Conservation, Climate, and Sustainable Use in Alpine Ecosystems

Muhammad Nauman Khan
https://orcid.org/0000-0003-3790-0902
Department of Botany,University of Chakwal, Pakistan

IGI Global
Scientific Publishing
Publishing Tomorrow's Research Today

Vice President of Editorial Melissa Wagner
Director of Acquisitions Mikaela Felty
Director of Book Development Jocelynn Hessler
Production Manager Mike Brehm
Cover Design Jose Rosado

Published in the United States of America by
IGI Global Scientific Publishing
701 East Chocolate Avenue
Hershey, PA, 17033, USA
Tel: 717-533-8845 | Fax: 717-533-7115
Website: https://www.igi-global.com E-mail: cust@igi-global.com

Library of Congress Cataloging-in-Publication Data

(CIP Data Pending)
ISBN13: 9798260012079
Isbn13Softcover: 9798260012086
EISBN13: 9798260012093
British Cataloguing in Publication Data
A Cataloguing in Publication record for this book is available from the British Library.

Table of Contents

Preface xv

Chapter 1
Ecosystem Services and Driving Factors Based on Land Use of Alpine Zones .. 1
Barkat Ali, Sichuan Agricultural University, China
Javed Khan, Sichuan Agricultural University, China
Yi Tang, Sichuan Agricultural University, China
Huanxiu Li, Sichuan Agricultural University, China

Chapter 2
Climate Change and Its Impact on Alpine Vegetation 27
Sikandar Shah, Institute of Biology, Warsaw University of Life Sciences, Poland
Sheharyar Khan, College of Environmental Science and Engineering, Tianjin, China
Sadaf, University of Peshawar, Pakistan
Sulaiman, University Putra Malaysia, Malaysia

Chapter 3
Anthropogenic Activities and Alpine Ecosystem Degradation 51
Hansa Gul, University of Mianwali, Pakistan
Haris Ahmed Khan, University of Mianwali, Pakistan
Zahida Nasreen, University of Mianwali, Pakistan
Nasir Assad, University of Sargodha, Pakistan
Baber Ali, Western Sydney University, Australia
Sezai Ercişli, Ataturk University, Erzurum, Turkey

Chapter 4
Sustaining Alpine Landscape: Biodiversity, Climate Change, and Conservation Imperatives 79
Shivansu Sachan, CSJM University, Kanpur, India
Nouman Ahmad, University of Mianwali, Punjab, Pakistan
Aleena Gul, University of Mianwali, Punjab, Pakistan
Mamoona Uzma Kalsoom, University of Mianwali, Punjab, Pakistan
Mridulesh Singh, CSJM University, Kanpur, India
Arvind Chauhan, CSJM University, Kanpur, India

Chapter 5
Conservation of Alpine Endemism Under Climate and Anthropogenic Pressures Emphasizing Phytogeographic Dimensions and Future Directions .. 111
Syed Waseem Gillani, Quaid-i-Azam University, Islamabad, Pakistan
Mushtaq Ahmad, Quaid-i-Azam University, Islamabad, Pakistan
Muhammad Manzoor, Quaid-i-Azam University, Islamabad, Pakistan
Mumna Munir, Quaid-i-Azam University, Islamabad, Pakistan
Shazia Sultana, Quaid-i-Azam University, Islamabad, Pakistan
Amir Sohail, Quaid-i-Azam University, Islamabad, Pakistan
Najeeb Ullah, Quaid-i-Azam University, Islamabad, Pakistan
Ahmed Ali, The University of Azad Jammu and Kashmir, Muzaffarabad, Pakistan

Chapter 6
Biodiversity of the National Parks and Nurseries in the Alpine Zones 137
Yasmeen Bibi, Government Graduate College for Women, Mianwali, Pakistan
Asma Bibi, University of Mianwali, Punjab, Pakistan
Mujeeb ur Rahman, Shandong University, Weihai, China
Nouman Ahmad, University of Mianwali, Punjab, Pakistan
Syed Nasar Shah, Quaid-i-Azam University, Islamabad, Pakistan
Abdul Bari Shah, Al-Farabi Kazakh National University, Kazakhstan
Fethi Ahmet Ozdemir, Bingol University, Bingol, Turkey
Tewekel Melese Gemechu, Ambo University, Ethiopia

Chapter 7
The Ethnobotany of Alpine Vegetation: Traditional Practices and Modern Science 165
Alevcan Kaplan, Sason Vocational School, Turkey
Sana Wahab, Quaid-i-Azam University, Islamabad, Pakistan
Barkat Ullah, Islamia College Peshawar, Pakistan
Abdul Razzaq, Qarshi Herb Research Center, Qarshi Industries (Pvt.) Ltd. Hattar, Haripur, Pakistan & Qarshi University, Lahore, Pakistan
Majid Iqbal, Institute of Geographic Sciences and Natural Resources Research, China
Maryam Bibi, Independent Researcher, Pakistan

Chapter 8
Alpine Medicinal Plant Strategies and Their Role in Conservation 191
Hansa Gul, University of Mianwali, Punjab, Pakistan
Zahida Nasreen, University of Mianwali, Punjab, Pakistan
Muhammad Nauman Khan, University of Chakwal, Punjab, Pakistan
Samina Siddiqui, University of Peshawar, Peshawar, Pakistan
Syed Nasar Shah, Quaid-i-Azam University, Islamabad, Pakistan
Sarah Abdul Razak, Universiti Malaya, Kuala Lumpur, Malaysia
Fethi Ahmet Ozdemir, Bingol University, Bingol, Turkey
Alevcan Kaplan, Sason Vocational School, Batman University, Batman, Turkey

Chapter 9
From Alpine Herbs to Nanotech: Biodiversity Conservation Through Medicinal Plants and Nanoparticle Research... 215
Arshad Ali, University of Sargodha, Pakistan
Maria Kanwal, University of Sargodha, Pakistan
Nasir Assad, University of Sargodha, Pakistan
Muhammad Mustaqeem, University of Sargodha, Pakistan
Rao Muhammad Faisal Iqbal, The University of Lahore, Sargodha, Pakistan
Laiba Manahil, University of Sargodha, Pakistan
Jamshed Iqbal, University of Genoa, Italy

Chapter 10
Proximate and Nutritional Study of the Wild Food Plants (WFPs) of the Alpine Zones 241
Nasir Assad, University of Sargodha, Pakistan
Muhammad Naeem Ul Hassan, University of Sargodha, Pakistan
Tooba, University of Sargodha, Pakistan
Yasir Assad, Hazara University, Mansehra, Pakistan
Marzia Batool Laila, University of Sargodha, Pakistan
Attia Khalid, University of Sargodha, Pakistan
Muhammad Khalid Mukhtar, University of Sargodha, Pakistan
Amjad Ali, Universita Cattolica del Sacro Cuore, Italy
Ummul Banin Zahra, Islamia College Peshawar, Pakistan
Muhammad Adnan, Bacha Khan University, Charsadda, Pakistan

Chapter 11
Biological Effects and Phytochemical Composition of Wild Food Plants (WFPs) in Alpine Zones: 279
Ahmet Ismail Ozkan, Artvin Coruh University, Turkey
Kemal Yüce, Ağrı İbrahim Çeçen University, Turkey
Nevin Aslan, Harran University, Turkey

Chapter 12
Alpine Biodiversity at the Crossroads: Medicinal Plants, Threats, and Nanoparticle-Based Intervention Strategies 341
Sajid Ali Khan Bangash, The University of Agriculture, Peshawar, Pakistan

Chapter 13
Honey Bee Flora: Identification, Characterization, Constraints, Pollen Substitutes, and Opportunity of Beekeeping in Alpine Zones........ 375
Hansa Gul, Department of Zoology, University of Mianwali, Punjab, Pakistan
Muhammad Nauman Khan, University of Chakwal, Pakistan
Muhammad Irfan, Department of Zoology, University of Mianwali, Punjab, Pakistan
Nouman Ahmad, University of Mianwali, Punjab, Pakistan
Majid Iqbal, Institute of Geographic Sciences and Natural Resources Research, Chinese Academy of Sciences, China

Compilation of References 397

About the Contributors 507

Index 513

Detailed Table of Contents

Preface xv

Chapter 1
Ecosystem Services and Driving Factors Based on Land Use of Alpine Zones .. 1
Barkat Ali, Sichuan Agricultural University, China
Javed Khan, Sichuan Agricultural University, China
Yi Tang, Sichuan Agricultural University, China
Huanxiu Li, Sichuan Agricultural University, China

Alpine zones deliver ecosystem services far beyond their spatial extent, including climate regulation, water supply, soil protection, biodiversity conservation, and cultural values. These services are highly sensitive to interactions between land-use change, rapid climate warming, and socioeconomic pressures. This chapter synthesizes existing evidence and develops a practical framework linking alpine land-use patterns such as grazing, afforestation, cropland expansion, protected areas, and settlements with changes in ecosystem service provision and trade-offs. Using remote sensing, ecological assessments, spatial statistics, and scenario analysis, we identify key drivers including elevation, climate regimes, grazing intensity, tourism, market access, and policy instruments. Results indicate that land conversion often degrades regulating and supporting services, while targeted restoration can recover them but may reduce short-term provisioning benefits. Distinct trade-offs across elevations highlight the need for context-specific, climate-resilient land-use planning.

Chapter 2
Climate Change and Its Impact on Alpine Vegetation 27
Sikandar Shah, Institute of Biology, Warsaw University of Life Sciences, Poland
Sheharyar Khan, College of Environmental Science and Engineering, Tianjin, China
Sadaf, University of Peshawar, Pakistan
Sulaiman, University Putra Malaysia, Malaysia

Alpine vegetation, which covers less than 3% of Earth's land surface, plays a vital role in supporting biodiversity, regulating water resources, and safeguarding soil integrity. These habitats are exceedingly susceptible to the impacts of climate change, owing to their narrow temperature tolerances, brief growing seasons, and distinctive plant communities adapted to extreme environmental conditions. Consequently, species that favor warmer conditions tend to migrate upward, leading to habitat loss for cold-adapted flora and heightened competition at lower elevations. Phenological changes, including earlier flowering and extended growing seasons, disrupt pollination and nutrient cycling. This situation underscores the pressing necessity for enhanced monitoring, conservation efforts and research initiatives, particularly in understudied regions and specialized communities to avert irreversible environmental degradation.

Chapter 3
Anthropogenic Activities and Alpine Ecosystem Degradation 51
Hansa Gul, University of Mianwali, Pakistan
Haris Ahmed Khan, University of Mianwali, Pakistan
Zahida Nasreen, University of Mianwali, Pakistan
Nasir Assad, University of Sargodha, Pakistan
Baber Ali, Western Sydney University, Australia
Sezai Ercişli, Ataturk University, Erzurum, Turkey

The alpine habitats reside on the highest points of the belts and are subject to the harsh climatic states and hence are among the most fragile but ecologically rich biomes of the planet. They are also sensitive to changes in climate and human activities that subject them to high levels of pressures rendering them extremely significant when it comes to providing early warnings of changes in the environment. The degradation of alpine vegetation has become the source of concern during the last decades of the high rates of the impact both of the natural and of the anthropogenic factors. Even though the climatic effects (temperature change, precipitation and surface albedo) have their physical consequences in the vegetation processes, human activity has also had its toll especially on highly populated or newly developed areas. In this chapter, the authors will analyze the complex of interplay between anthropogenic pressures and climate change to learn the forces behind alpine ecosystem degradation and define the efficient conservation policies.

Chapter 4
Sustaining Alpine Landscape: Biodiversity, Climate Change, and Conservation Imperatives 79
Shivansu Sachan, CSJM University, Kanpur, India
Nouman Ahmad, University of Mianwali, Punjab, Pakistan
Aleena Gul, University of Mianwali, Punjab, Pakistan
Mamoona Uzma Kalsoom, University of Mianwali, Punjab, Pakistan
Mridulesh Singh, CSJM University, Kanpur, India
Arvind Chauhan, CSJM University, Kanpur, India

The chapter is a thorough re-evaluation of the alpine conditions, the amazing biodiversity they host, the dramatic influence of the climate change that is currently inflicting on them, and the urgency behind its conservation. It discusses the bioclimatic conditions and global distribution of these high mountain ecosystems and explains the patterns and processes that control its flora and fauna. The chapter concludes the threats of habitat degradation, exotic species, and intense warming that these regions experience that in turn can fundamentally change ecological stability and vital ecosystem services. Particular focus is on contemporary conservation models which combine transboundary governance, community-based management, restoration ecology and leading-edge technological monitoring. By integrating ecological principles and socio-economic aspects, this review provides a synthesis that sets the stage for description of ecological and human cultural conservation challenges faced by these critical ecosystems at global scales.

Chapter 5
Conservation of Alpine Endemism Under Climate and Anthropogenic Pressures Emphasizing Phytogeographic Dimensions and Future Directions .. 111

Syed Waseem Gillani, Quaid-i-Azam University, Islamabad, Pakistan
Mushtaq Ahmad, Quaid-i-Azam University, Islamabad, Pakistan
Muhammad Manzoor, Quaid-i-Azam University, Islamabad, Pakistan
Mumna Munir, Quaid-i-Azam University, Islamabad, Pakistan
Shazia Sultana, Quaid-i-Azam University, Islamabad, Pakistan
Amir Sohail, Quaid-i-Azam University, Islamabad, Pakistan
Najeeb Ullah, Quaid-i-Azam University, Islamabad, Pakistan
Ahmed Ali, The University of Azad Jammu and Kashmir, Muzaffarabad, Pakistan

Endemic plant diversity of the Western Himalayas is a crucial component of global biodiversity but faces increasing environmental and anthropogenic threats. This study analyzed geographic distribution, compositional variation, elevational richness patterns, and β-diversity across 11 Himalayan sites between 3000 and 4300 m elevation. Chord diagrams revealed strong regional specificity with limited species overlap between Himalayan and Western Himalayan endemics. Multivariate analyses showed marked site-level heterogeneity, with a few sites driving overall dissimilarity. Species richness declined sharply with elevation, particularly among Himalayan endemics, while Western Himalayan endemics exhibited broader elevational ranges. PCA and PERMANOVA confirmed strong compositional segregation, and β-diversity was dominated by species turnover. Deforestation, grazing, NTFP extraction, tourism, and illegal trade emerged as major threats, emphasizing the need for region-specific, multi-site conservation strategies.

Chapter 6
Biodiversity of the National Parks and Nurseries in the Alpine Zones 137
Yasmeen Bibi, Government Graduate College for Women, Mianwali, Pakistan
Asma Bibi, University of Mianwali, Punjab, Pakistan
Mujeeb ur Rahman, Shandong University, Weihai, China
Nouman Ahmad, University of Mianwali, Punjab, Pakistan
Syed Nasar Shah, Quaid-i-Azam University, Islamabad, Pakistan
Abdul Bari Shah, Al-Farabi Kazakh National University, Kazakhstan
Fethi Ahmet Ozdemir, Bingol University, Bingol, Turkey
Tewekel Melese Gemechu, Ambo University, Ethiopia

Alpine zones are among the most fragile and ecologically significant regions on Earth, harboring unique biodiversity adapted to extreme climatic conditions. This study explored the biodiversity of national parks and nurseries established within alpine zones, emphasizing their role in the conservation of endemic, rare, and threatened species. National parks serve as in situ conservation hubs, preserving intact ecosystems and natural habitats, whereas nurseries act as ex situ conservation centers supporting the propagation, rehabilitation, and restoration of native alpine flora. Together, these initiatives contribute to safeguarding genetic diversity, sustaining ecosystem services, and promoting ecological resilience against climate change and anthropogenic pressures. These findings highlight the importance of integrated conservation strategies and effective management practices to maintain the biodiversity and ecological integrity of alpine regions.

Chapter 7
The Ethnobotany of Alpine Vegetation: Traditional Practices and Modern Science 165
Alevcan Kaplan, Sason Vocational School, Turkey
Sana Wahab, Quaid-i-Azam University, Islamabad, Pakistan
Barkat Ullah, Islamia College Peshawar, Pakistan
Abdul Razzaq, Qarshi Herb Research Center, Qarshi Industries (Pvt.) Ltd. Hattar, Haripur, Pakistan & Qarshi University, Lahore, Pakistan
Majid Iqbal, Institute of Geographic Sciences and Natural Resources Research, China
Maryam Bibi, Independent Researcher, Pakistan

Ethnobotany is valued for its interdisciplinary contributions to natural product research, particularly in, development, and standardization of pharmaceutical and pharmacological products. Alpine plants, often found above the tree line, contribute significantly to the biodiversity of mountain systems and represent an unexplored area that warrants ethnobotanical attention. Although much traditional knowledge has been lost due to various factors, traditional medicinal practices demonstrate that their complementarity with modern medicine should not be overlooked. In today's world, where interdisciplinarity has evolved into a post-disciplinary approach, pluralism in treatment is an undeniable reality. In line with these developments, this chapter examines processes defined and completed through the mutual exchange of information between ethnobotanists and pharmacologists, in the field. One or more aspects of the ethnobotanical use of plants growing in the alpine zone, and their valuable role in pharmacological research serving modern medicine, will be highlighted.

Chapter 8

Alpine Medicinal Plant Strategies and Their Role in Conservation 191
Hansa Gul, University of Mianwali, Punjab, Pakistan
Zahida Nasreen, University of Mianwali, Punjab, Pakistan
Muhammad Nauman Khan, University of Chakwal, Punjab, Pakistan
Samina Siddiqui, University of Peshawar, Peshawar, Pakistan
Syed Nasar Shah, Quaid-i-Azam University, Islamabad, Pakistan
Sarah Abdul Razak, Universiti Malaya, Kuala Lumpur, Malaysia
Fethi Ahmet Ozdemir, Bingol University, Bingol, Turkey
Alevcan Kaplan, Sason Vocational School, Batman University, Batman, Turkey

Alpine medicinal plants make an essential interchange between biodiversity, cultural traditions, and pharmacological potential. Prospering in alpine areas these species developed special adaptive features and bioactive compounds interesting to traditional healing systems. Their survival is, however, threatened by climate change, habitat fragmentation, over-harvesting, and unsustainable land use. In this chapter, the ecological and pharmacological importance of alpine vegetation is analyzed critically, with an understanding of the fact that there is an impending need of putting together conservation measure. It examines not only in-situ methods, including habitat protection, community-based stewardship and practices of sustainable harvest, but also ex-situ methods, including seed banks, botanic gardens, and technologies of tissue culture. Furthermore, it proposes a blend between the indigenous knowledge systems and scientific innovation such that these valuable plant resources may be long term preserved and wisely utilized.

Chapter 9
From Alpine Herbs to Nanotech: Biodiversity Conservation Through Medicinal Plants and Nanoparticle Research.. 215
Arshad Ali, University of Sargodha, Pakistan
Maria Kanwal, University of Sargodha, Pakistan
Nasir Assad, University of Sargodha, Pakistan
Muhammad Mustaqeem, University of Sargodha, Pakistan
Rao Muhammad Faisal Iqbal, The University of Lahore, Sargodha, Pakistan
Laiba Manahil, University of Sargodha, Pakistan
Jamshed Iqbal, University of Genoa, Italy

Alpine ecosystems are climatically extreme, biodiversity-rich, and among the most vulnerable yet pharmaceutically valuable habitats on Earth. Alpine herbs have long formed the basis of traditional medical systems across Eurasia, the Himalayas, and other mountain regions. However, the growing global demand for herbal medicines and bioactive compounds has placed unprecedented pressure on these fragile ecosystems, threatening both plant diversity and indigenous knowledge. Green nanotechnology has recently emerged as a sustainable link between biodiversity conservation and biotechnology. Green synthesis of nanoparticles using alpine medicinal plants offers opportunities to enhance bioactivity, reduce toxicity, develop value-added bioproducts, and minimize pressure on wild populations. This chapter explores the ecological, biochemical, and technological intersections between alpine medicinal flora and nanoscience, proposing nanotechnology as a conservation-oriented strategy for sustainable use of alpine biodiversity.

Chapter 10
Proximate and Nutritional Study of the Wild Food Plants (WFPs) of the Alpine Zones 241
Nasir Assad, University of Sargodha, Pakistan
Muhammad Naeem Ul Hassan, University of Sargodha, Pakistan
Tooba, University of Sargodha, Pakistan
Yasir Assad, Hazara University, Mansehra, Pakistan
Marzia Batool Laila, University of Sargodha, Pakistan
Attia Khalid, University of Sargodha, Pakistan
Muhammad Khalid Mukhtar, University of Sargodha, Pakistan
Amjad Ali, Universita Cattolica del Sacro Cuore, Italy
Ummul Banin Zahra, Islamia College Peshawar, Pakistan
Muhammad Adnan, Bacha Khan University, Charsadda, Pakistan

Alpine medicinal plants, adapted to high-altitude environments, represent an important source of bioactive compounds with therapeutic potential. Exposure to low temperatures, intense ultraviolet radiation, and nutrient-poor soils has promoted the synthesis of diverse secondary metabolites, including flavonoids, alkaloids, terpenoids, and phenolic compounds. This chapter explores the phytochemical diversity of alpine medicinal flora, highlighting their ecological adaptations, pharmacological significance, and traditional uses by indigenous communities. Key species such as Saussurea lappa, Bergenia ciliata, Picrorhiza kurroa, and Artemisia maritima exhibit antioxidant, anti-inflammatory, hepatoprotective, and antimicrobial activities. Modern analytical approaches, including HPLC, GC-MS, and NMR spectroscopy, are discussed for phytochemical profiling. The chapter emphasizes sustainable utilization and conservation amid climate-driven habitat loss.

Chapter 11
Biological Effects and Phytochemical Composition of Wild Food Plants (WFPs) in Alpine Zones: 279
Ahmet Ismail Ozkan, Artvin Coruh University, Turkey
Kemal Yüce, Ağrı İbrahim Çeçen University, Turkey
Nevin Aslan, Harran University, Turkey

All plants that we refer to as domesticated plants were once wild food plants (WFPs) before being cultivated by humans. These plants are species that can naturally grow in their natural habitats without direct human intervention. They have grown in various ecosystems and adapted to various environmental conditions. Recently, WFPs classified as functional foods have been found to alleviate nutritional deficiencies and even treat diseases. WFPs can be useful in combating malnutrition. They contain bioactive compounds including essential source of vitamins, minerals, and secondary metabolites such as essential oils, alkaloids, and phenolics. In addition, increasing the intake of these compounds can further improve human health. WFPs are also evaluated in terms of their pharmacological properties due to their significant medicinal importance. WFPs were noted to have the highest abundance in the alpine zone. This chapter was focused on the biological effects including antimicrobial, antioxidant and anticancer, and phytochemical composition of WFPs in the alpine zone.

Chapter 12
Alpine Biodiversity at the Crossroads: Medicinal Plants, Threats, and Nanoparticle-Based Intervention Strategies 341
Sajid Ali Khan Bangash, The University of Agriculture, Peshawar, Pakistan

Alpine countries possess a distinctive and varied collection of medicinal plants due to harsh environmental factors and centuries-old folk traditions. In recent decades, these species have become endangered because of climate change, habitats destruction and unsustainable harvesting practices. This chapter reviews the ecological value and pharmacological potential of major alpine medicinal plants, outlining the main threats to their survival and the resulting consequences for ecosystems and local populations. It then critically evaluates the potential role of emerging nanoparticle-based intervention strategies, such as targeted delivery systems and plant resilience enhancement, in conservation and sustainable use. By integrating ecological, ethnobotanical, and nano-biotechnological perspectives, the chapter shows how an interdisciplinary approach can help protect alpine medicinal biodiversity and support the development of sustainable solutions for future therapeutic applications.

Chapter 13
Honey Bee Flora: Identification, Characterization, Constraints, Pollen Substitutes, and Opportunity of Beekeeping in Alpine Zones 375

Hansa Gul, Department of Zoology, University of Mianwali, Punjab, Pakistan
Muhammad Nauman Khan, University of Chakwal, Pakistan
Muhammad Irfan, Department of Zoology, University of Mianwali, Punjab, Pakistan
Nouman Ahmad, University of Mianwali, Punjab, Pakistan
Majid Iqbal, Institute of Geographic Sciences and Natural Resources Research, Chinese Academy of Sciences, China

The alpine regions, with their own climatic characteristic and a variety of floral resources, present rather unexploited but promising perspectives of sustainable beekeeping. It is in these areas at high altitude that this chapter addresses in quantitative detail the identification and characterization of honey bee flora, its diversity and seasonal availability of both nectar and pollen providing sources. The special emphasis is made on the classification of the types of pollen, nutritional constituents of the pollen and the part played by the pollen in maintaining bee health, productivity and disease resistance. Ecological and agricultural importance of alpine bee flora is discussed also, especially its contribution to increased pollination of crops and maintenance of biodiversity. In spite of the possibilities, a few limitations which prevail on beekeeping in the alps are climatic extremes, shortage of forage, loss of habitat, and poor infrastructure.

Compilation of References 397

About the Contributors 507

Index 513

Preface

Alpine ecosystems, though occupying a relatively small proportion of the Earth's surface, are among the most ecologically significant and environmentally sensitive regions of the planet. Characterized by extreme climatic conditions, unique biodiversity, and fragile ecological balances, these high-altitude landscapes provide indispensable ecosystem services, including water regulation, carbon sequestration, soil stabilization, and cultural value. However, in the face of accelerating climate change and increasing anthropogenic pressures, alpine ecosystems are now at a critical juncture. *Conservation, Climate, and Sustainable Use in Alpine Ecosystems* seeks to address this urgency by presenting a comprehensive and interdisciplinary exploration of the challenges and opportunities associated with these vital environments.

This volume brings together contributions that integrate ecological science, conservation biology, ethnobotany, and emerging technological approaches to offer a multidimensional understanding of alpine systems. The chapters collectively highlight the intricate relationships between climate dynamics, land-use patterns, biodiversity, and human livelihoods. Importantly, the book emphasizes not only the ecological significance of alpine regions but also their socio-economic and cultural relevance, particularly for communities that depend on these ecosystems for sustenance and traditional knowledge.

A central theme of this collection is the need for sustainable and context-specific conservation strategies. The contributors explore how traditional practices, modern scientific methods, and innovative technologies—such as remote sensing and nanotechnology—can be combined to address complex environmental challenges. By bridging local knowledge systems with global scientific advancements, this volume advocates for integrative approaches that ensure both conservation and sustainable utilization of alpine resources.

CHAPTER OVERVIEW

The first chapter examines the ecosystem services provided by alpine zones and analyzes how different land-use patterns influence these services. It highlights the trade-offs between development and conservation, emphasizing the need for climate-resilient land-use planning.

Chapter 2 explores the impacts of climate change on alpine vegetation, including shifts in species distribution, phenological changes, and the vulnerability of specialized plant communities, underscoring the urgency of monitoring and conservation efforts.

Chapter 3 investigates the combined effects of anthropogenic activities and climate variability on alpine ecosystem degradation, offering a holistic understanding of the drivers of environmental change and strategies for mitigation.

The fourth chapter provides a comprehensive review of alpine biodiversity and the effects of climate change, while discussing conservation frameworks that integrate ecological principles with socio-economic considerations.

The fifth chapter focuses on the conservation of endemic plant species in alpine regions, analyzing patterns of species distribution and the threats posed by human activities, and proposing region-specific conservation strategies.

Chapter 6 examines the role of protected areas and plant nurseries in conserving alpine biodiversity, highlighting the importance of both in situ and ex situ approaches in maintaining ecological resilience.

Chapter 7 explores the ethnobotanical significance of alpine vegetation, emphasizing the value of traditional knowledge systems and their integration with modern scientific research for sustainable resource use.

The eighth chapter analyzes the ecological and pharmacological importance of alpine medicinal plants and outlines conservation strategies that combine habitat protection, sustainable harvesting, and technological interventions.

Chapter 9 investigates the intersection of alpine biodiversity and nanotechnology, proposing innovative approaches for sustainable utilization and conservation through green synthesis and biotechnological applications.

The tenth chapter examines the nutritional and phytochemical properties of wild food plants in alpine regions, highlighting their potential for health benefits and the importance of their sustainable use and conservation.

Chapter 11 explores the biological effects and phytochemical composition of wild food plants, emphasizing their role as functional foods and their potential in addressing nutritional and health challenges.

Chapter 12 evaluates the threats to alpine medicinal biodiversity and discusses advanced nanoparticle-based strategies for conservation and sustainable utilization, integrating ecological and technological perspectives.

The thirteenth and final chapter focuses on the diversity and ecological importance of honey bee flora in alpine regions, addressing both the opportunities and constraints for sustainable beekeeping and its role in supporting biodiversity and livelihoods.

Together, these chapters provide a comprehensive and forward-looking perspective on alpine ecosystems, emphasizing the interconnectedness of conservation, climate change, and sustainable development. This volume is intended to serve as a valuable resource for researchers, environmentalists, policymakers, and practitioners engaged in the study and preservation of high-altitude ecosystems.

It is our hope that this book will contribute to a deeper understanding of alpine environments and inspire collaborative, innovative, and sustainable approaches to safeguarding these critical ecosystems for future generations.

Muhammad Nauman Khan

Department of Botany, University of Chakwal, Pakistan

Chapter 1
Ecosystem Services and Driving Factors Based on Land Use of Alpine Zones

Barkat Ali
http://orcid.org/0009-0008-5324-2931
Sichuan Agricultural University, China

Javed Khan
http://orcid.org/0009-0008-0793-9906
Sichuan Agricultural University, China

Yi Tang
Sichuan Agricultural University, China

Huanxiu Li
Sichuan Agricultural University, China

ABSTRACT

Alpine zones deliver ecosystem services far beyond their spatial extent, including climate regulation, water supply, soil protection, biodiversity conservation, and cultural values. These services are highly sensitive to interactions between land-use change, rapid climate warming, and socioeconomic pressures. This chapter synthesizes existing evidence and develops a practical framework linking alpine land-use patterns such as grazing, afforestation, cropland expansion, protected areas, and settlements with changes in ecosystem service provision and trade-offs. Using remote sensing, ecological assessments, spatial statistics, and scenario analysis, we identify key drivers including elevation, climate regimes, grazing intensity, tourism, market access, and policy instruments. Results indicate that land conversion often degrades

DOI: 10.4018/979-8-2600-1207-9.ch001

regulating and supporting services, while targeted restoration can recover them but may reduce short-term provisioning benefits. Distinct trade-offs across elevations highlight the need for context-specific, climate-resilient land-use planning.

1.1 STUDY AREA DESCRIPTION

Alpine regions are some of the most biologically sensitive and climate-sensitive regions on the planet and these belonged to regions that are located over the natural tree line and under permanent snow cover (Buytaert et al., 2011). The features of these zones are short growth seasons, rugged terrain, low temperatures, and a large range of microclimates. The combination of these characteristics offers unique ecosystems with outstanding biodiversity and responsiveness (Humphries, 2020). Even though alpine areas are delicate, they provide very essential ecological services, which far exceed physical boundaries, particularly in terms of water management, cultural applicability, and carbon sequestration (Buytaert et al., 2011).

The Tibetan Plateau, European Alps, Andes, Himalayas, and Rocky Mountains are the best representatives of the alpine ecosystems of the world, and each one of them has a distinctive land use model dependent on the socioeconomic conditions of a region (Hong & Chongjian, 2024). To illustrate, in Andes and Tibetan Plateau, which are characterized by traditional pastoralism, soil stability and grassland productivity are highly impacted by grazing pressure (Dong et al., 2020). The European Alps, in contrast, show the struggle between the conservation of the environment and economic development in terms of a topography of safeguarded zones, tourist infrastructure, and dispersed population (Messerli, 2015). An example of such land use patterns in the Himalayas, which highlight how local people adapt to challenging conditions and heavily rely on ecosystem services, is crop rotation, terraced agriculture, and collection of medicinal plants (Hazarika et al., 2024).

Similar issues occur in many locations regardless of the different government structures, land-use patterns and cultures. No laws and services are secure due to overgrazing, deforestation, uncontrolled tourism, and infrastructural development, which cause land degradation and threat to the laws and services of the local community and the downstream. Moreover, climate change exacerbates these issues by raising the number of occurrences of extreme weather patterns and altering vegetation zones and snowmelt patterns. Determining the ecosystem service provision and evaluating the biophysical and socioeconomic processes involved in the provision of the service require a comprehensive understanding of this multifaceted and interconnected alpine ecosystem.

1. 2 INTRODUCTION

Alpine ecosystems are some of the most vulnerable but vital in the planet. Among the important ecological services that they provide are soil stability, freshwater management, carbon sequestration, biodiversity conservation and cultural and recreational values just to mention a few. Although these ecosystems have a limited global extent, these services have downstream effects and support local livelihood (Lavorel et al., 2006; Yang et al., 2023). Nonetheless, these systems are under a significant challenge, including land-use intensification, climate alteration, and socioeconomic upheavals, posing a threat to the ecological integrity of such systems and human well-being (Pauli & Halloy, 2019).

Agricultural and non-timber logging in high volume used to maintain the equilibrium within the local environment. Nonetheless, the ecosystem service trade-offs are evolving due to the alteration of land cover. The primary causes of these changes are increasing human demand, infrastructural development, increasing tourism, and increasing integration of the markets. An example is the alteration of alpine shrub meadows, which reduces grazing services drastically, modifying the regulatory processes such as carbon storage and water retention, as the results of research on the Tibetan Plateau show (Qian, Fan, et al., 2023). Meanwhile, as the European Alps have undergone a shift in the traditional forms of landscape and a growing agricultural intensive farming practice since the 19th century, it diminishes cultural services in most locations (Tasser et al., 2024). In-depth assessments of high attitude sites remain extremely infrequent, in spite of the extensive research in the field of land use and ecological services (Mengist et al., 2020). In order to understand the overall impact of these factors on the availability, distribution, and trade-offs of ecosystem services, it is necessary to analyze them at multiple scales topography, climate and socioeconomics (Shen et al., 2021). For instance, it is well known that seasonal dynamics changed by early snowmelt have a significant impact on the microbial and biogeochemical cycles in alpine soils, which in turn affects carbon storage and nutrient availability (Broadbent et al., 2021; Rindt et al., 2023). According to (Geneletti et al., 2018; Schirpke et al., 2013) leisure activities in the Alps and Alpine fortresses indicate that cultural services (biodiversity, aesthetic value) are seen as declining in regions with strong supply demand (feed, agriculture). To close these knowledge gaps, this chapter will: (i) review the main ecosystem services that alpine land use system provide; (ii) analyze land use changes patterns and their effects on the provision and spatial distribution of ecosystem services; (iii) identify the major biophysical and socioeconomics drivers behind these dynamics (climate change, topography, policy, population pressures); and (iv) explore management implications and policy mechanisms for maintaining ecosystem services in the context of ongoing global change.

2. CONCEPTUAL FRAMEWORK

2.1 Ecosystem services in alpine zones

The four major categories of ecosystem services are provisioning, regulatory, cultural, and support services (Table 1) (Costanza, 2016). Freshwater supplies, medicinal plants, feed resources, and, occasionally, limited agricultural products are examples of supply services in high-altitude regions (Ali et al., 2021). Because steep slopes and thin soils render the terrain extremely vulnerable to erosion and degradation, regulatory functions including soil protection, water purification, and climate regulation are especially crucial in these delicate situations (BöHmer, 2011). Cultural services including leisure and recreation, ecotourism, and spiritual values are becoming more and more important as mountain environments draw visitors looking for distinctive natural and cultural experiences (Kaltenborn et al., 2019; M. Schmitt et al., 2024; Schirpke et al., 2016). By guaranteeing the resilience and long-term sustainability of ecosystems, supportive services such as habitat supply and biodiversity conservation lay the groundwork for all other services.

Table 1. Alpine zones provide important environmental services and examples.

Ecosystem service category	Key services	Examples in alpine zones	References
Provisioning services	Freshwater, medicinal resources, forage	Glacier-fed rivers, alpine herbs, grazing pastures	(Costanza, 2016)
Regulating services	Climate regulation, soil protection, water purification	Carbon sequestration, erosion control on steep slopes	
Cultural services	Recreation, tourism, spiritual values	Mountain tourism, pilgrimage sites, scenic landscapes	
Supporting services	Biodiversity maintenance, habitat provision	Endemic flora and fauna, pollinator habitats	

2.2 Alpine regions land use dynamics

Traditional livelihoods, including as grazing, small-scale agriculture, and forestry product gathering, have historically influenced land use patterns in high-altitude locations (Garcés-Pastor et al., 2021; Tappeiner & Tasser, 2019). However, due to population pressures, infrastructural development, increased tourism, and ecological restoration initiatives, these patterns have drastically altered in recent decades (Kou et al., 2024). Grassland degradation often originates from numerous sources, including settlement development, afforestation of degraded slopes, expansion of valley farming, and overgrazing (Table 2) (Khazieva et al., 2022). These modifications impact the ecosystem's ability to deliver necessary services in addition to changing the landscape's physical structure. According to (Verhagen et al., 2016), this often results in intricate trade-offs and synergies between various service categories.

Table 2. Alpine land use changes and their effects on ecosystem services.

Land use change	Main drivers	Ecosystem service impacts	References
Overgrazing	Livestock intensification, population pressure	Reduced forage quality, soil erosion, biodiversity loss	(Garcés-Pastor et al., 2021)
Expansion of valley agriculture	Food demand, irrigation access	Increased provisioning services, reduced regulating services	(Tappeiner & Tasser, 2019)
Afforestation/reforestation	Ecological restoration policies	Enhanced carbon storage, possible water yield reduction	
Tourism infrastructure	Economic development	Increased cultural services, habitat fragmentation	
Settlement expansion	Urbanization, accessibility	Loss of supporting services, increased environmental pressure	

2.3 Ecosystem services driving forces

Numerous interacting elements influence the delivery of alpine ecosystem services. Plant distribution, hydrological cycles, and soil processes are all greatly impacted by climate influences, including rising temperatures and changed precipitation patterns (Y. Liu et al., 2023). Land use decisions are also influenced by socioeconomic factors such as population growth, market demand, tourism development, and agricultural intensification (Table 3) (Prabhakar, 2021). Furthermore, policy frameworks such as land use zoning, ecological compensation systems, and conservation initiatives play a vital role in shaping the outcomes of land management (dos Santos et al., 2015). These components work together to form a dynamic socio-ecological system in which decisions about land use must strike a balance between human demands and ecological sustainability (Meyfroidt et al., 2022). Forcing factors of ecosystems services and land use in elevated localities. This table is a comprehensive study of driving factors in high-altitude socio-ecology systems in the work of previous studies on climate, socioeconomic factors, policies, and management and their association.

Table 3. Key variables affecting ecosystem services in mountainous areas

Driving factor category	Key drivers	Primary land-use impacts	Affected ecosystem services	Typical outcomes in alpine zones	References
Climate drivers	Rising temperature, altered precipitation patterns, increased climate variability	Vegetation zone shifts, glacier retreat, changes in soil moisture and hydrological regimes	Provisioning (water supply), regulating (climate regulation, erosion control), supporting (habitat stability)	Altered runoff patterns, increased erosion risk, changes in species composition	(Prabhakar, 2021)
Socioeconomic drivers	Population growth, market demand, tourism expansion, agricultural intensification	Expansion of grazing and farmland, infrastructure development, land abandonment in marginal areas	Provisioning (food, fodder), cultural (tourism, recreation), regulating (soil and water regulation)	Increased pressure on fragile ecosystems, trade-offs between economic benefits	
Policy and institutional drivers	Conservation policies, ecological compensation schemes, land-use zoning, restoration programs	Restricted land conversion, afforestation, grassland restoration, protected area expansion	Regulating (soil protection, carbon sequestration), supporting (biodiversity conservation)	Improved ecosystem stability in protected areas, but potential livelihood constraints	
Technological and management drivers	Improved farming techniques, grazing management, irrigation practices	Intensification or optimization of land use, reduced land degradation	Provisioning (agricultural output), regulating (water-use efficiency)	Enhanced productivity with reduced environmental footprint when sustainably managed	
Integrated socio-ecological interactions	Interaction among climate, socioeconomic, and policy factors	Adaptive land-use decisions balancing development and conservation	All ecosystem service categories	Dynamic trade-offs and synergies among ecosystem services	

3 DATA AND METHODS

3.1 Data sources

A number of data sources are applicable in assessing the relationship between ecosystem services and the land use in high altitude regions. Remote sensing datasets such as Landsat, Sentinel or MODIS can be used to obtain long-term observational data of vegetation indices, snow cover, and land cover dynamics with varying spatial resolutions. Besides the collection of site-specific biological indicators such as biomass, soil organic matter and richness of the species, other field surveys can also be used to confirm land cover classifications. The socioeconomic data, which includes the demographics, livestock density, agricultural productivity and tourism activity, can be obtained through national statistical bureaus, local government reports or international databases such as the FAO and the World Bank. Combined, these data sets offer a powerful one on the ecological process and human stress at high-altitude.

3.2 Land use classification

The field verification and the imagery of remote sensors are used to classify land use and land cover (LULC). Forest, grassland, cultivated land, water bodies, built-up regions, and unused land are typical classifications. To differentiate between various land use types in varied mountainous terrain, classification approaches like supervised classification, object-oriented image analysis, or machine learning methods (random forests, support vector machines) are frequently employed. Reliable interpretation of land use patterns is ensured by accuracy evaluation utilizing high-resolution images or ground-based data.

3.3 Ecosystem service assessment

Four categories-provisioning, regulation, cultural, and sustaining services are used to quantify ecosystem services. Indicators including biomass yield, water supply, and the availability of medicinal plants are used to evaluate provisioning services. Ecological models (InVEST, RUSLE (for soil erosion), CASA (for net primary productivity) or analogous factor approaches are used to assess regulating services, such as soil conservation, water purification, and carbon sequestration. Indicators including tourism earnings, visitor numbers, and accessibility to recreational landscape spaces are used to evaluate cultural services. Species richness and habitat quality indicators were used to evaluate supporting services, especially biodiversity protection. Depending on the availability of data, a variety of valuation techniques were used, such as the market approach, replacement cost approach, and benefit transfer approach.

3.4 Analysis of driving factors

We looked at both biophysical and socioeconomic factors to determine the forces behind changes in ecosystem services. Global climate databases and meteorological records were used to determine the climate factors (temperature, precipitation, and snow cover). Digital elevation models (SRTM or ASTER) were used to extract topographic parameters such as elevation, slope, and aspect. Population increases, cattle density, land-use regulations, and the degree of tourism development were examples of socioeconomic factors. Regression models, correlation analysis, and structural equation modeling (SEM) were among the statistical methods used in this study to evaluate the relative importance of different factors. While GIS-based spatial analysis brought attention to regional variations, scenario analysis provided insights into possible future trajectories of ecosystem service supply under various land use plans.

4. LAND USE CHANGE PATTERNS IN ALPINE ZONES

Over the past few decades, land usage in high-altitude areas has altered dramatically, reflecting the combined effects of climate change and human activity. Natural grasslands are generally declining because of overgrazing, valley-level farming expansion, and settlement expansion brought on by population increase and infrastructure development, according to remote sensing study of various high-altitude systems. On the other hand, in many places, especially on degraded slopes, focused afforestation and ecological restoration initiatives have resulted in a partial restoration of vegetation cover. Land use patterns in the European Alps and Himalayas have also been altered by tourism-driven land use, such as ski resorts, hiking paths, and service facilities. In general, the spatial heterogeneity of land use emphasizes how sensitive alpine environments are to both external policy interventions and local populations' means of subsistence.

4.1 Ecosystem service provision

Complex patterns are seen in supply services in high-altitude areas. Although crop yields and animal output are essential sources of food and revenue, overexploitation frequently results in the loss of freshwater supplies, fuel, and fodder. In regions where overgrazing and deforestation are prevalent, regulation services-especially those related to soil conservation and water purification-decline; nevertheless, in regions where conservation or afforestation programs are being carried out, they improve. For instance, soil conservation efforts in the Andes have stabilized slopes, whereas afforestation programs on the Tibetan Plateau have increased carbon storage and decreased soil erosion. The steady growth of cultural services, especially ecotourism and recreation, frequently comes at the expense of habitat fragmentation and infrastructure expansion. Even though protected areas continue to protect high-value habitats, supporting services like biodiversity conservation are nevertheless seriously endangered by land-use change and climate impacts.

4.2 Synergies and trade-offs between ecosystem services

Decisions about land use have a big impact on how ecosystem services interact. Soil degradation and biodiversity loss result from the expansion of arable land and settlements, which frequently increases provisioning services in the near term but decreases regulating and supporting services. On the other hand, while they may occasionally diminish short-term provisioning benefits like grazing capacity, ecological restoration and afforestation can increase regulating functions like soil stability and carbon sequestration and frequently work together to enhance supporting services. Although tourism can lead to trade-offs through habitat degradation, increased waste, and effects on infrastructure, it also fosters synergies by enhancing cultural services and benefiting the livelihoods of locals. Altitude gradients affect these trade-offs and synergies; higher-altitude regions concentrate more on moderating and supporting functions, whereas lower-altitude mountain valleys exhibit larger supply gains.

4.3 Driving factors of ecosystem service change

The pattern of ecosystem services is exposed to a number of factors that are influenced by statistical and geographical analysis. The increase in temperatures and the shift in precipitation patterns, especially, result in the vegetation areas shifting higher, which changes the time of snowmelt and access to water. The socio-economic conditions such as population, intensification of livestock production, and tourism development are the direct causes of land use. Policy interventions like afforestation activities, creation of preserved areas and ecological compensation programs have become necessary to help curb the degradation of the ecosystems and to recover the ecosystem services. Moreover, the topographical limitations, including elevation, slope, and aspect, further limit the anthropogenic influence on the ecosystems, and these complex relations result in the high-altitude locations to have a significant diversity (Figure 1).The conceptual framework illustrates the effect of the complex interplay between the anthropogenic and the natural processes on the supply of ecosystem services. As Figure 1 shows, the outcomes of such interaction are highly determined by land use patterns and government systems. Topographical variables such as elevation, slope and aspect have significant geographical heterogeneity which limits the extent to which human activity can have an effect on the ecosystems. Finally, the results reveal that provision of the ecosystem services in the high-altitude locations is defined by such complex interactions and would rely on the proper land-use plans and governance frameworks.

The findings prove that land use patterns in high-altitude places have a significant impact on the quantity, quality, and dispersion of ecosystem services. Increased cultivated areas and communities normally boost the short-term service delivery and diminish the regulation and support functions, which is also in line with previous study on mountain systems (Fedrigotti et al., 2016; Grêt-Regamey et al., 2012; Körner, 2004; Qiu et al., 2021; Schirpke et al., 2013). Conversely, conservation and restoration, including afforestation on the Tibetan Plateau and slope stabilization in Andes, can increase regulatory services such as soil conservation and carbon storage as well as increase ecosystem long-term resilience (Huang et al., 2018; X. Liu et al., 2023). These results demonstrate that ecological variables and land management methods affect the trade-offs and synergies between ecosystem services. The problems in alpine ecosystems throughout the world are similar, but the development of the alpine ecosystem varies, Subsistence farming has been replaced by a tourism-driven economy in the European Alps, which has overstretched the ecology, but at the same time, it has caused synergies in the cultural activity (Huber et al., 2020; Klein et al., 2019). Population growth and agricultural intensification are increasingly putting a strain on the supply of ecosystem services in the Himalayas thereby reducing biodiversity and water availability (Ma et al., 2012; Tiwari & Joshi, 2014).). Qinghai-Tibet Plateau shows that when performed on a large scale, ecological restoration can partially undo the process of degradation, but this has a complicated effect on the livelihoods based on herding (Hopping et al., 2018; Zhen et al., 2018). These regional differences present the necessity of site-specific management plans that are aware of local socioeconomic and ecological factors. What is more, this analysis underlines the importance of the driving forces affecting the dynamics of ecosystem services. The increase in temperatures and alteration of precipitation patterns is increasing vegetation alterations and modulating the hydrological cycle of areas in high altitude regions across the globe (Mastrotheodoros et al., 2019; Yang et al., 2023; Zierl & Bugmann, 2005). With the topographical constraints, socioeconomic factors, including cattle husbandry, market integration, and tourism have different impacts on different regions. Though ecosystem services payment systems and setting up protected zones have proven effective, unless enforced properly or designed appropriately, they may increase inequality or drive the pressure elsewhere (Oldekop et al., 2016; Osano et al., 2013; Y. Zhang et al., 2024).

The trade-offs have discovered that to ensure that high-altitude regions have sustainable land management, then there has to be a balance between dogging the immediate human needs and the ecological health in the long-term. The promising strategies are ecological zoning, adaptive grazing management, and land use planning and participatory models of governance. Also, the policymakers can also choose priority locations to protect or restore based on quantifying ecosystem services with explicit spatial methods such as RUSLE, InVEST or habitat quality indices. Because the high-altitude areas have been one of the ecosystems that are most adversely affected by global warming, the future efforts should also incorporate climate adaptation methods.

5. PROVISIONING ECOSYSTEM SERVICES AND LAND USE

The supply of ecosystem services, which are directly related to land-use patterns, is the most evident benefit that alpine ecosystems provide to human communities. In high-altitude regions, provisioning services support subsistence lifestyles, downstream populations, and regional economy. However, because to their fragility, alpine ecosystems' capacity to deliver these services is highly susceptible to land-use intensity, climatic variability, and management strategies (Saleem et al., 2024). Therefore, knowing how terrain usage impacts service provision is necessary to achieve sustainable development in hilly places.

5.1 Grazing and livestock production

In alpine locations around the world, especially in places like the Tibetan Plateau, the Andes, the Himalayas, and Central Asian mountain ranges, grazing and livestock production provide one of the primary ecosystem services (Duchicela et al., 2024). The foundation of pastoral and agro-pastoral lifestyles, alpine meadows supply seasonal fodder for animals including yaks, sheep, goats, and cattle. Transhumant and nomadic grazing systems have historically maintained grassland production and ecological balance by allowing vegetation to recover through seasonal migration (Ingty, 2021).

Understanding the how much of fodder supply is essential for sustainable pastoralism, as proved by recent evaluations of livestock carrying capacity in the Kashmir Himalayas (Saleem et al., 2024). According to quantitative research, if stocking rates stay below carrying capacity criteria, alpine grasslands can sustainably support 39-62 million sheep units. However, grazing patterns have changed significantly in recent decades due to changes in livestock management, population increase, and intensification of land use. In many alpine grasslands, overgrazing has resulted from increased stocking densities, decentralization of herders, and decreased grazing mobility (Chen et al., 2025). These alterations endanger local people' food security and financial stability by causing vegetation degradation, decreased biomass output, soil compaction, and decreases in palatable fodder species.

Beyond the availability of fodder, overgrazing has negative ecological effects. Poor grasslands have lowered the soil organic matter and water infiltration capacity, and thus makes it difficult to regulate services such as soil conservation and water management (Qian, Li, et al., 2023). The study carried out in the Qinghai-Tibetan Plateau has shown that extreme grazing reduces soil enzymes activity by 3953, rises toxic weeds by 15, and reduces edible species biomass by 57 (Chen et al., 2025).

Alternatively, properly controlled grazing systems can enhance the resilience of the ecosystem through enhancement of plant diversity, nutrient cycles and carbon storage. There are many recognized adaptive grazing management strategies that can be used as a balance between the productivity level of livestock and the sustainability of the ecosystem, such as rotational grazing, seasonal resting, and adjusting the stocking rate (Yahdjian et al., 2026). Studies at rangelands in different parts of the world have indicated that moderate grazing levels may ensure the preservation of plant richness and quality of fodder as well as pollination and biodiversity services (Liu et al., 2025).

5.2 Medicinal and aromatic Plants

Because of their diverse climate, high endemism, and evolutionary adaptations, alpine ecosystems are vital global reservoirs of aromatic and medicinal plant variety (Kougioumoutzis et al., 2024). Several alpine plant species are crucial to primary healthcare for mountain communities and are used in traditional medical systems such as Tibetan, Ayurvedic, and Andean ethnomedicine. Furthermore, in isolated high-altitude areas, the commercialization of aromatic and therapeutic plants has formed significant revenue-generating prospects (Chandra, Singh, Lingwal, Jalal, et al., 2022).

The availability and sustainability of medicinal plant resources are significantly impacted by land-use practices. While land modification for infrastructure or agriculture directly destroys habitats, overgrazing can harm delicate plant species and hinder regrowth (Chandra, Singh, Lingwal, Bisht, et al., 2022). High-value species are further stressed by uncontrolled harvesting driven by market demand, which commonly results in local extinctions and population decreases. The main risk to species survival in Western Himalayan regions is habitat degradation due to extreme grazing and commercial collecting, according to documentation of vulnerable alpine medicinal plants (Chandra, Singh, Lingwal, Jalal, et al., 2022). However, land use policies that integrate conservation and sustainable usage have demonstrated encouraging outcomes. However, land-use policies that integrate conservation and sustainable usage have demonstrated encouraging outcomes (Gupta et al., 2023). Medicinal plant populations can be strengthened through the establishment of protected areas, rotational collection, harvesting quotes, and community-based resource management. In some regions, the cultivation of medicinal plants in lower-altitude zones or agroforestry systems have reduced the stress on wild populations while maintaining economic opportunities. Incorporating the conservation of medicinal plants into land-use planning is essential for preserving this valuable cultural and economic resource.

5.3 Fresh water resources

Among the most important ecosystem services that alpine areas offer is freshwater provision whose benefits go well beyond the local area. The high altitude ecosystems act as natural water towers and store the precipitation as wetlands, snow and glaciers and then discharge it slowly into the rivers and reservoirs situated lower down. This is an essential service that helps in the irrigation, hydropower generation, supply of drinking water and the health of the ecosystem of the lowland areas. Proper alpine water management systems are critical to the water security of regions and the entire world as the freshwater sources of over fifty percent of the population of the world are highly regulated by the land use (Feng et al., 2025). Elements such as soil structure, vegetation cover and land management techniques have a great impact on infiltration, runoff production and sediment transport. Overgrazing and vegetation degradation lead to increased surface runoff and reduced soil permeability, which leads to more apparent hydrological responses and reduced water availability in the dry season (Tian et al., 2025).

Nutrient loading and sedimentation are rising as a result of settlement growth and agricultural expansion, which may have an adverse effect on downstream users. Conversely, conservation-oriented land use strategies such afforestation, grassland restoration, and wetland preservation improve water filtration and retention (Qian, Fan, et al., 2023). These methods help preserve water quality, lower flood peaks, and stabilize streamflow. Water storage capacity is greatly increased by intact vegetation and efficient soil microclimate management, according to research on alpine catchments. Grasslands dominated by bryophytes store more water than places devoid of vegetation (Jaroszynska et al., 2023). Protecting freshwater provisioning services through sustainable land-use management in alpine regions has become a global priority due to rising downstream water demand and climate-induced changes in snowmelt dynamics.

5.4 Food and fuel resources

Alpine land-use patterns sustain subsistence livelihoods by offering a variety of food and fuel resources in addition to livestock products. Crop cultivation is mostly limited to terraced slopes and lower-altitude valleys where the climate allows for brief growing seasons. Staple crops and vegetables are produced by these agricultural systems to supplement pastoral efficiency and improve food security.

The stability between food provisioning and ecosystem sustainability has altered due to changes in land usage. In the near term, increasing cultivated land frequently lifts food production, but it can also result in soil degradation, water scarcity, and the loss of regulating services. Land abandonment brought on by regulatory constraints or outmigration may decrease local food supply in marginal areas while allowing the natural vegetation to rebound.

For many mountain households, fuel resources, especially fuelwood and biomass remain vital sources of energy. In forest-alpine transition zones, unsustainable fuelwood extraction can undermine slope stability and deteriorate plant cover. While preserving vital provisioning functions, promoting alternate energy sources and increasing energy efficiency might lessen the strain on alpine ecosystems.

6 REGULATING ECOSYSTEM SERVICES AND LAND USE

Due to the steep topography, poor soils, and sensitivity to extreme weather events, regulating ecosystem services is disproportionately important in alpine locations. These services support ecosystem stability and shield inhabitants downstream and in the mountains from environmental threats. In high-altitude environments, land-use decisions have a significant impact on how well services are regulated (Feng et al., 2025).

6.1 Climate regulation and carbon sequestration

By storing carbon in plants and soils, alpine ecosystems help control the climate (Bai et al., 2025). Alpine grasslands and shrub-lands store significant amounts of carbon belowground, with alpine meadows sequestering 26-79 Tg C yearly over the Tibetan Plateau, although generally having smaller biomass stocks than lowland forests (Wang et al., 2023). Carbon dynamics are significantly impacted by land-use activities such afforestation, vegetation clearance, and grazing intensity (Qu et al., 2022).

Plant biomass and soil organic carbon are reduced by overgrazing and land degradation, which releases greenhouse gases and impairs climate regulating processes (Q. Zhang et al., 2024). On the other hand, ecological restoration techniques, such as reduced disturbance and revegetation, greatly increase the capacity for carbon sequestration; revegetation increases carbon storage by quantifiable amounts and fodder supply by considerable percentages (Du et al., 2025). However, by speeding up the breakdown of soil carbon and changing the character of vegetation, climate change presents new difficulties (Bai et al., 2025). For alpine regions to continue providing carbon control services, land-use management must be integrated with climate adaptation plans.

6.2 Water regulation and flood control

Water regulation and flood control

Water regulation is one of the main regulatory roles of alpine ecosystems (Li et al., 2022). Soil composition, vegetation cover, and landscape design all affect how much water alpine catchments can hold and discharge over time. According to (Viviroli et al., 2011), intact wetlands, woodlands, and grasslands are natural buffers which reduce the speed of runoff and minimize the flood peaks. Nevertheless, land-use intensification and increase in infrastructure can disrupt the mechanisms by increasing impermeable surfaces, lowering infiltration, and compaction of soils (Qian, Fan, et al., 2023). Consequently, it can be expected that this can lead to the higher risk of floods, particularly in periods of high precipitation. Land-use planning is important to the management of alpine watersheds, as alpine river research has shown that vegetation restoration programs that enhance the plant cover and soil stability are effective in improving water management and reducing the risks of floods(Carolli et al., 2025).

6.3 Soil conservation and erosion control

In the alpine regions, the steep slopes and the harsh weather are the ones that increase the chances of soil erosion making soil protection critical in such areas. Land-use activities such as the clearing of vegetation, trampling, and cultivation have a direct effect on soil stability. Problems such as increased erosion, nutrient loss and reduced productivity are typical of degraded sites and result in feedback processes that further worsen the ecosystem services (Macicasan et al., 2024). Examples of sustainable land management practices that can significantly reduce soil erosion and raise land productivity include terracing, slope stabilization, and controlled grazing (Zhang et al., 2025). Moreover, efficient maintenance of soil conservation services must be provided in the long term provisioning services such as grazing and agriculture.

6.4 Natural hazard regulation (avalanches, landslides)

Alpine ecosystems are essential for preventing natural disasters including avalanches, landslides, and debris flows (Barredo et al., 2020). Vegetation cover restricts snow movement in avalanche-prone locations and stabilizes slopes by strengthening roots. Transportation lines, infrastructure, and towns are all protected by forests and shrublands. However, road construction, deforestation, and the development of tourism infrastructure are examples of land use changes that decrease plant cover and increase susceptibility to environmental risks. Furthermore, by changing precipitation intensity and snowfall patterns, climate change increases these hazards (Hurtado et al., 2024). Therefore, in order to improve the resilience of alpine socio-ecological systems in the face of increasing climate unpredictability, hazard regulation must be incorporated into land-use planning and conservation efforts. Alpine ecosystems are ultimately dynamic socio-ecological systems, and the intricate linkages between ecosystem services are influenced by land-use choices.

7. CONCLUSION AND FUTURE PERSPECTIVES

Even though they make up a very small portion of the Earth's surface, mountainous areas offer vital ecosystem services to both locals and people who live downstream. This chapter illustrates that alterations in land use which include expansion of arable land, grazing, tourism, and ecological restoration directly affect the provision of cultural, regulatory, and supporting service. Although regulatory and supporting roles may be interfered with occasionally by short-term enhancement of service delivery, conservation and restoration have demonstrated the ability to reestablish ecological equilibrium and generate long-term gains. The interaction of multiple driving factors also reveals the analysis of how the dynamic nature of ecosystem services is created. The impact of climate change, socioeconomic changes and governmental policies vary in dependence on local habitats and altitude gradients. These findings point to the necessity of site-based, flexible solutions instead of general solutions. Every three important strategies should take the first place going forward. To ensure the short-term development does not affect the long-term ecological stability, first, integrated land-use planning needs to consider the trades-offs and synergies of ecosystem services. Second the governance systems that involve the local communities, parliamentarians and scientists can be used to enhance the management through making decisions that are social and environmentally responsible. Lastly, scenario analysis, long-term monitoring and comprehensive spatial modeling should all be used in future studies to better predict how climate change and socioeconomic changes will impact alpine ecosystem services. The coordination of land-use patterns and ecological boundaries and climate adaption methods can help high-altitude areas to protect their essential ecosystem services and to sustainably serve mountain communities. The knowledge acquired in these high-altitude locations can also be used in strategies of other vulnerable ecosystems in the world that face similar problems because of global climate change.

REFERENCES

Ali, A., Khan, M. A., Ashraf, M. I., & Goheer, M. A. (2021). Economic valuation of ecosystem services in alpine pastures of western Himalayas: A case study of Kaghan valley, Pakistan. *Pakistan Journal of Botany, 53*(5), 1801–1806. DOI: 10.30848/PJB2021-5(18)

Bai, Y., Peng, Y., Zhang, D., Yang, G., Chen, L., Kang, L., Zhou, W., Wei, B., Xie, Y., & Yang, Y. (2025). Heating up the roof of the world: Tracing the impacts of in-situ warming on carbon cycle in alpine grasslands on the Tibetan Plateau. *National Science Review, 12*(2), nwae371. DOI: 10.1093/nsr/nwae371 PMID: 39872222

Barredo, J. I., Mauri, A., & Caudullo, G. (2020). Alpine tundra contraction under future warming scenarios in Europe. *Atmosphere (Basel), 11*(7), 698. DOI: 10.3390/atmos11070698

BöHmer. H. J. (2011). Vulnerability of tropical montane rain forest ecosystems due to climate change. In *Coping with Global Environmental Change, Disasters and Security: Threats, Challenges, Vulnerabilities and Risks* (pp. 789-802). Springer.

Broadbent, A. A., Snell, H. S., Michas, A., Pritchard, W. J., Newbold, L., Cordero, I., Goodall, T., Schallhart, N., Kaufmann, R., Griffiths, R. I., Schloter, M., Bahn, M., & Bardgett, R. D. (2021). Climate change alters temporal dynamics of alpine soil microbial functioning and biogeochemical cycling via earlier snowmelt. *The ISME Journal, 15*(8), 2264–2275. DOI: 10.1038/s41396-021-00922-0 PMID: 33619353

Buytaert, W., Cuesta-Camacho, F., & Tobón, C. (2011). Potential impacts of climate change on the environmental services of humid tropical alpine regions. *Global Ecology and Biogeography, 20*(1), 19–33. DOI: 10.1111/j.1466-8238.2010.00585.x

Carolli, M., Geneletti, D., Majone, B., Bellin, A., & Zolezzi, G. (2025). Effects of climate change and anthropic water uses on ecosystem services provided by an Alpine river. *Environmental Research Letters, 20*(8), 084036. DOI: 10.1088/1748-9326/ade902

Chandra, N., Singh, G., Lingwal, S., Bisht, M., Tewari, L. M., & Joshi, V. C. (2022). Ecological status of alpine medicinal and aromatic plants of Western Himalaya. *Journal of Herbs, Spices & Medicinal Plants, 28*(1), 73–88. DOI: 10.1080/10496475.2021.1963903

Chandra, N., Singh, G., Lingwal, S., Jalal, J., Bisht, M., Pal, V., Bisht, M., Rawat, B., & Tiwari, L. (2022). Ecological niche modeling and status of threatened alpine medicinal plant Dactylorhiza hatagirea D. Don in Western Himalaya. *Journal of Sustainable Forestry, 41*(10), 1029–1045. DOI: 10.1080/10549811.2021.1923530

Chen, B., Ma, X., Zhou, X., Zhang, X., Wang, X., Li, Z., Yang, X., Lu, S., & Du, W. (2025). Medium-Term Effect of Livestock Grazing Intensities on the Vegetation Dynamics in Alpine Meadow Ecosystems. *Land (Basel)*, *14*(3), 591. DOI: 10.3390/land14030591

Costanza, R. (2016). Ecosystem services in theory and practice. In *Routledge handbook of ecosystem services* (pp. 15–24). Routledge. DOI: 10.4324/9781315775302-3

Dong, S., Shang, Z., Gao, J., & Boone, R. B. (2020). Enhancing sustainability of grassland ecosystems through ecological restoration and grazing management in an era of climate change on Qinghai-Tibetan Plateau. *Agriculture, Ecosystems & Environment*, *287*, 106684. DOI: 10.1016/j.agee.2019.106684

dos Santos, R. F., Antunes, P., Ring, I., & Clemente, P. (2015). Engaging local private and public actors in biodiversity conservation: The role of agri-environmental schemes and ecological fiscal transfers. *Environmental Policy and Governance*, *25*(2), 83–96. DOI: 10.1002/eet.1661

Du, J., Wei, P., Bahadur, A., & Chen, S. (2025). The Impacts of Revegetation on Ecosystem Services in the Extremely Degraded Alpine Grassland of Permafrost Regions. *Sustainability (Basel)*, *17*(4), 1512. DOI: 10.3390/su17041512

Duchicela, S. A., Llambí, L. D., Bonnesoeur, V., & Román-Dañobeytia, F. (2024). Pastoralism in the high tropical Andes: A review of the effect of grazing intensity on plant diversity and ecosystem services. *Applied Vegetation Science*, *27*(3), e12791. DOI: 10.1111/avsc.12791

Fedrigotti, C., Aschonitis, V., & Fano, E. A. (2016). Effects of forest expansion and land abandonment on ecosystem services of alpine environments: Case study in ledro valley (Italy) for the period 1859–2011. *Global NEST Journal*, *18*(4), 875–884. DOI: 10.30955/gnj.001770

Feng, B., Zhou, J., Hu, L., Liu, Z., Yang, Y., Yang, S., Ni, J., Bai, W., & Zhao, S. (2025). Land Use and Ecosystem Service Value Spatiotemporal Dynamics, Topographic Gradient Effect and Their Driving Factors in Typical Alpine Ecosystems of the East Qinghai-Tibet Plateau: Implications for Conservation and Development. *Ecology and Evolution*, *15*(3), e71125. DOI: 10.1002/ece3.71125 PMID: 40099216

Garcés-Pastor, S., Coissac, E., Lavergne, S., Schwoerer, C., Theurillat, J.-P., Heintzman, P., Wangensteen, O., Tinner, W., Rey, F., & Heer, M. (2021). High resolution ancient sedimentary DNA shows that alpine plant biodiversity is a result of human land use.

Geneletti, D., Scolozzi, R., & Adem Esmail, B. (2018). Assessing ecosystem services and biodiversity tradeoffs across agricultural landscapes in a mountain region. *The International Journal of Biodiversity Science, Ecosystem Services & Management, 14*(1), 188–208. DOI: 10.1080/21513732.2018.1526214

Grêt-Regamey, A., Brunner, S. H., & Kienast, F. (2012). Mountain ecosystem services: Who cares? *Mountain Research and Development, 32*(S1), S23–S34. DOI: 10.1659/MRD-JOURNAL-D-10-00115.S1

Gupta, S., Jasrotia, S., & Purohit, M. (2023). Conservation, cultivation, and sustainable use of high altitude medicinal and aromatic plants for socio-economic development: A review. *Journal of Mountain Research, 18*(1), 201–209. DOI: 10.51220/jmr.v18i1.22

Hazarika, A., Nath, A. J., Reang, D., Pandey, R., Sileshi, G. W., & Das, A. K. (2024). Climate change vulnerability and adaptation among farmers practicing shifting agriculture in the Indian Himalayas. *Environmental and Sustainability Indicators, 23*, 100430. DOI: 10.1016/j.indic.2024.100430

Hong, T., & Chongjian, Y. (2024). Temporal and Spatial Evolution of Ecosystem Service Supply and Demand in the Tibetan Plateau: Implications for Land Use Patterns and Relationships. *Polish Journal of Environmental Studies, 33*(3), 2679–2691. DOI: 10.15244/pjoes/175059

Hopping, K. A., Knapp, A. K., Dorji, T., & Klein, J. A. (2018). Warming and land use change concurrently erode ecosystem services in Tibet. *Global Change Biology, 24*(11), 5534–5548. DOI: 10.1111/gcb.14417 PMID: 30086187

Huang, L., Cao, W., Xu, X., Fan, J., & Wang, J. (2018). Linking the benefits of ecosystem services to sustainable spatial planning of ecological conservation strategies. *Journal of Environmental Management, 222*, 385–395. DOI: 10.1016/j.jenvman.2018.05.066 PMID: 29870967

Huber, L., Schirpke, U., Marsoner, T., Tasser, E., & Leitinger, G. (2020). Does socioeconomic diversification enhance multifunctionality of mountain landscapes? *Ecosystem Services, 44*, 101122. DOI: 10.1016/j.ecoser.2020.101122

Humphries, H. C. (2020). Alpine ecosystems in temperate mountains of North America.

Hurtado, P., de Los Ríos, A., & Arias-Real, R. (2024). (Dis) connecting the Globe Through Water-Driven Ecological and Biogeochemical Corridors in the Polar-Alpine Biome. *Global Change Biology, 30*(12), e17606. DOI: 10.1111/gcb.17606 PMID: 39614421

Ingty, T. (2021). Pastoralism in the highest peaks: Role of the traditional grazing systems in maintaining biodiversity and ecosystem function in the alpine Himalaya. *PLoS One*, *16*(1), e0245221. DOI: 10.1371/journal.pone.0245221 PMID: 33411837

Jaroszynska, F., Althuizen, I., Halbritter, A. H., Klanderud, K., Lee, H., Telford, R. J., & Vandvik, V. (2023). Bryophytes dominate plant regulation of soil microclimate in alpine grasslands. *Oikos*, *2023*(12), e10091. DOI: 10.1111/oik.10091

Kaltenborn, B. P., Kaltenborn, E. F., & Linnell, J. D. (2019). It's All About the Scenery. *Arctic*, *72*(1), 1–12. DOI: 10.14430/arctic67944

Khazieva, E., Verburg, P. H., & Pazúr, R. (2022). Grassland degradation by shrub encroachment: Mapping patterns and drivers of encroachment in Kyrgyzstan. *Journal of Arid Environments*, *207*, 104849. DOI: 10.1016/j.jaridenv.2022.104849

Klein, J. A., Tucker, C. M., Nolin, A. W., Hopping, K. A., Reid, R. S., Steger, C., Grêt-Regamey, A., Lavorel, S., Müller, B., Yeh, E. T., Boone, R. B., Bourgeron, P., Butsic, V., Castellanos, E., Chen, X., Dong, S. K., Greenwood, G., Keiler, M., Marchant, R., & Yager, K. (2019). Catalyzing transformations to sustainability in the world's mountains. *Earth's Future*, *7*(5), 547–557. DOI: 10.1029/2018EF001024

Körner, C. (2004). Mountain biodiversity, its causes and function. *Ambio*, *33*(sp13), 11–17. DOI: 10.1007/0044-7447-33.sp13.11 PMID: 15575177

Kou, X., Zhao, J., & Sang, W. (2024). Impact of Typical Land Use Expansion Induced by Ecological Restoration and Protection Projects on Landscape Patterns. *Land (2012)*, *13*(9).

Kougioumoutzis, K., Tsakiri, M., Kokkoris, I. P., Trigas, P., Iatrou, G., Lamari, F. N., Tzanoudakis, D., Koumoutsou, E., Dimopoulos, P., Strid, A., & Panitsa, M. (2024). Assessing the vulnerability of medicinal and aromatic plants to climate and land-use changes in a Mediterranean biodiversity hotspot. *Land (Basel)*, *13*(2), 133. DOI: 10.3390/land13020133

Lavorel, S., Quetier, F., Thebault, A., Liancourt, P., Daigney, S., Davies, I., & de Chazal, J. (2006). Vulnerability to land use change of services provided by alpine landscapes. *Open Science Conference on Global Change in Mountain Regions (GLOCHAMORE 2005)*

Li, Y., Zhang, X., Shao, Q., Fan, J., Chen, Z., Dong, J., Hu, Z., & Zhan, Y. (2022). Community composition and structure affect ecosystem and canopy water use efficiency across three typical alpine ecosystems. *Frontiers in Plant Science*, *12*, 771424.

Liu, J., Wang, X., Miao, H., & Yang, Y. (2025). Community Beta Diversity as a Catalyst: Mediating Livestock Grazing Effects on Ecosystem Multifunctionality in the Qinghai-Tibet Plateau. *Land Degradation & Development*.

Liu, X., Ding, J., & Zhao, W. (2023). Divergent responses of ecosystem services to afforestation and grassland restoration in the Tibetan Plateau. *Journal of Environmental Management*, *344*, 118471. DOI: 10.1016/j.jenvman.2023.118471 PMID: 37364488

Liu, Y., Lü, Y., Zhao, M., & Fu, B. (2023). Multiple pressures and vegetation conditions shape the spatiotemporal variations of ecosystem services in the Qinghai-Tibet Plateau. *Frontiers in Plant Science*, *14*, 1127808. DOI: 10.3389/fpls.2023.1127808 PMID: 36743496

Ma, B., Singh, R. B., & Hietala, R. (2012). Human driving forces for ecosystem services in the Himalayan region. *Environment and Ecology*, (3, Iss. 1), 53–57.

Macicasan, V., Ocrain, A., Balc, R., Dicu, T., Bodmer, M. D., & Roba, C. (2024). How can land use management in traditional cultural landscapes become a policy instrument for soil organic carbon sequestration and climate change mitigation? A Transylvanian case study. *Applied Sciences (Basel, Switzerland)*, *14*(21), 9851. DOI: 10.3390/app14219851

Mastrotheodoros, T., Pappas, C., Molnar, P., Burlando, P., Hadjidoukas, P., & Fatichi, S. (2019). Ecohydrological dynamics in the Alps: Insights from a modelling analysis of the spatial variability. *Ecohydrology*, *12*(1), e2054. DOI: 10.1002/eco.2054

Mengist, W., Soromessa, T., & Legese, G. (2020). Ecosystem services research in mountainous regions: A systematic literature review on current knowledge and research gaps. *The Science of the Total Environment*, *702*, 134581. DOI: 10.1016/j.scitotenv.2019.134581 PMID: 31734614

Messerli, P. (2015). Die Alpen–Geschichte und Zukunft einer europäischen Kulturlandschaft [The Alps—History and Future of a European Cultural Landscape]. *Mountain Research and Development*, *35*(4), 428–429. DOI: 10.1659/mrd.mm171

Meyfroidt, P., De Bremond, A., Ryan, C. M., Archer, E., Aspinall, R., Chhabra, A., Camara, G., Corbera, E., DeFries, R., Díaz, S., Dong, J., Ellis, E. C., Erb, K.-H., Fisher, J. A., Garrett, R. D., Golubiewski, N. E., Grau, H. R., Grove, J. M., Haberl, H., & zu Ermgassen, E. K. H. J. (2022). Ten facts about land systems for sustainability. *Proceedings of the National Academy of Sciences of the United States of America*, *119*(7), e2109217118. DOI: 10.1073/pnas.2109217118 PMID: 35131937

Oldekop, J. A., Holmes, G., Harris, W. E., & Evans, K. L. (2016). A global assessment of the social and conservation outcomes of protected areas. *Conservation Biology, 30*(1), 133–141. DOI: 10.1111/cobi.12568 PMID: 26096222

Osano, P. M., Said, M. Y., de Leeuw, J., Ndiwa, N., Kaelo, D., Schomers, S., Birner, R., & Ogutu, J. O. (2013). Why keep lions instead of livestock? Assessing wildlife tourism-based payment for ecosystem services involving herders in the M aasai M ara, K enya. Natural Resources Forum, Pauli, H., & Halloy, S. R. (2019). High mountain ecosystems under climate change. In *Oxford research encyclopedia of climate science*.

Prabhakar, S. (2021). A succinct review and analysis of drivers and impacts of agricultural land transformations in Asia. *Land Use Policy, 102*, 105238. DOI: 10.1016/j.landusepol.2020.105238

Qian, D., Fan, B., Lan, Y., Si, M., Li, Q., & Guo, X. (2023). Ecosystem service relationships, drivers, and regulation strategies in a degraded alpine shrub meadow on the northeastern Qinghai-Tibetan plateau. *Diversity, 15*(5), 596. DOI: 10.3390/d15050596

Qian, D., Li, Q., Guo, X., Fan, B., Lan, Y., Si, M., & Cao, G. (2023). Ecosystem services relationship characteristics of the degraded alpine shrub meadow on the Qinghai-Tibetan Plateau. *Ecology and Evolution, 13*(7), e10351. DOI: 10.1002/ece3.10351 PMID: 37492459

Qiu, J., Queiroz, C., Bennett, E. M., Cord, A. F., Crouzat, E., Lavorel, S., Maes, J., Meacham, M., Norström, A. V., Peterson, G. D., Seppelt, R., & Turner, M. G. (2021). Land-use intensity mediates ecosystem service tradeoffs across regional social-ecological systems. *Ecosystems and People (Abingdon, England), 17*(1), 264–278. DOI: 10.1080/26395916.2021.1925743

Qu, R., He, L., He, Z., Wang, B., Lyu, P., Wang, J., Kang, G., & Bai, W. (2022). A study of carbon stock changes in the Alpine Grassland Ecosystem of Zoigê, China, 2000–2020. *Land (Basel), 11*(8), 1232. DOI: 10.3390/land11081232

Rindt, O., Rosinger, C., Bonkowski, M., Rixen, C., Brüggemann, N., Urich, T., & Fiore-Donno, A. M. (2023). Biogeochemical dynamics during snowmelt and in summer in the Alps. *Biogeochemistry, 162*(2), 257–266. DOI: 10.1007/s10533-022-01005-8

Saleem, S., Rather, J. A., Ahmad, S., Ahmed, R., & Hajam, L. A. (2024). Assessment of livestock carrying capacity in the alpine grasslands of the Kashmir Himalayas. *Pastoralism: Research. Policy & Practice, 14*, 13541.

Schirpke, U., Leitinger, G., Tasser, E., Schermer, M., Steinbacher, M., & Tappeiner, U. (2013). Multiple ecosystem services of a changing Alpine landscape: Past, present and future. *The International Journal of Biodiversity Science, Ecosystem Services & Management*, *9*(2), 123–135. DOI: 10.1080/21513732.2012.751936 PMID: 31491038

Schirpke, U., Timmermann, F., Tappeiner, U., & Tasser, E. (2016). Cultural ecosystem services of mountain regions: Modelling the aesthetic value. *Ecological Indicators*, *69*, 78–90. DOI: 10.1016/j.ecolind.2016.04.001 PMID: 27482152

Schmitt, M., T., Haensel, M., Kaim, A., Lee, H., Reinermann, S., & Koellner, T. (. (2024). Recreation and its synergies and trade-offs with other ecosystem services of Alpine and pre-Alpine grasslands. *Regional Environmental Change*, *24*(2), 57. DOI: 10.1007/s10113-024-02213-8

Shen, J., Li, S., Liu, L., Liang, Z., Wang, Y., Wang, H., & Wu, S. (2021). Uncovering the relationships between ecosystem services and social-ecological drivers at different spatial scales in the Beijing-Tianjin-Hebei region. *Journal of Cleaner Production*, *290*, 125193. DOI: 10.1016/j.jclepro.2020.125193

Tappeiner, U., & Tasser, E. I. (2019). *Historical Land Uses and Their Changes in the European Alps*. Oxford Bibliographies. DOI: 10.1093/obo/9780199363445-0114

Tasser, E., Leitinger, G., Tappeiner, U., & Schirpke, U. (2024). Shaping the European Alps: Trends in landscape patterns, biodiversity and ecosystem services. *Catena*, *235*, 107607. DOI: 10.1016/j.catena.2023.107607

Tian, Y., Liu, Z., Fan, Y., Li, Y., Tao, H., Han, C., Ao, X., & Chen, R. (2025). Patterns and Drivers of Surface Energy Flux in the Alpine Meadow Ecosystem in the Qilian Mountains, Northwest China. *Plants*, *14*(2), 155. DOI: 10.3390/plants14020155 PMID: 39861508

Tiwari, P. C., & Joshi, B. (2014). Environmental changes and their impact on rural water, food, livelihood, and health security in Kumaon Himalayas. *Journal of Urban and Regional studies on contemporary India, 1*(1), 1-12.

Verhagen, W., Van Teeffelen, A. J., Baggio Compagnucci, A., Poggio, L., Gimona, A., & Verburg, P. H. (2016). Effects of landscape configuration on mapping ecosystem service capacity: A review of evidence and a case study in Scotland. *Landscape Ecology*, *31*(7), 1457–1479. DOI: 10.1007/s10980-016-0345-2

Viviroli, D., Archer, D. R., Buytaert, W., Fowler, H. J., Greenwood, G. B., Hamlet, A. F., Huang, Y., Koboltschnig, G., Litaor, M., López-Moreno, J. I., Lorentz, S., Schädler, B., Schreier, H., Schwaiger, K., Vuille, M., & Woods, R. (2011). Climate change and mountain water resources: Overview and recommendations for research, management and policy. *Hydrology and Earth System Sciences, 15*(2), 471–504. DOI: 10.5194/hess-15-471-2011

Wang, Y., Xiao, J., Ma, Y., Ding, J., Chen, X., Ding, Z., & Luo, Y. (2023). Persistent and enhanced carbon sequestration capacity of alpine grasslands on Earth's Third Pole. *Science Advances, 9*(20), eade6875. DOI: 10.1126/sciadv.ade6875 PMID: 37196073

Yahdjian, L., Campana, S., Tognetti, P. M., Alberti, J., Graff, P., Molina, C., Borer, E. T., Seabloom, E. W., Prober, S. M., MacDougall, A. S., Risch, A. C., Price, J. N., Power, S. A., Barrio, I. C., Hersch-Green, E., Fay, P. A., Bagchi, S., Bakker, J. D., Blumenthal, D., & Virtanen, R. (2026). Insights on global rangeland ecosystem services shaped by grazing and fertilization. *Frontiers in Ecology and the Environment, 24*(3), 70022. DOI: 10.1002/fee.70022

Yang, H., Gou, X., Xue, B., Ma, W., Kuang, W., Tu, Z., Gao, L., Yin, D., & Zhang, J. (2023). Research on the change of alpine ecosystem service value and its sustainable development path. *Ecological Indicators, 146*, 109893. DOI: 10.1016/j.ecolind.2023.109893

Zhang, J., Ren, X., Xu, E., Evans, A. M., Jing, W., Wang, R., Jia, X., Bi, M., Amoah, I. D., Pohlmann, M., Mecha, C., & Smith, C. K. (2025). Soil Chemical Properties Along an Elevational Gradient in the Alpine Shrublands of the Northeastern Tibetan Plateau. *Soil Systems, 9*(3), 95. DOI: 10.3390/soilsystems9030095

Zhang, Q., Chen, X., Zhou, X., Nie, X., Liu, G., Zhuang, G., Zheng, G., Fortin, D., & Ma, A. (2024). Tibetan Plateau grasslands might increase sequestration of microbial necromass carbon under future warming. *Communications Biology, 7*(1), 686. DOI: 10.1038/s42003-024-06396-y PMID: 38834864

Zhang, Y., Wang, Y., Yang, H., Hull, V., Zhang, J., Wang, F., Zhao, Z., & Liu, J. (2024). Heterogeneous and interactive effects of payments for ecosystem services on household income across giant panda nature reserves. *Heliyon, 10*(15), e34866. DOI: 10.1016/j.heliyon.2024.e34866 PMID: 39145020

Zhen, L., Du, B., Wei, Y., Xiao, Y., & Sheng, W. (2018). Assessing the effects of ecological restoration approaches in the alpine rangelands of the Qinghai-Tibetan Plateau. *Environmental Research Letters, 13*(9), 095005. DOI: 10.1088/1748-9326/aada51

Zierl, B., & Bugmann, H. (2005). Global change impacts on hydrological processes in Alpine catchments. *Water Resources Research, 41*(2), 2004WR003447. DOI: 10.1029/2004WR003447

Chapter 2
Climate Change and Its Impact on Alpine Vegetation

Sikandar Shah
http://orcid.org/0009-0002-5503-8236
Institute of Biology, Warsaw University of Life Sciences, Poland

Sheharyar Khan
College of Environmental Science and Engineering, Tianjin, China

Sadaf
University of Peshawar, Pakistan

Sulaiman
University Putra Malaysia, Malaysia

ABSTRACT

Alpine vegetation, which covers less than 3% of Earth's land surface, plays a vital role in supporting biodiversity, regulating water resources, and safeguarding soil integrity. These habitats are exceedingly susceptible to the impacts of climate change, owing to their narrow temperature tolerances, brief growing seasons, and distinctive plant communities adapted to extreme environmental conditions. Consequently, species that favor warmer conditions tend to migrate upward, leading to habitat loss for cold-adapted flora and heightened competition at lower elevations. Phenological changes, including earlier flowering and extended growing seasons, disrupt pollination and nutrient cycling. This situation underscores the pressing

DOI: 10.4018/979-8-2600-1207-9.ch002

necessity for enhanced monitoring, conservation efforts and research initiatives, particularly in understudied regions and specialized communities to avert irreversible environmental degradation.

INTRODUCTION

The Intergovernmental Panel on Climate Change (IPCC) stated that the global average air temperature increased by 0.6 ± 0.2 °C during the 20th century (IPCC 2001). According to the World Meteorological Organization (WMO), 2024 was the warmest year on record since 1850, with the global average temperature around 1.5 °C above the preindustrial level, ranging from 1.48 to 1.52 °C (WMO, 2024). Moreover, global precipitation has generally become more intense and unpredictable, with much greater regional and seasonal variation. For example, the Middle East, parts of America, and North Africa show a declining trend, while the rest show an increasing trend (Mesquita et al., 2024). This unpredictability makes it difficult for the model to predict future changes in climate-driven disasters worldwide.

Due to a 1% annual increase in greenhouse gas (GHG) emissions, temperature fluctuations will also continue to rise (Rivera et al., 2024). Which causes many unpredictable events, such as forest fires in the mountains, Alpine degradation, rapid snowmelt, and early-flowering-induced pollinator mismatches. The rapid, unpredictable changes in the climate and global warming have sparked widespread concern among the global research community, and future predictions require advanced tools, additional effort, and funding to deploy them on mountains. Moreover, the rise in temperature was not uniform worldwide, and some areas of the alpine regions were experiencing air temperature increases twice as large to the global temperature changes (Böhm et al., 2001).

The spatiotemporal changes in rainfall and snowfall patterns were also not uniform, and according to the Intergovernmental Panel on Climate Change, some areas show around 10% increased in the precipitation between 1900 to 2000 (IPCC, 2002 and 2023). Frequent heat waves, rising temperatures and sea levels, melting glaciers, unpredictable rainfall, ozone exposure, intense and frequent flooding and droughts, along with pathogens and pests, are among the signs of climate-induced changes (Mack et al., 2008; NASA, 2024; European Commission, n.d.). These events, individually and in combination, pose significant risks to global biodiversity and to the structure, function, and composition of ecosystems, with solid evidence of extinction and negative impacts worldwide. Recent studies show that global climate change, particularly rising and fluctuating temperatures, has significantly affected many natural ecosystems (Field & Barros, 2014). One of the worst scenarios is that the effects of climate change are much more evident in mountain environments, because the alpine ecosystem responds very quickly to global warming, precipitation fluctuations, and rising temperatures. This shows the role of alpine communities as early warning indicators of global climate change and as important experimental systems to study the effects of climate change on plants, animals, and their interactions.

Alpine vegetation starts above the bioclimatic tree line, and its vegetation structure and taxonomic composition are primarily influenced by relatively short growing seasons and low temperatures (Körner, 2003, 2012). However, complex topography, periglacial disturbance regimes, microclimatic heterogeneity, and variable substrates create tree line ecotones that form a mosaic of low-growing alpine grasslands and heathlands supporting endemism and high biodiversity (Körner, 2004; Elsen & Tingley, 2015; Perrigo et al., 2020). Worldwide, 2.6% of Earth's land outside Antarctica is classified as alpine landscape, which still encompasses many glaciated or barren areas, with actual plant-covered regions closer to 2% (Körner et al., 2011). The Alpine World clearly supports high biodiversity. This partly explains why alpine terrain is a relatively safe place for adapting to climate change (Körner, 1999). Most alpine plants are adapted to survive for long periods and withstand harsh conditions. Some are even older than earth's oldest trees (Körner & Hiltbrunner, 2021). Alpine zones contain some of Earth's rarest and most resilient plants. From specialized lichens to tiny shrubs and herbaceous plants, alpine vegetation is uniquely adapted to survive in cracks, cold temperatures, thin soil, high winds, and UV exposure (Körner, 2021; Rundel & Millar, 2016). These tough, persistent organisms are essential for preserving alpine ecosystems. In alpine environments, abiotic factors, especially climate, primarily influence biotic interactions. Although short-term climatic fluctuations are expected to have minimal impact on alpine vegetation due to the slow growth rates and long-life cycles of species (Grabherr et al., 2010; Pauli et al., 2003), high-altitude vegetation is probably more susceptible to long-term climate changes (Pauli & Halloy, 2019; Theurillat & Guisan, 2001). Climate change significantly increases the frequency and severity of mass-movement hazards in the Alpine region by weakening permafrost, glaciers, snow, and rock slopes. Based on a review of more than 300 scientific studies, the research shows that processes such as rock avalanches, rockfalls, debris flows, and ice and snow avalanches are occurring more frequently and are becoming less predictable (Jacquemart et al., 2024). This emphasizes the vulnerability of mountains, especially the Alps, as highly sensitive indicators of ongoing climate change (Jacquemart et al., 2024). Therefore, as the climate continues to warm, the stability of alpine ecosystems and vegetation becomes increasingly vulnerable (Grabherr et al., 2010). Although numerous other factors such as water, light availability, and nutrients can influence alpine vegetation dynamics, climate change is widely regarded as the primary factor driving declines in snow cover, variable precipitation patterns, longer growing seasons, and warmer temperatures (Körner, 2003; Hock et al., 2019). Climate change has also apparently led to the loss of highly specialized, cryophilic vegetation maintained by periglacial processes, such as snow patch communities, due to the encroachment of thermophilic and generalist alpine species (Verrall, 2023).

As climate change progresses, we expect more frequent and severe weather extremes, with impacts on rainfall, snowfall, evaporation, cloud cover and temperatures (Figure 1). However, future planning is uncertain due to limited data, especially in alpine zones. Further research is needed to monitor changes to improve prediction and planning. This chapter compiles all the information on climate impacts without thoroughly considering elevation specific processes such as snow insulation, permafrost dynamics, and short growing seasons. By focusing specifically on alpine vegetation, this chapter helps counteract this bias and highlights ecosystems that are both highly vulnerable and ecologically irreplaceable. Furthermore, research on alpine vegetation is often fragmented by region (e.g., Alps, Himalayas, Andes, Rockies), which hampers the ability to draw generalizable conclusions. This chapter synthesizes evidence from various mountain systems to elucidate common patterns and region-specific responses, thereby enhancing the applicability of the findings.

Figure 1. Schematic Overview of the pathways by which climate change affects the alpine vegetation

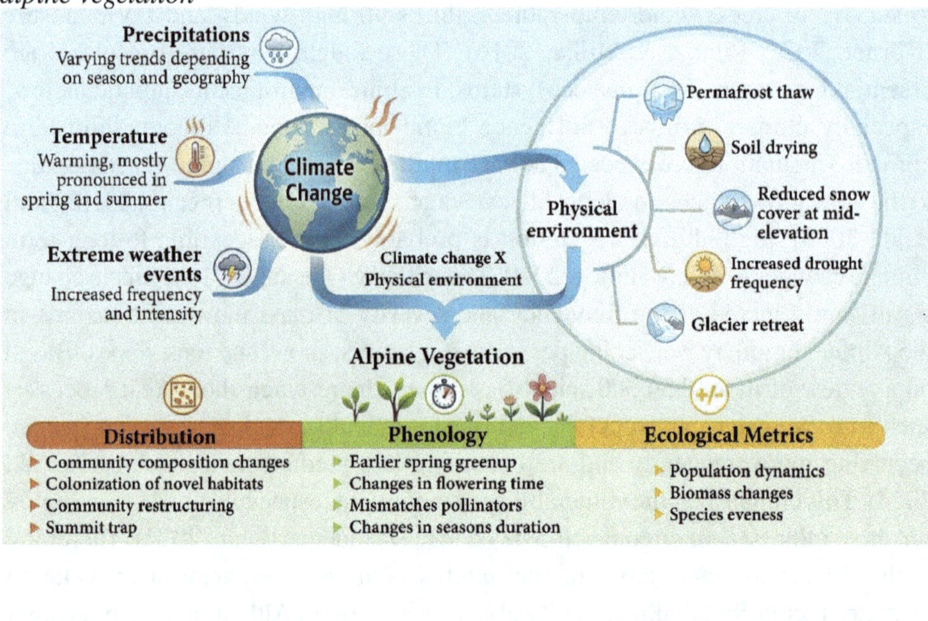

1. CLIMATE CHANGE IMPACTS ON ALPINE ECOSYSTEM FUNCTIONING

The Alpine region shows high plant endemism because its complex geography creates diverse environments and many ecological niches (Schickhoff et al., 2016). This habitat supports numerous unique species and is especially sensitive to global warming. This vulnerability arises from rapid climate change and the small size of high-elevation habitats, which makes it difficult for species to migrate to higher altitudes (Elsen & Tingley, 2015). Microclimatic and environmental factors contribute to greater species richness and endemism at higher elevations (Ashcroft & Gollan, 2013). Many species could lose up to 44% of their suitable habitat or undergo major shifts in their climatically suitable zones (Warren et al., 2018). Consequently, climate change is expected to substantially perturb alpine ecosystems, endangering their distinctive plant diversity and precipitating unforeseen taxonomic and functional alterations within alpine communities, alongside the possible extinction of endemic species (Colwell & Lees, 2000).

A study conducted in 2021 on ten rare vascular plant species across sixteen Northeastern alpine regions from 2006 to 2019 predicted declines due to climate change, atmospheric deposition, and recreational activities. Only two species demonstrated a statistically significant decline. Similar findings were observed in investigations carried out in 2010 and 2011. These trends are ascribed to increased temperatures, prolonged growing seasons, nitrogen deposition, and decreased trampling by recreational visitors (Howard et al., 2021). Long-term vegetation data from Master Schrankogel shows that climate change is causing an upward shift of warm-adapted species while reducing the abundance of cold-adapted alpine specialists, potentially delaying biodiversity losses at high elevations (Lamprecht et al., 2018).

Long-term observations in the Scottish Highlands clearly show that climate change is causing significant declines in isolated arctic-alpine plant populations. Over the past 30 years, populations of three specialists *Sagina nivalis, Sabulina rubella,* and *Saxifraga cernua* have declined by more than 50% since the mid-1990s and were eliminated at their lowest-elevation sites, demonstrating a clear contraction of their elevational range (Watts et al., 2022). Numerous studies in the Alpine environment, such as David (1993) in the French Alps, have shown that during the Boreal period (5000–9000 years BP), the tree line was at a higher elevation than it is today. It is likely that plants respond to warming by accelerating cell division, thereby increasing the rate at which they use assimilated carbon (Seth & Sebastian, 2024).

A four-year field study conducted at Finse Mountain in the alpine region of southern Norway showed that climate change could affect both species interactions and the population dynamics of Arctic plants. The removal of the dominant dwarf shrub, *Dryas octopetala,* significantly increased leaf production in the herb

Thalictrum alpinum and the sedge *Carex vaginata*. It also increased the flowering frequency of *Carex vaginata*, highlighting Dryas's substantial competitive influence on neighboring species (Klanderud, 2005). Bertness and Callaway (1994) predicted that as abiotic stress decreases, neighboring plant interactions would shift from facilitation to competition. Moreover, a study across eleven alpine sites by Callaway et al. (2002) found that species interactions changed from positive to negative with increasing summer temperatures along latitudinal and altitudinal gradients.

Climate change can directly and indirectly affect plants and pollinators, altering the frequency and nature of their interactions. These changes may consequently lead to diminished pollination and reproductive success in plants (Inouye, 2020). Predictive models suggest that areas of high conservation value, which currently have low invasion potential, could experience an average 77% increase in invasion areas as temperatures rise (Bellard et al., 2013).

In the European Alps, rising temperatures are causing a "greening" effect, where fast-growing, non-native plants colonize high-altitude areas that have become suitable for growth (Table 1). This change risks the loss of specialized native species (Rumpf et al., 2022). For example, the glacier buttercup (*Ranunculus glacialis*), one of the tallest-growing vascular plants in the Alps, faces increasing threats as competitors encroach on its habitat. Meanwhile, more adaptable species or those with better dispersal abilities may thrive and spread further. However, these species could also become invasive elsewhere, outcompeting native plants and disturbing existing ecosystems (Kulonen et al., 2018).

Table 1. Some recent spring phenological changes in plants in the European Alps.

Plants Taxa	Study period	Area	Number of species	Phenological variable	Trends (days decade^{-1})	Temperature sensitivity (days °C^{-1})	References
Woody plants	1970–2012	Alps and lowlands in Switzerland	14	Leaf-out or flowering	−3.4 ± 0.3	−5.1 ± 0.4	Güsewell (2014)
Herbaceous plants	1970–2012	Alps and lowlands in Switzerland	5	Flowering	−2.8 ± 0.4	−5.6 ± 0.6	Güsewell (2014)
Trees	1998–2020	Alps and lowlands in Switzerland	8	Leaf-out	−2.2 ± 0.7	NA	PhenoForest (2026)
Woody plants	2005–2016	French Alps	6	Leaf-out	−0.5 ± 2.3	NA	Bison et al. (2019)

Alpine flora is migrating to higher elevations as temperatures increase. This trend is particularly noticeable in European mountain ranges, where species that historically occurred at lower elevations are now observed at higher elevations. For instance, the viviparous knotweed (*Polygonum viviparum*) and the alpine forget-menot (*Myosotis alpestris*) are examples of species that have expanded their geographic ranges upward (Nieuwschepen, 2024).

2. PHYSIOLOGICAL AND PHENOLOGICAL RESPONSES

Alpine ecosystems are amongst the biomes that are climate sensitive since they experience long growing seasons, low temperatures, and are highly dependent on snow cover (Rixen et al., 2022). To survive under these extreme conditions, alpine plant species have developed very specialized physiological, morphological, and reproductive features. Nevertheless, climate change is irregularly influencing these adaptations, and the rise in temperature, changes in precipitation patterns, and various changes in snow dynamics expose alpine vegetation to multiple stressors (Körner, 2023; Lu et al., 2025). The alpine vegetation response to these changes can be classified into physiological, phenological, and ecological dimensions that have great consequences to the stability of the ecosystems, conservation of biodiversity and ecosystem services (Choler et al., 2025).

2.1. Temperature stress and Plant metabolism

2.1.1. Effects on Photosynthesis

Temperature is an essential controller of plant metabolism, and even small changes in temperature are especially sensitive to alpine plants due to their adjustment to low temperatures (Li et al., 2022). Low temperatures often restrict photosynthesis in alpine plants to the extent that it affects the activity of enzymes in the Calvin cycle and lowers stomatal conductance (Bhattacharya, 2022). Consequently, there would be an initial increase in the rate of photosynthesis with moderate warming due to the proximity of temperatures to physiological assays (Hou et al., 2024). According to some experimental studies of the alpine grasslands by using open-top chambers, that moderate increase in temperature results in invariable short-term improvements of net photosynthesis, leaf expansion, and aboveground biomass (Morales et al., 2025). Also, research on tundra and alpine warming experiments indicated that an increase of temperature from 1-3 °C resulted in an average growth of plant productivity of about 19% (Banerjee et al., 2025). This growth is greatly explained by high photosynthesis and prolonged growing periods (Figure 2).

However, at temperatures that are higher than species-specific thermal optima, photosynthesis is reduced because enzymes like Rubisco become denatured, thylakoid membranes are disrupted, and photorespiration is enhanced (Aslam et al., 2022). Alpine plants, which are adapted to low thermal conditions, have low thermal safety margins, so additional warming can quickly decrease the capacity to assimilate carbon (Chandra et al., 2025). Photosynthesis is usually less temperature sensitive than respiration which rises exponentially with temperature (Arnold et al., 2022). This imbalanced cumulative growth in respiratory carbon loss is capable of counteracting the increase in photosynthesis to give lower net carbon balance. It is also revealed that increases in plant and soil respiration caused by warming could greatly decrease net ecosystem productivity in alpine systems (Posch et al., 2025). As a result, alpine ecosystems which are now weak carbon sinkers might transform to carbon neutral or even carbon sources with the further warming (Wang et al., 2023).

Figure 2. Mechanisms of climate change impacts on alpine vegetation

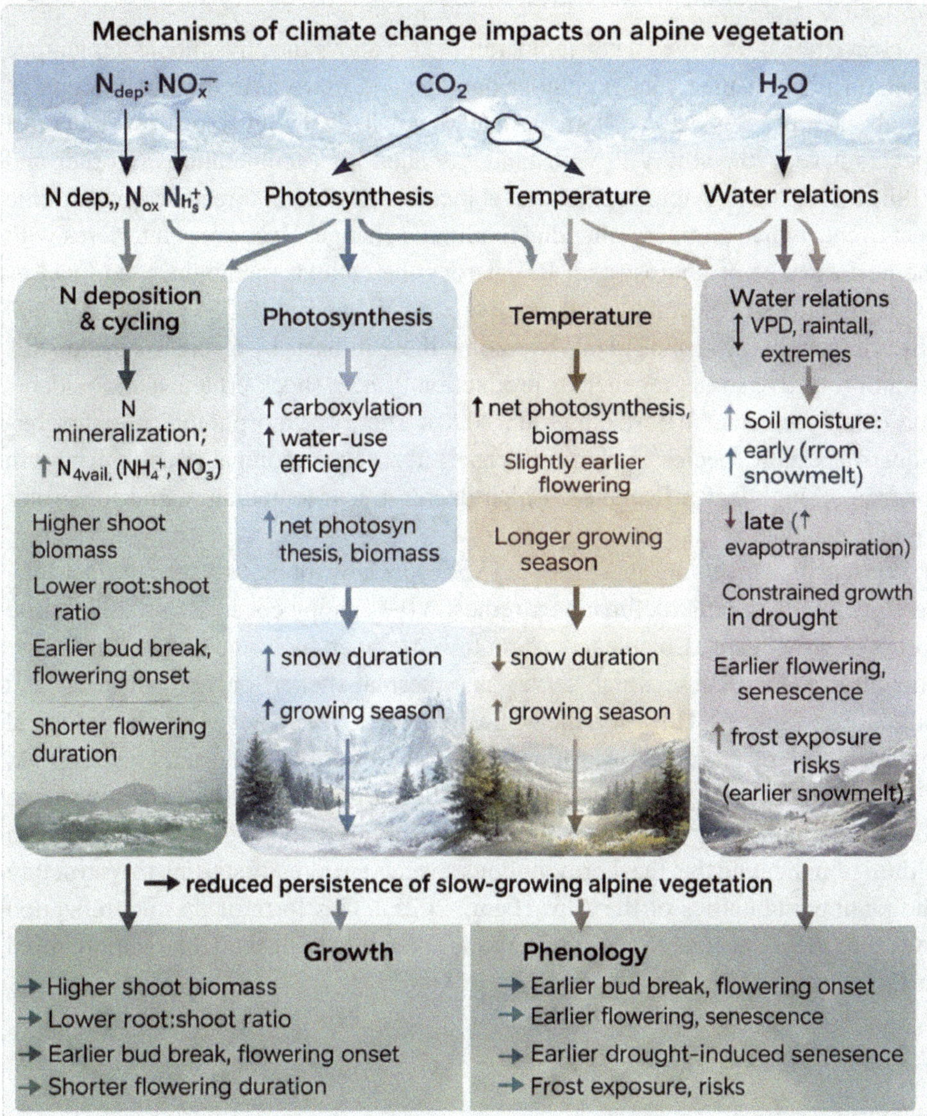

Moreover, warming also has an impact on nutrient cycling, especially mineralization of nitrogen that may also impact photosynthetic performance. Higher soil temperatures are known to increase the rate at which microbes live and increase the availability of nitrogen that can boost the growth of the plant in the short run (Du et al., 2025).

2.1.2. Heat Stress Tolerance

Heat stress is observed when temperatures are above the physiological tolerance levels of plants which causes cellular damage and may cause the malfunction of metabolic processes. Alpine flora has a tendency to deal with cold stress, and most species have a low ability to withstand extended high temperatures (Körner and Hiltbrunner, 2021). Alpine vegetation is increasingly being threatened by extreme heat events which are increasing due to climate change. Heat stress interferes with the performance of photosystem II, destroys the cellular membranes, and causes oxidative stress by accumulating reactive oxygen species (ROS) (von Büren and Hiltbrunner, 2022). Alpine plants overcome these impacts by trying to develop different protective strategies such as production of heat shock proteins, antioxidants and osmo-protectants (Rosbakh et al., 2020). There are morphological characteristics of certain species, including dense pubescence, compact cushion growth forms, and high leaf reflectance, which decrease leaf temperature and low water loss (Jaganathan et al., 2020).

The cushion plants are also believed to be especially robust to warming since they can ameliorate the microclimatic extremes in the canopy cover (Zhao, 2023). The forms of growth are also able to sustain lower temperatures and elevated moisture content in the structure, which serves as a thermal shelter for the plant itself and other related species. However, the well-evolved species do not avoid physiological stress in case of prolonged warming and drought periods (Magaña Ugarte et al., 2019). Studies on long-term monitoring of the European Alps have already shown that thermophilous species are simply replacing cold-adapted specialists at high altitudes, indicating that the thermophilous species are already starting to restructure the plant communities of the Alps (Feng, 2025). This form of thermophilization indicates the reduced ability of most alpine species to withstand the rising thermal stress (Hou et al., 2024).

2.2. Changes in Phenology

2.2.1. Alterations in flowering and fruiting times

Phenology is one of the most sensitive indicators of climate change impacts on vegetation, which refers to the timing of biological events (Inouye, 2022). Snowmelt, bud break, flowering, and seed maturation are phenological events in alpine ecosystems, which are strongly linked with temperatures and snow processes (Jabis et al., 2020). Phenological timing has been widespread in the mountain areas due to climate warming. Observational studies that are recorded over a long period of time in the Rocky Mountains have demonstrated that flowering of numerous alpine species has moved forward with several days to weeks in the past decades (Jerome et al., 2021). Likewise, European alpine monitoring networks have reported extensive improvement in flowering phenology as a result of an increase in temperature (Kudo, 2020). These modifications are mainly influenced by the general earlier snow melting and the rising temperatures during spring that speed up the growth of plants.

The earlier flowering may be beneficial as well as harmful to alpine plants. On the one hand, the growth season might be longer, enhancing growth and reproduction opportunities (Adhikari & Kumar, 2020). Conversely, early flowering leads to the high exposure of plants to late frost, which may destroy flowers and lead to low reproductive success (Inouye, 2022). Reproductive tissue damage has been noted as one of the key limitations on the alpine plant fitness during the changing climatic conditions.

2.2.2. Mismatch with Pollinators

One of the most significant effects of the changes in phenology is the possible mismatch in the flowering of plants and their pollinators. The alpine ecosystems with relatively short and well-timed biological activity are more vulnerable to the phenological decoupling (Kudo & Cooper, 2019). Alpine systems have been reported in empirical studies to be experiencing phenological mismatch. For example, Inouye (2022) found that the number of early flowering species in the Rocky Mountains is facing increasing pollinator limitation due to an increase in flowering, without a corresponding enhancement in pollinator emergence. Similarly, Kudo and Ida (2019) established that early flowering under warming conditions caused a decrease in pollination success of a number of alpine plant species due to the lack of pollinators.

Reduced seed set can decrease population viability of rare or endemic species in the Alps, which already have limited distributions (Wang et al., 2024). Furthermore, the change in plant-pollinator networks may disrupt mutualistic interactions, thus leading to co-extinctions in systems with high specialization (Li et al., 2024). Plants and generalist pollinators that have adaptable reproductive methods might be more resilient to phenological alterations, and highly specialized taxa are at high risk (Rose-Person et al., 2024). Such reactions also lead to changes in community make-up during climate change.

2.3. Drought stress and Plant response

2.3.1. Soil Moisture Changes

One of the key determinants of the performance of plants in the alpine ecosystem is water availability. Though the climate phenomena surrounding the alpine regions are usually observed as humid because of the presence of snow, the effects of climate change transform hydrology significantly. Increase in temperature reduces the depth of snowpack, increases snow melt, and evapotranspiration, making soils drier in the growing season (Sumner & Venn, 2022). The earlier meltdown of the snow may extend the growing season, but very frequently ends with a prolonged late-summer drought, especially in areas where the precipitation does not enhance enough to counter additional evaporative load (Tello-García et al., 2020). The seasonal change of water supply subject water stress to alpine plants when they are most in need of water during periods of critical growth and reproduction.

The experimental drought measurements in alpine grasslands have reported a significant decrease in the biomass production, leaf area, and the rate of photosynthesis under the reduced moisture levels in the soil (Rose-Person et al., 2024). Water stress reduces the stomatal conductance, which reduces the uptake of CO_2 and further reduces photosynthesis. Long-term drought can also cause hydraulic breakdown, tissue dehydration and increased mortality particularly in shallow-rooted plants (Kudo & Cooper, 2019).

2.3.2. Tolerance to Water Scarcity

Alpine plants have been shown to have several adaptations to increase drought tolerance despite adverse conditions. The morphological characteristics of small, thick leaves, large mass of leaf per area, and large root systems are helpful in enhancing water use efficiency (Sumner et al., 2022). Plants can maintain the functioning of their cells under conditions of water limitation through physiological as well as osmotic adjustment, accumulation of compatible solutes, and control of stomatal conductance (Haynes et al., 2021).

Some of the alpine species are characterized by a strong phenotypic plasticity, which allows the adaptation to the environmental variability to change the patterns of growth (Binter et al., 2025). For example, plants might make more biomass root systems during drought, which enhances the ability to take up water. Some seek drought avoidance measures by completing their life cycles in a short time span when conditions are favorable (Sumner & Venn, 2021).

3. PROJECTIONS AND SCENARIOS MODELLING

3.1. Present Models and Uncertainties

The effects of climate change on Alpine vegetation are discussed in terms of species distribution models (SDMs), ecological niche models (ENMs), and dynamic vegetation models (Jan, 2024). These models use the inter-relations between species occurrence and climatic variables in order to predict future distributions under different climatic scenarios. SDMs are always indicative of an increase in the environmental suitability of alpine flora as warming changes (Pecchi et al., 2020). But these changes are limited by topography, where species eventually utilize all available space, and emerge.

Such models contain serious uncertainties, despite their usefulness. Most SDMs are based on the assumption of species/climate equilibrium, thus poor consideration of biotic interactions, dispersal constraints, microclimatic refugia, or even adaptive capacity (Xue et al., 2024). Further, small-scale heterogeneity of the alpine environments can establish microhabitats that cushion species against climate trends over a region, making predictions on a large scale difficult.

There is also uncertainty due to differences in climate models and scenarios with regard to emissions. The projections based on low-emission scenarios (e.g., SSP1 2.6) demonstrate significantly lower habitat loss than the projections based on high-emission scenarios (e.g., SSP58.5), which highlights why mitigation is important to the conservation of biodiversity (Liao et al., 2020).

3.2. Estimating Future Loss of Habitats

According to numerous research works, a significant portion of the alpine flora could be deprived of most of the favorable habitat by the end of the 21st century (Lee et al., 2020). The European Alps have modeling research that predicts that 60 to 80% of alpine plant species may see considerable range shrinkage in the case of high-emission conditions (Pecchi et al., 2020). Similar trends have been observed in the other mountain systems. The climate models in the Himalayas, Tibetan Plateau, Andes, and Rocky Mountains indicate strong upward shifts in the vegetation areas as the alpine meadows begin to be increasingly replaced by shrublands and forest encroachment (Xue et al., 2024).

Lee et al. (2020) reported great changes in the numbers of species per square meter of flat expanse on the alpine summits in Europe significantly because of the uphill relocation of species. Although this is likely to increase local diversity in the short term, it eventually results in the extinction of cold-adapted species and biotic homogenization. Such alterations could reduce the functionality diversity and ecosystem stability over a period of time (Lee et al., 2020).

4. KNOWLEDGE GAPS AND FUTURE RESEARCH

Climate change is likely to cause a substantial decline in alpine plant populations. The scientific community is increasingly aware of the threats facing mountainous ecosystems and the potential ripple effects on alpine and surrounding environments as these plants diminish. Therefore, new conservation strategies and novel methods are needed to develop and implement in these areas, such as advanced monitoring tools, plant translocation moments, and seed reseeding to preserve these plants (Ohler et al., 2020; Denver, 2024). However, there are no long-term, sustainable solutions or predictive modeling tools to collect data from these mountains, which makes it difficult to predict future changes. The alpine mountains are stretched across various countries and terrains, and the cross-border conflicts and lack of collaboration to share data make it very difficult to perform research on these areas.

Till now, there are no specific and in-depth studies conducted on identifying the role of snow cover, its duration and depth on alpine plants because it is challenging to monitor the changes in alpine plants at very cold temperatures. Therefore, some understudied regions, such as the Himalayas, required multidecadal demographic data and cross-border modeling to estimate the full extent of changes in alpine communities amid upslope migration. Moreover, the effects of interactive stress factors, such as nitrogen, pests, and pathogens, cannot be predicted because each stress has a unique effect, and there are few studies to predict their simultaneous effects.

The north-facing slopes (Microrefugia) that buffer aridization are usually unexplored, making it difficult to predict changes.

The idea that native alpine plant species may face difficulties to adapting to rising temperatures may face competition for food and space with immigrant species, and as a result, lead to widespread extinction (Rammig, 2023). One suggestion to address this problem is to translocate the species out of its usual growth and elevation range. However, this solution is temporary and may not be a long-term feasible solution because of the difficulties for the alpine plants to survive in new environments. The high alpine areas may be overtaken by lower-elevation plants and subalpine trees as their range expands. Besides this, several areas currently suitable for the translocation of alpine plants could become too warm, making conditions unfavorable for the survival of alpine species (Lesica, 2023).

Some researchers claim that climate change may prolong the viability of alpine seeds, with some seeds lasting more than 20 years (White et al., 2023). One of these climate adaptive goals of the seeds is to remain capable of germination wherever the climatic conditions become more favorable again for the alpine vegetation. According to the Intergovernmental Panel on Climate Change (IPCC), if global emissions are cut, the weather will remain slightly warmer before returning to cooler temperatures

(IPCC, 2023). However, at the moment, it remains uncertain whether the seeds will retain full viability to support germination in the alpine ecosystem.

There is also a limited understanding of the impacts of microclimate within alpine environments. So much research is needed on the small-scale effects and changes, including the observation to microclimatic variations.

Several studies are conducted in labs to examine the effects of climate change on plant interactions and pollination; however, in predictive modeling, these aspects are completely neglected, making it difficult to understand the changing alpine ecosystem. The resource allocation between reproductive and vegetative growth is not properly studied, like how climate change will affect carbohydrate allocation in vegetative and reproductive parts under changing climate scenario.

5. CONCLUSION

If greenhouse gas emissions continue at this level, the alpine plant ecosystem is at high risk to endangerment due to the changing climate. The effects of climate change are already visible in these communities due to significant changes in flowering times, pollination timing, and growth patterns. The proposed solutions can only slow the decline of the alpine plant communities but cannot fully stop the impacts of climate change and global warming. This chapter highlights and collects historical data on events in Alpine communities due to climate change, presenting them in the form of figures. In addition, the physiological, agronomic, and phenological changes through which climate change induces changes in alpine plants are also discussed in detail. However, this Alpine ecosystem consists not only of plants but also of animals, microbes, and other factors, making it difficult to predict the full extent of changes that have occurred in the ecosystem over the last couple of decades. Therefore, changes induced by climate change are not limited to a single factor within the community; they also slowly influence the entire alpine community. So, protecting and restoring these communities is very important for human survival. Based on our literature review and analysis, we established that, due to extreme weather conditions, high altitude, the risk of accidents, and limited funds, there is limited research on this type of community, which makes it difficult to predict future changes with confidence.

REFERENCES

Adhikari, B. S., & Kumar, R. (2020). Effect of snowmelt regime on phenology of herbaceous species at and around treeline in Western Himalaya, India. *Notulae Scientia Biologicae, 12*(4), 901–919. DOI: 10.15835/nsb12410716

Arnold, P. A., Wang, S., Catling, A. A., Kruuk, L. E., & Nicotra, A. B. (2022). Patterns of phenotypic plasticity along a thermal gradient differ by trait type in an alpine plant. *Functional Ecology, 36*(9), 2412–2428. DOI: 10.1111/1365-2435.14128

Ashcroft, M. B., & Gollan, J. R. (2013). The sensitivity of topoclimatic models to fine-scale microclimatic variability and the relevance for ecological studies. *Theoretical and Applied Climatology, 114*(1), 281–289. DOI: 10.1007/s00704-013-0841-0

Aslam, M., Fakher, B., Ashraf, M. A., Cheng, Y., Wang, B., & Qin, Y. (2022). Plant low-temperature stress: Signaling and response. *Agronomy (Basel), 12*(3), 702. DOI: 10.3390/agronomy12030702

Banerjee, S., Rathore, N., Semerád, J., Cajthaml, T., Münzbergová, Z., & Thakur, D. (2025). Soil moisture and its interaction with temperature determine root metabolomes of a Himalayan alpine shrub. *Physiologia Plantarum, 177*(4), e70444. DOI: 10.1111/ppl.70444 PMID: 40798898

Bellard, C., Thuiller, W., Leroy, B., Genovesi, P., Bakkenes, M., & Courchamp, F. (2013). Will climate change promote future invasions? *Global Change Biology, 19*(12), 3740–3748. DOI: 10.1111/gcb.12344 PMID: 23913552

Bertness, M. D., & Callaway, R. (1994). Positive interactions in communities. *Trends in Ecology & Evolution, 9*(5), 191–193. DOI: 10.1016/0169-5347(94)90088-4 PMID: 21236818

Bhattacharya, A. (2022). Effect of low temperature stress on photosynthesis and allied traits: A review. In *Physiological processes in plants under low temperature stress* (pp. 199–297). Springer. DOI: 10.1007/978-981-16-9037-2_3

Binter, J., Macek, M., & Dolezal, J. (2025). Integrating morphological, anatomical, and physiological traits to explain elevational distributions in Himalayan steppe and alpine plants. *Journal of Integrative Plant Biology, 67*(10), 2643–2657. DOI: 10.1111/jipb.13971 PMID: 40662507

Bison, M., Asse, D., Vitasse, Y., Chuine, I., Delpierre, N., Yoccoz, N. G., & Randin, C. F. (2019). Warmer winters reduce the advance of tree spring phenology induced by warmer springs in the Alps. *Agricultural and Forest Meteorology, 252*, 220–230.

Böhm, R., Auer, I., Brunetti, M., Maugeri, M., Nanni, T., & Schöner, W. (2001). Regional temperature variability in the European Alps: 1760–1998 from homogenized instrumental time series. *International Journal of Climatology*, *21*(14), 1779–1801. DOI: 10.1002/joc.689

Bueno de Mesquita, C. P., Elmendorf, S. C., Smith, J. G., & Suding, K. N. (2024). Shifting alpine plant distributions with global change: Testing the environmental matching hypothesis. *Arctic, Antarctic, and Alpine Research*, *56*(1), 2393443. DOI: 10.1080/15230430.2024.2393443

Callaway, R. M., Brooker, R. W., Choler, P., Kikvidze, Z., Lortie, C. J., Michalet, R., Paolini, L., Pugnaire, F. I., Newingham, B., Aschehoug, E. T., Armas, C., Kikodze, D., & Cook, B. J. (2002). Positive interactions among alpine plants increase with stress. *Nature*, *417*(6891), 844–848. DOI: 10.1038/nature00812 PMID: 12075350

Chandra, S., Trivedi, V., Purohit, V., Nautiyal, M., & Chandra, S. (2025). Climatic change and physiological adaptations in alpine medicinal plants. In *High altitude medicinal plants: Botany, conservation and cultivation* (pp. 11–35). Springer.

Choler, P., Bayle, A., Fort, N., & Gascoin, S. (2025). Waning snowfields have transformed into hotspots of greening within the alpine zone. *Nature Climate Change*, *15*(1), 80–85. DOI: 10.1038/s41558-024-02177-x

Colwell, R. K., & Lees, D. C. (2000). The mid-domain effect: Geometric constraints on the geography of species richness. *Trends in Ecology & Evolution*, *15*(2), 70–76. DOI: 10.1016/S0169-5347(99)01767-X PMID: 10652559

Du, E., de Vries, W., Collalti, A., & De Marco, A. (2025). Climate warming alters nutrient cycling and its constraint on CO2 fertilization in global forests. *Current Climate Change Reports*, *11*(1), 1–11. DOI: 10.1007/s40641-025-00201-6

Elsen, P. R., & Tingley, M. W. (2015). Global mountain topography and the fate of montane species under climate change. *Nature Climate Change*, *5*(8), 772–776. DOI: 10.1038/nclimate2656

European Commission. (n.d.). *Consequences of climate change*. Climate Action. Retrieved January 18, 2026, from https://climate.ec.europa.eu/climate-change/consequences-climate-change_en

Feng, M. (2025). *Functional diversity and metabolic capabilities of cold-adapted microbial communities in the Swiss Alps* [Doctoral dissertation, University of Zurich].

Field, C. B., & Barros, V. R. (Eds.). (2014). *Climate change 2014–impacts, adaptation and vulnerability: Global and sectoral aspects*. Cambridge University Press. DOI: 10.1017/CBO9781107415379

Grabherr, G., Gottfried, M., & Pauli, H. (2010). Climate change impacts in alpine environments. *Geography Compass*, *4*(8), 1133–1153. DOI: 10.1111/j.1749-8198.2010.00356.x

Güsewell, S. (2014). Long-term trends in spring phenology of plants in Switzerland. *International Journal of Biometeorology*, *58*(1), 1–14. PMID: 23800836

Haynes, K. R., Friedman, J., Stella, J. C., & Leopold, D. J. (2021). Assessing climate change tolerance and the niche breadth-range size hypothesis in rare and widespread alpine plants. *Oecologia*, *196*(4), 1233–1245. DOI: 10.1007/s00442-021-05003-9 PMID: 34331105

Hock, R., Rasul, G., Adler, C., Cáceres, B., Gruber, S., Hirabayashi, Y., Jackson, M., Kääb, A., Kang, S., Kutuzov, S., Milner, A., Molau, U., Morin, S., Orlove, B., & Steltzer, H. (2019). High mountain areas. In Pörtner, H.-O., Roberts, D. C., Masson-Delmotte, V., Zhai, P., Tignor, M., Poloczanska, E., Mintenbeck, K., Alegría, A., Nicolai, M., Okem, A., Petzold, J., Rama, B., & Weyer, N. M. (Eds.), *IPCC special report on the ocean and cryosphere in a changing climate* (pp. 131–202). IPCC.

Hou, M., Leng, C., Zhu, J., Yang, M., Yin, Y., Xing, Y., & Chen, J. (2024). Alpine and subalpine plant microbiome mediated plants adapt to the cold environment: A systematic review. *Environmental Microbiome*, *19*(1), 82. DOI: 10.1186/s40793-024-00614-0 PMID: 39487507

Howard, T. G., White, K., & Goren, J. (2021). Thirteen years of rare plant population changes in the Adirondack alpine. *Northeastern Naturalist*, *28*(sp11), 29–46. DOI: 10.1656/045.028.s1103

Inouye, D. W. (2020). Effects of climate change on alpine plants and their pollinators. *Annals of the New York Academy of Sciences*, *1469*(1), 26–37. DOI: 10.1111/nyas.14104 PMID: 31025387

Inouye, D. W. (2022). Climate change and phenology. *Wiley Interdisciplinary Reviews: Climate Change*, *13*(3), e764. DOI: 10.1002/wcc.764

Intergovernmental Panel on Climate Change. (2001). *Climate change 2001: Synthesis report*. Cambridge University Press. https://www.ipcc.ch/2001/

Intergovernmental Panel on Climate Change. (2013). Summary for policymakers. In Stocker, T. F., Qin, D., Plattner, G.-K., Tignor, M., Allen, S. K., Boschung, J., Nauels, A., Xia, Y., Bex, V., & Midgley, P. M. (Eds.), *Climate change 2013: The physical science basis. Contribution of Working Group I to the fifth assessment report of the Intergovernmental Panel on Climate Change*. Cambridge University Press.

Intergovernmental Panel on Climate Change. (2023). *Climate change 2023: Synthesis report. Contribution of Working Groups I, II and III to the sixth assessment report of the Intergovernmental Panel on Climate Change* [Core Writing Team, H. Lee & J. Romero (Eds.)]. IPCC. https://www.ipcc.ch/report/ar6/syr/

Jabis, M. D., Winkler, D. E., & Kueppers, L. M. (2020). Warming acts through earlier snowmelt to advance but not extend alpine community flowering. *Ecology, 101*(9), e03108. DOI: 10.1002/ecy.3108 PMID: 32455489

Jacquemart, M., Weber, S., Chiarle, M., Chmiel, M., Cicoira, A., Corona, C., Deline, P., & Stoffel, M. (2024). Detecting the impact of climate change on alpine mass movements in observational records from the European Alps. *Earth-Science Reviews, 258*, 104886. DOI: 10.1016/j.earscirev.2024.104886

Jaganathan, G. K., Dalrymple, S. E., & Pritchard, H. W. (2020). Seed survival at low temperatures: A potential selecting factor influencing community level changes in high altitudes under climate change. *Critical Reviews in Plant Sciences, 39*(6), 479–492. DOI: 10.1080/07352689.2020.1848277

Jan, A. (2024). *Predicting species distribution in freshwaters facing climate change and biological invasions using ecological niche models.*

Jerome, D. K., Petry, W. K., Mooney, K. A., & Iler, A. M. (2021). Snow melt timing acts independently and in conjunction with temperature accumulation to drive subalpine plant phenology. *Global Change Biology, 27*(20), 5054–5069. DOI: 10.1111/gcb.15803 PMID: 34265142

Klanderud, K. (2005). Climate change effects on species interactions in an alpine plant community. *Journal of Ecology, 93*(1), 127–137. DOI: 10.1111/j.1365-2745.2004.00944.x

Körner, C. (1999). *Alpine plant life: Functional plant ecology of high mountain ecosystems* (Vol. 4). Springer. DOI: 10.1007/978-3-642-98018-3

Körner, C. (2003). The alpine life zone. In *Alpine plant life: Functional plant ecology of high mountain ecosystems* (pp. 9–20). Springer. DOI: 10.1007/978-3-642-18970-8_2

Körner, C. (2004). Mountain biodiversity, its causes and function. *Ambio, 33*(sp13), 11–17. DOI: 10.1007/0044-7447-33.sp13.11 PMID: 15575177

Körner, C. (2012). *Alpine treelines: Functional ecology of the global high elevation tree limits*. Springer. DOI: 10.1007/978-3-0348-0396-0

Körner, C. (2021). The alpine life zone. In *Alpine plant life: Functional plant ecology of high mountain ecosystems* (pp. 23–51). Springer.

Körner, C. (2023). Concepts in alpine plant ecology. *Plants*, *12*(14), 2666. DOI: 10.3390/plants12142666 PMID: 37514280

Körner, C., & Hiltbrunner, E. (2021). Why is the alpine flora comparatively robust against climatic warming? *Diversity*, *13*(8), 383. DOI: 10.3390/d13080383

Körner, C., Paulsen, J., & Spehn, E. M. (2011). A definition of mountains and their bioclimatic belts for global comparisons of biodiversity data. *Alpine Botany*, *121*(2), 73–78. DOI: 10.1007/s00035-011-0094-4

Kudo, G. (2020). Dynamics of flowering phenology of alpine plant communities in response to temperature and snowmelt time: Analysis of a nine-year phenological record collected by citizen volunteers. *Environmental and Experimental Botany*, *170*, 103843. DOI: 10.1016/j.envexpbot.2019.103843

Kudo, G., & Cooper, E. J. (2019). When spring ephemerals fail to meet pollinators: Mechanism of phenological mismatch and its impact on plant reproduction. *Proceedings of the Royal Society B*, *286*(1904), 20190573. DOI: 10.1098/rspb.2019.0573

Kulonen, A., Imboden, R. A., Rixen, C., Maier, S. B., & Wipf, S. (2018). Enough space in a warmer world? Microhabitat diversity and small-scale distribution of alpine plants on mountain summits. *Diversity & Distributions*, *24*(2), 252–261. DOI: 10.1111/ddi.12673

Lamprecht, A., Semenchuk, P. R., Steinbauer, K., Winkler, M., & Pauli, H. (2018). Climate change leads to accelerated transformation of high-elevation vegetation in the central Alps. *The New Phytologist*, *220*(2), 447–459. DOI: 10.1111/nph.15290 PMID: 29938796

Lee, S., Jung, H., & Choi, J. (2020). Projecting the impact of climate change on the spatial distribution of six subalpine tree species in South Korea using a multi-model ensemble approach. *Forests*, *12*(1), 37. DOI: 10.3390/f12010037

Lesica, P. (2023). What good are herbaria? *Sage (Atlanta, Ga.)*, *45*(3), 2.

Li, K.-J., Liu, X.-F., Yang, L., & Shen, S.-K. (2024). Alpine Rhododendron population contractions lead to spatial distribution mismatch with their pollinators under climate change. *The Science of the Total Environment*, *926*, 171832. DOI: 10.1016/j.scitotenv.2024.171832 PMID: 38521263

Li, S., Dong, S., Fu, Y., Zhou, B., Liu, S., Shen, H., Xu, Y., Gao, X., Xiao, J., Wu, S., & Li, F. (2022). Air or soil temperature matters the responses of alpine plants in biomass accumulation to climate warming. *The Science of the Total Environment, 844*, 157141. DOI: 10.1016/j.scitotenv.2022.157141 PMID: 35798113

Liao, Z., Zhang, L., Nobis, M. P., Wu, X., Pan, K., Wang, K., Dakhil, M. A., Du, M., Xiong, Q., Pandey, B., & Tian, X. (2020). Climate change jointly with migration ability affect future range shifts of dominant fir species in Southwest China. *Diversity & Distributions, 26*(3), 352–367. DOI: 10.1111/ddi.13018

Lu, X., Zheng, X., Liang, E., Piao, S., Babst, F., Elliott, G. P., Sigdel, S. R., Wang, T., Wang, Y., Li, X., Gao, S., Zhang, L., Sun, J., Li, J., Zhu, H., Rossi, S., Peñuelas, J., & Camarero, J. J. (2025). Patterns, dynamics and drivers of alpine treelines and shrublines. *Nature Reviews. Earth & Environment, 6*(8), 489–502. DOI: 10.1038/s43017-025-00703-9

Mack, A., Choffnes, E. R., Hamburg, M. A., & Relman, D. A. (Eds.). (2008). *Global climate change and extreme weather events: Understanding the contributions to infectious disease emergence: Workshop summary*. National Academies Press.

Magaña Ugarte, R., Escudero, A., & Gavilán, R. G. (2019). Metabolic and physiological responses of Mediterranean high-mountain and alpine plants to combined abiotic stresses. *Physiologia Plantarum, 165*(2), 403–412. DOI: 10.1111/ppl.12898 PMID: 30536685

Morales, L. V., Sierra-Almeida, A., Sandoval-Urzúa, C., & Arroyo, M. T. (2025). Thermal decoupling may promote cooling and avoid heat stress in alpine plants. *Plants, 14*(13), 2023. DOI: 10.3390/plants14132023 PMID: 40648032

National Aeronautics and Space Administration. (2024, October 23). *The effects of climate change*. NASA Science. https://science.nasa.gov/climate-change/effects/

Nieuwschepen, M. (2024, July 3). *Climate change and shifting ecosystems: Understanding the impact of altitudinal and latitudinal shifts*. European Wilderness Society. https://wilderness-society.org/climate-change-and-shifting-ecosystems-understanding-the-impact-of-altitudinal-and-latitudinal-shifts/

Ohler, L. M., Lechleitner, M., & Junker, R. R. (2020). Microclimatic effects on alpine plant communities and flower-visitor interactions. *Scientific Reports, 10*(1), 1366. DOI: 10.1038/s41598-020-58388-7 PMID: 31992825

Pauli, H., Gottfried, M., & Grabherr, G. (2003). Effects of climate change on the alpine and nival vegetation of the Alps. *Journal of Mountain Ecology, 7*(Suppl), 9–12.

Pauli, H., & Halloy, S. R. (2019). High mountain ecosystems under climate change. In *Oxford research encyclopedia of climate science*. Oxford University Press. DOI: 10.1093/acrefore/9780190228620.013.764

Pecchi, M., Marchi, M., Moriondo, M., Forzieri, G., Ammoniaci, M., Bernetti, I., Bindi, M., & Chirici, G. (2020). Potential impact of climate change on the forest coverage and the spatial distribution of 19 key forest tree species in Italy under RCP4.5 IPCC trajectory for 2050s. *Forests, 11*(9), 934. DOI: 10.3390/f11090934

Perrigo, A., Hoorn, C., & Antonelli, A. (2020). Why mountains matter for biodiversity. *Journal of Biogeography, 47*(2), 315–325. DOI: 10.1111/jbi.13731

PhenoForest. (2026). *PhenoForest database: Phenological trends of forest tree species in Switzerland* [Data set]. Swiss Federal Institute for Forest, Snow and Landscape Research (WSL). Retrieved January 15, 2026.

Posch, B. C., Amoanimaa-Dede, H., Aparecido, L. M., Atkin, O. K., Bison, N. N., Blonder, B. W., Coast, O., Doughty, C. E., Guo, J. S., & Van Haren, J. (2025). High-temperature acclimation of photosystem II in land plants. *The New Phytologist*. Advance online publication. PMID: 41178112

Rammig, S. P. D. A. (2023). *Local and regional responses of alpine flora to climate change induced warming*.

Rivera, A., Movalia, S., & Rutkowski, E. (2024). *Global greenhouse gas emissions: 1990–2022 and preliminary 2023 estimates*. Rhodium Group.

Rixen, C., Høye, T. T., Macek, P., Aerts, R., Alatalo, J. M., Anderson, J. T., Arnold, P. A., Barrio, I. C., Bjerke, J. W., Björkman, M. P., Blok, D., Blume-Werry, G., Boike, J., Bokhorst, S., Carbognani, M., Christiansen, C. T., Convey, P., Cooper, E. J., Cornelissen, J. H. C., & Zong, S. (2022). Winters are changing: Snow effects on Arctic and alpine tundra ecosystems. *Arctic Science, 8*(3), 572–608. DOI: 10.1139/as-2020-0058

Rosbakh, S., Margreiter, V., & Jelcic, B. (2020). Seedlings of alpine species do not have better frost-tolerance than their lowland counterparts. *Alpine Botany, 130*(2), 179–185. DOI: 10.1007/s00035-020-00237-4

Rose-Person, A., Spasojevic, M. J., Forrester, C., Bowman, W. D., Suding, K. N., Oldfather, M. F., & Rafferty, N. E. (2024). Early snowmelt advances flowering phenology and disrupts the drivers of pollinator visitation in an alpine ecosystem. *Alpine Botany, 134*(2), 141–150. DOI: 10.1007/s00035-024-00315-x

Rumpf, S. B., Gravey, M., Brönnimann, O., Luoto, M., Cianfrani, C., Mariethoz, G., & Guisan, A. (2022). From white to green: Snow cover loss and increased vegetation productivity in the European Alps. *Science, 376*(6597), 1119–1122. DOI: 10.1126/science.abn6697 PMID: 35653482

Rundel, P. W., & Millar, C. I. (2016). Alpine ecosystems. In Zavaleta, E., & Mooney, H. (Eds.), *Ecosystems of California* (pp. 613–634). University of California Press.

Schickhoff, U., Bobrowski, M., Böhner, J., Bürzle, B., Chaudhary, R. P., Gerlitz, L., & Schwab, N. (2016). Climate change and treeline dynamics in the Himalaya. In *Climate change, glacier response, and vegetation dynamics in the Himalaya: Contributions toward future earth initiatives* (pp. 271–306). Springer. DOI: 10.1007/978-3-319-28977-9_15

Seth, P., & Sebastian, J. (2024). Plants and global warming: Challenges and strategies for a warming world. *Plant Cell Reports, 43*(1), 27. DOI: 10.1007/s00299-023-03083-w PMID: 38163826

Sumner, E., & Venn, S. (2021). Plant responses to changing water supply and availability in high elevation ecosystems: A quantitative systematic review and meta-analysis. *Land (Basel), 10*(11), 1150. DOI: 10.3390/land10111150

Sumner, E. E., & Venn, S. E. (2022). Thermal tolerance and growth responses to in situ soil water reductions among alpine plants. *Plant Ecology & Diversity, 15*(5-6), 297–308. DOI: 10.1080/17550874.2022.2160674

Sumner, E. E., Williamson, V. G., Gleadow, R. M., Wevill, T., & Venn, S. E. (2022). Acclimation to water stress improves tolerance to heat and freezing in a common alpine grass. *Oecologia, 199*(4), 831–843. DOI: 10.1007/s00442-022-05245-1 PMID: 35974110

Tello-García, E., Huber, L., Leitinger, G., Peters, A., Newesely, C., Ringler, M.-E., & Tasser, E. (2020). Drought-and heat-induced shifts in vegetation composition impact biomass production and water use of alpine grasslands. *Environmental and Experimental Botany, 169*, 103921. DOI: 10.1016/j.envexpbot.2019.103921

Theurillat, J. P., & Guisan, A. (2001). Potential impact of climate change on vegetation in the European Alps: A review. *Climatic Change, 50*(1), 77–109. DOI: 10.1023/A:1010632015572

Verrall, B. (2023). *Assessing alpine vegetation dynamics using long-term ecological monitoring amidst rapid climate change.*

von Büren, R. S., & Hiltbrunner, E. (2022). Low winter temperatures and divergent freezing resistance set the cold range limit of widespread alpine graminoids. *Journal of Biogeography, 49*(8), 1562–1575. DOI: 10.1111/jbi.14455

Wang, W., Du, J., He, Z., Miao, C., Wu, J., Ma, D., & Zhao, P. (2024). Pollinator peaking earlier than flowering is more detrimental to plant fecundity. *Science of the Total Environment, 917*, 170458.[1]

Wang, Y., Xiao, J., Ma, Y., Ding, J., Chen, X., Ding, Z., & Luo, Y. (2023). Persistent and enhanced carbon sequestration capacity of alpine grasslands on Earth's Third Pole. *Science Advances, 9*(20), eade6875. DOI: 10.1126/sciadv.ade6875 PMID: 37196073

Warren, R., Price, J., VanDerWal, J., Cornelius, S., & Sohl, H. (2018). The implications of the United Nations Paris Agreement on climate change for globally significant biodiversity areas. *Climatic Change, 147*(3), 395–409. DOI: 10.1007/s10584-018-2158-6

Watts, S. H., Mardon, D. K., Mercer, C., Watson, D., Cole, H., Shaw, R. F., & Jump, A. S. (2022). Riding the elevator to extinction: Disjunct arctic-alpine plants of open habitats decline as their more competitive neighbours expand. *Biological Conservation, 272*, 109620. DOI: 10.1016/j.biocon.2022.109620

White, F. J., Hay, F. R., Abeli, T., & Mondoni, A. (2023). Two decades of climate change alters seed longevity in an alpine herb: Implications for ex situ seed conservation. *Alpine Botany, 133*(1), 11–20. DOI: 10.1007/s00035-022-00289-8

World Meteorological Organization. (2025, January 10). *WMO confirms 2024 as warmest year on record at about 1.55 °C above pre-industrial level.* https://wmo.int/news/media-centre/wmo-confirms-2024-warmest-year-record-about-155degc-above-pre-industrial-level

Xue, N., Li, K., Chen, K., Li, P., Ji, X., Ma, Z., & Ji, W. (2024). Predicting climate change impacts on distribution and conservation of critically endangered *Picea neoveitchii* using MaxEnt. *Frontiers in Forests and Global Change, 7*, 1472857. DOI: 10.3389/ffgc.2024.1472857

Zhao, R. (2023). The influence of cushion plants on soil fertility islands under their canopies with ontogenesis in an extreme alpine ecosystem. *Plant and Soil, 484*(1), 379–392. DOI: 10.1007/s11104-022-05798-3

Chapter 3
Anthropogenic Activities and Alpine Ecosystem Degradation

Hansa Gul
http://orcid.org/0009-0001-5554-617X
University of Mianwali, Pakistan

Nasir Assad
http://orcid.org/0009-0004-9435-1414
University of Sargodha, Pakistan

Haris Ahmed Khan
http://orcid.org/0000-0002-5093-9476
University of Mianwali, Pakistan

Baber Ali
Western Sydney University, Australia

Zahida Nasreen
http://orcid.org/0009-0002-0995-5041
University of Mianwali, Pakistan

Sezai Ercişli
Ataturk University, Erzurum, Turkey

ABSTRACT

The alpine habitats reside on the highest points of the belts and are subject to the harsh climatic states and hence are among the most fragile but ecologically rich biomes of the planet. They are also sensitive to changes in climate and human activities that subject them to high levels of pressures rendering them extremely significant when it comes to providing early warnings of changes in the environment. The degradation of alpine vegetation has become the source of concern during the last decades of the high rates of the impact both of the natural and of the anthropogenic factors. Even though the climatic effects (temperature change, precipitation and surface albedo) have their physical consequences in the vegetation processes, human activity has also had its toll especially on highly populated or newly developed areas. In this

DOI: 10.4018/979-8-2600-1207-9.ch003

chapter, the authors will analyze the complex of interplay between anthropogenic pressures and climate change to learn the forces behind alpine ecosystem degradation and define the efficient conservation policies.

1. INTRODUCTION

The Himalayas, the most prominent and densely populated mountain range in Asia, stand as both a global biodiversity hotspot and a region of immense ecological, cultural, and economic importance. These landscapes are critical for sustaining biodiversity, sequestering carbon, regulating hydrological cycles, and supporting millions of livelihoods in downstream regions (Dhar, 2002; Negi et al., 2018; Rana et al., 2021). But these vulnerable systems of the alpine environment are becoming more and more threatened by a combination of natural and anthropogenic challenges. Among the most important ones is climate change, which has changed the patterns of distributions of species and the productivity of ecosystems, increasing the vulnerability of endemic species (Behera et al., 2019; Wani et al., 2021). In addition to these abiotic stressors, biotic influences including species invasion and habitat loss, have further contributed to ecological instability in the Himalayas (Negi et al., 2022).

Insensitive actions by humans, especially deforestation, unsustainable fuelwood and timber exploitation, agricultural expansion, and a high rate of infrastructure development keep undermining the stability of alpine ecosystems (Negi & Maikhuri, 2017; Thakur et al., 2022). The same pattern of degradation is realized in the realm of alpine areas worldwide, where anthropogenic disturbances increase the erosion rate, lead to the fragmentation of habitats, and encourage the extinction of biodiversity. Despite the fact that the establishment of the so-called protected area (PA) networks has become a successful tool to safeguard biodiversity and to a substantial extent have contributed to the minimization of the risk of extinction (Karanth et al., 2010; Oldekop et al., 2016), the continued loss of biodiversity despite the spread of the use of PAs worldwide highlights major issues related to the implementation of the strategy, ecological surveillance, and the involvement of the local population (Jenkins & Joppa, 2009).

Alpine forest and ecosystems in the Indian Himalayan Region (IHR) are especially vulnerable, and demand strong ecological databanks and innovative conservation approaches. The development of Conservation Priority Indices (CPI), which incorporate parameters such as species richness, nativity, and endemism, has proven effective for identifying priority habitats and communities requiring urgent management interventions (Samant, 2007; Thakur et al., 2020). Also, the incorporation of vulnerability mapping and community-based conservation in the protected spaces, as illustrated in the Gulmarg Wildlife Sanctuary, is emphasized as an essential adaptive intervention to respond to the anthropogenic and climate-driven stressors.

An alpine ecosystem is an exclusive biome which is found above the tree line in high altitude areas where trees cannot grow because of the severe environmental conditions. Such ecosystems are typified by highly specialized plant and animal communities that are designed to survive under extreme conditions such as low temperatures, high UV radiation and small growing periods. Alpine ecosystems are widely distributed over the major mountains around the world including the Himalayas, Rockies, Andes, and Alps with elevation starting at different levels based on the closeness to the equator.

Alpine ecosystems are not degraded by chance but are a direct consequence of global climate change- a long-term change in weather patterns, mostly due to a warming atmosphere. Although natural processes also play a role in climate variability, overwhelming scientific evidence has proven that anthropogenic activities are the most important cause of the warming that has taken place since the mid-twentieth century. An in-depth assessment of scientific literature showed that 90-100% of the literature conclude on human-caused causes of global warming (Cook & Lewandowsky, 2016). The impacts of this anthropogenic climate change are now being felt in all of the major ecosystems- forests, oceans, and most importantly, the delicate, high-altitude alpine settings of mountains around the globe.

This chapter consequently examines the interaction of anthropogenic pressures and climate change that strengthens the pace of alpine ecosystem degradation and the Himalayas is an important example of such cases. It explores ecological, social and policy aspects of degradation, conservation methods, including ecological restoration, fortification of protected zones and biodiversity corridors and focuses on the application of scientific frameworks to prioritization of species, communities, and habitats to protect alpine biodiversity in the face of the accelerating impacts of global environmental change.

2. CHARACTERISTICS OF ALPINE ECOSYSTEMS

Alpine ecosystems can be found in mountain ranges all over the world. Globally, this biome covers approximately 2.6% of the terrestrial area outside Antarctica, yet it harbors a surprisingly high concentration of biodiversity (Körner & Hiltbrunner, 2021; Körner et al., 2011). While often perceived as harsh and depauperate, alpine zones are hotspots of biodiversity, hosting about 4% of the world's known angiosperm species. This abundance does not come as a result of homogenous climatic conditions but is shaped by geodiversity the complicated interaction of topography, solar radiation, and wind. It is this interaction that forms a fine-grained mosaic of microhabitats in which the conditions of life may differ sharply just a few meters, as Körner and Hiltbrunner (2021) explain. The alpine plants because of their low height create their own microclimate, which captures the sun heat and works at much higher temperatures when compared to the weather stations. The resulting topographically regulated thermal differentiation generates a series of thermal niches adjacent, that can supersede the most deplorable climate warming conditions, and offer a potent hedge to the regional temperature changes (C. Körner & E. J. D. Hiltbrunner, 2021).

2.1 Flora and Fauna Diversity

Alpine systems of the Western Himalayas form a biodiversity hotspot of the world as one of the few places on Earth possessing a unique collection of flora that is specially adapted to the harsh environmental factors. There was a study of the rich floristic richness of the region in the alpine highlands of the Kashmir a study recorded 251 species of vascular plants representing 51 families (Shaheen et al., 2023).

Flora is defined by the prevalence of several main families which are especially well-adapted to high-altitude conditions. The most prevalent of these are Asteraceae (daisy family), Poaceae (grasses) and Rosaceae (rose family). This prevalence of herbaceous families is typical of alpine areas, which have morphology and physiology specifically adapted to endure a short growing period, extreme UV radiation, and low temperatures (Wesche et al., 2000) (Shaheen et al., 2023). In terms of life-forms, the alpine vegetation is highly biased with Hemicryptophytes (soil-surface-budded plants) about 49 percent of the local flora. This hegemony is one of the principal adaptive mechanisms, which help plants to live through long winters, as well as reproduce fast within a short summer growing season (Noroozi et al., 2008). However, in ecosystems under intense anthropogenic pressure, this life-form spectrum is often inverted. A detailed study of the grass family (Poaceae) in the heavily utilized Charsadda district, found that therophytes (annuals that survive harsh seasons as seed) were the overwhelmingly dominant life form, accounting for 62.74% of all grass species (Khan et al., 2019). The dominance of annuals is a key bio-indicator of frequent disturbance from factors like overgrazing and agriculture, as they rapidly colonize bare ground where perennial species struggle to persist

Even with this variety, the flora of the Alps is very delicate. Due to microhabitat specificity and limited distribution, the same study identified 51 vulnerable and 17 threatened species. Others, like *Aconitum heterophyllum*, *Saussurea costus*, are overexploited as sources of medicinal products, and others are at risk due to overgrazing and habitat destruction. Since they occupy small ecological niches, the loss of even a few species can cause cascading ecological effects, highlighting the urgent need for conservation action (Shaheen et al., 2023).

Snow plays a primary role in ecosystem structure and function in extratropical alpine regions. Korerner and Hiltbrunner, also state that snowpack is an important insulator, which disconnects the plants with the severe air temperature of winter and, in most cases, leaves the soil surface unfrozen even when the air temperature is below -20 C. Interactions between wind and topography create a mosaic of snow distribution, producing habitats with contrasting growing season lengths. Consequently, while climate change is expected to alter the timing of snowmelt, the fundamental spatial pattern of snow beds and wind-exposed ridges and the specialized species they support will likely persist as long as winter precipitation remains consistent (C. Körner & E. J. D. Hiltbrunner, 2021). As a comparative example from another major mountain system, the distinct flora and fauna of California's alpine ecosystems are summarized in Table 1, highlighting the high degree of endemism that characterizes these environments (Rundel & Millar, 2016).

Table 1. Summary of Flora and Fauna Diversity in California Alpine Ecosystems

Category	Key Species and Characteristics
Plant Species	
Sierra Nevada	• Species Count: 385 native species (≥3,500 m) • Dominant Families: Asteraceae, Poaceae, Brassicaceae • Key Genera: *Carex, Draba, Lupinus* • Endemism: High (36 endemic species).
White Mountains	• Species Count: 163 native species (≥3,500 m) • Dominant Families: Asteraceae, Brassicaceae, Poaceae • Endemism: Low (3 endemic species).
Sweetwater Mountains	• Species Count: 173 native species • Characteristics: Shares 94% of its flora with the Sierra Nevada.
Animal species	
Large Mammals	• Sierra Nevada Bighorn Sheep (*Ovis canadensis sierrae*) • Mule Deer (*Odocoileus hemionus*) • Mountain Lion (*Puma concolor*)
Small & Meso-Mammals	• American Pika (*Ochotona princeps*) • Yellow-bellied Marmot (*Marmota flaviventris*) • Sierra Nevada Red Fox (*Vulpes vulpes necator*) • American Marten (*Martes americana sierra*)
Amphibians	• Sierra Nevada Yellow-legged Frog (*Rana muscosa*) • Yosemite Toad (*Anaxyrus canorus*) *(Both species have experienced dramatic population declines).*
Birds	• Gray-crowned Rosy Finch (*Leucostichte tephrocotis*) • Water Pipit (*Anthus spinoletta alticola*) • Sage Grouse (*Centrocercus urophasianus*)
Invasive Fauna	• Non-native Trout: Harmful to native amphibians • White-tailed Ptarmigan (*Lagopus leucurus*): Established in alpine grasslands.

1.3 Anthropogenic Activities Affecting Alpine Ecosystems

The primary human-driven pressures on alpine ecosystems, which will be discussed throughout this section, are summarized in the conceptual diagram in Figure 1.

Figure 1. A conceptual model illustrating the primary anthropogenic activities that impact and degrade alpine ecosystems

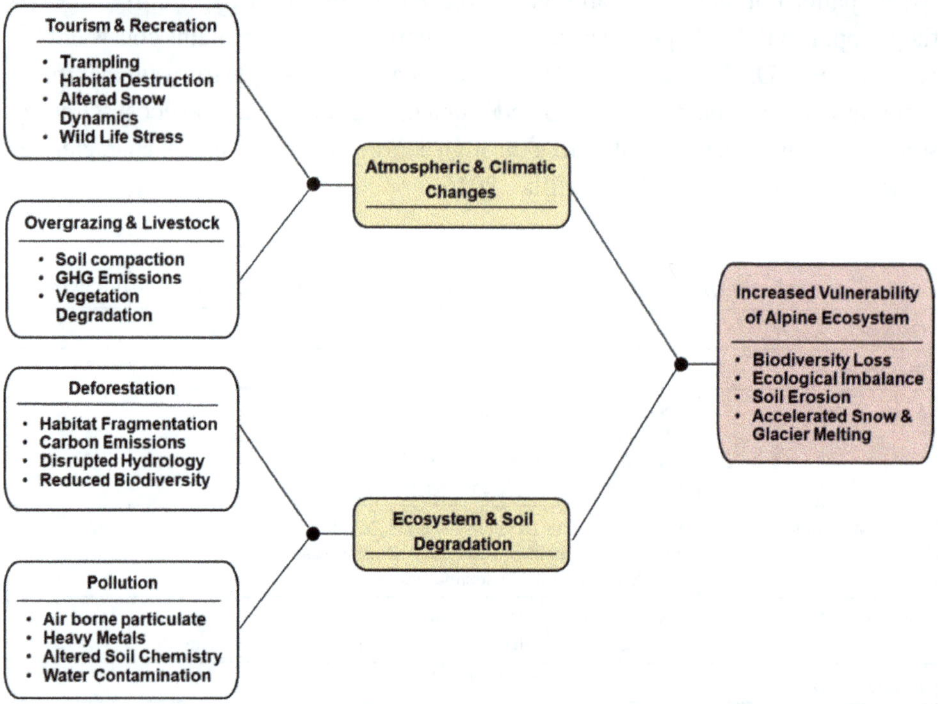

3. ANTHROPOGENIC ACTIVITIES AFFECTING ALPINE ECOSYSTEMS

3.1 Tourism and Recreational Activities

Direct physical disruptions to alpine vulnerable vegetation are brought about by tourism and recreational activities like hiking and mountaineering. Another major effect is human trampling which can affect human species in plant communities. The results of a study by Chardon (Chardon et al., 2023), in the Garibaldi Provincial Park in Canada showed that trampling lowers plant cover in all the species studied.

The reaction however, was mixed, as the size of dominant shrubs such as *Phyllodoce empetriformis* decreased, but in *Cassiope mertensiana*, trampling was the major cause of reduced reproductive output. This study indicates the necessity of management specific to species since various species have different susceptibilities to the same disruption.

In addition to trampling, there are other forms of tourism related disturbances that increase ecosystem vulnerability. Skiing, camping and tourism infrastructure like ski lifts, trails and lodges destroy habitats, alter movement of wildlife and augment chances of invasive species establishment. Snow sports and other seasonal activities may slow the growth of plants and cause losses in species richness by changing the dynamics of snow cover and hydrology. Most wildlife species, such as alpine ungulates and ground-nesting birds, have been found to be stressed, less efficient foraging, and less successful in breeding as a result of more human presence. Taken together, these effects underscore the point that uncontrolled tourism has the potential to hasten the process of the biodiversity destruction and ecosystem harm, and means of managing eco-tourism like controlled trail infrastructure, visitor education and seasonal access rights can significantly reduce ecological devastation within the alpine areas.

3.2 Overgrazing and Livestock Pressure

Overgrazing and soil compaction are not the only causes of livestock pressure, but also massive contributors of climate change on a global scale. At the global level, the livestock industry contributes to 18% of total greenhouse gas (GHG) emissions, which is more significant than the transport industry across the globe. With the increased demand, the livestock emissions in the world are expected to be doubled by 2050, which continues to raise the climatic stressors on vulnerable ecosystems such as alpine meadows (Ilea, 2009). Overgrazing in alpine areas causes displacement of palatable plants by other plants, proliferation of non-palatable plants and non-palatable or invasive plants, and eventual loss of soil fertility. Such developments do not only decrease the level of biodiversity, but also disrupt the ecological balance, making alpine ecosystem more susceptible to erosion and change in climate conditions.

3.3 Deforestation and Fuel wood Collection

Deforestation, whether for agriculture, fuelwood, or infrastructure, is a major contributor to changes in both local and global climatic conditions. The deforestation process alters the local hydrological patterns and plays a significant role in the carbon burden in the atmosphere that causes global warming. This connection is a key factor in the anthropogenic climate change, where intensive deforestation does have a quantifiable consequence to the global climate system that eventually affects all ecosystems, even alpine (Ramankutty et al., 2007). Reliance on the use of fuel wood as a source of energy in Himalayas has been one of the most crucial causes of the destruction of forests especially in small villages where there are few sources of other energies. Fragmentation of alpine and subalpine forests not only decreases biodiversity, but also disrupts the ecosystem services like sequestration of carbon, soil stabilization and water control. These compounding effects have an aggravating effect on the susceptibility of alpine ecosystems to both anthropogenic and climate stressors (Qazi et al., 2023).

3.4 Pollution (Air, Water, and Soil Contamination)

In addition to the physical disturbances, industrial and urban Centre's have their chemical pollution which has extensive impact on the alpine ecosystems.

Air Pollution: The emissions of industries and vehicle fumes may move long distances and be deposited in the high altitudes. The direct physiological effects of these pollutants on plants are interference with stomatal functioning, chlorophyll breakdown, slowed photosynthesis, foliar necrosis (tissue death), and a general decrease in growth and yield (Iqbal et al., 2018). The huge land surrounding the industrial sites which at one time was green is now dusty and barren because of the massive amount of dust released by plants such as cement factories and stone quarries.

Soil and Water Contamination: The content of industrial effluents and municipal sewage is usually highly contaminated with heavy metals (including lead, cadmium, copper and zinc) and other toxic substances. When such polluted water is irrigated or discharged into natural systems, the pollutants build up on the soil. This has the ability to change the soil chemistry and render it toxic to the native plant life. The ongoing rise in the level of metals has been found to alter the soil pH and induce excessive absorption by plants, resulting in detrimental outcomes and even eventual extinction of sensitive species (Iqbal et al., 1999; Iqbal et al., 2018). Vegetation adjacent to polluted effluent channels also tends to be seriously impacted, with a severe reduction in the diversity of the species and the preponderance of a few highly resistant species, which are usually non-native (Iqbal et al., 1998; Iqbal et al., 2018).

Human induced pollution is a major cause of the atmospheric alterations, which undermine the alpine ecosystem. Transportation pollution, through burning fossil fuels, contributes about 26% of global carbon dioxide emission (Bandh et al., 2021; Chapman, 2007). Industrial processes are also major contributors, with cement production alone emitting 5–8% of all human-made GHGs (Aïtcin, 2000; Bandh et al., 2021). Furthermore, pollutants like black carbon, produced from biomass and fossil fuel burning, are recognized as the second-largest contributor to global warming after CO_2, accelerating atmospheric warming and the melting of snow and ice in high-altitude environments (Ramanathan & Carmichael, 2008) (Bandh et al., 2021).

4. IMPACTS OF ANTHROPOGENIC ACTIVITIES ON ALPINE ECOSYSTEMS

4.1 Habitat Fragmentation and Loss

The anthropogenic factor that has the most immediate and irreversible effect is the physical destruction and fragmentation of natural habitats. This is especially harmful in the case of alpine ecosystems. The Western Himalayan area consists of interrelated habitat layers which are a continuation of the foothill grasslands to the high alpine meadow. These adjacent habitats are vital to many well-known Himalayan bird and mammal species and are where they migrate to at various times of the year to up and down the steep mountain slopes according to the changing conditions. These migratory ways are interrupted when any of these layers (especially the sub-alpine conifer forests) are destroyed or altered by human activities like deforestation to obtain farmland, build infrastructure, or collect fuelwood (Iqbal et al., 2018). This division separates populations, lessens genetic exchange and may, in the long term, contribute to the local extinction of species that will no longer be able to reproduce their life cycles.

4.2 Soil Erosion and Degradation of the Land

Alpine areas are steep slopes with shallow growing soils, and thus, they are prone to erosion. The activities of humans greatly increase this process. Iqbal et al. (2018) identifies the influence of high precipitation on steep erodible soils without sufficient protective vegetation cover as the factors contributing to massive soil erosion and agricultural productivity decline in areas such as the Pothwar upland (Iqbal et al., 2018). The same principle is more strongly applied in alpine regions. The grazing activities of animals destroy the stabilizing cover of herbaceous cover, exposing the soil to wind and water. The trampling that takes place continuously compacts the soil making it unable to absorb water and resulting in high surface runoff which carries away the thin rich nutrient filled topsoil. This soil erosion does not only stop growth of indigenous plants, but also affects downstream hydrology (Khan et al., 2022).

Furthermore, the shift in vegetation life-forms induced by human activity directly accelerates soil degradation. Perennial grasses, typical of stable ecosystems, maintain year-round root structures that bind soil. The observed dominance of annual therophytes in heavily grazed regions like Charsadda (62.74% of grasses) means that for a significant portion of the year after these plants die, the soil surface is left bare and unprotected (Khan et al., 2019). This lack of continuous vegetative cover and root mass makes the soil highly vulnerable to erosion from wind and seasonal rains, a process that is greatly magnified on the steep slopes characteristic of mountain regions.

4.3. Alteration of Hydrological Cycles

The well-being of the Pakistani alpine ecosystems cannot be separated in any way with the water resources of the territory as it is the main source of most of the rivers in the country. Rapid melting and receding of feeding glaciers is shaking up the mass equilibrium of the high-altitude regions due to accelerated rise in temperature. This has two far reaching implications. One, it puts a direct threat to the very existence of the high-altitude wetlands which rely on a certain supply of the glacial meltwater. Second, and more disastrously, it significantly predisposes the occurrence of natural catastrophes such as landslides, avalanches and most notably Glacial Lake Outburst Floods (GLOFs). GLOFs are the situations when the natural dams within the lakes of meltwater fail, and huge amounts of water and debris flow downstream. Over one hundred lakes have already been identified in the alpine regions of Pakistan alone, and this presents a long-term and continually increasing hazard to the downstream farms, industrial and residential communities (Qadir, 2025).

Deforestation and soil erosion cause larger and faster runoffs and cause changes in the flow regimes and raise the amount of sediments carried by rivers. Also, by dumping untreated industrial and municipal waste into these river systems they cause a plethora of pollutants. This extensive pollution of water has posed life-threatening conditions to the water life with some fish species in major rivers such as the Ravi almost becoming extinct (Iqbal et al., 2018; Javed & Hayat, 1999). An environmental imbalance and extreme lack of clean water have also been observed to have affected the population of migratory birds visiting Pakistan meaning that there is a wide scale interference with the aquatic ecosystems formed in the mountains. Furthermore, natural hydrological processes are also disrupted by the alteration of the snowmelt period caused by deforestation and climate warming that jeopardizes irrigation systems, agriculture, and drinking water of the downstream populations that are millions of people. These cascading effects accentuate the dire necessity of having sustainable watershed management in the alpine areas.

4.4. Decline in Biodiversity and Endemic Species

The overall effect of anthropogenic stresses has caused disturbing biodiversity loss, with certain scientists stating that the world is going through a sixth mass extinction because of the ongoing human-made stress factors (Khan et al., 2016). This has already been identified to be a near 12% of the total flora in Pakistan alone, which has been classified as threatened or endangered, a figure which is likely to increase as the destruction of habitat, deforestation and over exploitation are allowed to take place without any effort at controlling them.

Natural ecological gradients are usually characterized by a decrease in the richness of species with altitude but human activities normally interfere with this pattern. Research has found that species richness is often minimal at low elevation, which is both the most reachable by humans and where the anthropogenic pressure is the highest, especially in transitory sub-alpine environments where the sub-alpine habitat is a crucial buffer to the higher alpine ecosystems. The findings of this type are the reason to mark the disproportional susceptibility of such critical ecological areas to human disturbances (Iqbal et al., 2018).

The risk to endemic species is particularly serious as their whole population is restricted to very small geographical areas; in many instances to one mountain-valley or mountain-range. When they are destroyed, there is no other home left to the endemic taxa hence the taxa are prone to extinction. Pakistan is one of the countries where 306 species of plants are endemic and the threats include overgrazing, fragmentation of habitats and pollution which are existential. Such a weakness is why several researchers have sounded warnings that we have entered a sixth mass extinction that is largely due to human-made changes in natural landscapes (Cowie et al., 2022). Extinction of endemic species is not merely a local loss in biodiversity, but a loss in the global genetic pool that is irreversible. Endemic plants usually disappear first under environmental pressure due to their habitat needs which cannot be moved and thus should be considered as an important indicator of the suitability of the regional and world biodiversity management plans.

Given their specialized habitat requirements and limited adaptability, endemic plants are often the first to vanish under environmental stress, making their conservation a crucial benchmark for assessing the effectiveness of regional and global biodiversity management strategies.

4.5 Invasive Species Introduction

Degraded and disturbed ecosystems are particularly vulnerable to invasion by non-native species. Anthropogenic activities create openings in the natural vegetation that aggressive, generalist alien species can exploit, often outcompeting the native, specialized alpine flora. Introduction of exotic species, such as Mesquite (*Prosopis juliflora*), has been cited as one of the greatest threats, replacing the landscapes at a huge economic and social price. Such species may be broadly distributed, which causes the destruction of indigenous trees such as Himalayan Ash and Himalayan Elm, as well as radically changes the structure and functioning of ecosystem (Iqbal et al., 2018). Having established, invasive species tend to develop very fast because there are no natural predators, and it is rather expensive and extremely hard to eradicate them. In mountainous areas, where the native species are ecologically sensitive, invasives may disrupt pollination, decrease the quality of forage, and increase the rate of biodiversity loss.

4.6 Disentangling Human vs. Natural Impacts on a Global Scale

Although the effects of localized anthropogenic activities are many, and they are plainly devastating, it is vital to comprehend them in the context of greater forces of nature. In a large-scale analysis of residual analysis to isolate the factors of vegetation change, it was discovered that the weight of the significant natural factors affecting alpine vegetation was higher than the effect of anthropogenic activities. This is probably due to the fact that most of the alpine areas are isolated and sparsely populated and thus the broad climatic trends prevail in vegetation changes.

Interestingly, the study also found that the positive impacts of anthropogenic activities (leading to greening) were greater than the negative impacts (leading to browning) on an annual basis globally. This suggests that while activities like overgrazing and infrastructure cause degradation in specific areas, other human-driven factors such as CO_2 fertilization, nitrogen deposition, or targeted conservation and revegetation projects have had a net positive effect on vegetation greenness at this broad scale (Zhang et al., 2022). This highlights that human intervention is not uniformly destructive and underscores the potential for positive action.

5. SYNERGISTIC EFFECTS WITH CLIMATE CHANGE

Anthropogenic activities never take place in isolation but their effects are enhanced by their general impacts of the global climate change. One of the most vulnerable to climatic changes is mountain ecosystem which is frequently considered the initial sign of global environmental degradation. A combination of direct pressures (grazing, deforestation and tourism) and indirect ones (warmer climate) exerts a synergistic effect on the rate of ecosystem degradation (Ishaq et al., 2016). Among the most documented impacts of the change in climate in the alpine regions is the migration of species upwards. The increase in temperatures compels vegetation zones to wander up into areas of higher elevation in search of conditions that are favorable and this is a common occurrence, commonly referred to as, upslope migration. Although such a reaction can be observed in mountain systems around the globe, it also poses a severe issue of the so-called summit trap as the species that have already been pushed to the highest altitudes will become extinct as no new habitats can be found (Ishaq et al., 2016).

However, it is also necessary to note the adaptive possibilities of alpine plants. Mountainous terrain brings about characteristic mosaic patterns of microclimates, which allows certain species to survive by simply relocating a few meters to better-adapted niches, as opposed to comprehensive elevational changes (Körner & Hiltbrunner, 2021). Therefore, although there is happening compositional turnover in summit communities, there is a possibility of regional extinction being reduced in large mountain ranges which retain this level of geodiversity.

The other significant impact of warming is the progressive increase of climatic treeline. Despite the fact that as glaciers are retreating, they show new substrates to growth, the net increase in habitat is low when weighed against the huge losses as a result of encroaching forests to the alpine areas. International scenarios indicate that a permanent increase in temperature of 2.2 K may cause half of the land area that could support alpine vegetation to be destroyed since the treelines will move upwards by 200-400 meters. This occurs because mountains narrow with elevation, meaning that even modest vertical shifts translate into dramatic reductions in habitable area at the broader base of the alpine belt (Körner & Hiltbrunner, 2021).

Besides habitat loss, warming is changing even the structure of alpine ecosystems. One of the most notable processes is the shrubification of arctic and alpine habitats, where ecologically unproductive species like the green alder (*Alnus alnobetula*) spread to taller altitudes, altering the vegetation competition and the communities. Even though it is difficult to study the growth of shrubs because they do not develop symmetrically in rugged areas, this is the key to making predictions about the future vegetation patterns (Oberhuber et al., 2023). These are accompanied by extensive changes in community structure, such as those in the Tibetan Plateau a warming region with twice the global mean where alpine grasslands have switched to sledge communities to grass communities (Liu et al., 2018). These examples are used to emphasize that climate change, together with human stressors, is causing radical restructuring of the alpine biodiversity, and the end result may be irreversible in terms of ecosystem functionality.

5.1 Increased Vulnerability to Natural Disasters

The vulnerability of natural disasters in the alpine is a direct worsening of human-induced climate change. A good example is the European Alps. A huge retreat of the glaciers was caused by the summer heatwave of 2003 which caused the highest temperatures to be experienced in Central Europe in half a millennium (Luterbacher et al., 2004). This was due to minimal snow cover that subjected the underlying ice to warming of the atmosphere and hastened melting. The combination of this with serious drought also caused forest fires. Only two years later, in August 2005, catastrophic floods hit the same Alpine region, due to the record rates of rainfall resulting in a great loss of life and infrastructure (Keiler et al., 2010). This is evidenced by these occurrences where climatic change results in a unstable environment where extreme heat, dryness, fire and floods can occur closely. This trend of increasing climate extremes is not limited to Europe, but analogous effects, such as the accelerated glacial retreat, heat waves and devastating floods are being witnessed in the Himalayan alpine region and along with them are causing analogous threats to its ecosystems and communities.

5.2 Climate-Induced Migration and Social Pressures

The degradation of alpine ecosystems has profound social consequences, including the forced migration of human populations. As climate change undermines the resources that mountain communities depend on such as water, food from agriculture, and stable land it can render their traditional homes uninhabitable (Stern, 2006). For populations in fragile mountain areas, environmental degradation and the loss of livelihoods are becoming significant drivers of displacement, adding a critical human dimension to the ecological crisis (Bandh et al., 2021).

5.3 Plant Functional Traits and Adaptation

The environmental degradation and livelihood loss is proving to be a key factor that causes displacement in populations in the weak mountainous regions which gives a human dimension of seriousness to the ecological crisis. The outstanding biodiversity of the alpine areas is supported by a set of functional characteristics that allow the plant to survive in the most severe conditions. Binter et al. (2025) propose a recent comprehensive study of the Himalayan steppe and alpine zones, which gives potent mechanistic information about these adaptations. The authors found that a species with a specific ecological optimum is defined by a set of distinct trait syndromes by studying 17 functional traits in 310 species over 3,500 meters of elevation gradient (Binter et al., 2025).

In the most extreme and high alpine areas, species have a conservative life history strategy that entails stress-tolerance and conservation of resources. This effective trait syndrome that accounted 61% of the variation in species elevational optima represents the combination of short stature and compact growth forms with large portion of storage tissue (parenchyma, avg. 63%). Physiologically, these plants have high water-use efficiency (as indicated by delta 13C), high freezing resistance (as indicated by high levels of fructans), and high nutrient retention (as indicated by high levels of root nitrogen and leaf phosphorus). Meanwhile, species that grow well in the lower, drier steppe environments are more competitive growth and avoid drought, and their competitive success is better explained by aspects of life span and clonal reproduction. The average steppe vegetation is more likely to be taller (42 cm vs. 20 cm), faster-growing, and it has more structural (lignified) tissues than the high-alpine vegetation (Binter et al., 2025) (Binter et al., 2025). Such hierarchical combinations of morphological, anatomical and physiological aspects prove the fact that survival in alpine conditions depends on a complex of strategies, and such an approach is consistent with the original ecological ideas about stress-tolerant and competitive life histories (Grime, 1974; Körner & Hiltbrunner, 2021).

6. CONSERVATION AND MANAGEMENT STRATEGIES

6.1 In-situ Conservation (On-site Protection)

This approach focuses on conserving species within their natural habitats, allowing evolutionary processes to continue. It is considered the most fundamental method for preserving biodiversity.

Protected Areas: The establishment and management of National Parks, Wildlife Sanctuaries, and Biosphere Reserves form the backbone of in-situ conservation. These areas safeguard entire ecosystems and the communities of species within them. For alpine plants, protected high-altitude regions like the Valley of Flowers National Park in the Indian Himalayas and Nanda Devi Biosphere Reserve play a critical role in preventing habitat degradation and conserving endemic species adapted to harsh alpine environments (Niesenbaum, 2019; Saxena & Brahman, 1996).

Sacred Groves: Tracts of forest protected by local communities due to cultural or religious beliefs often act as vital repositories of Rare, Endangered, and Threatened (RET) and endemic plant species. In many Himalayan regions, alpine meadows and high-altitude sacred landscapes are traditionally protected, serving as community-managed conservation areas for rare medicinal plants such as *Saussurea obvallata* and *Picrorhiza kurroa* (Kisku & Barik, 2015; Panda et al., 2014).

On-farm Conservation: For domesticated or cultivated alpine species, maintaining traditional farming systems helps preserve genetic diversity. Alpine crops such as buckwheat, barley, and amaranth are traditionally cultivated by mountain communities and provide both food and genetic resources for resilience to climate change. Despite being a cornerstone of conservation, in-situ approaches face challenges such as climate-induced habitat shifts, grazing pressures, and human encroachment, which pose a serious risk to alpine plant populations (Heywood, 2014).

6.2 Ex-situ Conservation (Off-site Protection)

Ex-situ strategies serve as a backup for species too threatened to survive in their natural habitat, or for research and restoration purposes.

Botanical Gardens: Botanical gardens worldwide collectively manage about 30% of all known plant species and over 41% of threatened species in their living collections (Mounce et al., 2017). Alpine botanical gardens such as the High Altitude Plant Physiology Research Centre (HAPPRC) in Uttarakhand, India, focus specifically on cultivating and preserving alpine species under controlled conditions, thereby aiding in their restoration.

Seed Banks and Germplasm Banks: These facilities provide long-term storage of alpine species seeds, which is particularly valuable given that many alpine plants produce orthodox seeds that can be stored under low-temperature and low-moisture conditions. The Millennium Seed Bank has prioritized several high-altitude medicinal plants for conservation (Breman et al., 2021).

Cryopreservation: Many alpine species bear recalcitrant seeds or propagate vegetatively, making cryopreservation indispensable. Plant materials such as embryos, shoot tips, and pollen can be stored at -196°C, ensuring long-term genetic conservation of alpine species like *Aconitum heterophyllum* and *Podophyllum hexandrum* (Corlett, 2016; Engelmann, 2011).

6.3 Modern Biotechnological Methods

Biotechnology offers powerful tools to accelerate propagation and ensure the long-term survival of endangered species, particularly alpine plants that are slow-growing and difficult to reproduce conventionally.

In Vitro Propagation (Micropropagation): Tissue culture enables the growth of whole plant under sterile conditions of small samples of the plant tissues. This technique can multiply in bulk the RET alpine species which normally are very susceptible to overharvesting, like *Gentiana kurroo* and *Nardostachys jatamansi* (Sarasan et al., 2006).

Cryobiotechnologies: It is possible to preserve alpine plant cells cultures, shoot tips and embryos by combining cryopreservation with in vitro methods. This guarantees that there is a wider range of ex-situ conservation over time and that reintroduction programs to their original habitats are supported (Pence et al., 2020).

6.4 Policy Frameworks and Community Engagement

This cannot be done just by technical solutions. Conservation needs to be engrained through social, legal and educational frameworks that are participatory at the local levels.

Priorities and Laws: Aids such as the IUCN Red List would be vital in determining the species within the Alps that are at greatest risk, based on habitat loss, and climate change, hence setting conservation priorities. Global alignment is offered by international policies like the UN Sustainable Development Goals (SDG 13: Climate Action and SDG 15: Life on Land) and the Global Strategy of Plant Conservation (GSPC) (Mounce et al., 2017; Pence et al., 2020).

Documentation and Indigenous Knowledge (IK): Mountain communities possess ethnobotanical knowledge of alpine plants of immense scale, as utilized in traditional medicine, food and rituals. The key in documenting this experience is the issue of preservation and the approach of harvesting the plant material in the alpine like *Arnebia benthamii,* and *Rheum australe* sustainably (Smita et al., 2012; Tailang, 2014). This is essential in the process of carrying out comprehensive ethnobotanical surveys, including that of District Charsadda. The research reported the numerous applications with the identification of 66 medicinal species, 28 vegetable species, 27 fruit species, and 28 ornamental species among others (Khan & Badshah, 2019). It is not a list of plants; it is an essential storehouse of indigenous knowledge. This kind of data forms the scientific foundation of community-based conservation initiatives to incorporate the needs of local communities and sustainable harvesting activities so that the knowledge will be added to but not subtracted to the biodiversity conservation.

Public Awareness and Sustainable Practices: In the case of Alpine plant diversity unsustainable harvesting is placing a tremendous amount of stress on the ecosystem and more than 90% of medicinal alpine plants are harvested wild (Panda et al., 2023; Schippmann et al., 2002). The pressure on natural populations can be reduced through community participation in sustainable harvesting, social forestry and eco-tourism projects. Examples of programs, through which fragile ecosystems are preserved by grassroots involvement, include programs like participatory monitoring of alpine meadow in the Himalayas.

Policy Interventions and International Frameworks

Transboundary collaboration among Himalayan countries is particularly critical, as alpine ecosystems extend across national boundaries. Effective implementation of these frameworks, supported by local participation, financial mechanisms, and scientific monitoring, is key to ensuring the long-term survival of alpine ecosystems in a rapidly changing climate (Figure 2).

Figure 2. Comprehensive Conservation and Management Strategies for Alpine Plants

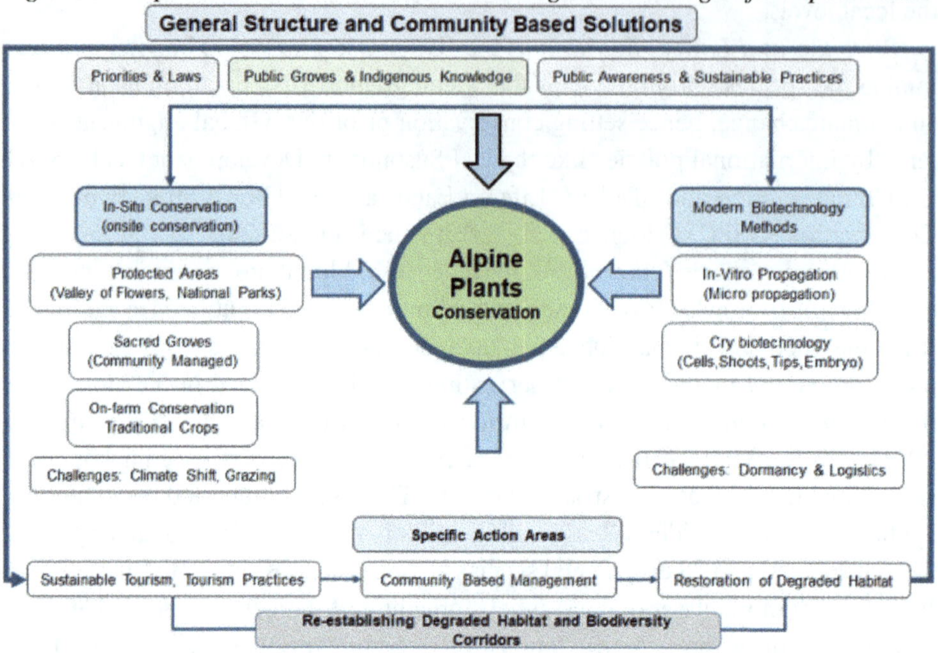

6.5 Sustainable Tourism Practices

The alpine regions have become a threat as well as an opportunity in tourism. Although it does bring income to mountain communities, uncontrolled tourism usually causes habitat destruction, littering and disruption of delicate alpine environments. Sustainable tourism aims at reducing such effects by enhancing environmentally friendly facilities, controlling the number of visitors and implementation of trekking, camping and interaction with wildlife codes of conduct. Ecotourism projects that take place in the community like in the Himalayas have shown that environmental education when used hand in hand with tourism can bring awareness to the tourists and at the same time they provide economic benefit to the people of the community, thus encouraging them to preserve the biodiversity. Sustainable tourism when well controlled does not only lessen the ecological pressure but also offers alternative means of livelihood which lessens relying on unsustainable harvesting of the alpine resources.

6.6 Restoration of Degraded Habitats

Overgrazing and the development of infrastructure and changes caused by climate jeopardize both biodiversity and ecosystem services in the alpine habitats. The work to restore such vulnerable landscapes must take specialized methods, such as reseeding native alpine grasses, managing the encroachment of invasive shrubs and soil stabilization through bioengineering. Planting hardy pioneer species in the disturbed sites is an example of how the ecological succession is hastened in glacier-fed valleys and soil erosion is prevented. Going on, participatory restoration projects with local communities are put into practice so that there is a maintenance of the long run and ownership to the conservation. Effective habitat recovery does not only restore the biodiversity but also maintains the important ecosystem processes like water management, carbon capture, and pollination which are very crucial to the mountain inhabitants and downstream individuals.

6.7 Strengthening Protected Areas and Biodiversity Corridors

The alpine conservation strategies are based on the conserved areas which are often isolated as ecological islands that can be easily fragmented. The improvement of these areas entails increasing the number of networks of National Parks, Biosphere Reserves and Community Reserves, as well as proper management and enforcement. It is also important to create biodiversity corridors that would connect the view of alpine habitats which have been fragmented because of climatic changes and allow the species to move. As an illustration, the use of corridors within the Eastern Himalayas has helped in the exchange of fauna and flora between the reserves allowing genetic interchange and survival in the long term. By including the local community stewardship in the running of these corridors, conservation benefits and sustainable development will balance.

6.8 Disaster Risk Reduction and Early Warning Systems

Since frequency of climate related hazards has been on the rise in alpine areas, conservation should be considered along with the disaster risk reduction. The increasing danger of the Glacial Lake Outburst Floods (GLOFs) requires the active approach to the development of the traditional habitat protection. One major initiative is the implementation of the use of the modern sensor-based early warning systems in the prone glacial regions. These systems may be used to check the level of the lakes and seismic activity, which will give the downstream communities an early warning about the possibility of flooding (Qadir) at el, 2025). This is a method that is directly related to the monitoring of ecosystems and the safety of people and the infrastructure. Additionally, training disaster teams in the areas and building climate-adaptive infrastructure in the ones in risky areas is the key aspect of the complex approach to the total management, which will eliminate the whole range of climate change effects on the mountain environment and people that are dependent on it.

7. FUTURE PERSPECTIVES

With the increasing effects of anthropogenic activity and climate change, the future of alpine ecosystems is one of the key problems to be predicted correctly. According to the review of Kozlowski (2024), one of the most important points that are shared by the prominent alpine ecologists is that the use of broad-scale species distribution models (SDMs), despite its usefulness, does not always reflect the situation on the ground. According to Kozlowski (2024), nothing makes sense in alpine vegetation biology, until one considers micro-climate (Kozlowski, 2024). This highlights the fact that fine grained, micro topographical information is critical in any sound modeling endeavor. The way forward in future studies can therefore be seen to be beyond the large scale climate forecasts and instead to the detailed interactions within the mosaic of micro habitats that characterize alpine terrain. The mechanisms by which plants negotiate their way through this fine-grained environment is important in understanding their persistence and in the formulation of effective site specific conservation strategies

8. CONCLUSION

The alpine ecosystems in the Himalayas are among the most delicate and important landscapes on Earth. These high-altitude biomes, once thought remote and untouched, are now facing unprecedented stresses due to local anthropogenic factors and global climate change. Their endemic and extreme conditions have produced unique biodiversity that is slowly being eroded, with cascading effects on global biodiversity, regional climate stability, and the millions of people who rely on mountain-water resources. Human activities remain major sources of degradation. Unregulated tourism, overgrazing, fuelwood cutting, and poorly planned infrastructure fragment habitats, disrupt migratory routes, and increase extinction risks for species with narrow ecological niches. Soil erosion and land degradation reduce ecosystem productivity, while altered hydrological cycles increase flood risks and threaten water security. Long-range pollution from distant industrial centers adds further strain, as heavy metals and black-carbon deposits weaken alpine flora and fauna. These localized threats are intensified by climate change, which is warming alpine areas faster than the global average. Species are being driven upslope and often fall victim to a "summit trap," with no space left to migrate. Vegetation is shifting due to rapid shrubification of alpine meadows, altering ecosystem interactions and functions. Meanwhile, melting ice caps and unpredictable rainfall disrupt hydrological processes. Together, these pressures create a feedback loop in which a degraded ecosystem becomes less resilient to climate change, while climate change amplifies local stress. Conservation must therefore shift from passive to proactive strategies. These should be supported by ex-situ measures such as cryopreservation and micropropagation for highly endangered species. Yet conservation cannot succeed without strong community participation. Traditional knowledge, sustainable livelihoods like ecotourism, and community-based resource management are essential to reducing destructive practices. At broader scales, effective governance, national commitment, and international collaboration are required to integrate conservation into regional development and align it with global goals like the Sustainable Development Goals. The fate of these alpine ecosystems is a warning for global ecological health. Protecting these "water towers of the world" safeguards biodiversity, climate security, and human well-being. Urgent, coordinated action linking science, policy, and communities is essential to ensure their survival for future generations.

REFERENCES

Aïtcin, P.-C. (2000). Cements of yesterday and today: Concrete of tomorrow. *Cement and Concrete Research*, *30*(9), 1349–1359. DOI: 10.1016/S0008-8846(00)00365-3

Bandh, S. A., Shafi, S., Peerzada, M., Rehman, T., Bashir, S., Wani, S. A., & Dar, R. (2021). Multidimensional analysis of global climate change: A review. *Environmental Science and Pollution Research International*, *28*(20), 24872–24888. DOI: 10.1007/s11356-021-13139-7 PMID: 33763833

Behera, M. D., Behera, S. K., & Sharma, S. (2019). Recent advances in biodiversity and climate change studies in India. *Biodiversity and Conservation*, *28*(8), 1943–1951. DOI: 10.1007/s10531-019-01781-0

Binter, J., Macek, M., & Dolezal, J. (2025). Integrating morphological, anatomical, and physiological traits to explain elevational distributions in Himalayan steppe and alpine plants. *Journal of Integrative Plant Biology, n/a*(n/a). https://doi.org/https://doi.org/10.1111/jipb.13971

Breman, E., Ballesteros, D., Castillo-Lorenzo, E., Cockel, C., Dickie, J., Faruk, A., & Sharrock, S. (2021). Plant diversity conservation challenges and prospects—The perspective of botanic gardens and the Millennium Seed Bank. *Plants*, *10*(11), 2371. DOI: 10.3390/plants10112371 PMID: 34834734

Chapman, L. (2007). Transport and climate change: A review. *Journal of Transport Geography*, *15*(5), 354–367. DOI: 10.1016/j.jtrangeo.2006.11.008

Chardon, N. I., Stone, P., Hilbert, C., Maclachlan, T., Ragsdale, B., Zhao, A., Goodwin, K., Collins, C. G., Hewitt, N., & Elphinstone, C. (2023). Species-Specific Responses to Human Trampling Indicate Alpine Plant Size Is More Sensitive than Reproduction to Disturbance. *Plants*, *12*(17), 3040. DOI: 10.3390/plants12173040 PMID: 37687287

Cook, J., & Lewandowsky, S. (2016). Rational irrationality: Modeling climate change belief polarization using Bayesian networks. *Topics in Cognitive Science*, *8*(1), 160–179. DOI: 10.1111/tops.12186 PMID: 26749179

Corlett, R. T. (2016). Plant diversity in a changing world: Status, trends, and conservation needs. *Plant Diversity*, *38*(1), 10–16. DOI: 10.1016/j.pld.2016.01.001 PMID: 30159445

Cowie, R. H., Bouchet, P., & Fontaine, B. (2022). The Sixth Mass Extinction: Fact, fiction or speculation? *Biological Reviews of the Cambridge Philosophical Society*, *97*(2), 640–663. DOI: 10.1111/brv.12816 PMID: 35014169

Dhar, U. (2002). Conservation implications of plant endemism in high-altitude Himalaya. *Current Science*, •••, 141–148.

Engelmann, F. (2011). Use of biotechnologies for the conservation of plant biodiversity. *In Vitro Cellular & Developmental Biology. Plant*, *47*(1), 5–16. DOI: 10.1007/s11627-010-9327-2

Grime, J. P. (1974). Vegetation classification by reference to strategies. *Nature*, *250*(5461), 26–31. DOI: 10.1038/250026a0

Heywood, V. H. (2014). An overview of in situ conservation of plant species in the Mediterranean. *Flora Mediterranea*, *24*, 5–24. DOI: 10.7320/FlMedit24.005

Ilea, R. C. (2009). Intensive livestock farming: Global trends, increased environmental concerns, and ethical solutions. *Journal of Agricultural & Environmental Ethics*, *22*(2), 153–167. DOI: 10.1007/s10806-008-9136-3

Iqbal, M., Sherwani, A., & Shafiq, M. (1999). Vegetation characteristics and trace metals (Cu, Zn and Pb) in soils along the super highways near Karachi, Pakistan. *Studia Botanica Hungarica*, *29*, 79–86.

Iqbal, M. Z., Gill, D., & Shafiq, M. (1998). Plant communities along the sewage effluents channels of Lyari river in Pakistan. *Taiwania*, *43*(1), 1–11.

Iqbal, M. Z., Shafiq, M., & Kabir, M., & Farooqi, Z.-u.-R. (2018). Impact of anthropogenic activities on biodiversity in Pakistan: A review. *Bioscience Research*, *15*(1), 481–490.

Ishaq, S., Khan, M., Begum, F., Hussain, K., Amir, R., Hussain, A., & Ali, S. (2016). Climate change impact on mountain biodiversity: A special reference to Gilgit-Baltistan of Pakistan. *Journal of Mountain Area Research*, *1*, 53–63. DOI: 10.53874/jmar.v1i0.12

Javed, M., & Hayat, S. (1999). Heavy metal toxicity of river Ravi aquatic ecosystem. *Pakistan Journal of Agricultural Sciences*, *36*, 1–9.

Jenkins, C. N., & Joppa, L. (2009). Expansion of the global terrestrial protected area system. *Biological Conservation*, *142*(10), 2166–2174. DOI: 10.1016/j.biocon.2009.04.016

Karanth, K. K., Nichols, J. D., Karanth, K. U., Hines, J. E., & Christensen, N. L., Jr. (2010). The shrinking ark: patterns of large mammal extinctions in India. *Proceedings of the Royal Society B: Biological Sciences*, *277*(1690), 1971-1979. DOI: 10.1098/rspb.2010.0171

Keiler, M., Knight, J., & Harrison, S. (2010). Climate change and geomorphological hazards in the eastern European Alps. *Philosophical Transactions of the Royal Society A: Mathematical, Physical and Engineering Sciences, 368*(1919), 2461-2479.

Khan, M. N., & Ali, S. RAZAK, S. A., Zaman, A., Iqbal, M., & Shah, S. N. J. B. J. o. B. D. (2022). Assessment of floristic diversity in the mountain ecosystem of Marghazar Valley, Hindukush Range, Swat, Pakistan. *23*(2).

Khan, M. N., Ali, S., Yaseen, T., Ullah, S., Zaman, A., Iqbal, M.,...Sciences, A. (2019). Eco-taxonomic study of family Poaceae (Gramineae). *10*(2), 63-75.

Khan, M. N., & Badshah, L. (2019). Floristic diversity and utility of flora of district Charsadda, Khyber Pakhtunkhwa. *Acta Ecologica Sinica, 39*(4), 306–320. https://doi.org/https://doi.org/10.1016/j.chnaes.2018.10.003. DOI: 10.1016/j.chnaes.2018.10.003

Khan, S., Nabi, G., Ullah, M. W., Yousaf, M., Manan, S., Siddique, R., & Hou, H. (2016). Overview on the role of advance genomics in conservation biology of endangered species. *International Journal of Genomics, 2016*(1), 3460416. DOI: 10.1155/2016/3460416 PMID: 28025636

Kisku, S., & Barik, D. K. L. (2015). Ecofloristic Studies on Sacred Groves of Bamanghaty Subdivision in Mayurbhanj District, Odisha and its Conservation Strategies. *Indian Journal of Research*, *4*(6), 482–483.

Körner, C. (2004). Mountain biodiversity, its causes and function. *Ambio*, *33*(sp13), 11–17. DOI: 10.1007/0044-7447-33.sp13.11 PMID: 15575177

Körner, C., & Hiltbrunner, E. (2021). Why is the alpine flora comparatively robust against climatic warming? *Diversity*, *13*(8), 383. DOI: 10.3390/d13080383

Körner, C., & Hiltbrunner, E. J. D. (2021). Why is the alpine flora comparatively robust against climatic warming?, *13*(8), 383.

Körner, C., Paulsen, J., & Spehn, E. M. (2011). A definition of mountains and their bioclimatic belts for global comparisons of biodiversity data. *Alpine Botany*, *121*(2), 73–78. DOI: 10.1007/s00035-011-0094-4

Kozlowski, G. (2024). Understanding Arctic-Alpine Plants from Ecological and Evolutionary Perspectives. *Plants*, *13*(11), 1560. Advance online publication. DOI: 10.3390/plants13111560 PMID: 38891368

Liu, H., Mi, Z., Lin, L., Wang, Y., Zhang, Z., Zhang, F., & Cao, G. (2018). Shifting plant species composition in response to climate change stabilizes grassland primary production. *Proceedings of the National Academy of Sciences of the United States of America, 115*(16), 4051–4056. DOI: 10.1073/pnas.1700299114 PMID: 29666319

Luterbacher, J., Dietrich, D., Xoplaki, E., Grosjean, M., & Wanner, H. (2004). European seasonal and annual temperature variability, trends, and extremes since 1500. *Science, 303*(5663), 1499–1503. DOI: 10.1126/science.1093877 PMID: 15001774

Mounce, R., Smith, P., & Brockington, S. (2017). Ex situ conservation of plant diversity in the world's botanic gardens. *Nature Plants, 3*(10), 795–802. DOI: 10.1038/s41477-017-0019-3 PMID: 28947807

Negi, V. S., Joshi, B. C., Pathak, R., Rawal, R. S., & Sekar, K. C. (2018). Assessment of fuelwood diversity and consumption patterns in cold desert part of Indian Himalaya: Implication for conservation and quality of life. *Journal of Cleaner Production, 196*, 23–31. DOI: 10.1016/j.jclepro.2018.05.237

Negi, V. S., & Maikhuri, R. (2017). Forest resources consumption pattern in Govind wildlife sanctuary, western Himalaya, India. *Journal of Environmental Planning and Management, 60*(7), 1235–1252. DOI: 10.1080/09640568.2016.1213707

Negi, V. S., Tiwari, D. C., Singh, L., Thakur, S., & Bhatt, I. D. (2022). Review and synthesis of climate change studies in the Himalayan region. *Environment, Development and Sustainability, 24*(9), 10471–10502. DOI: 10.1007/s10668-021-01880-5

Niesenbaum, R. A. (2019). *The integration of conservation, biodiversity, and sustainability* (Vol. 11). MDPI.

Noroozi, J., Akhani, H., & Breckle, S.-W. (2008). Biodiversity and phytogeography of the alpine flora of Iran. *Biodiversity and Conservation, 17*(3), 493–521. DOI: 10.1007/s10531-007-9246-7

Oberhuber, W., Dobler, A. L., Heinzle, T., Scandurra, F., Gruber, A., & Wieser, G. (2023). Climate Overrides the Influence of Microsite Conditions on Radial Growth of the Tall Multi-Stemmed Shrub Alnus alnobetula at Treeline. *Plants, 12*(8), 1708. Advance online publication. DOI: 10.3390/plants12081708 PMID: 37111935

Oldekop, J. A., Holmes, G., Harris, W. E., & Evans, K. L. (2016). A global assessment of the social and conservation outcomes of protected areas. *Conservation Biology, 30*(1), 133–141. DOI: 10.1111/cobi.12568 PMID: 26096222

Panda, D., Bisoi, S. S., & Palita, S. (2014). Floral diversity conservation through sacred groves in Koraput district, Odisha, India: a case study.

Panda, T., Mishra, N., Rahimuddin, S., Pradhan, B. K., Apollo, M., Kar, M. K., & Mohanty, R. B. (2023). Diversity and Conservation Significance of Angiosperm Climbers in Bhadrak District of Odisha, India. [JTBC]. *Journal of Tropical Biology & Conservation, 20*, 135–160. DOI: 10.51200/jtbc.v20i.4652

Pence, V. C., Ballesteros, D., Walters, C., Reed, B. M., Philpott, M., Dixon, K. W., Pritchard, H. W., Culley, T. M., & Vanhove, A.-C. (2020). Cryobiotechnologies: Tools for expanding long-term ex situ conservation to all plant species. *Biological Conservation, 250*, 108736. DOI: 10.1016/j.biocon.2020.108736

Qadir, A. (2025) VULNERABILITY CHALLENGES TO HIGH ALTITUDE WETLANDS AND THEIR FEEDING GLACIERS IN PAKISTAN REGARDING CLIMATE CHANGE.

Qazi, A. W., Saqib, Z., Zaman-ul-Haq, M. J. P. o. t. P. A. o. S. B. L., & Sciences, E. (2023). Diversity and distribution of endemic flora in Pakistan. *60*(2), 165-179.

Ramanathan, V., & Carmichael, G. (2008). Global and regional climate changes due to black carbon. *Nature Geoscience, 1*(4), 221–227. DOI: 10.1038/ngeo156

Ramankutty, N., Gibbs, H. K., Achard, F., Defries, R., Foley, J. A., & Houghton, R. (2007). Challenges to estimating carbon emissions from tropical deforestation. *Global Change Biology, 13*(1), 51–66. DOI: 10.1111/j.1365-2486.2006.01272.x

Rana, S. K., Rawal, R. S., Dangwal, B., Bhatt, I. D., & Price, T. D. (2021). 200 years of research on Himalayan biodiversity: Trends, gaps, and policy implications. *Frontiers in Ecology and Evolution, 8*, 603422. DOI: 10.3389/fevo.2020.603422

Rundel, P., & Millar, C. (2016). Alpine ecosystems. *In: Zavaleta, E.; Mooney, H., eds. Ecosystems of California. Berkeley, California: University of California Press: 613-634. Chapter 29.*, 613-634.

Samant, S. P. S. (2007). Assessment of plant diversity and prioritization of communities for conservation in Mornaula Reserve Forest. *Applied Ecology and Environmental Research, 5*(2), 123–138. DOI: 10.15666/aeer/0502_123138

Sarasan, V., Cripps, R., Ramsay, M. M., & Atherton, C. McMICHEN, M., Prendergast, G., & Rowntree, J. K. (2006). Conservation in vitro of threatened plants—progress in the past decade. *In Vitro Cellular & Developmental Biology-Plant, 42*(3), 206-214.

Saxena, H., & Brahman, M. (1996). The Flora of Orissa" 1-4 Orissa Forest Development Corporation Ltd. *Bhubaneswar, India.*

Schippmann, U., Leaman, D. J., & Cunningham, A. B. (2002). Impact of cultivation and gathering of medicinal plants on biodiversity: global trends and issues. *Biodiversity and the ecosystem approach in agriculture, forestry and fisheries.*

Shaheen, H., Aziz, S., Nasar, S., Waheed, M., Manzoor, M., Siddiqui, M. H., Alamri, S., Haq, S. M., & Bussmann, R. W. (2023). Distribution patterns of alpine flora for long-term monitoring of global change along a wide elevational gradient in the Western Himalayas. *Global Ecology and Conservation, 48*, e02702. https://doi.org/https://doi.org/10.1016/j.gecco.2023.e02702. DOI: 10.1016/j.gecco.2023.e02702

Smita, R., Sangeeta, R., Kumar, S. S., Soumya, S., & Deepak, P. (2012). An ethnobotanical survey of medicinal plants in Semiliguda of Koraput District, Odisha, India. *Research Journal of Recent Sciences, 5*, 97–107.

Stern, N. (2006). Stern Review: The economics of climate change.

Tailang, M. (2014). Ethnomedicinal plants used by some Indian tribe. *World Journal of Pharmaceutical Research, 4*, 549–558.

Thakur, S., Negi, V. S., Dhyani, R., Bhatt, I., & Yadava, A. (2022). Influence of environmental factors on tree species diversity and composition in the Indian western Himalaya. *Forest Ecology and Management, 503*, 119746. DOI: 10.1016/j.foreco.2021.119746

Thakur, S., Negi, V. S., Pathak, R., Dhyani, R., Durgapal, K., & Rawal, R. S. (2020). Indicator based integrated vulnerability assessment of community forests in Indian west Himalaya. *Forest Ecology and Management, 457*, 117674. DOI: 10.1016/j.foreco.2019.117674

Wani, Z. A., Pant, S., & Sharma, V. (2021). Diversity, Distribution, and Status of Weed Species of Northwest Himalaya. In *Bioremediation using weeds* (pp. 3–34). Springer. DOI: 10.1007/978-981-33-6552-0_1

Wesche, K., Miehe, G., & Kaeppeli, M. (2000). The significance of fire for afroalpine ericaceous vegetation. *Mountain Research and Development, 20*(4), 340–347. DOI: 10.1659/0276-4741(2000)020[0340:TSOFFA]2.0.CO;2

Zhang, Y., Sun, J., Lu, Y., & Song, X. J. L. (2022).. . *Influence of Anthropogenic Activities and Major Natural Factors on Vegetation Changes in Global Alpine Regions., 11*(7), 1084.

Chapter 4
Sustaining Alpine Landscape:
Biodiversity, Climate Change, and Conservation Imperatives

Shivansu Sachan
http://orcid.org/0000-0002-9893-8018
CSJM University, Kanpur, India

Nouman Ahmad
http://orcid.org/0009-0009-3370-2862
University of Mianwali, Punjab, Pakistan

Aleena Gul
University of Mianwali, Punjab, Pakistan

Mamoona Uzma Kalsoom
University of Mianwali, Punjab, Pakistan

Mridulesh Singh
http://orcid.org/0009-0000-6365-4392
CSJM University, Kanpur, India

Arvind Chauhan
http://orcid.org/0000-0002-3647-2988
CSJM University, Kanpur, India

ABSTRACT

The chapter is a thorough re-evaluation of the alpine conditions, the amazing biodiversity they host, the dramatic influence of the climate change that is currently inflicting on them, and the urgency behind its conservation. It discusses the bioclimatic conditions and global distribution of these high mountain ecosystems and explains the patterns and processes that control its flora and fauna. The chapter concludes the threats of habitat degradation, exotic species, and intense warming that these regions experience that in turn can fundamentally change ecological stability and vital ecosystem services. Particular focus is on contemporary conservation models which combine transboundary governance, community-based management, restoration ecology and leading-edge technological monitoring. By integrating

DOI: 10.4018/979-8-2600-1207-9.ch004

ecological principles and socio-economic aspects, this review provides a synthesis that sets the stage for description of ecological and human cultural conservation challenges faced by these critical ecosystems at global scales.

1. INTRODUCTION TO ALPINE LANDSCAPE

The alpine ecosystems are high region ecosystems at elevations that rise above the climatic tree-line and are too hostile to support the growth of trees. These ecosystems occupy an area of nearly 3.56 million km^2 in the world, which is around 2.64% of total terrestrial land area of the earth with the exception of Antarctica (Testolin et al., 2020). They can be found in mountainous areas in all the continents, including temperate areas such as the European Alps and Rocky Mountains, boreal, tundra, and tropical high mountains such as the Andes (Testolin et al., 2020; Suárez et al., 2023). Alpine ecosystems despite latitude and climate differences do have common bioclimatic traits, usually found at temperatures less than an average yearly temperature of approximately 5.9 °C, whose vegetation and productivity are determined by the regional conditions, including seasonality and continentally (Testolin et al., 2020). Alpine zones are studied for low-growing vegetation with sparse species adapted to cold, windy, and nutrient-starved conditions, the richness of plant species can be determined by historical factors, area size, and the local environment. These ecosystems have an ecological significance in the water regulation by snow melting and are sensitive signals of the climatic change because they are sensitive to warming and species dispersal changes (Grabherr et al., 2010; Testolin et al., 2020). The alpine growing season is short, and root zone temperatures are usually approximately 4.5 °C and vegetation is adapted to cold with low nutrient conditions (Noroozi & Körner, 2018). Topography, slope aspect, and soil depth cause high microclimatic variability, resulting in extensive species buffering to larger climate warming because of large temperature variance even on small spatial scales (Malanson et al., 2024). These bioclimatic features determine the vegetation patterns, productivity and species distributions, local and regional contingencies are also significant in addition to macroclimatic variables (Testolin et al., 2020) Figure 1.

Figure 1. A Conceptual Model of an Alpine Socio-Ecological System

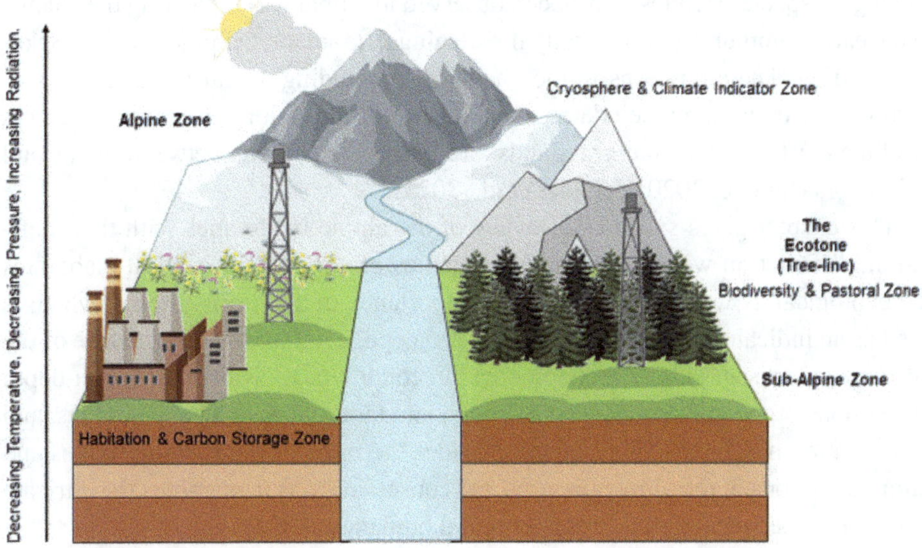

Alpine ecosystems are also important ecological and socio-cultural resources that provide various ecosystem services, vital to both the immediate societies and the population at large. These landscapes provide biodiversity, regulate water resources by snowmelt, as well as provide habitats to symbolic species that favor cultural identity (Ebner et al., 2022). Alpine regions iconically possess beauty of the scenery, recreational value, and spiritual resonance, and traditionally governed landscapes are viewed as hotspots of cultural ecosystem services, e.g. leisure and heritage value (Zoderer, Tasser, et al., 2016). These landscapes have been historically formed by the interaction of human activities with alpine ecosystem, the socio-economic processes of which influence the sustainability of resilience and stability, and which contradict the concept of the traditional fixed use of the land (Carrer et al., 2020). The role of tourism in shaping perceptions of alpine cultural services is influenced by land use, social-demographic context, and cultural background that influence perception of the services, which ultimately influences the landscape planning and management (Zoderer, Tasser, et al., 2016).

Recent issues like climate changes and the melting of glaciers demand a different principle of governance combining the experience of local practices with wider policy structures to ensure the preservation of ecological and cultural roles of alpine landscapes (Schirpke et al., 2021; Zimmer et al., 2022). Alpine ecosystems are also very delicate and effective indicator of climate change because they require low temperature and a snow cover that directly influence vegetation, soil, and freshwater dynamics (Rogora et al., 2018; Inouye, 2020; Sekar et al., 2025).

Changes towards more warm-adapted vegetation communities (thermophilization) and higher species richness have been observed in Himalayas indicating the change in the early communities potentially threatening cold-adapted alpine species (Sekar et al., 2025). The alterations in the phenology, including an earlier green-up that correlates with the snowmelt dates, and temperature alterations, are other indicators of climate effects in alpine grasslands and in the interactions between plants and pollinators (Inouye, 2020; Zehnder et al., 2025).

This chapter gives systematic review of the alpine landscapes with their definition, distribution worldwide and bioclimatics of alpine landscape. It elaborates the ecological as well as socio-cultural importance of such ecosystems, which is used as an indicator of climatic change. The scope, objectives, and structure of the book are also discussed in the chapter that preconditions transformation to in-depth discussions of the alpine biodiversity, flora, and fauna, climate change effects, and conservation measures. This chapter provides the basis of the pressing issues and future directions of the alpine research and conservation as it examines the intricate interactions between alpine ecosystems and human societies.

2. ALPINE BIODIVERSITY PATTERNS AND PROCESSES

Geological and evolutionary process form species richness and endemism of alpine regions such that biodiversity hotspots with numerous small-range species are created, particularly in tropical mountains. Their bedrock in these regions is frequently unusual, (e.g. mafic and ultramafic rocks), and presents the plants with special physiological requirements, and high levels of endemism. Deep-time evolutionary and ecological histories also make mountains affect surrounding lowlands in terms of biotic exchange and regional climate impacts (Rahbek et al., 2019).

Floristic and faunal diversity in the alpine ecology vary along both elevation and climatic gradients, plant species are most abundant in the mid-latitudes, especially in the Northern Hemisphere. The degree of alpine area, the history of the glaciers, and the local environment conditions such as pH level and macroclimate have an effect on this richness and not the mere latitudinal gradients. Change in temporal and spatial diversity of alpine biodiversity is also demonstrated by alpine bird communities and other fauna based on climate and land use, which also reflects the dynamic character of alpine biodiversity (García-Navas et al., 2020; Testolin et al., 2020, 2021). There are also different elevational and seasonal trends of below-ground microbial communities due to soil temperature and moisture, highlighting the multifaceted nature of alpine ecosystem functions and stability (Losapio et al., 2021; Ruiz-Gonzalez, 2024; Schroeder et al., 2024). A biodiversity distribution patterns steepened by natural climatic conditions, soil characteristics, and past events including a glacial history and biome changes are drivers of biodiversity distribution along elevation gradients. The changes in upslope biomes, particularly in plant lineages such as Saxifraga, have accelerated in response to a cooling of climates in the late-Cenozoic period, and have determined the alpine plant diversity over millions of years. Temperature and moisture are powerful environmental filters that affect the species composition and community assembly, and vegetation dynamics and biodiversity homogenization continues to occur in part of the alpine areas due to changes in land use and climate (Carruthers et al., 2024; Ramskogler et al., 2025).

The comparison of continental studies indicates that alpine biodiversity patterns are influenced by regional peculiarities, evolutionary history and environmental gradients, but not by the latitudinal gradient patterns. High biodiversity of alpine vegetation in temperate Eurasia is able to compete with tropical ecosystems and microbial biodiversity in the soil reflects both a sensitive reaction to alterations in elevation and time of year. These results support the need to protect alpine biodiversity as an outcome of a complex and long-term ecological and evolutionary events (Rahbek et al., 2019; Testolin et al., 2021; Carruthers et al., 2024; Ruiz-Gonzalez, 2024).

3. PLANT LIFE IN ALPINE ENVIRONMENTS

Alpine vegetation has a complex of physiological and morphological changes that allow them to survive in the most adverse conditions (low temperature, high levels of UV radiation, and scarce water supply). Such adaptations encompass small stature, shrub-like growth patterns, increased storage structures (tissues) and characteristics that aid in water retention and frost resistance, which jointly clarify most of the variation in elevation distributions of species in the Himalayas (Binter et al., 2025). The characteristics of leaves also change with altitude: the area of the leaf reduces, and its thickness, stomatal density, and mesophyll layers become greater, which diminishes the amount of transpiration and increases photosynthesis and the ability to retain water in cold and dry climate (Liu et al., 2020).

3.1. Life cycle tactics and growth forms in alpine plants

Life-cycle tactics and growth forms in alpine plants are characterized by resource conservation and stress tolerance instead of competitive dominance and a high proportion of species may be clonally propagated and have long lifespans to survive in adverse environments (De Villemereuil et al., 2018; Binter et al., 2025).

Floral distribution according to continent is as follow

The alpine plant diversity in the Himalayas is very high with the species richness taking the hump-shaped distribution along the gradient of elevations. There is higher diversity in the eastern Himalaya compared to central and western areas with numerous endemic and specialized species that are suited to the alpine conditions (Ahmad et al., 2025).

The diversity of the alpine vegetation is distinguished by a large number of specialized, and, in most cases, endemic species, which adapted to certain substrates and microhabitats, and the distribution of which was heavily determined by geographic boundaries and past climatic phenomena. Such diversity reveals the significance of conservation activities based in the regions to ensure the preservation of these exclusive alpine plant communities throughout the whole world (Boucher et al., 2021; Testolin et al., 2021; Ahmad et al., 2025; Voisin et al., 2025).

The invasive occurrence of non-native plants in the alpine settings is not very extensive as compared to the lowland habitat, mainly because of low propagule pressure as well as the extreme environmental limitations in high altitudes. Alpine areas have almost 200 non-native species, yet only several of these are cold-adapted and have great influences on the native habitats. However, future invasions by pre-adapted species are possible due to climate warming and human changes, which will endanger the native alpine biodiversity. Alpine ecosystems not only need to be relatively resistant to invaders but also to be monitored and prevented in order to retain their unique plant assemblages (Alexander et al., 2016).

4. ALPINE FAUNA AND TROPHIC INTERACTIONS

4.1. Invertebrates

Complex spatial and temporal distributions of invertebrates, especially insects and beetles are curved by habitat and climate. In alpine and temperate regions, there are pollinators *Diptera, Hymenoptera,* and *Coleoptera* and above 1500 m, flies predominate (Schuldt et al., 2009). In tropical savannas, the family *Formicidae, Carabidae*, and many *Diptera* have a weak fit to vertebrate and highlight the importance of invertebrate data on conservation (Oberprieler et al., 2019a). There are also other minor invertebrates such as spiders, beetles, and flies which can be found in a variety of microhabitats. Aquatic insects, Hydrophilidae beetles, gastropods associated with certain aquatic plants such as Typha and Nelumbo species, affect their spatial distribution (Son et al., 2021). The taxa are pollinating, decomposing, and food web contributors that are threatened by climatic changes and habitat loss (Schuldt et al., 2009; Son et al., 2021).

4.2. Aquatic life

The aquatic life on the alpine and other freshwater ecosystems consists of different types of vertebrates and invertebrates that have unique tropic interactions. The Arctic char (*Salvelinus alpinus*), brown trout (*Salmo trutta*) and brook trout (*Salvelinus fontinalis*), occupy the highest predator sites in alpine lakes in Europe, North America, and Asia (Valdez, 2020). The Odonata larvae (e.g., *Aeshnidae, Libellulidae*), *Ephemeroptera* (mayflies: *Baetidae, Heptageniidae*), Plecoptera (stoneflies: *Perlidae*), and *Trichoptera* (caddisflies: *Hydropsychidae*) are the aquatic macroinvertebrates which constitute the important trophic connections in streams and wetlands (Balian et al., 2008). These wetlands are also vulnerable to hydrologic shifts and warming, as well as the food web stability (Balian et al., 2008; Son et al., 2021).

4.3. Amphibians

Amphibians form part of the alpine and tropical ecosystems, and are usually predated by arthropod predators. Globally, frogs (*Anura*) comprise more than a third of vertebrate prey in arthropod predation events, and frogs of the families *Ranidae* (Rana), *Hylidae* (Hyla), *Bufonidae* (Bufo) and *Dendrobatidae* (Dendrobates) are often prey (Valdez, 2020). The amphibians supported by the alpine wetlands are the following: common frog (*Rana temporaria*), the alpine newt (*Ichthyosaura alpestris*) and fire salamander (*Salamandra salamandra*), that rely on salinity and habitat conditions to encounter food webs with microbes and invertebrates (Balian et al., 2008; Titley et al., 2017). These interactions indicate the significance of the amphibians in the multitrophic alpine systems.

4.4. Reptiles

Reptiles live in alpine and tropical ecosystems having different trophic interactions. Viviparous lizard (*Zootoca vivipara*), alpine salamander (*Salamandra atra*) and the common European adder (*VIpera berus*), are examples of alpine reptiles, whereas common tropical reptiles are all sorts of lizards (e.g., Anolis, Varanus), snakes (e.g., Bothrops, Naja) and turtles (e.g., Chelonia, Trachemys). These reptiles play vital role in the working of these alpine ecosystems involving predators and prey combining trophic levels (Schuldt et al., 2009; Valdez, 2020).

4.5. Predators

European Alps abode a number of predators, such as golden eagles (*Aquila chrysaetos*), lynx (*Lynx lynx*) and wolves (*Canis lupus*), which consume roe deer (*Capreolus capreolus*), wild boar (*Sus scrofa*), ibex (*Capra ibex*), red deer (*Cervus elaphus*) and chamois (*Rupicapra rupicapra*), with the help of spatial and temporal patterns of predator distributions (Oberprieler et al., 2019a). Mountain lions (*Puma concolor*), black bears (*Ursus americanus*), and coyotes (*Canis latrans*) serve the same function in the alpine and montane ranges of North America. In Asia, snow leopards (*Panthera uncia*) attack blue sheep (*Pseudois nayaur*) and Himalayan tahr (*Hemitragus jemlahicus*). Trophic regulation is also provided by birds of prey like peregrine falcon (*Falco peregrinus*) and bearded vultures (*Gypaetus barbatus*). These giant vertebrates ensure ecosystem sustainability and biodiversity due to their trophic functions (Valdez, 2020).

4.6. Mammals

The alpine and tropical regions have different trophic interactions and distributions of mammals. The herbivores inhabiting the Alps are the Alpine ibex (*Capra ibex*), the chamois (*Rupicapra rupicapra*), marmots (*Marmota marmota*), and mountain hares (*Lepus timidus*), whereas the carnivores are the Eurasian lynx (*Lynx lynx*), the red fox (*Vulpes vulpes*), and the stoats (*Mustela erminea*) (Oberprieler et al., 2019b; Mammola et al., 2021). Examples of tropical mammals are bats (*Chiroptera: Pteropus, Myotis*), primates (e.g., *Macaca, Ateles*), and carnivores (e.g., jaguars (*Panthera onca*) and civets (*Viverridae*). Prey-predator interactions between spiders (Araneae: *Theridiidae*), centipedes (*Chilopoda*), and beetles (Coleoptera: *Carabidae*) are reported to be complicated predators of mammals, particularly bats, worldwide (Valdez, 2020). On the whole, the alpine fauna, including invertebrates as well as large vertebrates, have intricate, continent-scale trophic relationships determined by environmental gradients, habitat organization, and climate.

5. CLIMATE CHANGE AND ALPINE ECOSYSTEMS

Climatic change is one of the important changes in the alpine ecosystems, which are mainly expressed in the increase in temperature and precipitation distribution. In the mountain areas, warming takes place two times faster than the global average, which results in lower levels of snow, sooner snowmelt, and the longer growing seasons (Huss et al., 2017; Rumpf et al., 2022). These fluctuations modify the narrow-range phenology of the alpine flora and their pollinators, leading to discrepancies in flowering and pollinators levels, and potentially reducing the reproductive success and changing the species interactions (Inouye, 2020). Besides, warming supports the rising of a low-lying species, as it brings new competitors, and changes the community composition (Inouye, 2020; Verrall & Pickering, 2020). Alterations in the snow cover and permafrost thaw also influence soil stability, hydrology and nutrient cycling with implications on vegetation cover and ecosystem benefits i.e. water regulation and carbon sequestration (Huss et al., 2017; Rumpf et al., 2022; Piccinelli & Cannone, 2025).

There is also altered precipitation patterns and nitrogen deposition that further affect alpine ecosystems by changing the microbial communities in soils and vegetation cover, which results in the growth of forests and shrubs at the cost of the alpine meadows and steppes, and the net primary productivity is expected to rise significantly (Gao et al., 2016). Nevertheless, the changes may decrease the availability of habitats to specific alpine species and stability of ecosystem services (Z. Yang et al., 2025). A shorter duration of snow cover raises the soil temperature and microbial activity in the growing season and adjustments in streamflow and water chemistry influence freshwater biota and aquatic food webs (Hotaling et al., 2017; Rogora et al., 2018). The disappearance of permanent snow and ice diminishes the albedo, increasing warming and impacting human and ecological water supply. They are harmful to the survival of species adapted to cold, and these physical changes interfere with the trophic interaction within terrestrial and aquatic ecosystems (Huss et al., 2017; Verrall & Pickering, 2020; Rumpf et al., 2022). Such complex effects and integrative management to maintain the services of alpine ecosystem in a scenario of climatic change require ecological long-term monitoring and research methods alterations (Rogora et al., 2018; Verrall & Pickering, 2020).

6. SOIL, NUTRIENT CYCLING, AND CARBON DYNAMICS

Like in lowland regions, the primary processes of soil formation in the alpine landscapes are climate, biological factors, parent material, topography and time. Like in the arctic regions, mountain regions achieve most of their development as a result of soil limited with the severe weather conditions (Donhauser & Frey, 2018). Temperature and other climatic variables. And moisture influence the type and rate of weathering. While in physical fracturing of rocks prevails that is characteristic of cold and arid areas colder and drier climate is unfavorable to biological and chemical processes (Thorn et al., 2011). The Alpine glaciers are melting at a very high rate in response to climatic changes, with lots of unproductive bare rocks (Donhauser & Frey, 2018).

Microbial life is important to Carbon cycle and Nitrogen cycle in alpine landscapes. Biogeochemical is of utmost importance in soils through soil microorganisms cycles on the regional and on the supra-regional level (Rousk & Bengtson, 2014; Donhauser & Frey, 2018). Being exposed to high temperature, all biogeochemical cycles are likely to be speeded up to give an augmentation in both output and input in the ecosystem (e.g. respiration and photosynthesis). Nonetheless, microbial responses are dependent on factors other than temperature on an excess of climate conditions like other climatic factors (e.g. moisture), substrates and nutrients availability, biological processes, and physicochemical characteristics of soil (Donhauser & Frey, 2018). The complex interactions between warming, microbial activity, and the C- and N- cycles in alpine soils are summarized in Figure 2.

Figure 2. The Impact of Global Warming on C- and N- Cycles in Alpine Soils.

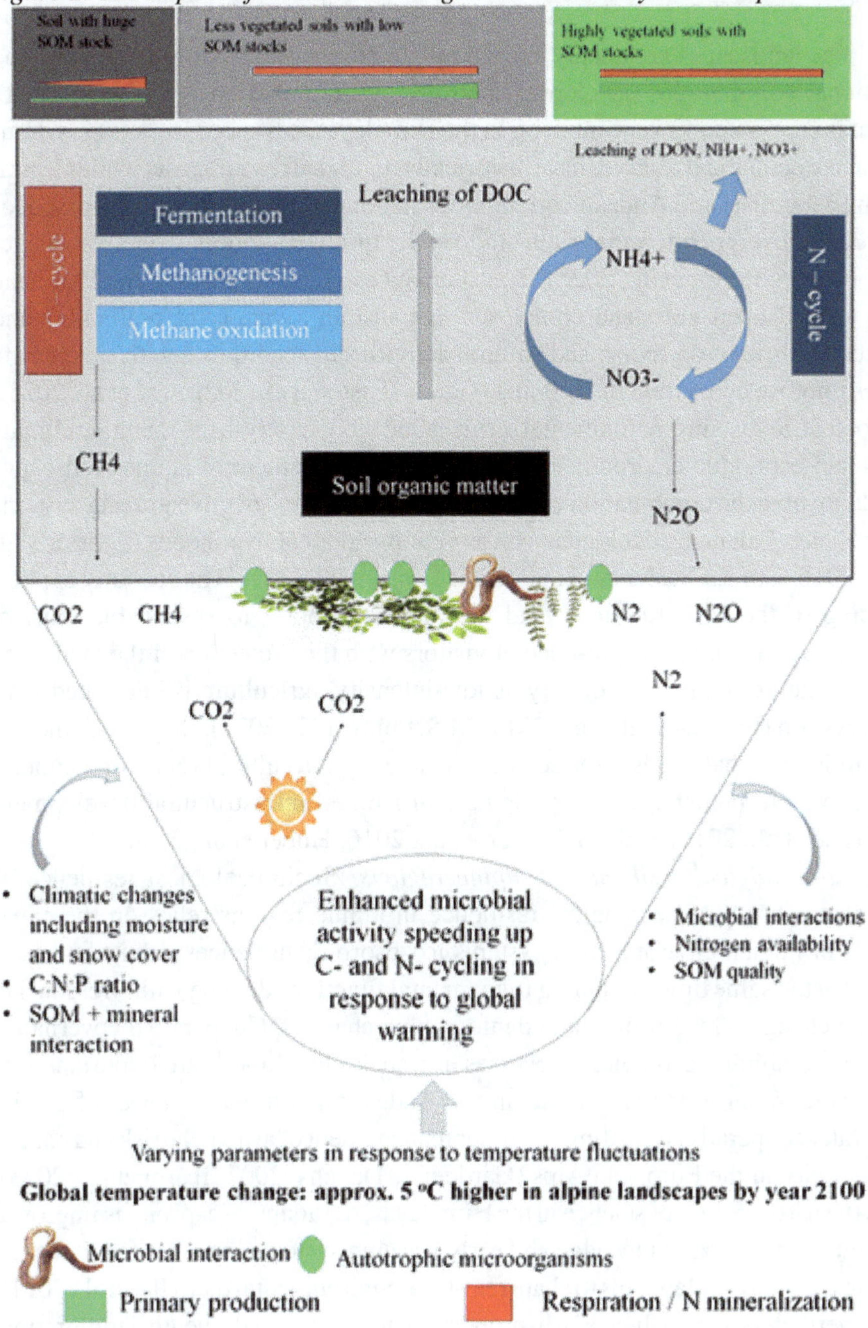

7. HUMAN DIMENSIONS OF ALPINE LANDSCAPES

Alpine landscapes are deeply shaped by human presence, with Indigenous and traditional communities developing land use systems that have co-evolved with mountain ecosystems over centuries. In the Himalayas, advanced irrigation systems and sacred ecological reserves have historically regulated resource use, while Central Asian pastoralists and Andean communities maintain communal grazing practices that prevent overgrazing and sustain biodiversity through rotational pasture management (Garcés-Pastor et al., 2022). Transhumance and pastoralism in the European Alps formed open grassland complexes that sustain a variety of pollinators and species-specific bird groups, and cultural activities in accordance with species life cycles improve ecosystem functioning (Garcés-Pastor et al., 2022; Caf et al., 2025).

For that, to sustain environmental services and biodiversity in the changing climate and socio-economic environments, sustainable management of alpine landscapes should involve the combination of traditional knowledge, adaptive grazing systems and policies, balancing ecological conservation and local livelihoods (Egarter Vigl et al., 2016; Garcés-Pastor et al., 2022; H. Yang et al., 2023). The spatially explicit mapping of the CES resources and demand contributes to sustainable tourism planning by balancing the presence of visitors with the protection of the landscape and facilitating multi-functionality as low-intensity agriculture is integrated with conservation (Egarter Vigl et al., 2017; M.Schmitt et al., 2024). In general, the opportunities and challenges to the ecological integrity and cultural values of mountain landscapes are predetermined by alpine tourism and infrastructural development (Schirpke et al., 2016; Zoderer, Tasser, et al., 2016; Huber et al., 2020).

Socio-ecological resilience of alpine regions: Socio-ecological resilience of alpine regions: Socio-ecological resilience in alpine regions relies on the capabilities of human-environmental systems to absorb disturbances, adjust and later change at the same time sustaining fundamental functions dealing with stressors i.e. climate change, globalization and demographic changes. Cooperative governance systems combining ecological processes and social livelihoods are important, e.g. Indigenous water user associations in the Andes to address the impact of glacial meltwater cooperatively, and cross-border management of avalanche risk and habitat connectivity in the European Alps (Gardner & Dekens, 2007; Ibarra et al., 2023). With the participation of science in the Himalayas, the adaptive capacity is improved with the help of local knowledge and early warning systems, which is facilitated by mobile technology platforms to share real-time environmental data (Jha et al., 2021).

Nevertheless, the resilience is disproportionate because of the youth outmigration, loss of the traditional ecological knowledge, and intensification of the climate that question the pre-existing governance and the collective resources management (Jha et al., 2021; Ibarra et al., 2023). Research in the Tibetan Plateau and Chinese pastoral regions in the alpine has shown that with the help of coordinating grasslands, pastoralist organizations, and restoration technologies, it is possible to maintain the ecological quality and pastoral returns which will encourage their socio-ecological adaptability (Ji et al., 2024; Cai et al., 2025). Multi-scalar governance, economic diversification, ecosystem services payments, and transdisciplinary strategies which incorporate Indigenous and local worldviews would be needed to address the future resilience-building requirement (Lavorel et al., 2019; Stotten et al., 2021; Ibarra et al., 2023).

8. THREATS TO ALPINE BIODIVERSITY

The threats facing alpine biodiversity are complex and can operate in synergy where the effects of these threats are devastating in terms of habitat degradation and loss of species diversity. There is growing threat to alpine biodiversity in the form of habitat degradation and fragmentation due to infrastructure development, tourism, mining, hydropower projects and road building which further isolates populations and ecological connectivity which is essential to the climate-driven ranges of species. Ski resorts and access roads in the European Alps have changed the habitat, posing a threat to cold-adapted species, such as alpine butterflies and ptarmigans, and altered the river valley and pasture layout in the Himalayas and Hindu Kush-Karakoram, which attracts migratory ungulates and predator prey interplay (Schmeller et al., 2022; Roseo et al., 2025). Overgrazing and extraction of natural resources on fragile alpine pastures with low regenerative potential exacerbate the pressure on the vegetation, which is eroded, and the soil is eroded, resulting in changes in the balance of nutrients; overexploitation of medicinal plants also endangers biodiversity and local livelihoods (Schmeller et al., 2022).

Higher growth of intrusive species, which is aided by climate warming, tourism, and livestock movement, alters native alpine communities due to competition with specialized, slow-growing species, e.g. invasive grasses in the Alps and cheat-grass in the Rockies, and invasive weeds and pathogens in the Andes and Himalayas (Hotaling et al., 2017; Schmeller et al., 2022). Nitrogen, heavy metals, and microplastics pollute the air and reach even remote alpine areas, resulting in nutrient enrichment and development of fast-growing species and pollute aquatic ecosystems, influencing biodiversity, microbes, and higher trophic levels (Schmeller et al., 2022; Fahy et al., 2024). The climate change plays a critical role of a multiplier, augmenting physiological pressure on cold-adapted species, facilitating the escalation of intrusive species, augmenting grazing pressure, and diminishing the upcast migration of species because of habitat fragmentation, augmenting the risk of extinction and altering major biodiversity hotspots such as alpine wetlands and meadows in the Himalayas and the Andes (Wilkes et al., 2023; Lamouille-Hébert et al., 2024).

Such interacting stressors decrease the resilience of the ecosystem and raise the chance of occurrence of sudden ecological transitions, which contradicts traditional conservation strategies dealing with threats separately. The conservation of the alpine biodiversity and ecosystem services it provides needs integrated and multi-sectoral strategies of protection on a landscape scale, sustainable management of resources, climate-resilient policies, and high-quality engagement with the local community (Chauvier-Mendes et al., 2024; Schmeller et al., 2022). Protecting climate refugia and enhancing connectivity across elevations and latitudes are also essential to maintain genetic diversity and species persistence under future climate scenarios (Chauvier-Mendes et al., 2024; Roseo et al., 2025). Overall, alpine ecosystems are among the most vulnerable globally, necessitating urgent, coordinated efforts to mitigate compounded threats and preserve their unique biodiversity (Meza-Joya et al., 2023).

9. CONSERVATION STRATEGIES AND POLICY FRAMEWORKS

9.1. Transboundary and Protected Area Governance

The planning of alpine conservation is gradually modifying its approaches to incorporate the transboundary governance schemes in order to protect biodiversity and ecosystems services. When discussing the European continent, the Alpine Convention offers a collaboration platform between nations, including Austria, France, Italy, and Switzerland, harmonizing land-use planning, nature conservation, and mitigation of climatic change (Egarter Vigl et al., 2021). Natura 2000 network is additionally preserving alpine habitats throughout the EU, but climate-induced changes in the distribution of species endanger the connectivity of ecosystems, necessitating adaptive conservation to safeguard up to 70% of alpine flora estimated to be under threat by 2080 (Chauvier-Mendes et al., 2024). Similar examples of cross-border planning, including the Maloti Drakensberg Conservation and Development Area in Africa, can illustrate the role of regional planning in mountain landscapes (Egarter Vigl et al., 2021).

9.2. Community-Based Conservation and Indigenous participation

The project will mobilize the community in the running of the project and confront the local issues or challenges in implementation. Community conservation is very crucial in the alpine regions where communities rely on natural resources. Some of the protected areas including Qomolangma National Nature Preserve in Asia-Pacific involve the local volunteers in maintaining forests, and wildlife conservation. The Annapurna Conservation Area of Nepal is also a biodiversity conservation area that incorporates tourism, pasture control, and benefit sharing to address the local livelihoods (Manzoor et al., 2025). In South Asia, zonation strategies (core, buffer and transition zone) are used in high-altitude protected areas including the Great Himalayan national park and the Spiti biosphere reserve to protect fragile ecologies, cultural wealth and flagship species such as snow leopards and musk deer (Ji et al., 2024; Manzoor et al., 2025).

9.3. Restoration Ecology in Alpine Systems

The alpine ecosystems that have been destroyed by the overgrazing, infrastructural development, recreations and climatic changes have to be restored through restoration ecology. In Central Himalaya, controlled grazing, reintroduction of native species and stabilization of soil has also increased the resilience of the ecosystems by restoring alpine meadows in Dayara Bugyal. One example is the restoration in the Rocky Mountains, which aims to restore the native vegetation and rehabilitate tourism-disturbed soils, whereas in Andes, erosion control efforts are combined with watershed rehabilitation to preserve the downstream biodiversity and water security (Maiti et al., 2022). New tools including robotics-assisted habitat monitoring are emerging in the European Alps to enhance data collection to assist adaptive management in challenging environments (Del Curto et al., 2022).

9.4. Climate-Smart Conservation and Adaptation Strategies

The focus of climate-smart conservation is landscape connectivity, climate refugia protection, and adaptive governance. The Alpine Climate Target System 2050 harmonizes signatories of the Alpine Convention in terms of adaptation and mitigation policy, increasing transboundary ecological connectivity and resilience (Chauvier-Mendes et al., 2024). Andes Community-based watershed management and agro forestry buffer climatic variability and support traditional agriculture and biodiversity. Hindu Kush-Himalayan example of glacier grafting and ice stupas, which combine traditional knowledge and climate change adaptation, are used to sustain the water resources of the region in the face of glaciation (Maiti et al., 2022).

9.5. Science Policy Integration and Institutional Frameworks

The instruments of international policy support alpine conservation through the development of coordinated governance and adaptive management. Convention on biological diversity promotes regional and cross-border conservation and almost 40% of the reporting countries are active in transboundary mountain activities. The United Nations Framework Convention on Climate Change prioritizes water security of the mountain ecosystem and resilience as a part of national adaptation strategies, and the Man and the Biosphere Programme of UNESCO balances conservation and sustainability in the mountain area through the integration of core-protected zones and research-oriented buffer areas as examples of Spiti (Ji et al., 2024). Successful conservation of the alps requires the combining of scientific investigation, management and the involvement of the locals. Science also enlightens biodiversity dynamics, climate effects, and restoration performances, which can be used in adaptive management (Ding et al., 2023). The legitimacy and stewardship are increased through the governance structures linking the multi-level institutions with local and Indigenous stakeholders. Practices throughout the Alps, Rockies, Andes, and Himalayas have repeatedly shown that with the growing rate of environmental change, inclusive governance, science-policy integration, and community involvement are key to the sustainability of alpine biodiversity and livelihoods (Chauvier-Mendes et al., 2024; Previšić, 2025).

10. MONITORING, MODELING, AND TECHNOLOGICAL ADVANCES

The long-term ecological monitoring is the basis of the knowledge of the mechanics of the alpine ecosystem and climatic influence. In the European Alps, such networks as ILTER and GLORIA offer strong multi-decadal species composition and snow cover and phenology data, which help to identify climate-induced alterations (Hotaling et al., 2017; Rogora et al., 2018). The Canadian Rocky Mountain plots determine changes in elevation of flora and fauna in response to warming and changing disturbance regimes, and the Andean surveys combine the changes in the mass balance of the glaciers with the indicators of biodiversity in order to record the impact of the rapid glacier retreat in the tropical regions. Integrated glacier, hydrology, and biodiversity monitoring aids the adaptation planning and water management in climate uncertainty in the Hindu Kush-Himalayan region (Hotaling et al., 2017; Che et al., 2019).

The alpine monitoring has been transformed by remote sensing and GIS technologies which allow repetitive, spatially broad observations to be made. The extent of the snow, the phenomenon of vegetation, and disturbances are registered using a high-resolution satellite image (e.g., Sentinel-2, Landsat), whereas fine-scale alterations in the topography that are important in permafrost and landslide risk assessment are observed using LiDAR and photogrammetry (Kiese et al., 2018; Okamoto et al., 2024). Multi spectral remote sensing supports the detection of avalanches and tree-line monitoring in the Alps, and GIS is used to determine the priorities in conservation by basing their results on the combination of microclimate and topography data (Rogora et al., 2018; Verrall & Pickering, 2020).

Species distribution and dynamic vegetation models are ecological and climate modeling tools which project alpine species range shifts in the future under different emission scenarios. To illustrate, Andean alpine specialists are predicted to shrink by a long distance and tree-line to shift upslope in the Himalayas, which will impact on endemic grasslands and water supplies. The projections guide the national adaptation plan of countries such as Switzerland, Norway, Nepal, and Chile to strike the right balance amid conservation, development, and community needs (Hotaling et al., 2017; Rogora et al., 2018).

Improved molecule and genomic discovery improve biodiversity monitoring in the alps by detecting cryptic species, as well as measuring genetic diversity and connectivity. Environmental DNA (eDNA) sampling is a method where the rare or difficult to locate species are detected without actually seeing them, whereas genomic results show the response of evolution to microclimatic refugia and landscape fragmentation in the Alps and Andes, guiding both the structure of the corridors and their conservation units. Participatory monitoring and citizen science increase the data gathering and involvement of the community. The European Mountain watch is a volunteer phenology and wildlife-monitoring initiative and supplements the scientific study and helps in the stewardship of the environment. In the Himalayas, community-based networks combine traditional knowledge with scientific protocols to follow the glacial lakes and wildlife and enhance the early warning on threats i.e. the glacial lake outburst floods (GLOFs). The local and global decision-making is facilitated by the democratization of data, which is supported by mobile apps and open platforms (Hotaling et al., 2017; Thuiller et al., 2024). Combined, these combined monitoring, modeling and technological innovations make possible adaptive, data-driven conservation strategies to respond to the accelerated environmental change in alpine ecosystems to protect biodiversity and ecosystem services that are important to both mountain and downstream populations.

11. CASE STUDIES FROM MAJOR ALPINE REGIONS; THE HIMALAYAS

The youngest and the largest high-elevation mountain system in the world, the Himalayas live in Nepal, India, Pakistan, Bhutan, and China, and offers important ecological and cultural services to more than 240 million people. A good example of an effective community-based conservation is the Annapurna Conservation Area Project (ACAP)-Nepal, that combines local governance with eco-friendly tourism, sustainable agriculture, and protection of the watershed, which has been effective in protecting species such as snow leopard, musk deer and has boosted the livelihoods of the surrounding community through revenue sharing in eco-tourism (Khan et al., 2013). The management of visitor effects and preservation of culturally important sceneries have been sought to be achieved in Sagarmatha National Park through long-term ecological observation and community partnerships. Climate change accelerates the meltdown of the glaciers and destabilizations of the permafrost that pose a threat to the water security of the major rivers; community-based glacier guardianship and participatory monitoring of the glacial lakes have enhanced early warning systems of the hazards like glacial lake outburst floods (GLOFs). Cold-desert biosphere projects in the Spiti Valley and Ladakh of India combine the management of traditional irrigation patterns with climate adaptation in order to maintain the alpine pastures and increase drought resistance. Table 1 gives other case studies of key alpine regions.

Table 1. Case Studies from Major Alpine Regions

Region	Key Alpine Systems / Sites	Governance & Management Approaches	Major Conservation Focus	Climate Change Responses & Tools	Key References
European Alps	Vanoise NP; Hohe Tauern NP; Natura 2000 network	Transboundary governance under the Alpine Convention; strong science–policy integration	Habitat mosaics; alpine flora and fauna (ibex, Apollo butterfly); carbon sequestration	Long-term monitoring (GLORIA); GIS and LiDAR for erosion, avalanche risk, and habitat connectivity	(Khan et al., 2013; Vincent et al., 2019; Sun et al., 2023; Canella et al., 2025).
Andes	Páramo (Colombia); Cordillera Blanca (Peru); Sajama NP (Bolivia); Podocarpus–El Cóndor Corridor	Protected areas combined with community-based research and conservation corridors	Endemic biodiversity; freshwater provision; soil carbon storage	Glacial retreat monitoring; ecological modeling of upslope vegetation shifts; climate-adaptive pasture restoration	(Vincent et al., 2019; Sun et al., 2023)
Rockies	Rocky Mountain NP; Banff; Jasper; Yellowstone–Yukon Corridor	Federal park agencies; Indigenous partnerships; NGO-led connectivity initiatives	Wildlife connectivity (grizzly bears, wolves); ecosystem services; recreation management	Snowpack, phenology, and streamflow monitoring; wildfire and invasive species restoration	(Vincent et al., 2019)
Central Asian & Hindu Kush–Karakoram	Pamir; Tien Shan; Sary-Chelek Biosphere Reserve; Karakoram Range; Hindu Kush	Cross-border protected areas; community pasture and forest management	Snow leopard conservation; pasture sustainability; climate hazard mitigation	Remote sensing; UAV and eDNA surveys; research on the Karakoram glacier anomaly	(Khan et al., 2013; Sun et al., 2023)

12. FUTURE PERSPECTIVES AND RESEARCH PRIORITIES

Subsequent studies of the alpine environments and conservation have to concern important knowledge gaps in below-ground activity, microbial dynamics, and multi-trophic interactions that form the nutrient cycling and ecosystem resilience, particularly in underrepresented areas such as the Himalayas, Andes, and Central Asia (Hotaling et al., 2017; Verrall & Pickering, 2020). It is urgently needed to scale the long-term datasets to include more microorganisms, invertebrates, and cryptic species and in order to better grasp the tipping points of alpine landscapes environment i.e. vegetation meltdown after permafrost thaw or hydrological transformation after the loss of glacier (Brighenti et al., 2019; Wilkes et al., 2023). It is important to have adaptive management frameworks which include scenario planning, iterative learning, and flexible policies to adapt to the accelerating ecological reorganization and climatic changes of alpine systems (Verrall & Pickering, 2020; Meza-Joya et al., 2025). Real-time monitoring and climate-informed ecological modeling can be use to forecast the change in a species range, modified disturbance regimes and new stressors, such as invasive species or new pathogens (Chamberlain et al., 2013; Meza-Joya et al., 2025). It is suggested that protecting climate refugia, improving elevational and latitudinal connectivity, and focusing on ecosystem functions instead of fixed species assemblages should be promoted, and uncertainty has to be explicitly included into decision-making (Chamberlain et al., 2013; Brambilla et al., 2022). Conservation of biodiversity and sustainable livelihoods take a primary agenda because millions of people rely on alpine ecosystems as a source of grazing, farming, water and tourism. Transboundary cooperation is becoming an essential tool since the issues of alpine ecosystems and climate effects are at the cross of political boundaries. Effective models such as the Alpine Convention and Yellowstone-to-Yukon initiative show the advantages of the shared governance, coordinated monitoring, and policies, yet such mechanisms are still somewhat ineffective in certain regions of Asia, South America, and Central Asia (Ramel et al., 2020; Rowland et al., 2023).

All in all, the conservation of the alpines in the future should consider these landscapes as dynamic socio-ecological systems, which are subject to climate change, human activities, and ecological feedbacks. Long-term monitoring purposes, strong modeling, an emerging technology, inclusive governance, and local participation will be relevant to creating adaptive, resilient, and equitable stewardship of the alpine ecosystems (Brighenti et al., 2019; Verrall & Pickering, 2020; Rowland et al., 2023; Meza-Joya et al., 2025). The interdisciplinary gap between ecology, climatology, social sciences and policy will assist in dealing with the complexity and uncertainty that accompany these sensitive mountain settings.

13. CONCLUSION

The Alpine landscapes are a very valuable yet a fragile socio-ecological system that provides some of the essential ecosystem services that include water regulation and cultural heritage. As has been demonstrated in this chapter, a complex of increased climatic change and habitat fragmentation and unviable human actions is undermining their unique biodiversity. This will demand a different approach to work - it will be time to shift to joined up conservation at scale, rather than isolated acts. In critical imperatives, there is a need to bridge transboundary governance, empower the local and Indigenous community when it comes to the management of resources, and the use of science-based restoration ecology. The continued existence of such critical areas in the future lies in the establishment of climate-wise policies and advanced monitoring systems to assist in informing adaptive management. Finally, the future of the alpine landscapes around the globe will be founded on the systems-based vision that integrates the iterative combination of scientific knowledge, inclusive governance, and local acting stewardship to assist the societies in accepting the change in the rapidly changing world.

REFERENCES

Ahmad, M., Luo, Y.-H., Rathee, S., Spicer, R. A., Zhang, J., Wambulwa, M. C., Zhu, G.-F., Cadotte, M. W., Wu, Z.-Y., Khan, S. M., Maity, D., Li, D.-Z., & Liu, J. (2025). Multifaceted plant diversity patterns across the Himalaya: Status and outlook. *Plant Diversity*, *47*(4), 529–543. DOI: 10.1016/j.pld.2025.04.003 PMID: 40734839

Alexander, J. M., Lembrechts, J. J., Cavieres, L. A., Daehler, C., Haider, S., Kueffer, C., Liu, G., McDougall, K., Milbau, A., Pauchard, A., Rew, L. J., & Seipel, T. (2016). Plant invasions into mountains and alpine ecosystems: Current status and future challenges. *Alpine Botany*, *126*(2), 89–103. DOI: 10.1007/s00035-016-0172-8

Balian, E. V., Segers, H., Lévèque, C., & Martens, K. (2008). The Freshwater Animal Diversity Assessment: An overview of the results. *Hydrobiologia*, *595*(1), 627–637. DOI: 10.1007/s10750-007-9246-3

Binter, J., Macek, M., & Dolezal, J. (2025). Integrating morphological, anatomical, and physiological traits to explain elevational distributions in Himalayan steppe and alpine plants. *Journal of Integrative Plant Biology*, *67*(10), 2643–2657. DOI: 10.1111/jipb.13971 PMID: 40662507

Boucher, F. C., Dentant, C., Ibanez, S., Capblancq, T., Boleda, M., Boulangeat, L., Smyčka, J., Roquet, C., & Lavergne, S. (2021). Discovery of cryptic plant diversity on the rooftops of the Alps. *Scientific Reports*, *11*(1), 11128. DOI: 10.1038/s41598-021-90612-w PMID: 34045566

Brambilla, M., Rubolini, D., Appukuttan, O., Calvi, G., Karger, D. N., Kmecl, P., Mihelič, T., Sattler, T., Seaman, B., Teufelbauer, N., Wahl, J., & Celada, C. (2022). Identifying climate refugia for high-elevation Alpine birds under current climate warming predictions. *Global Change Biology*, *28*(14), 4276–4291. DOI: 10.1111/gcb.16187 PMID: 35441422

Brighenti, S., Tolotti, M., Bruno, M. C., Wharton, G., Pusch, M. T., & Bertoldi, W. (2019). Ecosystem shifts in Alpine streams under glacier retreat and rock glacier thaw: A review. *The Science of the Total Environment*, *675*, 542–559. DOI: 10.1016/j.scitotenv.2019.04.221 PMID: 31030160

Caf, N., Sabatier, P., Cucinotta, J., Rapuc, W., Dolenec, M., & Šmuc, A. (2025). Human influence at the Holocene treeline in the Julian Alps, Slovenia. *The Holocene*, *35*(4), 446–459. DOI: 10.1177/09596836241307300

Cai, L., Zhao, J., & Chen, J. (2025). Coevolutionary dynamics in the grass-livestock social-ecological system of China's alpine pastoral areas: A case study of the Qilian Mountains region in China. *PLoS One*, *20*(1), e0317769. DOI: 10.1371/journal.pone.0317769 PMID: 39883742

Canella, M., Beltran-Sanz, N., Gröger, A., Pungaršek, Š., Rome, M., Natale, S., Bonomi, C., Wiesinger, H. C., Mainetti, A., Scapin, A., Valecic, M., Sensato, S., Sommacal, M., Piutti, E., Bottelli, F., Pizzato, M. T., Senn, J., Monod, A., La Rocca, N., & Dal Grande, F. (2025). The role of Alpine botanical gardens in integrating germplasm bank collections and mission. *PLANTS, PEOPLE, PLANET*, ppp3.70120. DOI: 10.1002/ppp3.70120

Carrer, F., Walsh, K., & Mocci, F. (2020). Ecology, Economy, and Upland Landscapes: Socio-Ecological Dynamics in the Alps during the Transition to Modernity. *Human Ecology: an Interdisciplinary Journal*, *48*(1), 69–84. DOI: 10.1007/s10745-020-00130-y

Carruthers, T., Moerland, M. S., Ebersbach, J., Favre, A., Folk, R. A., Hawkins, J. A., Muellner-Riehl, A. N., Röser, M., Soltis, D. E., Tkach, N., Baker, W. J., De Vos, J. M., & Eiserhardt, W. L. (2024). Repeated upslope biome shifts in Saxifraga during late-Cenozoic climate cooling. *Nature Communications*, *15*(1), 1100. DOI: 10.1038/s41467-024-45289-w PMID: 38321017

Chamberlain, D. E., Negro, M., Caprio, E., & Rolando, A. (2013). Assessing the sensitivity of alpine birds to potential future changes in habitat and climate to inform management strategies. *Biological Conservation*, *167*, 127–135. DOI: 10.1016/j.biocon.2013.07.036

Chauvier-Mendes, Y., Pollock, L. J., Verburg, P. H., Karger, D. N., Pellissier, L., Lavergne, S., Zimmermann, N. E., & Thuiller, W. (2024). Transnational conservation to anticipate future plant shifts in Europe. *Nature Ecology & Evolution*, *8*(3), 454–466. DOI: 10.1038/s41559-023-02287-3 PMID: 38253754

Che, T., Li, X., Liu, S., Li, H., Xu, Z., Tan, J., Zhang, Y., Ren, Z., Xiao, L., Deng, J., Jin, R., Ma, M., Wang, J., & Yang, X. (2019). Integrated hydrometeorological, snow and frozen-ground observations in the alpine region of the Heihe River Basin, China. *Earth System Science Data*, *11*(3), 1483–1499. DOI: 10.5194/essd-11-1483-2019

De Villemereuil, P., Mouterde, M., Gaggiotti, O. E., & Till-Bottraud, I. (2018). Patterns of phenotypic plasticity and local adaptation in the wide elevation range of the alpine plant *Arabis alpina. Journal of Ecology*, *106*(5), 1952–1971. DOI: 10.1111/1365-2745.12955

Del Curto, D., Garzulino, A., Menini, G., & Schiesaro, C. (2022). Sustainable Conservation and Management of a 20th-Century Landscape in the Alps: The Former Sanatorium Village of Sondalo. *Sustainability (Basel)*, *14*(12), 7424. DOI: 10.3390/su14127424

Ding, J., Ge, W., Liu, Q., Wang, Q., Kong, D., & Yin, H. (2023). Temperature drives the coordination between above-ground nutrient conservation and below-ground nutrient acquisition in alpine coniferous forests. *Functional Ecology*, *37*(6), 1674–1687. DOI: 10.1111/1365-2435.14330

Donhauser, J., & Frey, B. (2018). Alpine soil microbial ecology in a changing world. *FEMS Microbiology Ecology*, *94*(9), fiy099. Advance online publication. DOI: 10.1093/femsec/fiy099 PMID: 30032189

Ebner, M., Fontana, V., Schirpke, U., & Tappeiner, U. (2022). Stakeholder perspectives on ecosystem services of mountain lakes in the European Alps. *Ecosystem Services*, *53*, 101386. DOI: 10.1016/j.ecoser.2021.101386

Egarter Vigl, L., Depellegrin, D., Pereira, P., De Groot, R., & Tappeiner, U. (2017). Mapping the ecosystem service delivery chain: Capacity, flow, and demand pertaining to aesthetic experiences in mountain landscapes. *The Science of the Total Environment*, *574*, 422–436. DOI: 10.1016/j.scitotenv.2016.08.209 PMID: 27639477

Egarter Vigl, L., Marsoner, T., Schirpke, U., Tscholl, S., Candiago, S., & Depellegrin, D. (2021). A multi-pressure analysis of ecosystem services for conservation planning in the Alps. *Ecosystem Services*, *47*, 101230. DOI: 10.1016/j.ecoser.2020.101230

Egarter Vigl, L., Schirpke, U., Tasser, E., & Tappeiner, U. (2016). Linking long-term landscape dynamics to the multiple interactions among ecosystem services in the European Alps. *Landscape Ecology*, *31*(9), 1903–1918. DOI: 10.1007/s10980-016-0389-3

Fahy, J. C., Demierre, E., & Oertli, B. (2024). Long-term monitoring of water temperature and macroinvertebrates highlights climate change threat to alpine ponds in protected areas. *Biological Conservation*, *290*, 110461. DOI: 10.1016/j.biocon.2024.110461

Gao, Q., Guo, Y., Xu, H., Ganjurjav, H., Li, Y., Wan, Y., Qin, X., Ma, X., & Liu, S. (2016). Climate change and its impacts on vegetation distribution and net primary productivity of the alpine ecosystem in the Qinghai-Tibetan Plateau. *The Science of the Total Environment*, *554–555*, 34–41. DOI: 10.1016/j.scitotenv.2016.02.131 PMID: 26950617

Garcés-Pastor, S., Coissac, E., Lavergne, S., Schwörer, C., Theurillat, J.-P., Heintzman, P. D., Wangensteen, O. S., Tinner, W., Rey, F., Heer, M., Rutzer, A., Walsh, K., Lammers, Y., Brown, A. G., Goslar, T., Rijal, D. P., Karger, D. N., Pellissier, L., Pouchon, C., & Alsos, I. G. (2022). High resolution ancient sedimentary DNA shows that alpine plant diversity is associated with human land use and climate change. *Nature Communications*, *13*(1), 6559. DOI: 10.1038/s41467-022-34010-4 PMID: 36333301

García-Navas, V., Sattler, T., Schmid, H., & Ozgul, A. (2020). Temporal homogenization of functional and beta diversity in bird communities of the Swiss Alps. *Diversity & Distributions*, *26*(8), 900–911. DOI: 10.1111/ddi.13076

Gardner, J. S., & Dekens, J. (2007). Mountain hazards and the resilience of social–ecological systems: Lessons learned in India and Canada. *Natural Hazards*, *41*(2), 317–336. DOI: 10.1007/s11069-006-9038-5

Grabherr, G., Gottfried, M., & Pauli, H. (2010). Climate Change Impacts in Alpine Environments. *Geography Compass*, *4*(8), 1133–1153. DOI: 10.1111/j.1749-8198.2010.00356.x

Hotaling, S., Finn, D. S., Joseph Giersch, J., Weisrock, D. W., & Jacobsen, D. (2017). Climate change and alpine stream biology: Progress, challenges, and opportunities for the future. *Biological Reviews of the Cambridge Philosophical Society*, *92*(4), 2024–2045. DOI: 10.1111/brv.12319 PMID: 28105701

Huber, L., Schirpke, U., Marsoner, T., Tasser, E., & Leitinger, G. (2020). Does socioeconomic diversification enhance multifunctionality of mountain landscapes? *Ecosystem Services*, *44*, 101122. DOI: 10.1016/j.ecoser.2020.101122

Huss, M., Bookhagen, B., Huggel, C., Jacobsen, D., Bradley, R. S., Clague, J. J., Vuille, M., Buytaert, W., Cayan, D. R., Greenwood, G., Mark, B. G., Milner, A. M., Weingartner, R., & Winder, M. (2017). Toward mountains without permanent snow and ice. *Earth's Future*, *5*(5), 418–435. DOI: 10.1002/2016EF000514

Ibarra, J. T., Caviedes, J., Marchant, C., Mathez-Stiefel, S.-L., Navarro-Manquilef, S., & Sarmiento, F. O. (2023). Mountain social-ecological resilience requires transdisciplinarity with Indigenous and local worldviews. *Trends in Ecology & Evolution*, *38*(11), 1005–1009. DOI: 10.1016/j.tree.2023.07.004 PMID: 37574394

Inouye, D. W. (2020). Effects of climate change on alpine plants and their pollinators. *Annals of the New York Academy of Sciences*, *1469*(1), 26–37. DOI: 10.1111/nyas.14104 PMID: 31025387

Jha, S. K., Negi, A. K., Alatalo, J. M., & Negi, R. S. (2021). Socio-ecological vulnerability and resilience of mountain communities residing in capital-constrained environments. *Mitigation and Adaptation Strategies for Global Change*, *26*(8), 38. DOI: 10.1007/s11027-021-09974-1 PMID: 34744484

Ji, Z., Zou, S., Zhang, W., Song, F., Yuan, T., & Xu, B. (2024). Optimizing zoning for ecological management in alpine region by combining ecosystem service supply and demand with ecosystem resilience. *Journal of Environmental Management*, *365*, 121508. DOI: 10.1016/j.jenvman.2024.121508 PMID: 38924889

Khan, S. M., Page, S. E., Ahmad, H., & Harper, D. M. (2013). Sustainable utilization and conservation of plant biodiversity in montane ecosystems: The western Himalayas as a case study. *Annals of Botany*, *112*(3), 479–501. DOI: 10.1093/aob/mct125 PMID: 23825353

Kiese, R., Fersch, B., Baessler, C., Brosy, C., Butterbach-Bahl, K., Chwala, C., Dannenmann, M., Fu, J., Gasche, R., Grote, R., Jahn, C., Klatt, J., Kunstmann, H., Mauder, M., Rödiger, T., Smiatek, G., Soltani, M., Steinbrecher, R., Völksch, I., & Schmid, H. P. (2018). The TERENO Pre-Alpine Observatory: Integrating Meteorological, Hydrological, and Biogeochemical Measurements and Modeling. *Vadose Zone Journal*, *17*(1), 1–17. DOI: 10.2136/vzj2018.03.0060

Lamouille-Hébert, M., Arthaud, F., & Datry, T. (2024). Climate change and the biodiversity of alpine ponds: Challenges and perspectives. *Ecology and Evolution*, *14*(2), e10883. DOI: 10.1002/ece3.10883 PMID: 38327685

Lavorel, S., Colloff, M. J., Locatelli, B., Gorddard, R., Prober, S. M., Gabillet, M., Devaux, C., Laforgue, D., & Peyrache-Gadeau, V. (2019). Mustering the power of ecosystems for adaptation to climate change. *Environmental Science & Policy*, *92*, 87–97. DOI: 10.1016/j.envsci.2018.11.010

Liu, W., Zheng, L., & Qi, D. (2020). Variation in leaf traits at different altitudes reflects the adaptive strategy of plants to environmental changes. *Ecology and Evolution*, *10*(15), 8166–8175. DOI: 10.1002/ece3.6519 PMID: 32788969

Losapio, G., Schöb, C., Staniczenko, P. P. A., Carrara, F., Palamara, G. M., De Moraes, C. M., Mescher, M. C., Brooker, R. W., Butterfield, B. J., Callaway, R. M., Cavieres, L. A., Kikvidze, Z., Lortie, C. J., Michalet, R., Pugnaire, F. I., & Bascompte, J. (2021). Network motifs involving both competition and facilitation predict biodiversity in alpine plant communities. *Proceedings of the National Academy of Sciences of the United States of America*, *118*(6), e2005759118. DOI: 10.1073/pnas.2005759118 PMID: 33526655

Maiti, P., Kuniyal, J. C., Sekar, K. C., Satish, K. V., Singh, D., Bisht, N., Kumar, A., Arya, S. C., Nand, M., & Sundriyal, R. C. (2022). Landscape level ecological assessment and eco-restoration strategies for alpine and sub-alpine regions of the Central Himalaya. *Ecological Engineering*, *180*, 106674. DOI: 10.1016/j.ecoleng.2022.106674

Malanson, G. P., Fagre, D. B., Butler, D. R., & Shen, Z. (2024). Alpine plant communities and current topographic microrefugia vary with regional climates. *Geomorphology*, *458*, 109241. DOI: 10.1016/j.geomorph.2024.109241

Mammola, S., Pétillon, J., Hacala, A., Monsimet, J., Marti, S., Cardoso, P., & Lafage, D. (2021). Challenges and opportunities of species distribution modelling of terrestrial arthropod predators. *Diversity & Distributions*, *27*(12), 2596–2614. DOI: 10.1111/ddi.13434

Manzoor, M., Ahmad, M., Gillani, S. W., Waheed, M., Shaheen, H., Mehmood, A. B., Fonge, B. A., & Al-Andal, A. (2025). Population dynamics, threat assessment, and conservation strategies for critically endangered Meconopsis aculeata in alpine zone. *BMC Plant Biology*, *25*(1), 358. DOI: 10.1186/s12870-025-06361-9 PMID: 40102752

Meza-Joya, F. L., Morgan-Richards, M., Koot, E. M., & Trewick, S. A. (2023). Global warming leads to habitat loss and genetic erosion of alpine biodiversity. *Journal of Biogeography*, *50*(5), 961–975. DOI: 10.1111/jbi.14590

Meza-Joya, F. L., Morgan-Richards, M., & Trewick, S. A. (2025). Forecasting Range Shifts in Terrestrial Alpine Insects Under Global Warming. *Ecology and Evolution*, *15*(1), e70810. DOI: 10.1002/ece3.70810 PMID: 39803196

Noroozi, J., & Körner, C. (2018). A bioclimatic characterization of high elevation habitats in the Alborz mountains of Iran. *Alpine Botany*, *128*(1), 1–11. DOI: 10.1007/s00035-018-0202-9 PMID: 29576762

Oberprieler, S. K., Andersen, A. N., Gillespie, G. R., & Einoder, L. D. (2019a). Vertebrates are poor umbrellas for invertebrates: Cross-taxon congruence in an Australian tropical savanna. *Ecosphere*, *10*(6), e02755. DOI: 10.1002/ecs2.2755

Oberprieler, S. K., Andersen, A. N., Gillespie, G. R., & Einoder, L. D. (2019b). Vertebrates are poor umbrellas for invertebrates: Cross-taxon congruence in an Australian tropical savanna. *Ecosphere*, *10*(6), e02755. DOI: 10.1002/ecs2.2755

Okamoto, R., Ide, R., & Oguma, H. (2024). Automatically drawing vegetation classification maps using digital time-lapse cameras in alpine ecosystems. *Remote Sensing in Ecology and Conservation*, *10*(2), 188–202. DOI: 10.1002/rse2.364

Piccinelli, S., & Cannone, N. (2025). Divergent Responses of Alpine Rock Glaciers to Climate Change: A Review of Ecological and Abiotic Dynamics. *Permafrost and Periglacial Processes, 36*(3), 438–450. DOI: 10.1002/ppp.2278

Previšić, B. (2025). Cultures of the Alps: The intersection of sectors as the main tool for future. *Archalp, 13*(2024). DOI: 10.30682/aa2413c

Rahbek, C., Borregaard, M. K., Antonelli, A., Colwell, R. K., Holt, B. G., Nogues-Bravo, D., Rasmussen, C. M. Ø., Richardson, K., Rosing, M. T., Whittaker, R. J., & Fjeldså, J. (2019). Building mountain biodiversity: Geological and evolutionary processes. *Science, 365*(6458), 1114–1119. DOI: 10.1126/science.aax0151 PMID: 31515384

Ramel, C., Rey, P.-L., Fernandes, R., Vincent, C., Cardoso, A. R., Broennimann, O., Pellissier, L., Pradervand, J.-N., Ursenbacher, S., Schmidt, B. R., & Guisan, A. (2020). Integrating ecosystem services within spatial biodiversity conservation prioritization in the Alps. *Ecosystem Services, 45*, 101186. DOI: 10.1016/j.ecoser.2020.101186

Ramskogler, K., Lepesant, L., & Tasser, E. (2025). Understanding the long-term dynamics of vegetation since 1953 in high-mountain regions. *Journal of Ecology, 113*(2), 485–503. DOI: 10.1111/1365-2745.14472

Rogora, M., Frate, L., Carranza, M. L., Freppaz, M., Stanisci, A., Bertani, I., Bottarin, R., Brambilla, A., Canullo, R., Carbognani, M., Cerrato, C., Chelli, S., Cremonese, E., Cutini, M., Di Musciano, M., Erschbamer, B., Godone, D., Iocchi, M., Isabellon, M., & Matteucci, G. (2018). Assessment of climate change effects on mountain ecosystems through a cross-site analysis in the Alps and Apennines. *The Science of the Total Environment, 624*, 1429–1442. DOI: 10.1016/j.scitotenv.2017.12.155 PMID: 29929254

Roseo, F., Celada, C., & Brambilla, M. (2025). Ski resorts threaten climate refugia for high-elevation biodiversity under current and future conditions in the Alps. *Biological Conservation, 301*, 110890. DOI: 10.1016/j.biocon.2024.110890

Rousk, J., & Bengtson, P. (2014). Microbial regulation of global biogeochemical cycles. *Frontiers in Microbiology, 5*. Advance online publication. DOI: 10.3389/fmicb.2014.00103 PMID: 24672519

Rowland, J. A., Walsh, J. C., Beitzel, M., Brawata, R., Brown, D., Chalmers, L., Evans, L., Eyles, K., Gibbs, R., Grover, S., Grundy, S., Harris, R. M. B., Haywood, S., Hilton, M., Hope, G., Keaney, B., Keatley, M., Keith, D. A., Lawrence, R., & Moore, J. L. (2023). Setting research priorities for effective management of a threatened ecosystem: Australian alpine and subalpine peatland. *Conservation Science and Practice, 5*(3), e12891. DOI: 10.1111/csp2.12891

Ruiz-Gonzalez, A. (2024). Ion-Selective Electrodes in the Food Industry: Development Trends in the Potentiometric Determination of Ionic Pollutants. *Electrochem, 5*(2), 178–212. DOI: 10.3390/electrochem5020012

Rumpf, S. B., Gravey, M., Brönnimann, O., Luoto, M., Cianfrani, C., Mariethoz, G., & Guisan, A. (2022). From white to green: Snow cover loss and increased vegetation productivity in the European Alps. *Science, 376*(6597), 1119–1122. DOI: 10.1126/science.abn6697 PMID: 35653482

Schirpke, U., Timmermann, F., Tappeiner, U., & Tasser, E. (2016). Cultural ecosystem services of mountain regions: Modelling the aesthetic value. *Ecological Indicators, 69*, 78–90. DOI: 10.1016/j.ecolind.2016.04.001 PMID: 27482152

Schirpke, U., Wang, G., & Padoa-Schioppa, E. (2021). Editorial: Mountain landscapes: Protected areas, ecosystem services, and future challenges. *Ecosystem Services, 49*, 101302. DOI: 10.1016/j.ecoser.2021.101302

Schmeller, D. S., Urbach, D., Bates, K., Catalan, J., Cogălniceanu, D., Fisher, M. C., Friesen, J., Füreder, L., Gaube, V., Haver, M., Jacobsen, D., Le Roux, G., Lin, Y.-P., Loyau, A., Machate, O., Mayer, A., Palomo, I., Plutzar, C., Sentenac, H., & Ripple, W. J. (2022). Scientists' warning of threats to mountains. *The Science of the Total Environment, 853*, 158611. DOI: 10.1016/j.scitotenv.2022.158611 PMID: 36087665

Schmitt, M., T., Haensel, M., Kaim, A., Lee, H., Reinermann, S., & Koellner, T. (. (2024). Recreation and its synergies and trade-offs with other ecosystem services of Alpine and pre-Alpine grasslands. *Regional Environmental Change, 24*(2), 57. DOI: 10.1007/s10113-024-02213-8

Schroeder, L., Robles, V., Jara-Arancio, P., Lapadat, C., Hobbie, S. E., Arroyo, M. T. K., & Cavender-Bares, J. (2024). Drivers of plant diversity, community composition, functional traits, and soil processes along an alpine gradient in the central Chilean Andes. *Ecology and Evolution, 14*(2), e10888. DOI: 10.1002/ece3.10888 PMID: 38343572

Schuldt, A., Wang, Z., Zhou, H., & Assmann, T. (2009). Integrating highly diverse invertebrates into broad-scale analyses of cross-taxon congruence across the Palaearctic. *Ecography, 32*(6), 1019–1030. DOI: 10.1111/j.1600-0587.2009.05973.x

Sekar, S., Arumugasamy, S. K., Lee, S., Sekar, S., Sadhasivam, S., Vaithilingam, S., Srinivasan, N., Krishnan, E., Chang, S.-C., & Manikandan, R. (2025). Magnetosome-Based Immunosensors for Foodborne Pathogen Detection. *Chemosensors (Basel, Switzerland), 13*(9), 355. DOI: 10.3390/chemosensors13090355

Son, S.-H., Kwon, S.-J., Im, J.-H., Kim, S.-K., Kong, D., & Choi, J.-Y. (2021). Aquatic Macrophytes Determine the Spatial Distribution of Invertebrates in a Shallow Reservoir. *Water (Basel)*, *13*(11), 1455. DOI: 10.3390/w13111455

Stotten, R., Ambrosi, L., Tasser, E., & Leitinger, G. (2021). Social-ecological resilience in remote mountain communities: Toward a novel framework for an interdisciplinary investigation. *Ecology and Society*, *26*(3), art29. DOI: 10.5751/ES-12580-260329

Suárez, E., Encalada, A. C., Chimbolema, S., Jaramillo, R., Hofstede, R., & Riveros-Iregui, D. (2023). On the Use of "Alpine" for High-Elevation Tropical Environments. *Mountain Research and Development*, *43*(1). Advance online publication. DOI: 10.1659/mrd.2022.00024

Sun, J., Ni, C., & Wang, M. (2023). Analysis of Water Conservation Trends and Drivers in an Alpine Region: A Case Study of the Qilian Mountains. *Remote Sensing (Basel)*, *15*(18), 4611. DOI: 10.3390/rs15184611

Testolin, R., Attorre, F., Borchardt, P., Brand, R. F., Bruelheide, H., Chytrý, M., De Sanctis, M., Dolezal, J., Finckh, M., Haider, S., Hemp, A., Jandt, U., Kessler, M., Korolyuk, A. Y., Lenoir, J., Makunina, N., Malanson, G. P., Montesinos-Tubée, D. B., Noroozi, J., & Jiménez-Alfaro, B. (2021). Global patterns and drivers of alpine plant species richness. *Global Ecology and Biogeography*, *30*(6), 1218–1231. DOI: 10.1111/geb.13297

Testolin, R., Attorre, F., & Jiménez-Alfaro, B. (2020). Global distribution and bioclimatic characterization of alpine biomes. *Ecography*, *43*(6), 779–788. DOI: 10.1111/ecog.05012

Thorn, C. E., Darmody, R. G., & Dixon, J. C. (2011). Rethinking weathering and pedogenesis in alpine periglacial regions: Some Scandinavian evidence. *Special Publication - Geological Society of London*, *354*(1), 183–193. DOI: 10.1144/SP354.11

Thuiller, W., Saillard, A., Abdulhak, S., Augé, V., Birck, C., Bonet, R., Choler, P., Delestrade, A., Kunstler, G., Leccia, M.-F., Lienard, B., Poulenard, J., Valay, J.-G., Bayle, A., Bonfanti, N., Brousset, L., Bizard, L., Calderón-Sanou, I., Dentant, C., & Renaud, J.The Orchamp Consortium. (2024). ORCHAMP: An observation network for monitoring biodiversity and ecosystem functioning across space and time in mountainous regions. *Comptes Rendus Biologies*, *347*(G1), 223–247. DOI: 10.5802/crbiol.165 PMID: 39774921

Titley, M. A., Snaddon, J. L., & Turner, E. C. (2017). Scientific research on animal biodiversity is systematically biased towards vertebrates and temperate regions. *PLoS One*, *12*(12), e0189577. DOI: 10.1371/journal.pone.0189577 PMID: 29240835

Valdez, J. (2020). *Arthropods as Vertebrate Predators: A Review of Global Patterns*. BIOLOGY. https://doi.org/DOI: 10.20944/preprints202002.0119.v1

Verrall, B., & Pickering, C. M. (2020). Alpine vegetation in the context of climate change: A global review of past research and future directions. *The Science of the Total Environment*, *748*, 141344. DOI: 10.1016/j.scitotenv.2020.141344 PMID: 32814293

Vincent, C., Fernandes, R. F., Cardoso, A. R., Broennimann, O., Di Cola, V., D'Amen, M., Ursenbacher, S., Schmidt, B. R., Pradervand, J.-N., Pellissier, L., & Guisan, A. (2019). Climate and land-use changes reshuffle politically-weighted priority areas of mountain biodiversity. *Global Ecology and Conservation*, *17*, e00589. DOI: 10.1016/j.gecco.2019.e00589

Voisin, C., Choler, P., Vacher, C., Perrier, C., Renaud, J., Rioux, D., & Boucher, F. C. (2025). Deciphering the drivers of speciation in Alpine Noccaea (Brassicaceae) reveal an evolution predominantly shaped by geography. *Molecular Phylogenetics and Evolution*, *211*, 108381. DOI: 10.1016/j.ympev.2025.108381 PMID: 40456477

Wilkes, M. A., Carrivick, J. L., Castella, E., Ilg, C., Cauvy-Fraunié, S., Fell, S. C., Füreder, L., Huss, M., James, W., Lencioni, V., Robinson, C., & Brown, L. E. (2023). Glacier retreat reorganizes river habitats leaving refugia for Alpine invertebrate biodiversity poorly protected. *Nature Ecology & Evolution*, *7*(6), 841–851. DOI: 10.1038/s41559-023-02061-5 PMID: 37142743

Yang, H., Gou, X., Xue, B., Ma, W., Kuang, W., Tu, Z., Gao, L., Yin, D., & Zhang, J. (2023). Research on the change of alpine ecosystem service value and its sustainable development path. *Ecological Indicators*, *146*, 109893. DOI: 10.1016/j.ecolind.2023.109893

Yang, Z., You, J., Li, J., Zhao, W., Xing, M., Zhang, Y., Cui, M., Gong, Y., Zhao, Y., & Chen, X. (2025). Divergent functional traits and gene expression profiles in native and encroaching plant species across an alpine elevational gradient. *Frontiers in Plant Science*, *16*, 1656812. DOI: 10.3389/fpls.2025.1656812 PMID: 41127076

Zehnder, M., Pfund, B., Svoboda, J., Marty, C., Vitasse, Y., Alexander, J., Hille Ris Lambers, J., & Rixen, C. (2025). Snow Height Sensors Reveal Phenological Advance in Alpine Grasslands. *Global Change Biology*, *31*(5), e70195. DOI: 10.1111/gcb.70195 PMID: 40346003

Zimmer, A., Beach, T., Klein, J. A., & Recharte Bullard, J. (2022). The need for stewardship of lands exposed by deglaciation from climate change. *Wiley Interdisciplinary Reviews: Climate Change*, *13*(2), e753. DOI: 10.1002/wcc.753

Zoderer, B. M., Lupo Stanghellini, P. S., Tasser, E., Walde, J., Wieser, H., & Tappeiner, U. (2016). Exploring socio-cultural values of ecosystem service categories in the Central Alps: The influence of socio-demographic factors and landscape type. *Regional Environmental Change*, *16*(7), 2033–2044. DOI: 10.1007/s10113-015-0922-y

Zoderer, B. M., Tasser, E., Erb, K.-H., Lupo Stanghellini, P. S., & Tappeiner, U. (2016). Identifying and mapping the tourists perception of cultural ecosystem services: A case study from an Alpine region. *Land Use Policy*, *56*, 251–261. DOI: 10.1016/j.landusepol.2016.05.004

Chapter 5

Conservation of Alpine Endemism Under Climate and Anthropogenic Pressures Emphasizing Phytogeographic Dimensions and Future Directions

Syed Waseem Gillani
http://orcid.org/0009-0009-2926-0723
Quaid-i-Azam University, Islamabad, Pakistan

Mushtaq Ahmad
Quaid-i-Azam University, Islamabad, Pakistan

Muhammad Manzoor
http://orcid.org/0000-0002-5118-3957
Quaid-i-Azam University, Islamabad, Pakistan

Mumna Munir
Quaid-i-Azam University, Islamabad, Pakistan

Shazia Sultana
Quaid-i-Azam University, Islamabad, Pakistan

Amir Sohail
Quaid-i-Azam University, Islamabad, Pakistan

Najeeb Ullah
Quaid-i-Azam University, Islamabad, Pakistan

Ahmed Ali
http://orcid.org/0009-0008-8786-2952
The University of Azad Jammu and Kashmir, Muzaffarabad, Pakistan

DOI: 10.4018/979-8-2600-1207-9.ch005

ABSTRACT

Endemic plant diversity of the Western Himalayas is a crucial component of global biodiversity but faces increasing environmental and anthropogenic threats. This study analyzed geographic distribution, compositional variation, elevational richness patterns, and β-diversity across 11 Himalayan sites between 3000 and 4300 m elevation. Chord diagrams revealed strong regional specificity with limited species overlap between Himalayan and Western Himalayan endemics. Multivariate analyses showed marked site-level heterogeneity, with a few sites driving overall dissimilarity. Species richness declined sharply with elevation, particularly among Himalayan endemics, while Western Himalayan endemics exhibited broader elevational ranges. PCA and PERMANOVA confirmed strong compositional segregation, and β-diversity was dominated by species turnover. Deforestation, grazing, NTFP extraction, tourism, and illegal trade emerged as major threats, emphasizing the need for region-specific, multi-site conservation strategies.

1. INTRODUCTION

Global warming is one of the major threats to the biodiversity around the globe (Cahill et al., 2013; Sax et al., 2013). Mountain ecosystems are important hotspots of biological diversity, which sustain the species and ecological communities, which are frequently at high risk (Nogues- Nogués-Bravo et al., 2007; Hoorn et al., 2018). Climate change is especially sensitive to such systems but the vulnerability to climate change differs significantly across mountain regions (Engler et al., 2011). Plant species endemic to the mountainous regions should be particularly susceptible to climate-induced changes in habitats (Dullinger et al., 2012), since they usually inhabit small ecological niches, have a restricted geographic range, and are highly dependent on a particular habitat (Essl et al., 2009). Moreover, this has been strengthen by the fact that most of the endemic mountain plants typically have a low dispersal potential and thus their present distributions are not sufficient to monitor the previous climatic changes (Essl et al., 2011). This constraint will further impair their ability to adapt to climate change in future (Engler et al., 2009; Ozinga et al., 2009). This has consequently turned the determination of the future distributional extremes of endemic mountain plant species to be a significant conservation biological concern. Such measurements are important to inform active conservation planning and in decision making in an effort to mitigate the adverse impacts of climate change on biodiversity (Pereira et al., 2010; Parmesan et al., 2011).

Human activities put extensive pressures on global biodiversity other than climate change (Tittensor et al., 2014). Out of this group of drivers, land-use change (especially habitat conversion and degradation) has become the most dominant cause of biodiversity loss all over the world (Newbold et al., 2015; Banks-Leite et al., 2020; Kuipers et al., 2021). The recent evaluations suggest that these processes have led to an average 13.6% decrease in species richness and a decrease of up to 10.7% in total biological abundance on the planet (Newbold et al., 2015). These land-use change effects are compounded by habitat fragmentation that also leads to localized losses of biodiversity (13-75%) and interferes with critical ecosystem processes, such as nutrient cycling (Haddad et al., 2015). These interacting pressures collectively have led to changes in distributions of many alpine species (moving towards higher elevations or latitude) and also the probability of local extinction in others (Wang et al., 2021).

The Himalaya is globally recognized as one of the most important biodiversity hotspots in the world, owing to its vast altitudinal range, diverse climatic conditions, and complex geological history. This region supports exceptional species richness and endemism, providing critical habitats for numerous plant and animal species of ecological, medicinal, and conservation significance (Mittermeier et al., 2011) and is regarded as the area especially susceptible to the effects of climate change among many other regions (Kaur et al., 2022). The Himalaya being the largest mountain range on the planet is the home of an amazing endemism and a great deal of nearly untouched natural environments with a high level of biological diversity (Rather et al., 2022; Wani et al., 2023). However, over the last decades, anthropogenic land-use pressures such as deforestation, agricultural expansion, infrastructure development, and unregulated tourism have intensified across the region. These activities pose a serious threat to biodiversity by accelerating habitat loss, fragmentation, ecosystem degradation, and the decline of native and endemic species (Pandit, 2013; Rathore et al., 2022). At the same time, climate-related processes which include ice retreat and the rise of temperatures have been associated with an upslope movement of the treelines at higher altitudes in the Himalaya (Singh et al., 2015; Hamid et al., 2023). Such changes in elevation will pose risks to extinction of alpine flora existing above the treeline, where they might receive more competition due to the expansion of thermophilic, lower-elevation taxa into a previously cooler habitat (Sah et al., 2023; Hamid et al., 2020). In turn, the analysis of the diversity and spatial distribution of endemic species of alpine zone in the Himalayan area under the combined effects of climate change and the anthropogenic factors is vital. This study will contribute to a better understanding of the alpine ecosystem reaction and it will be a solid scientific basis on which conservation priorities can be established and future strategies to manage biodiversity successfully. Therefore this study aims to address following main objectives; (1). To record the diversity and the floristic structure of endemic vegetation in the alpine sites of the Himalayas in the Kashmir region. (2).To examine the patterns of spatial distribution of endemic plant species along an elevational and topographical gradient. (3).To investigate habitat association, anthropogenic threats, and ecological preference of alpine endemic plants in the Kashmir Himalayas. (4).To targets the areas of priority areas to be used to conserve the endemic plant diversity in the alpine environment in the Kashmir Himalayas.

2. MATERIALS AND METHODS

2.1. Study Area

This study has been conducted in Muzaffarabad Division of Azad Jammu and Kashmir (Pakistan) which comprises of three districts namely, Muzaffarabad, Neelum Valley, and Jehlum Valley. The area has a large variance of elevations that range between about 360 m above the sea level in the low land areas of the south to 6,325 m in the highlands of the north (Gillani et al., 2026) (Fig. 1). Such a variation in altitude creates a variety of climatic zones, including subtropical monsoon in the lower altitude, moist temperate in the middle altitude, and subalpine and alpine in the upper altitude. Precipitation of the region is highly seasonal with the highest rainfall in the monsoon months of July and August and a second wet season between February and April during spring. On the other hand, the lowest rainfall is usually recorded during November. Temperature changes in the extremes are clearly seasonal, with the warmest month of July, and the coldest month of January (Ashraf et al., 2012; GoAJK, 2017). The latest developments have shown that the Muzaffarabad Division is becoming more susceptible to climatic change as the temperatures are growing higher and the precipitation patterns are also changing. It has been estimated that climate change will continue to rise in the future, and by 2060, the rise would be 1.4 °C and as high as 3.0 °C by 2100 (GoAJK, 2017). The region where the study is taking place is a coniferous area which is mostly composed of *Abies pindrow*, *Taxus wallichiana*, *Pinus roxburghii*, *Picea smithiana*, *Cedrus deodora* and *Pinus wallichina* (Gillani et al., 2025a). The area is home to a number of significant medicinal and endemic species of plants, including *Saussurea costus*, *Podophyllum hexandrum*, *Arnebia benthamii*, *Aconitum chasmanthum*, *Rheum webbianum*, *Rheum australe*, *Viburnum grandiflorum*, *Meconopsis aculeata*, *Geranium wallichianum*, *Trillium govanianum* and *Polygonatum verticillatum* (Gillani et al., 2025b). The main indicator species also indicate the specific ecological areas on the altitudinal gradient of the Kashmir Himalayas. *Pinus roxburghii* and *Dodonaea viscosa* are the most common in the lowland forests in the subtropical zone. There are thick coniferous forests in the temperate zone which include *Pinus wallichiana*, *Cedrus deodara* and *Abies pindrow* (Alam et al., 2023). Farther upwards, the subalpine region is characterized by *Betula utilis* and *Quercus semecarpifolia* which are species well adapted to cooler and harsher conditions (Manzoor et al., 2024). *Aconitum chasmanthum*, *Rheum webbianum*, and *Arnebia benthamii* are the alpine zone, which grows at the highest altitude of the habitats and is exposed to harsh climatic conditions (Manzoor et al., 2026).

Figure 1. Study area map showing the sampling sites in the Kashmir alpine Himalayan region.

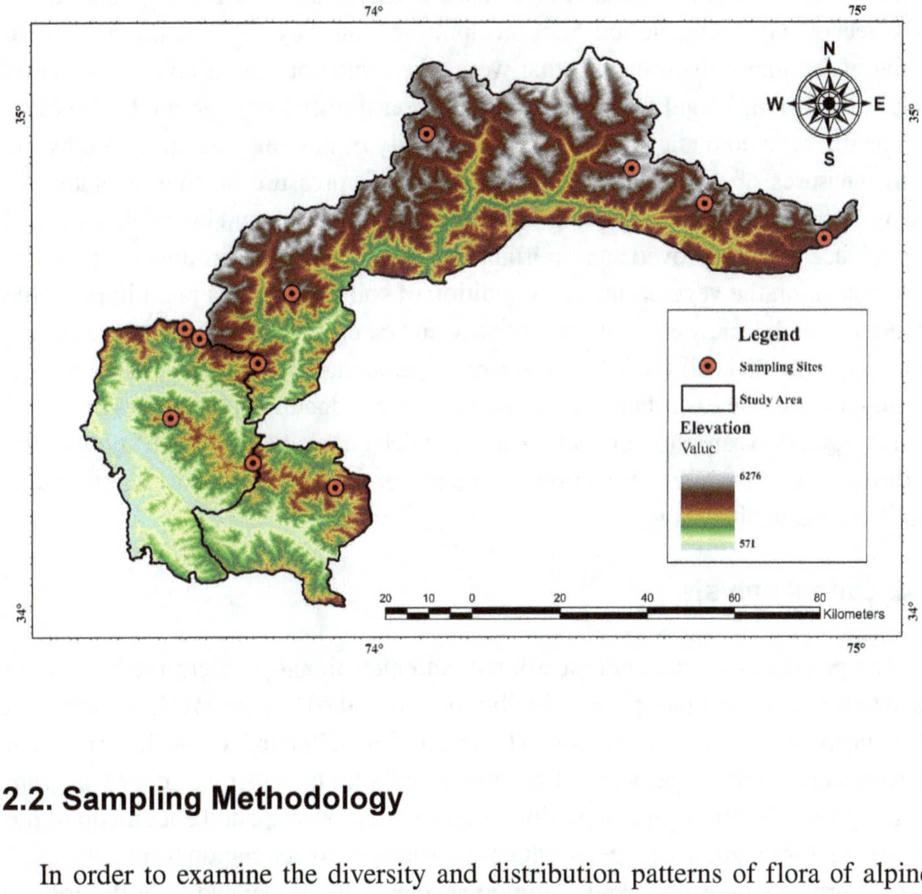

2.2. Sampling Methodology

In order to examine the diversity and distribution patterns of flora of alpine grassland in Azad Jammu and Kashmir (AJK), a sample site was chosen on the Muzaffarabad Division. Eleven sites which cover three districts: Neelum Valley, Jhelum Valley, and Muzaffarabad, were surveyed. Field sampling was conducted in 2022- 2024 between June to September to record species diversity, vegetation properties (species composition and types of vegetation) as well as structure (density, cover of species and growth forms in the alpine area). We used a systematic quadrat-based system where quadrat size for shrubs 25 m^2 (5 m x 5 m) and grasses and herbs had 1 m^2 (1 m x 1 m) quadrats (Cox, 1996). Both vegetation structure and microclimatic conditions were used in quadrat selection to cover all the alpine habitats in a representative sample. Basic vegetation data like cover, frequency and density were measured as per the standard phytosociological procedures (Cox, 1996). Geographic data (latitude, longitude) and topographic data such as height and aspect were measured at every site by the use of a GPS device.

2.3. Evaluation of Disturbance Indicators

A set of disturbance indicators was used to assess the systematic evaluation of the effect of anthropogenic activities on alpine endemic vegetation in the study area. Some of the major disturbances that were taken into consideration were grazing, fire, soil erosion, illegal trade, extraction of non-timber forest products (NTFPs), and nearness to nomadic settlements. The quality of grazing was measured by the field measures of the visual cues of the livestock pressure on the vegetation in terms of the hoof tracks, trampling, and browsed vegetation and bare soil spots and animal feces, which proved an important assessment of the anthropogenic pressure of livestock on the vegetation. The condition of soil erosion was rated in the same manner with the analysis of the soil surface and categorization of the levels of erosion into three: controlled (little or no erosion), moderate (topsoil erosion observed), and intense. These disturbance measurements were documented at the same level as the vegetation sampling of each location which enabled a combined study of the anthropogenic pressures effects on the plant community composition, structural features, and habitat integrity.

2.4. Data Analysis

The presence-absence, compositional, and elevational gradient methods were used to examine endemic species distribution across the Himalayas (H) and Western Himalayas (WH) and were integrated to reveal results. Records of the occurrence of species were banded together at 11 representative sites to construct a chord diagram which gives a simple visual depiction of specificity, overlap, and endemism of the region. In this diagram, the species that were specific to each region were associated to the corresponding node and common species were associated to both regions, resulting in quick evaluation of spatial patterns of distribution and exclusivity to the region. SIMPER (Similarity Percentage) analysis was applied to calculate compositional differences between sites, using Bray-Curtis dissimilarity, which showed the contribution made by sites to total community compositional variance. The locations that contributed the most were identified so as to identify the locations that convey dissimilarity across the study area. In order to measure the statistical significance of the group-level compositional differences, the use of permutational multivariate analysis of variance (PERMANOVA) with 9,999 permuations was used to offer a robust test of between-group divergence, taking into consideration within-group variance. The 3,000-4,300 m elevational gradient was examined using both Spearman rank correlation analysis and Poisson generalized linear models (GLMs) and compared the trends between the total species richness and Himalayan (H) and Western Himalayan (WH) endemics. This analysis helped to identify the patterns that were limited to elevation with emphasis on taxa that were focused at lower altitude compared to those that were widely distributed throughout the gradient. The Principal Component Analysis (PCA) was conducted to check the clustering trends of the H and WH endemics, where convex hulls represent boundaries of groups and vectors of site point to particular locations that make significant contributions to the division of groups. This multivariate method was able to visualize species weighted structure and ecological affinities between sites. Spatial heterogeneity and correspondence between patterns of species replacement and nesting were interpreted using a hierarchical clustering and a dendrogram with tanglegram visualization of β-diversity partitioned into turnover (βsim) and nestedness-resultant dissimilarity (βsne) components with the Sørensen index, which is implemented via the betapart package in R (Manzoor et al., 2025).

3. RESULTS

3.1. Geographic Distribution Patterns of Endemic Species

The chord diagram was used for the illustration of the geographic distribution of endemic species of plants in the Himalayas and the Western Himalayas and presented a simplified form of the specificity of the region and species overlap (Fig. 2). Out of the 41 species of endemics present, 25 were confined to the Himalayas. They were *Aconitum chasmanthum*, *Aconitum heterophyllum*, and *Anaphalis triplinervis*, which means that the endemism level was high in the area of the Himalayas (Table 1). The abundance of the species indicates the ecological diversity, elevation gradients, and a wide range of microhabitats, which define the Himalayan hotspot biodiversity. In comparison, the number of species being endemic to the Western Himalayas was 15, which highlight the floristic nature of the subregion. *Aconitum violaceum*, and *Berberis kashmirana* are present in this group and so the importance of local climatic regimes and geological controls in local distributions of plants. The Himalayas and the Western Himalayas shared six species which indicated a wider ecological tolerance and adaptability. It was interesting to note that both regions had the *Anaphalis triplinervis* and *Aconitum violaceum* meaning that these species can survive under different environmental conditions.

Figure 2. Species found only in the Himalayas in the chord diagram were depicted by blue chords and Western Himalayas ones were depicted by red chords. The species that were present in both of the territories were linked to both nodes, and the distribution overlap of the species was highlighted graphically.

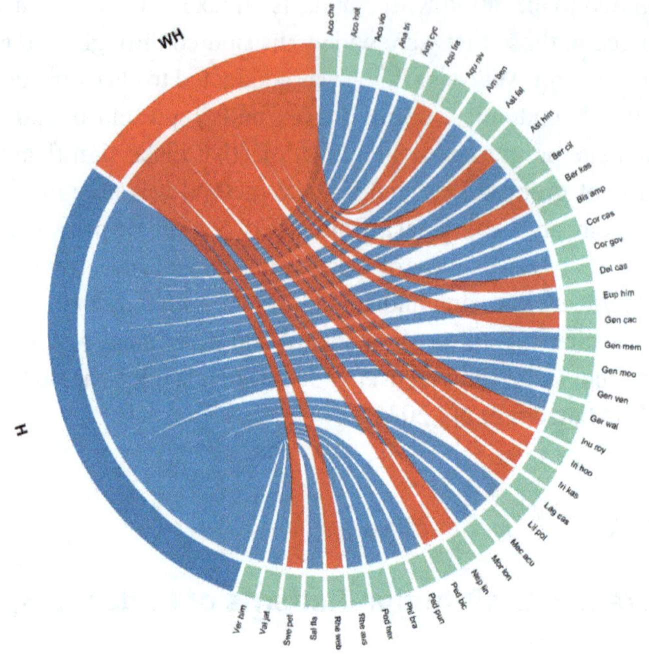

Table 1. Showing the all studied species with species names, family, species code and endemic status.

Sr. No.	Species Name	Family	Species Code	Endemic Status
1.	*Aconitum chasmanthum* Stapf ex Holmes	Ranunculaceae	Aco cha	H
2.	*Aconitum heterophyllum* Wall. ex Royle	Ranunculaceae	Aco het	H
3.	*Aconitum violaceum* Jacquem. ex Stapf	Ranunculaceae	Aco vio	H
4.	*Anaphalis triplinervis* (Sims) C.B.Clarke	Asteraceae	Ana tri	H
5.	*Angelica cyclocarpa* (C.Norman) M.Hiroe	Apiaceae	Ang cyc	H
6.	*Aquilegia fragrans* Benth.	Ranunculaceae	Aqu fra	WH
7.	*Aquilegia nivalis* (Falc. ex Brühl) J.R.Drumm. & Hutch.	Ranunculaceae	Aqu niv	WH

continued on following page

Table 1. Continued

Sr. No.	Species Name	Family	Species Code	Endemic Status
8.	*Arnebia benthamii* (Wall. ex G.Don) I.M.Johnst.	Boraginaceae	Arn ben	H
9.	*Aster falconeri* (C.B.Clarke) Hutch.	Asteraceae	Ast fal	H
10.	*Aster himalaicus* C.B.Clarke	Asteraceae	Ast him	H
11.	*Astragalus himalayanus* Klotzsch	Fabaceae	Ast him	WH
12.	*Berberis kashmirana* Ahrendt	Berberidaceae	Ber kas	WH
13.	*Bergenia ciliata* (Haw.) Sternb.	Saxifragaceae	Ber cil	H
14.	*Bistorta amplexicaulis* (D.Don) Greene	Polygonaceae	Bis amp	H
15.	*Corydalis cashmeriana* Royle	Papaveraceae	Cor cas	H
16.	Corydalis govaniana Wall.	Papaveraceae	Cor gov	H
17.	*Delphinium cashmerianum* Royle	Ranunculaceae	Del cas	WH
18.	*Euphrasia himalayica* Wettst.	Orobanchaceae	Eup him	H
19.	*Gentiana cachemirica* Decne.	Gentianaceae	Gen cac	WH
20.	*Gentiana membranulifera* T.N.Ho	Gentianaceae	Gen mem	H
21.	*Gentiana venusta* (G.Don) Wall. ex Griseb.	Gentianaceae	Gen ven	H
22.	*Gentianella moorcroftiana* (Wall. ex G.Don) Airy Shaw	Gentianaceae	Gen moo	H
23.	*Geranium wallichianum* D.Don ex Sweet	Geraniaceae	Ger wal	H
24.	*Inula royleana* DC.	Asteraceae	Inu roy	WH
25.	*Iris hookeriana* Foster	Iridaceae	Iri hoo	WH
26.	*Iris kashmiriana* Baker	Iridaceae	Iri kas	WH
27.	*Lagotis cashmeriana* (Royle ex Benth.) Rupr.	Plantaginaceae	Lag cas	WH
28.	*Lilium polyphyllum* D.Don	Liliaceae	Lil pol	H
29.	Meconopsis aculeata Royle	Papaveraceae	Mec acu	H
30.	*Morina longifolia* Wall. ex DC.	Caprifoliaceae	Mor lon	H
31.	*Nepeta linearis* Royle ex Benth.	Lamiaceae	Nep lin	WH
32.	*Pedicularis bicornuta* Klotzsch	Orobanchaceae	Ped bic	WH
33.	*Pedicularis punctata* Decne.	Orobanchaceae	Ped pun	H
34.	*Phlomoides bracteosa* (Royle ex Benth.) Kamelin & Makhm.	Lamiaceae	Phl bra	H
35.	*Podophyllum hexandrum* Royle	Berberidaceae	Pod hex	H
36.	*Rheum australe* D.Don	Polygonaceae	Rhe aus	H
37.	*Rheum webbianum* Royle	Polygonaceae	Rhe web	WH
38.	*Salix flabellaris* Andersson	Salicaceae	Sal fla	H
39.	*Swertia petiolata* D.Don	Gentianaceae	Swe pet	WH

continued on following page

Table 1. Continued

Sr. No.	Species Name	Family	Species Code	Endemic Status
40.	*Valeriana jatamansi* Jones ex Roxb.	Caprifoliaceae	Val jat	H
41.	*Veronica himalensis* D.Don	Plantaginaceae	Ver him	H

H= Himalayas and WH= Western Himalayas.

3.2. Trends in Compositional Change among Sites

The compositional variation in the 11 Himalayan sites using SIMPER (Similarity Percentage) which is based on Bray-Curtis dissimilarity indicated a high degree of heterogeneity in species composition with the total average dissimilarity of 70.41% of site groups. The SIMPER findings found important locations that significantly contributed to these differences in composition. The site with the greatest effect, which had 11.91% of the observed dissimilarity (average dissimilarity = 8.387) was Site 2 and then Site 5 (11.20%; 7.888) and Site 3 (11.01%; 7.753). These three sites accounted together as 34.12 of the compositional variance, meaning that site-level specific differences were clear in organizing community divergence. Further significant contributions came out of Sites 6, 11 and 10 that combined to increase the cumulative explained dissimilarity to 64.42. Distinctions between the mean species abundances of groups were especially significant at Sites 5 (Mean1 = 0.48; Mean0 = 0.125) and 6 (0.44 vs 0.188) indicating a high level of site specific patterns to drive between-group differentiation. Additional input of Sites 4, 7 and 8 even raised the cumulative dissimilarity to 93.64 and indicate that the compositional variability in a few key sites accounted most of the variability, and Site 9 itself had the least effect (6.36%), which pointed to relative compositional similarity to other sites. Although site-level contributions were high, the results of PERMANOVA showed no statistically significant overall differences in the compositions of the groups ($F = 1.513$, $p = 0.1388$; 9999 permutations). This non-significance indicates that there existed a high within-group variability probably due to homogeneous species distributions and stochastic variability between sites. The integrated results demonstrate the significance of assessing both local contributions and general group trends in understanding ecological dissimilarity because a large site-level heterogeneity may be found amid much broader group difference that is statistically blurred. These findings indicate that although individual locations can be viewed as the primary areas of community differentiation, there is a great deal of variability in the total compositional patterns between Himalayan sites, which can be attributed to the complicated interactions of elevation, heterogeneity of habitats, and distribution of endemic species.

3.3. Richness-Elevation Patterns of Endemic Species

The species richness of the surveyed sites in the Himalayan region had a strong negative trend with the elevational gradient between 3000m and 4300m. The relationship between total species richness and elevation was statistically significant and had a negative value (Spearman, $\rho = -0.602$, $p = 0.050$; Poisson GLM: $\beta = -0.00046$, $p = 0.016$), and more species were supported in lower altitude sites. This trend highlights the role of elevation limits in the total biodiversity in the area in terms of variation in microclimatic conditions, habitat structure, and resource availability across the gradient. The endemic species were especially altitude sensitive (H) and showed a strong negative relationship with the increase in the altitude (Spearman: $\rho = -0.803$, $p = 0.0029$; GLM: $\beta = -0.00047$, $p = 0.044$). Such a trend indicates that the H species are concentrated in the lower altitude bands and this implies that they have very narrow ecological ranges and therefore may be more susceptible to environmental change. Western Himalayan endemics (WH), in contrast, had a weakly and non-significantly positive increase in elevation (Spearman $\rho = -0.411$, $p = 0.21$; GLM: $\beta = -0.00044$, $p = 0.184$) and a more uniform distribution along the gradient thus ecologically tolerant (Fig. 3). The relative reaction of H species and WH species indicates that the endemic assemblages have strong elevational limitation and ecological strategies. Species of H seem to be localized in low altitude and middle altitude areas but those of WH are distributed more extensively and may be more adaptable to various environmental factors. The patterns have noteworthy interests to conservation planning since the H species that are restricted in their elevation can be vulnerable in terms of loss of habitats, climate change, and anthropogenic pressures, whilst the WH species can have a higher adaptive capacity. These trends are shown by graphical representation of richness per site, which shows a sharp decrease on H species compared to WH species across the elevation gradient. All these above findings together highlight the importance of elevation in organizing endemic plant diversity in the Western Himalayas.

Figure 3. Species richness along 3000–4300 m elevation gradient for Total, H (Himalayas), and WH (Western Himalayas) species.

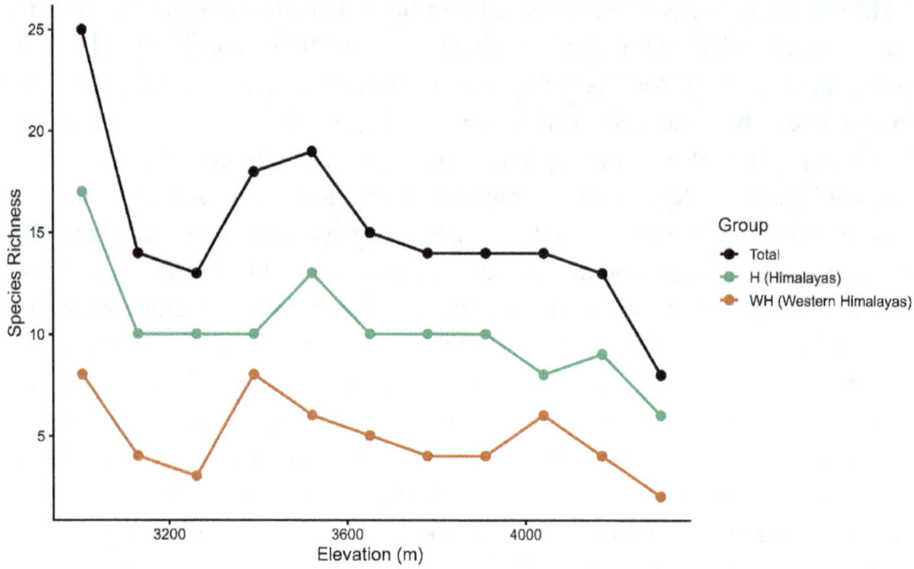

3.4. Different Endemic Clusters Differentiation

Analysis of the endemic species in the Himalayan and Western Himalayan sites in the form of Principal Component Analysis (PCA) demonstrated evident compositional segregation of the two regions. The two major components explained 34.55 percent of the total variance, with 19.89 percent and 14.66 percent explained by PC 1 and PC 2 respectively (Fig. 4). The cumulative variance was moderate; nevertheless, the ordination clearly grouped Himalayan (H) and Western Himalayan (WH) endemic species in 2 non-overlapping clusters, which showed high levels of differentiation in species composition and ecological affinities. Himalayan endemics were clumped into a small group, which indicated a fairly homogeneous occupation of the sites, and an invariable relation to the definite environmental conditions. Conversely, the Western Himalaya endemics showed more spread implying heterogeneous sites associations and more variation in term of ecological preference. The orientation of the site vectors with the ordination axes also determined the sites which provided maximum contribution to group separation, and thus indicate that specific locations were important contributors of a compositional divergence. Statistical test (PERMANOVA) revealed that the distance between species of H and WH was very significant ($p < 0.01$), which supported the trends of ordinations in the PCA.

Boundaries of the convex hulls enclosing the clusters further strengthened the ecological segregation between the groups, no overlap was found, which highlights the uniqueness between the assemblages of species between the two localities. These findings showed that endemic species of the Himalaya and the Western Himalaya are not simply geographically separated but ecologically as well, with site-specific affinities and distributions. The reduced dispersion of H species is an indication that it is highly dependent on particular habitat types or microclimatic conditions and that the relatively more dispersed WH species may be indicative of broader ecological tolerance or adaptation to dissimilar environmental conditions. All in all, the PCA puts significant emphasis on the biogeographic segregation of Himalayan endemics, and the necessity to preserve the representative sites in both areas to ensure the entire range of endemic diversity.

Figure 4. PCA biplot illustrating significant segregation of Himalayan (H) and Western Himalayan (WH) endemic species with influential site vectors.

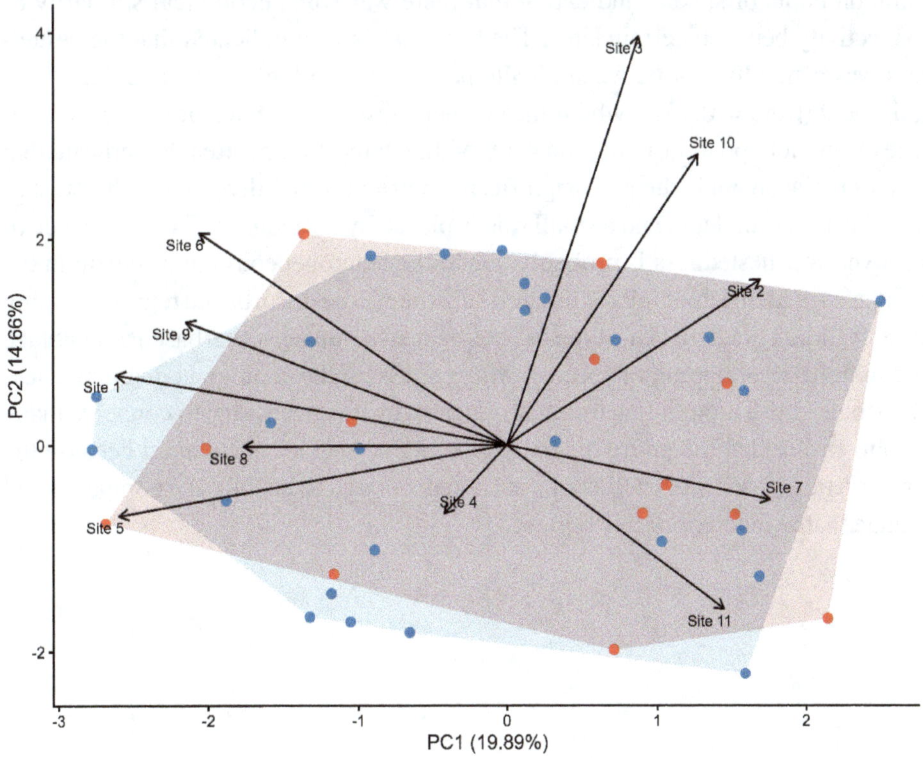

3.5. β -Diversity Patterns and Heterogeneity of the Sites

The β-diversity analysis of the 11 sites in the Himalaya showed that species turnover (βsim) was the largest element of compositional dissimilarity with a mean value of βsim of 0.553 meaning that there was a lot of replacement of species across sites. This high turnover indicates that a lot of sites contain exclusive species assemblages, which depicts strong spatial heterogeneity of the species composition along the elevational gradient that was moved during the survey. Comparatively, however, the resultant dissimilarity (βsne) was relatively low, with an average of 0.077, indicating that variations in species richness that arose as a result of species loss or gain made minimal contributions to total β -diversity. Analysis of βsim matrix showed that there were some site pairs that had very high turnover i.e. Site 5 and Site 7 (βsim = 0.786), Site 6 vs Site 7 (βsim = 0.857), which revealed that compositional differentiation was high and species overlap was minimal. On the other hand, lower values of turnover like Site 1 and Site 5 βsim = 0.143 indicated common pools of species indicating that there was some ecological similarity or connectivity between certain sites. The βsne matrix also indicated that the nestedness was generally low between all site pairs, with the highest value being Site 1 and Site 9 (βsne = 0.451), which means that subsets of species of richer communities were not apparent in the majority of site pairs (Fig. 5). Results indicate that species replacement is the most significant contributor to β-diversity in this west of Himalayan assemblage, and a small role is played by nestedness. The prevalence of turnover over nestedness is probably due to the heterogeneous environment in the habitat, such as different microclimates, substrates, and disturbance regime crossing the elevational gradient. This type of heterogeneity enhances site-based assemblage and minimal overlapping of species, with a stress on the ecological distinctiveness of a single site. Through this trend, it is clear why it is necessary to conserve many sites to ensure that the entire diversity of endemic species is captured because the conservation of few sites will not be sufficient enough to capture the compositional change that is inherent to this region.

Figure 5. Tanglegram depicting β-diversity patterns among 11 Himalayan sites. Left dendrogram represents species turnover (βsim) and right dendrogram represents nestedness-resultant dissimilarity (βsne). Branch colors correspond to individual sites, and connecting lines illustrate correspondence between site clusters in turnover vs nestedness. High βsim values indicate strong species replacement, while low βsne values reflect minimal nestedness across sites.

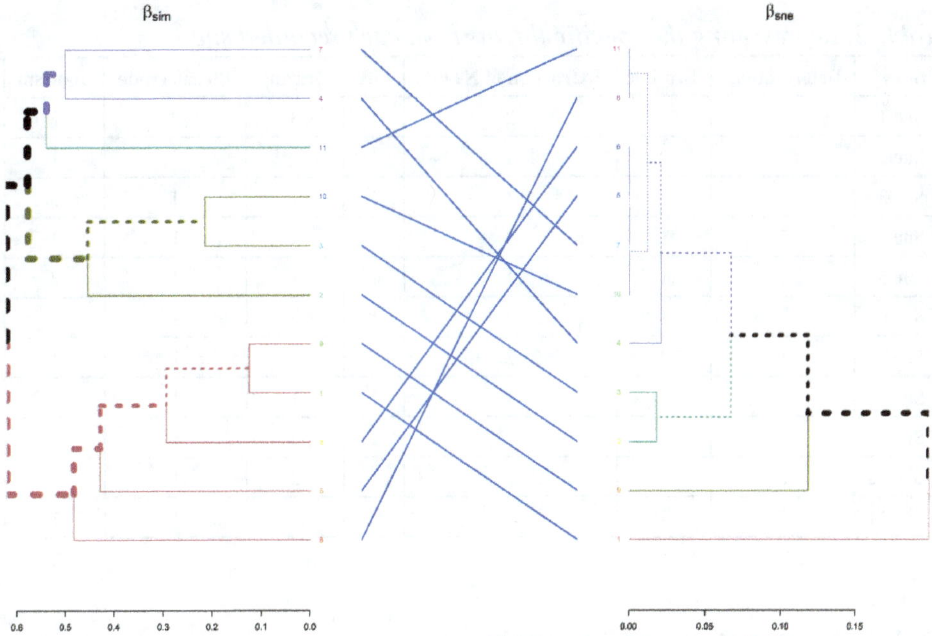

3.6. Anthopogenic Threat Assesment

Analysis of the anthropogenic threats in the 11 sites surveyed throughout the Himalayas showed that pressures affecting the endemic species and forest health were not uniformly distributed. Sites 3, 6 and 9 (score 3) had moderate to high deforestation which indicated the possibility of a localized habitat loss whereas other sites had low to moderate clearing. The patterns of soil erosion were variable with Sites 1, 2, 7 and 9 having slightly higher risk (score 2) which was due to steep slope and human interruption (Table 2). The extraction of non-timber forest products (NTFPs) was very strong at the Sites 3 and 6 (score 3), which made it clear that people relied on medicinal and edible plants. Incidents of fire were generally low and only moderate in the Site 4 (score 2). Site 1 (score 3) had the greatest grazing pressure followed by other sites, which had low to medium scores. The sites that were affected by the illegal trade of wildlife and plants were mainly Sites 1, 6, 10, which means that the exploitation is ongoing. Effects of tourism were generally moderate and Sites 3 and 10 recorded the best scores (3), indicating recreational disturbance. Sites 3, 6 and 9 were identified as hotspots of cumulative threats and there is need to use specific conservation strategies to reduce degradation areas and preserve endemic biodiversity.

Table 2. Representing the specific threats from each recorded site.

Sites	Deforestation	Erosion	Extraction of NTFPs	Fire	Grazing	Illegal Trade	Tourism
Site 1	1	2	2	1	3	2	2
Site 2	1	2	1	1	1	2	1
Site 3	3	1	3	1	1	1	3
Site 4	2	1	2	2	2	1	1
Site 5	1	1	1	1	1	1	2
Site 6	3	1	3	1	2	2	2
Site 7	1	2	2	1	1	1	1
Site 8	2	1	1	1	1	1	2
Site 9	3	2	2	1	2	1	1
Site 10	2	1	1	1	2	1	3
Site 11	2	1	1	1	1	1	1

3.7. Indicator Species Analysis

Indication Species Analysis was used to identify the top 20 species which were most associated with each site in the Western Himalayas according to the values of IndVal percent and permutation significance. This analysis showed that *Aconitum hetrophyllum* (IndVal = 87.5%, p = 0.015), *Aquilegia fragrans* (63%, p = 0.089), and *Corydalis cashmeriana* (50%, p = 0.111) had the highest values of the indicator showing a strong site fidelity and possibly significant ecological value. Others that were significant were *Anaphalis triplinervis* (31-42%, p = 0.472-0.480), *Angelica cyclocarpa* (31.8%, p = 0.472), and *Geranium wallichianum* (26.9%, p = 0.719), which, although not strongly linked, had compositional differentials at their respective locations. The significance test showed that some species were statistically significant ($p < 0.05$) such as *Aconitum hetrophyllum* and *Bergenia ciliata*, meaning that they can serve as reliable predictors of a particular habitat. The rest of the species with moderate percentages of IndVal were not found to be significant with non-significative p-values, indicating a weaker association with the site or less consistent findings. As a set, the most 20 indicator species indicate spatial heterogeneity and ecological differentiation among sites, since the indicator species uniquely relate with sites as well as provide potential bioindicators to be used in conservation monitoring. These findings put a strong emphasis on the necessity to consider the level of both IndVal magnitude and statistical significance in identifying species that best capture site specific ecological phenomena in the study area.

3.8. Conservation implications

The geographic isolation and compositional distinction of Himalayan and Western Himalayan endemic species emphasize the importance of the region-specific conservation approaches. The species that are restricted to a certain area, are the part of the biodiversity that is irreplaceable and that has a very limited ecological range. Their high clustering and their lower elevational confinement denotes that they are very sensitive to changes in habitat and climate fluctuations. The planning of conservation therefore should focus on the lower to mid-elevation habitats which enable concentrated endemic richness to be maintained so that microhabitats and environmental conditions that are vital to these species are preserved. The fact that crucial locations producing disproportionate compositional change is also a strong argument in favor of site-level management in that the conservation value of a small number of ecologically significant locations can be disproportionately high. A high β-diversity that is driven by species turnover means that no site can fairly represent endemic diversity in the region. This geographic diversity requires networked approach of conservation as opposed to closed areas of protection. In the areas that qualify as threat hotspots, especially those that are under heavy deforestation, grazing, NTFP harvesting, and tourism, urgent mitigation has to be undertaken by the use of regulated resource utilization, community involvement, and habitat restoration. Strongly fidelitic indicator species like *Aconitum heterophyllum* are convenient devices used to monitor an ecological system over the long term and provide an early warning about ecological change. Taken together, the findings highlight that conservation of endemic plant biodiversity in the Western Himalayas requires the protection of the various complementary locations, minimization of the local anthropogenic stressors, and incorporation of the ecological indicators into the adaptive management initiatives.

4. MANAGEMENT PRIORITIES FOR SUSTAINING HIMALAYAN ENDEMIZM

The Himalayas need a strong spatial, ecological and elevational pattern, which this study indicated as the basis of endemic plant diversity management priorities.

The evident geographic isolation of Himalayan and Western Himalayan endemic species shows that the conservation activities must be region specific instead of homogeneous. The species that are limited to a specific area, need special protection since due to their small distribution range, species are more susceptible to habitat destruction and environmental alteration. The management ought to thus focus on habitats that conducive to the regionally endemic species to ensure the Himalayan and the Western Himalayan floras have enough representation in conservation planning. The high level of heterogeneity at the site level in 11 sites that have been studied shows the necessity of conserving more than just a few sites with several species of plants. There were a few sites that contributed disproportionately to the total compositional variation such that loss or degradation of these important sites would cause enormous decreases in the diversities of the entire region. Management approaches ought to recognize and emphasize on such influential locations to have them more strictly protected, restored and monitored long term. Simultaneously, ecological connectivity between sites with overlapping species can be maintained, which can be used to maintain population persistence and genetic mixing. The patterns of species richness at elevation also indicate that preference should be given to lower and mid elevation areas. The management interventions policy in such areas should aim at reducing deforestation, overgrazing, and extraction of non-timber forest products in an unsustainable way. Wider endemics species on the other hand exhibited a wider geographic range in elevation, but need protection throughout the gradient to achieve ecological stability. There was high beta diversity with species turnover, the focus of which is that the location has a unique assemblage. This observation endorses the network based conservation strategy that safeguards a mosaic of habitats in elevations and sites. There were some places, where the pressure hot spots were identified, and controlled grazing, regulated tourism, and community based resource management are badly needed. Indicator species offer useful information in terms of tracking habitat quality and management performance. These species serve as ecological indicators that can be used to inform adaptive management to ensure Himalayan endemic plant diversity has a long term survival.

5. LIMITATIONS OF THE STUDY

Our study has a number of limitations even though it offers in-depth information on endemic plant diversity. It was also limited to sampling 11 sites across a given elevational gradient and this could not present the overall spatial and environmental variance of the wider Himalayan region. The range of temporal replication was restricted, and they could not evaluate interannual differences and seasonal changes in species composition. Other results of the analysis, including non-significant site group PERMANOVA, might indicate high within-group dispersion and low sample size but not ecological similarity. Also the semi-quantitative scoring that was used to make threat assessments partially depended on the result that might not reflect the full intensity or temporal dynamics of anthropogenic pressures. These limitations must be taken into account when making generalization of findings.

6. FUTURE DIRECTIONS

Further studies are required in the future to cover more area with more sites across broader elevational, climatic, and biogeographic gradients to enhance representativeness of endemic patterns of diversity. The plots of long-term monitoring would enable assessing the temporal variations in the species structure, β-diversity, and indicative species functionality under a continued climate and land-use alteration. High-resolution environmental variables (soil properties, microclimate, and disturbances measures based on remote sensing) would be useful in intensifying insights into the factors that cause species turnover and site specificity. The distribution modeling of species may also help anticipate changes in favorable habitats of endemics that are constrained by altitude to facilitate proactive conservation planning. More socio-ecological studies should also be focused on to measure the human dependence on the endemic plants and to develop community-based conservation programs. The use of functional and phylogenetic metrics of diversity in addition to taxonomic diversity would be a more comprehensive picture of ecosystem resiliency and vulnerability, and conservation strategies would be able to protect not only number of species, but ecological processes and evolutionary legacy.

7. CONCLUSION

Our study has shown that the endemic plant assemblage of the Himalayas and Western Himalayas are distinct geographically and ecologically, the geography, site specific environmental factors and the anthropogenic pressures differ. The gradient of decreasing species richness with elevation, indicates how the narrow-range taxa are prone to habitat change and climate change. The β-diversity is high due to species turnover, and this shows that the individual sites support unique species assemblage, which supports the insufficiency of single-sites conservation strategies. Rather, biodiversity conservation of the region will entail conservation of several, ecologically complementary sites. These measures of important sites which contribute to compositional change disproportionately, and indicator species with high habitat fidelity, can give practical advice on conservation priorities and monitoring.

Threat analyses also show that human activities that are localized, such as resource extraction and tourism, when located in sensitive areas can have a significant negative impact on the quality of habitats. Taken together, the results herein point to the fact that to achieve a successful conservation in the Western Himalayas, it is necessary to incorporate both spatial heterogeneity, elevation-sensitive planning, and threat mitigation on local scales. This work provides a solid framework of protection of endemic plant diversity and ecosystem resilience in one of the most vulnerable regions of the world offering a combination of quantitative ecological research and threat assessment.

REFERENCES

Alam, N. M., Shaheen, H., Manzoor, M., Tinghong, T., Arfan, M., & Idrees, M. (2023). Spatial distribution and population structure of Himalayan Fir (*Abies pindrow* (Royle ex D. Don) Royle) in moist temperate forests of the Kashmir region. *Forests*, *14*(3), 482. DOI: 10.3390/f14030482

Ashraf, A., Naz, R., & Roohi, R. (2012). Monitoring and estimation of glacial resource of Azad Jammu and Kashmir using remote sensing and GIS techniques. *Pakistan Journal of Meteorology*, *8*(16), 31–41.

Banks-Leite, C., Ewers, R. M., Folkard-Tapp, H., & Fraser, A. (2020). Countering the effects of habitat loss, fragmentation, and degradation through habitat restoration. *One Earth*, *3*(6), 672–676. DOI: 10.1016/j.oneear.2020.11.016

Cahill, A. E., Aiello-Lammens, M. E., Fisher-Reid, M. C., Hua, X., Karanewsky, C. J., Yeong Ryu, H., Sbeglia, G. C., Spagnolo, F., Waldron, J. B., Warsi, O., & Wiens, J. J. (2013). How does climate change cause extinction? *Proceedings of the Royal Society B: Biological Sciences*, 280(1750), Article 20121890. DOI: 10.1098/rspb.2012.1890

Cox, G. W. (1996). Laboratory manual of general ecology (7th ed.). Wm. C. *Brown Publishers*.

Dullinger, S., Gattringer, A., Thuiller, W., Moser, D., Zimmermann, N. E., Guisan, A., Willner, W., Plutzar, C., Leitner, M., Mang, T., Caccianiga, M., Dirnböck, T., Ertl, S., Fischer, A., Lenoir, J., Svenning, J.-C., Psomas, A., Schmatz, D. R., Silc, U., & Hülber, K. (2012). Extinction debt of high-mountain plants under twenty-first-century climate change. *Nature Climate Change*, *2*(8), 619–622. DOI: 10.1038/nclimate1514

Engler, R., Randin, C. F., Thuiller, W., Dullinger, S., Zimmermann, N. E., Araújo, M. B., Pearman, P. B., Le Lay, G., Piedallu, C., Albert, C. H., Choler, P., Coldea, G., De Lamo, X., Dirnböck, T., Gégout, J.-C., Gómez-García, D., Grytnes, J.-A., Heegaard, E., Høistad, F., & Guisan, A. (2011). 21st century climate change threatens mountain flora unequally across Europe. *Global Change Biology*, *17*(7), 2330–2341. DOI: 10.1111/j.1365-2486.2010.02393.x

Engler, R., Randin, C. F., Vittoz, P., Czáka, T., Beniston, M., Zimmermann, N. E., & Guisan, A. (2009). Predicting future distributions of mountain plants under climate change: Does dispersal capacity matter? *Ecography*, *32*(1), 34–45. DOI: 10.1111/j.1600-0587.2009.05789.x

Essl, F., Dullinger, S., Plutzar, C., Willner, W., & Rabitsch, W. (2011). Imprints of glacial history and current environment on correlations between endemic plant and invertebrate species richness. *Journal of Biogeography*, *38*(4), 604–614. DOI: 10.1111/j.1365-2699.2010.02425.x

Essl, F., Staudinger, M., Stöhr, O., Schratt-Ehrendorfer, L., Rabitsch, W., & Niklfeld, H. (2009). Distribution patterns, range size and niche breadth of Austrian endemic plants. *Biological Conservation*, *142*(11), 2547–2558. DOI: 10.1016/j.biocon.2009.05.027

Gillani, S. W., Ahmad, M., Manzoor, M., Al-Andal, A., Khan, R. W. A., Ayoola, J. O., & Waheed, M. (2025a). Integrating environmental variables and biotic interactions to predict the future distribution of the endangered *Aconitum heterophyllum* Wall. ex Royle. *BMC Plant Biology*, *26*(1), 45. DOI: 10.1186/s12870-025-07859-y PMID: 41339780

Gillani, S. W., Ahmad, M., Manzoor, M., Waheed, M., Khan, R. W. A., & Sohail, A. (2026). Geographic distribution patterns, population dynamics, and habitat vulnerability of endangered and endemic *Trillium govanianum*. *Environmental Monitoring and Assessment*, *198*(2), 162. DOI: 10.1007/s10661-026-15013-w PMID: 41569443

Gillani, S. W., Ahmad, M., Manzoor, M., Zafar, M., Shaheen, H., Hussain, J., Al Farraj, D. A., Elshikh, M. S., & Abbas, N. (2025b). Local perceptions of ecosystem services and anthropogenic disturbances on endangered *Taxus wallichiana*: Implications for conservation and sustainable management. *Polish Journal of Environmental Studies*, *34*(4), 1–14. DOI: 10.15244/pjoes/188236

GoAJK. (2017). *District disaster management plan*. District Disaster Management Authority.

Haddad, N. M., Brudvig, L. A., Clobert, J., Davies, K. F., Gonzalez, A., Holt, R. D., Lovejoy, T. E., Sexton, J. O., Austin, M. P., Collins, C. D., Cook, W. M., Damschen, E. I., Ewers, R. M., Foster, B. L., Jenkins, C. N., King, A. J., Laurance, W. F., Levey, D. J., Margules, C. R., & Townshend, J. R. (2015). Habitat fragmentation and its lasting impact on Earth's ecosystems. *Science Advances*, *1*(2), e1500052. DOI: 10.1126/sciadv.1500052 PMID: 26601154

Hamid, M., Gulzar, A., Dar, F. A., Singh, C. P., Malik, A. H., Kamili, A. N., & Khuroo, A. A. (2023). Microclimate heterogeneity modulates fine-scale edaphic and vegetation patterns on the Himalayan treelines: Implications under climate change. *Agricultural and Forest Meteorology*, *341*, 109688. DOI: 10.1016/j.agrformet.2023.109688

Hamid, M., Khuroo, A. A., Malik, A. H., Ahmad, R., & Singh, C. P. (2020). Assessment of alpine summit flora in Kashmir Himalaya and its implications for long-term monitoring of climate change impacts. *Journal of Mountain Science*, *17*(8), 1974–1988. DOI: 10.1007/s11629-019-5924-7

Hoorn, C., Perrigo, A., & Antonelli, A. (Eds.). (2018). *Mountains, climate and biodiversity*. Wiley Blackwell.

Kaur, D., Tiwana, A. S., Kaur, S., & Gupta, S. (2022). Climate change: Concerns and influences on biodiversity of the Indian Himalayas. In Kaur, D., Tiwana, A. S., Kaur, S., & Gupta, S. (Eds.), *Climate change: Impacts, responses and sustainability in the Indian Himalaya* (pp. 265–281). Springer. DOI: 10.1007/978-3-030-92782-0_13

Kuipers, K. J. J., May, R., & Verones, F. (2021). Considering habitat conversion and fragmentation in characterisation factors for land-use impacts on vertebrate species richness. *The Science of the Total Environment*, *801*, 149737. DOI: 10.1016/j.scitotenv.2021.149737 PMID: 34525717

Manzoor, M., Ahmad, M., Gillani, S. W., Waheed, M., Shaheen, H., Haq, S. M., & Mustafa, A. (2026). Alpine endemism under threat: Ecological patterns and conservation strategies in the changing Himalayas. *Global Ecology and Conservation*, *60*, e04051. DOI: 10.1016/j.gecco.2025.e04051

Manzoor, M., Ahmad, M., Gillani, S. W., Waheed, M., Shaheen, H., Mehmood, A. B., Fonge, B. A., & Al-Andal, A. (2025). Population dynamics, threat assessment, and conservation strategies for critically endangered *Meconopsis aculeata* in alpine zone. *BMC Plant Biology*, *25*(1), 358. DOI: 10.1186/s12870-025-06361-9 PMID: 40102752

Manzoor, M., Shaheen, H., Attique, A., Khan, R. W. A., Gillani, S. W., & Khan, M. S. (2024). Distribution pattern, population structure and habitat geography of *Betula utilis* subalpine forests of Kashmir region in the western Himalayas: Implications for conservation and sustainable management. *Environment, Development and Sustainability*, *26*(1), 1–26. DOI: 10.1007/s10668-024-05049-8

Mittermeier, R. A., Turner, W. R., Larsen, F. W., Brooks, T. M., & Gascon, C. (2011). Global biodiversity conservation: The critical role of hotspots. In Zachos, F. E., & Habel, J. C. (Eds.), *Biodiversity hotspots: Distribution and protection of conservation priority areas* (pp. 3–22). Springer. DOI: 10.1007/978-3-642-20992-5_1

Newbold, T., Hudson, L. N., Hill, S. L. L., Contu, S., Lysenko, I., Senior, R. A., Börger, L., Bennett, D. J., Choimes, A., Collen, B., Day, J., De Palma, A., Díaz, S., Echeverria-Londoño, S., Edgar, M. J., Feldman, A., Garon, M., Harrison, M. L. K., Alhusseini, T., & Purvis, A. (2015). Global effects of land use on local terrestrial biodiversity. *Nature*, *520*(7545), 45–50. DOI: 10.1038/nature14324 PMID: 25832402

Nogués-Bravo, D., Araújo, M. B., Errea, M. P., & Martínez-Rica, J. P. (2007). Exposure of global mountain systems to climate warming during the 21st century. *Global Environmental Change*, *17*(3–4), 420–428. DOI: 10.1016/j.gloenvcha.2006.11.007

Ozinga, W. A., Römermann, C., Bekker, R. M., Prinzing, A., Tamis, W. L. M., Schaminée, J. H. J., Hennekens, S. M., Thompson, K., Poschlod, P., Kleyer, M., Bakker, J. P., & van Groenendael, J. M. (2009). Dispersal failure contributes to plant losses in NW Europe. *Ecology Letters*, *12*(1), 66–74. DOI: 10.1111/j.1461-0248.2008.01261.x PMID: 19016826

Pandit, M. K. (2013). The Himalayas must be protected. *Nature*, *501*(7467), 283. DOI: 10.1038/501283a PMID: 24048033

Parmesan, C., Duarte, C., Poloczanska, E., Richardson, A. J., & Singer, M. C. (2011). Overstretching attribution. *Nature Climate Change*, *1*(1), 2–4. DOI: 10.1038/nclimate1056

Pereira, H. M., Leadley, P. W., Proença, V., Alkemade, R., Scharlemann, J. P. W., Fernandez-Manjarrés, J. F., Araújo, M. B., Balvanera, P., Biggs, R., Cheung, W. W. L., Chini, L., Cooper, H. D., Gilman, E. L., Guénette, S., Hurtt, G. C., Huntington, H. P., Mace, G. M., Oberdorff, T., Revenga, C., & Walpole, M. (2010). Scenarios for global biodiversity in the 21st century. *Science*, *330*(6010), 1496–1501. DOI: 10.1126/science.1196624 PMID: 20978282

Rather, Z. A., Ahmad, R., & Khuroo, A. A. (2022). Ensemble modelling enables identification of suitable sites for habitat restoration of threatened biodiversity under climate change: A case study of *Himalayan Trillium. Ecological Engineering*, *176*, 106534. DOI: 10.1016/j.ecoleng.2021.106534

Rathore, P., Roy, A., & Karnatak, H. (2022). Predicting the future of species assemblages under climate and land use land cover changes in Himalaya: A geospatial modelling approach. *Climate Change Ecology*, *3*, 100048. DOI: 10.1016/j.ecochg.2022.100048

Sah, P., Sharma, S., Latwal, A., & Shaik, R. (2023). Timberline and climate in the Indian western Himalayan region: Changes and impact on timberline elevations. In Shukla, B., Chakraborty, A., & Katiyar, N. K. (Eds.), *Climate change and urban environment sustainability* (pp. 205–225). Springer. DOI: 10.1007/978-981-19-7618-6_12

Sax, D. F., Early, R., & Bellemare, J. (2013). Niche syndromes, species extinction risks, and management under climate change. *Trends in Ecology & Evolution*, *28*(9), 517–523. DOI: 10.1016/j.tree.2013.05.010 PMID: 23769712

Singh, C. P., Mohapatra, J., & Dharaiya, N. (2015). Remote sensing of alpine treeline dynamics. *International Journal of Remote Sensing*, *21*(8), 106–113.

Tittensor, D. P., Walpole, M., Hill, S. L. L., Boyce, D. G., Britten, G. L., Burgess, N. D., Butchart, S. H. M., Leadley, P. W., Regan, E. C., Alkemade, R., Baumung, R., Bellard, C., Bouwman, L., Bowles-Newark, N. J., Chenery, A. M., Cheung, W. W. L., Christensen, V., Cooper, H. D., Crowther, A. R., & Ye, Y. (2014). A mid-term analysis of progress toward international biodiversity targets. *Science*, *346*(6206), 241–244. DOI: 10.1126/science.1257484 PMID: 25278504

Wang, W.-T., Guo, W.-Y., Jarvie, S., & Svenning, J.-C. (2021). The fate of Meconopsis species in the Tibeto-Himalayan region under future climate change. *Ecology and Evolution*, *11*(2), 887–899. DOI: 10.1002/ece3.7096 PMID: 33520173

Wani, S. A., Mugal, M. A., Dar, F. A., Reddy, C. S., Rashid, I., & Khuroo, A. A. (2023). Biodiversity data synthesis on trees of Indian Himalayan Region: Policy and management implications. *Global Ecology and Conservation*, *48*, e02698. DOI: 10.1016/j.gecco.2023.e02698

Chapter 6
Biodiversity of the National Parks and Nurseries in the Alpine Zones

Yasmeen Bibi
http://orcid.org/0009-0007-2366-7501
Government Graduate College for Women, Mianwali, Pakistan

Asma Bibi
University of Mianwali, Punjab, Pakistan

Mujeeb ur Rahman
Shandong University, Weihai, China

Nouman Ahmad
http://orcid.org/0009-0009-3370-2862
University of Mianwali, Punjab, Pakistan

Syed Nasar Shah
http://orcid.org/0000-0002-9002-6110
Quaid-i-Azam University, Islamabad, Pakistan

Abdul Bari Shah
http://orcid.org/0000-0002-2918-7556
Al-Farabi Kazakh National University, Kazakhstan

Fethi Ahmet Ozdemir
Bingol University, Bingol, Turkey

Tewekel Melese Gemechu
Ambo University, Ethiopia

ABSTRACT

Alpine zones are among the most fragile and ecologically significant regions on Earth, harboring unique biodiversity adapted to extreme climatic conditions. This study explored the biodiversity of national parks and nurseries established within alpine zones, emphasizing their role in the conservation of endemic, rare, and threatened species. National parks serve as in situ conservation hubs, preserving intact ecosystems and natural habitats, whereas nurseries act as ex situ conserva-

DOI: 10.4018/979-8-2600-1207-9.ch006

tion centers supporting the propagation, rehabilitation, and restoration of native alpine flora. Together, these initiatives contribute to safeguarding genetic diversity, sustaining ecosystem services, and promoting ecological resilience against climate change and anthropogenic pressures. These findings highlight the importance of integrated conservation strategies and effective management practices to maintain the biodiversity and ecological integrity of alpine regions.

1. INTRODUCTION

Mountainous areas are known as biodiversity hotspots worldwide, with a large number of species and a high density of endemic taxa (Antonelli et al., 2018; Hoorn et al., 2018). Much of this biodiversity is limited to alpine regions characterized by cold weather and unfriendly environmental conditions (Körner & Kèorner, 1999; Rahbek et al., 2019). Alpine ecosystems experience a high concentration of range-restricted species compared to low-elevation ecosystems, which explains their outstanding significance in terms of biodiversity conservation (Barthlott et al., 1996; Spehn et al., 2011). Habitat isolation, environmental heterogeneity, and historical climatic processes have been major factors driving the alpine pattern of biodiversity, leading to high endemism and distinctive biogeographical differentiation (Hobohm, 2014; Tribsch & Schönswetter, 2003).

The current global environmental change has a powerful impact on Alpine biodiversity, and numerous species are under the threat of extinction without the proper improvement of conservation activities (Miller et al., 2006; Pereira et al., 2012). Alpine habitats are ranked among the most threatened ecosystems on a global scale because they are sensitive to climate warming, and the temperature increase experienced, particularly at high altitudes, is disproportionately high (Barry, 2013; Dirnböck et al., 2011; Dullinger et al., 2012). Climate changes, including the extension of growing seasons and the decrease in the intensity of frost, enable the causal movement of species towards the mountains, which creates more competitive pressure in the alpine areas (Benito Alonso & Villar Pérez, 2012; Dullinger et al., 2007; Steger et al., 2013). In addition, the gradual movement of treelines decreases the spatial range of alpine habitats, further endangering cold-adapted species (Hagedorn et al., 2014; Harsch et al., 2009; Kullman, 2002).

Because a very high fraction of endemic and range-restricted species occur in alpine ecosystems, defining biodiversity hotspots and associated conservation gaps has been one of the primary goals of conservation biology (Myers, 1988; Scott et al., 1993). Several studies have indicated that alpine biodiversity hotspots in many cases are not adequately safeguarded, even in those biodiversity hotspots of the world, and are exposed to intense anthropogenic pressures (Mittermeier et al., 2011). One of the key features of this chapter is to underscore the ecological importance of alpine biodiversity, the susceptibility of alpine ecosystems to climate change and land-use pressures, and the need to enhance conservation planning in alpine regions.

The safeguarded zones are important in the preservation of alpine biodiversity, especially in mountain areas with a high elevation amplitude, which may cushion the effects of climate change on endemic and narrowly distributed animal species (Ludovicy et al., 2022). Nevertheless, even though the share of areas under protection in alpine hotspots is higher than in the rest of the regions, the existing level of protection is still below the internationally recommended level and can prove to be inadequate to protect the highly endangered alpine ecosystems (Benito Alonso & Villar Pérez, 2012; Dirnböck et al., 2011; Dullinger et al., 2012). Additional conservation strategies, such as the establishment and control of nurseries, could further support conservation by sustaining and recovering the diversity of alpine flora, particularly in areas where natural systems have been destroyed or damaged. Combined, safeguarded zones and facilitating conservation efforts are necessary to fill conservation gaps and preserve alpine biodiversity in the face of escalating environmental and human impacts (Steinbauer et al., 2018).

2. ECOLOGICAL FEATURES OF ALPINE ZONE

The naturally treeless areas above the montane belt and above the climatic treeline are called alpine regions (Körner & Hoch, 2023; Körner et al., 2011). In contrast to the usual usage of the term alpine to refer to any mountainous region, the biogeographic definition is defined by vegetation and climatic conditions. Alpine regions cover approximately 2.6 percent of the terrestrial surface of the Earth, excluding Antarctica (Körner et al., 2017). Temperature was also found to play a major role in the location of the treeline, which can be determined through climatic data, and the alpine delineation is connected to elevation and the environment (Paulsen & Körner, 2014). These areas occur worldwide, including in the Arctic and temperate mountains and highlands of tropical regions, and show significant differences in climate and biodiversity (Table 1).

Table 1. Ecological Drivers, Biodiversity Trends, and Threats

Ecological Feature	Description	Impact on Alpine Biodiversity	Specific Location Context	Reference
Aerodynamic Decoupling	Low-stature plants trap solar heat, creating a warmer microclimate than the ambient air.	Allows plants to survive and reproduce in air temperatures that would otherwise be too cold.	Global Alpine Zones	(Körner, 2023)
Endemism Rate	The percentage of species found only in a specific region.	76% of subalpine-alpine species in SW Asia are endemic; 80% of these are range-restricted (narrow distribution).	Irano-Anatolian Region	(Noroozi et al., 2023)
Thermal Warming (Aquatic)	Rapid increase in water temperature in high-elevation water bodies.	Summer water temperatures rose by 2.7°C per decade, threatening cold-adapted aquatic life.	Macun Pondscape (Switzerland)	(Fahy et al., 2024)
Cold Stenotherms	Organisms that can only survive in narrow, cold temperature ranges (e.g., *Agabus bipustulatus*).	Currently stable in high elevations but face a "tipping point" if warming continues or lowland species migrate up.	Alpine Ponds	(Fahy et al., 2024)
Land-Use Polarization	The simultaneous intensification of valley agriculture and abandonment of high slopes.	Both trends lead to biodiversity loss: intensification removes habitats; abandonment leads to forest encroachment.	European Alps (1860–2010)	(Tasser et al., 2024)
Clonal Persistence	The ability of alpine plants to reproduce vegetatively and survive for centuries.	Ensures long-term survival of populations even if sexual reproduction fails due to harsh weather.	Global Alpine Flora	(Körner, 2023)
Hydroperiod Fluctuation	Changes in how long ponds hold water (drying out vs. permanent).	Climate change causes permanent ponds to become temporary, stressing species not adapted to desiccation.	Alpine Ponds	(Lamouille-Hébert et al., 2024)
Traditional Ecological Knowledge	Local human knowledge regarding plant properties and management.	Preserves distinct plant communities associated with traditional mowing and grazing regimes.	South Tyrol (Italy)	(Petelka et al., 2020)
Summit Traps	The physical limit of mountain height.	Species migrating upwards to escape heat eventually run out of terrain and face extinction.	General Alpine Context	(Fahy et al., 2024; Noroozi et al., 2023)

Among other features, the climatic conditions of the Alps include low average temperatures, wide diurnal temperature changes, severe winds, and short growing seasons. Nevertheless, the microclimate conditions to which plants are frequently exposed vary significantly from standard weather data because of aerodynamic decoupling and topographic sheltering (Körner, 2021; Scherrer & Koerner, 2010). In sunny conditions, compact forms of plants in cushions or tussocks may reach tens of degrees higher than the ambient air leaf temperature (Dietrich & Körner, 2014; Gauslaa, 1984). Plants are further protected against extreme winter temperatures by snow cover, which enables them to survive at higher elevations than those at which bare tissues would otherwise freeze (Körner & Hiltbrunner, 2018; Scherrer & Koerner, 2010). These microclimatic impacts are critical for interpreting plant life and yield in alpine regions.

Alpine soils are extremely heterogeneous and are influenced by topography, glacial processes, erosion, and snow distribution (Stoecklin & Armbruster, 2016). The amount of moisture, nutrients, and organic matter changes drastically over a few centimeters, forming a mosaic of microhabitats that also select certain plant communities. These microhabitats, combined with the different microbial communities in the soil, greatly assist in the survival of plants under extreme conditions (Körner et al., 2022; Oehl & Körner, 2014). The depth of soil is usually low; hence, most species have to develop deep underground structures to store nutrients and avoid freezing of meristems.

Alpine vegetation has a collection of morphological, physiological, and life-history characteristics that allow it to endure extreme environments.

Life Form and Morphology: Most alpine species are low-growing in form, such as dwarf shrubs, tussock-forming graminoids, rosette herbs, and cushion plants (Billings, 1973; Körner, 1999; Körner, 2021). Shorter heights limit exposure to wind, leaves heat close to the ground, and protect buds and meristems against frost. Clonal prosperity guarantees mechanical durability in regions where reproduction is hazardous (Crivellaro & Büntgen, 2020; Körner, 2021).

- *Microclimate Exploitation:* Topographic positioning and plant architecture allow the exploitation of warmer microclimates closer to the soil surface, elevating leaf temperatures, and proving to be better at photosynthesis even at elevated altitudes (Dietrich & Körner, 2014; Gauslaa, 1984; Körner, 2021).
- *Developmental Controls and Phenology:* The phenological patterns of alpine species are genetically controlled. Growth, flowering, and seed set have a very small seasonal range, so reproduction is achieved despite fluctuating amounts of snowmelt and temperature (Körner et al., 2019; Möhl et al., 2022). Other species, such as *Soldanella pusilla*, begin to grow in the snow months prior to emergence, thus exhibiting an opportunistic response to take advantage of the limited growing season.
- *Clonal Persistence:* A large number of alpine species prefer longevity over fast growth. Clonal taxa can endure centuries as they are present in generations despite harsh environmental changes (Hiltbrunner et al., 2021; Steinger

et al., 1996). This consistency cushions the population against seasonal failures of sexual reproduction.

- *Functional Diversity:* The Alpine flora is morphologically, anatomically, and physiologically diverse. The biomass distribution (between ground and above ground) is only 10 to 90%, photosynthetic efficiency is diverse, and the leaf structure in each species is different, but they are all united by the fact that reduced stature is an adaptive trait (Körner, 2021; Körner & Renhardt, 1987; Patty et al., 2010). Diverse functional species can use different microhabitats, leading to greater ecological stability.
- *Productivity and Resource Use:* Annual productivity can be low because of short growing seasons, but the productivity of the land in years of active growth time can be equal to or even higher than that of the lowland ecosystems (Hiltbrunner et al., 2021; Körner, 2021; Monteiro & Körner, 2013). Effective nutrient recycling, carbohydrate storage, and media-favorable microbial populations allow fast accumulation of biomass in good times. High species diversity also leads to ecosystem stability, stabilizing microhabitats, and becoming resistant to erosion or weather variations (Onipchenko et al., 1998; Vorkauf et al., 2021).

In general, alpine vegetation is not only resistant to stress. They proliferate by exploiting microclimates, clonal endurance, quick yet timed growth, and excellent functional heterogeneity into robust ecosystems in harsh surroundings (Körner, 1999, 2003; Vorkauf et al., 2021).

Alpine ecosystems are diverse ecosystems whose biodiversity differs across different regions of the Alps.

3. BIODIVERSITY OF ALPINE ECOSYSTEMS

Alpine areas are known to be biodiversity hotspots in terms of plants, as they contain a staggering number of species despite occupying only a small percentage of the landmass. An example of this is the European Alps, which boast approximately 4,500 species of plants, constituting approximately 40% of all plant species in Europe (Morschel et al., 2004; Nagy et al., 2012). It has a wonderful diversity of flowers with many endemic species that are adapted to the severe conditions of the mountains, and plants with medicinal qualities that are appreciated by local populations (Zachos & Habel, 2011).

Alpine plants can survive extreme conditions owing to their structural and physiological adaptations. Such adaptations include small growth forms, reproduction via clonal means, and specialized leaves, which minimize the loss of water and endure high UV radiation. Flexible phenology is also observed in many species, including early flowering or delayed seed set, which enables them to take advantage of short growth periods and microclimatic changes (Brondízio et al., 2021; Stöcklin et al., 2007). However, in recent decades (the last 150 years), the variety of alpine plants has decreased, especially in areas with favorable agricultural conditions, where high land-use intensity and mechanization of agriculture have led to a decrease in habitat heterogeneity (Kirchner et al., 2016; Pe'er et al., 2014).

Alpine ecosystems are also home to a diverse number of animals, such as mammals, birds, insects, and microfauna that can be found in steep gradients, low temperatures, and seasonal availability of resources. Alpine pastures are inhabited by mammals such as chamois and marmots, and birds such as alpine accentor and golden eagle subsist in open and rocky environments. Insect populations, such as pollinators and soil arthropods, play important roles in the operation of ecosystems and help in the cycling of nutrients and plant reproduction. Microfauna (nematodes and soil microorganisms) help keep the soil fertile and the ecosystem resilient (Brondízio et al., 2021; Nagy et al., 2012).

Historically, faunal habitats have been transformed by human activities such as agriculture, forestry, and urbanization, and thus fragmented and reduced in certain localities. Animal movement and breeding are further interfered with by infrastructure development, such as roads and ski resorts, which specialize in climate change and invasive species pressures (Kowarik, 2011; Moser et al., 2007).

Alpine ecosystems are based on biodiversity that supports a large scale of ecosystem services (ESs) that are of value to the local and to the rest of the society. High plant and animal diversity has been useful in sustaining provisioning services, including food, fodder, medicinal plants, and timber. Interestingly, in the case of an overall loss of species diversity, the provisioning services in highly managed landscapes have frequently become more intense because of agricultural specialization and technological intensification (Kirchner et al., 2016; Schmithüsen, 2013).

In contrast, regulating and cultural services have tended to decrease. These include soil stabilization, water control, pollination, recreation, and aesthetic values. Scenes with widespread or unproductively utilized land use are more likely to offer more diverse benefits to ecosystems than intensively utilized land, which points to the trade-offs between economic exploitation and environmental sustainability (Brondízio et al., 2021; Pe'Er et al., 2022).

High biodiversity is thus important not only for maintaining the resilience of an ecosystem but also for ensuring the long-term well-being of humans. Economic incentives are supposed to be matched with biodiversity conservation through policy frameworks, including the 2030 Biodiversity Strategy of the European Union and the reforms of the Common Agricultural Policy, which encourage sustainable land use and nature-based solutions (Mann & Lanz, 2013; Pe'Er et al., 2022).

4. ROLE OF NATIONAL PARKS IN ALPINE BIODIVERSITY CONSERVATION

National parks are critical in situ conservation areas that safeguard ecosystems and species within their natural habitats. The establishment and management of protected areas represent one of the most direct strategies to counteract habitat loss, fragmentation, overexploitation of natural resources, climate change, and the spread of invasive species, all of which elevate extinction risk (Bona et al., 2022). In alpine regions, these areas serve as refugia for species adapted to harsh environmental conditions, including low temperatures, seasonal high flow, and unstable substrates. The integrity of headwater and low-order streams, often encompassed by mountain parks, is crucial for maintaining ecological connectivity and protecting sensitive taxa (Bona et al., 2022).

The Gran Paradiso National Park (GPNP) in Italy exemplifies how long-term protection can preserve habitat quality and ecological integrity. Surveys conducted in 2005 and 2020 along 12 high-altitude streams demonstrated minimal changes in hydromorphology. Indices such as the Habitat Quality Assessment (HQA), Habitat Modification Score (HMS), and lentic–Lotic River Descriptor (LRD) showed no significant differences between sampling years (Bona et al., 2022; Bona et al., 2008). The physical and chemical water parameters remained largely stable, with only minor variations in nitrate and pH. Importantly, sites with higher habitat heterogeneity supported richer diatom communities, illustrating that habitat protection directly sustains the species diversity (Bowler et al., 2012; Falasco et al., 2016).

Protected areas, such as the GPNP, also mitigate anthropogenic pressures on lotic ecosystems, which are particularly vulnerable because of their dendritic structure and susceptibility to upstream and downstream impacts. By preserving river headwaters and maintaining natural flow regimes, parks help maintain the functional richness of aquatic communities, fostering resilience to environmental changes (Doretto et al., 2020; Viktória et al., 2021).

National parks play a pivotal role in conserving rare, endemic, and threatened species. In GPNP, benthic communities, including diatoms and macro invertebrates, host numerous taxa listed in the Red List of diatoms (Hofmann et al., 2018). Between 2005 and 2020, sites such as Leviona, Roc, and Vaudaletta contributed the most to local β-diversity, indicating their ecological uniqueness and conservation priority (Legendre & De Cáceres, 2013; Ruhí et al., 2017). Sensitive species such as *Fragilaria arcus, Psammothidium subatomoides*, and several stenoecious macro invertebrates remained present over time, suggesting that park protection effectively buffers them from anthropogenic disturbances.

Community resilience to climate change and extreme events occurs on functional richness, which is promoted by habitat protection and environmental heterogeneity, and national parks are indispensable habitats of the alpine taxa (Mason et al., 2005; Viktória et al., 2021). These parks have headwater areas that could be future refugia in case of warming, especially of cold-adapted species (Sáinz-Bariáin et al., 2016).

An intelligent management of the national parks incorporates the habitat monitoring, long-term data gathering, and interventions which are adaptive. Repeat surveys of stream reaching more than 15 years at GPNP have made it possible to assess hydromorphological change, aquatic biodiversity, and ecological quality indices to make evidence-based management decisions (Bona et al., 2008). Measures like Shannon diversity, local contribution to β- diversity (LCBD) and species contribution to β- diversity (SCBD) are used to determine where to focus conservation efforts and species to focus on conservation (Legendre & De Cáceres, 2013).

Parks conserve the quality of habitat and structure of community by ensuring that the watercourses have a minimal human influence and are connected. The combined approach of benthic monitoring and habitat assessment highlights the value of functional richness as an indicator of ecosystem stability. All these measures show that the national parks are not only the locations of biodiversity but also the instruments that are used in maintaining the processes of ecology in the alpine regions (Bona et al., 2022).

5. ROLE OF NURSERIES IN EX SITU CONSERVATION

Particularly alpine botanical gardens (ABGs) are key centers of ex situ conservation, living collections of plant species that are kept out of their natural environment. These centers supplement in situ plans ensuring the protection of the threatened species of the climate change, habitat fragmentation, and land-use change (Canella et al., 2025). ABGs have merged horticultural skills, ecological studies and propagation experiments in order to grow native, endemic and threatened alpine taxa under predetermined settings that recapitulate their natural micro environments. By so doing, nurseries serve as active repositories of the genetic resources of plants, which can be used both in material and knowledge bases to support the larger conservation programs in future. The Alpine nurseries provide the possibility to propagate the native species with the help of the specific methods which should be developed according to the requirements of their ecology. Live collections permit germination cue experimentation, dormancy-breaking treatments and propagation of species with complicated reproductive needs. The endemics which are vitally endangered (*Isoetes malinverniana* and *Campanula morettiana*) have particular propagation requirements that nurseries are able to reproduce to preserve viable populations in the long run (Abeli et al., 2012; Canella et al., 2025). The resultant propagated people can be used in reintroduction and habitat restoration, which will guarantee the survival of the indigenous species in the restored and damaged locations.

Nurseries are also important in restoration and reinforcing of the ecology by supplying genetically suitable plant materials when conducting translocation and population enhancement. ABGs usually have small demonstration gardens and nurseries with specific alpine conditions, where it is possible to assess the performance of plants in various environmental conditions. This methodology allows restoration practitioners to use evidence-based interventions to strengthen populations in guarded regions or destroyed habitats and decrease the potential of maladaptation (Canella et al., 2025). By so doing, nurseries enhance the gap between the ex situ controlled cultivation and ecological restoration in the field hence effective conservation strategies with changing climatic conditions (Table 2).

In addition to species conservation, nurseries play an important role in protecting genetic diversity within and between populations. ABGs contain living collections of rare, endemic, and locally extinct taxa, many of which are also underrepresented in world germplasm banks (Canella et al., 2025). Nurseries preserve allelic diversity by culturing numerous accessions of a given species; minimize the chances of genetic swamping due to hybridization by selectively restricting pollination, and offer seeds, spores and vegetative tissue that can be stored ex situ over prolonged periods. The combination with seed banks, cryopreservation, and tissue cultures also make sure of the adaptability and the resilience of the alpine flora. Moreover, nurseries promote the study of germination ecology, storage patterns, and propagation regimes, producing the most important data, reinforcing ex situ and in situ conservation models.

Table 2. Conservation Strategies and Management Outcomes in Alpine Zones

Conservation Strategy	Focus Function	Key Outcome Observation	Limitation Challenge Identified	Reference
National Park Protection (Stream Morphology)	Preserving physical habitat structure (hydro-morphology) of high-altitude streams.	In Gran Paradiso National Park (Italy), stream morphology remained excellent with no significant artificial alterations over a 15-year period.	Protection from physical alteration did not prevent biotic homogenization in diatom communities.	(Bona et al., 2023)
Strict Nature Reserve (Pond Monitoring)	Monitoring abiotic conditions in strictly protected, high-elevation pondscapes.	In the Swiss National Park (Macun), human disturbance was successfully minimized (no fishing/pollution).	Strict legal protection failed to stop water temperatures from rising by 4°C due to global atmospheric warming.	(Fahy et al., 2024)
Alpine Botanical Gardens (ABGs)	Ex Situ Conservation: Acting as "living seed banks" and cultivation centers.	14 ABGs across the Alps currently cultivate ~1,800 alpine taxa (32% of the flora), including many rare endemics.	Gaps exist in the collection; some threatened taxa are still missing from global germplasm banks.	(Canella et al., 2025)
Propagation Protocols	Developing methods to grow wild species outside their natural habitat.	Nurseries serve as laboratories to determine germination requirements for species with complex dormancy.	Essential for "recalcitrant" seeds that cannot survive standard freezing/drying in traditional seed banks.	(Canella et al., 2025)
Identification of Hotspots	Mapping biodiversity density to prioritize protected area boundaries.	Identified that 32–98 grid cells in the Irano-Anatolian region qualify as diversity hotspots.	60% of these identified hotspots currently fall outside of protected areas (Conservation Gaps).	(Noroozi et al., 2023)
Cultural Keystone Species (CKS)	Aligning conservation with local cultural and medicinal values.	Focusing on culturally valued plants (e.g., *Arnica montana*) increases local community support for conservation.	Requires balancing commercial demand (harvesting) with ecological sustainability.	(Petelka et al., 2020)
Agri-environmental Subsidies	Using economic policy to influence land-use patterns.	"Green subsidies" in the Alps helped reduce agricultural intensification and maintained species-rich traditional pastures.	Higher subsidies can sometimes lead to land abandonment if not carefully managed.	(Tasser et al., 2024)
Connectivity Management	Maintaining pathways for species dispersal between water bodies.	Essential for allowing species to migrate upward or between ponds as climates warm.	Alpine ponds are naturally fragmented; loss of connectivity threatens species with low dispersal abilities.	(Lamouille-Hébert et al., 2024)

6. INTEGRATED CONSERVATION APPROACHES

The concept of biodiversity conservation demands a multi-layered level of intervention that incorporates the ecological, cultural and social levels. The European Alps and South Tyrol, in particular, offer a very good demonstration of how conservation strategies can work in protecting species, habitats, and local communities simultaneously (Petelka et al., 2020; Schunko et al., 2019).

In situ conservation In South Tyrol, it is primarily focused on the preservation of species in their natural environments (forests, grasslands, and wetlands) (Petelka et al., 2022). Nevertheless, there are culturally and ecologically valuable species which need complementary ex situ measures e.g. *Arnica montana* and *A. montana* is partially covered by wild populations, and the presence of the plant in herb gardens or controlled meadows guarantees its sustainable use without endangering the existence of wild ones (Blachnik & Saller, 2015; Schunko et al., 2019). Through a combination of in situ and ex situ, conservation projects ensure the ecology is not compromised, and locals continue to live well.

Conservation success requires alignment of various stakeholders, such as herb farmers, conventional farmers, environment NGOs, local governments, and pharmaceutical industry (Reed, 2008; Sterling et al., 2017). The power-interest mapping indicates that herb farmers and gastronomy users are very interested and influential at the current time, and the local authorities and residents have lower influence at the moment (Petelka et al., 2022). Research institutions and environmental NGOs can frequently serve as mediators, and in that way, help get conservation goals and cultural and commercial interests to work closely with each other.

The focus on culturally important species (CKS) discovery and management is accompanied by community involvement (Petelka et al., 2022). Herb farmers, pharmacists, historians, and botanists have been interviewed semi-structurally to provide local information on the use of plants, their distribution, and their cultural significance (Petelka et al., 2020; Schunko et al., 2019). Participatory strategies are effective in raising awareness among people, encouraging sustainable harvests, and incorporating traditional ecological knowledge in the conservation planning and restoration of habitats (Berkes et al., 2000; Conservancy et al., 2006).

Regional policies also make biodiversity easier through regulation of wild plant harvesting, promotion of organic farming and by over harvesting endangered species (Petelka et al., 2022). The autonomous system of South Tyrol promotes a balance between cultural, ecological, and economic interests and enables the introduction of interventions that do not violate local traditions but promote the preservation of biodiversity (Alber, 2017; Carlà, 2015). With these types of governance, conservation initiatives structure ecological security with cultural activities, which offer incentives for sustainable management of the ecosystem (Figure 1).

Figure 1. An Integrated Model for Alpine Biodiversity Conservation. This model ties In Situ conservation in National Parks (preservation of intact ecosystems) and Ex Situ conservation in Alpine Nurseries (preservation of genetic diversity). Community involvement provides the cycle with a reinforcement loop that ensures ecological resilience in the face of climate changes and anthropogenic pressures due to the sustainable management of Cultural Keystone Species.

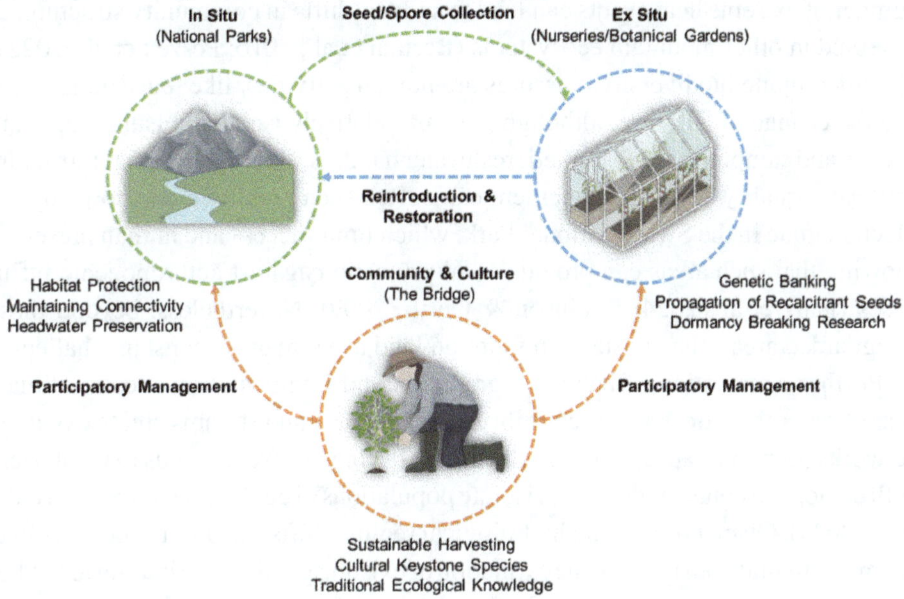

7. MENACE TO ALPINE BIODIVERSITY

Threats to Alpine biodiversity are a complex array of threat agents that exist on a variety of spatial and temporal scales. These dangers are a result of global processes, including climate change, as well as local human-induced stressors, including habitat alteration and the transfer of alien species.

Climate change is one of the major threats to alpine ecosystems. Warming is even faster in high-elevation regions than in the rest of the world, and it is unlikely to avoid causing changes in temperature and precipitation patterns (Dokulil, 2022; Vitasse et al., 2021). Alpine ponds and wetlands, which rely on snowmelt and precipitation, are particularly susceptible. For example, in the Swiss Alps Macun pondscape, the summer water temperatures in high-altitude ponds are rising by 4°C over 15 years, a rapid warming trend that is directly connected to the growth of air temperatures (Fahy et al., 2024; Robinson & Oertli, 2009).

An increase in temperature causes range shifts and changes in the phenology of alpine species. Cold stenothermal species are those species that are limited to cold environments, and as the temperature zones rise, so does their suitable habitat, which reduces. Extinctions in the area could occur when species are confined to peaks and are unable to expand to higher habitats, and specialist species spread to the regions, which may cause biotic homogenization (Fahy et al., 2024). Moreover, the increasing number of extreme heat events can lead to sudden shifts in community structure, as observed in other mountain ecosystems (Bertani et al., 2016; Peszek et al., 2022).

Other alpine biodiversity pressures are human activities, like tourism, grazing, and the change of land-use, although they are relatively isolated. Fishing, off-path hiking, and camping have disrupted freshwater habitats, altering the habitat structure and water quality. Local management can soften such pressures, as shown by the Macun cirque in the Swiss National Park, which limits access and human presence, showing that such areas can provide local protection against anthropogenic influences (Fahy et al., 2024; Robinson & Oertli, 2009). Nevertheless, beyond these safeguarded areas, the constant pressure on land use remains a constant challenge.

Further aggravation of habitat degradation and fragmentation adds to climate change and other stressor susceptibility. Alpine ponds and streams create extended networks (or pondscapes) in which the loss or drying of several ponds may interfere with ecological connectedness and isolate populations of cold-adapted species (Fahy et al., 2024). Glacier retreat and hydrological regime shifts can provide new habitats but may simultaneously cause the extinction of existing ponds, contributing to further population fragmentation and increasing the risk of extinction (Fahy et al., 2024).

Pathogens and invasive species further complicate the problem. Although high-elevation isolation initially constrains colonization, warming leads to the upslope movement of generalist and non-native taxa (Li et al., 2016; Oertli, 2010). These new arrivals can be extremely competitive, change the structure of the community, and bring with them new illnesses. The interaction of the two processes of climate-induced range shifts and invasive biological processes creates a twofold threat to freshwater ecosystems in the Alpines (Figure 2).

Figure 2. Cascading effects of climate change on Alpine biodiversity. This flow chart illustrates the pathway from global warming to local extinction. As temperatures rise, abiotic changes such as treeline advancement and pond drying force lowland generalist species upward. This creates a "Summit Trap" for cold-adapted endemic species, which face increased competition and habitat loss with no remaining higher elevation for refuge, ultimately leading to biotic homogenization.

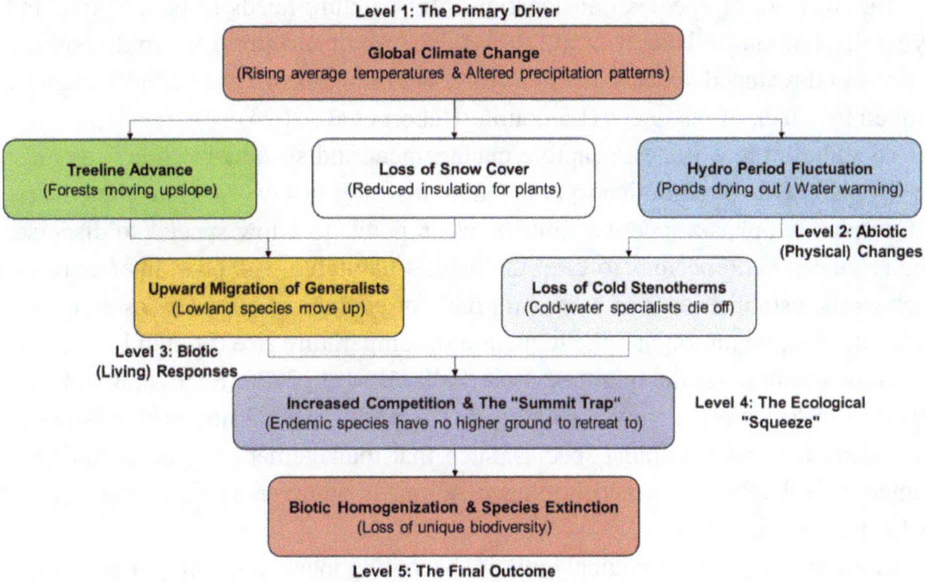

8. CONSERVATION CHALLENGES IN ALPINE PONDSCAPES

Alpine ponds are sensitive to climate change and therefore present a great challenge in terms of conservation because of their inaccessibility and unique ecology. However, there are factors that limit successful management. The most notable constraints are monitoring and research because the alpine ponds are small, remote, and understudied, especially in locations above an altitude of 4000 (m) (Khan & Baig, 2017; Lindholm et al., 2012). The existing literature is not geographically dispersed; the majority of the studies have been carried out in Europe and the Americas, while covered no less than a vast area, including Asia and Africa. Furthermore, the measurements of biodiversity vary in scale and process and it is difficult to generalize the findings and forecast species distribution, community structure and functioning (Lamouille-Hébert et al., 2024).

Resource and infrastructure restraints also limit conservation. Alpine ponds can occur in very rugged and remote areas, which make long-term monitoring, habitat restoration, or assisted movement logistically and financially challenging. These habitats are small, with variable hydroperiods, and as such, water management, habitat creation, or species translocation infrastructure needs to be adaptive and dynamic (Lamouille-Hébert et al., 2024). The ability to provide rapid response actions to threatened species due to drying, warming, or disease outbreaks can be limited by a lack of resources (Lamouille-Hébert et al., 2024).

To address these issues, adaptive management and sustainable practices must be adopted. This practice incorporates new research discoveries into the conservation plan, emphasizes connections between ponds to allow species to disperse, and regulates hydroperiods to keep the habitat habitable. Adaptive interventions, such as the establishment of artificial ponds or ecological corridors as a counter to habitat fragmentation and the reduction of temperature rise through local interventions, where possible, might be used (Pallarés et al., 2020; Sandvik & Odland, 2014). Conservation planning should also consider the dissimilarities between cold specialist and generalist species, such that management practices minimize competitive displacement and homogenization of biodiversity (Lindholm et al., 2012; Redmond, 2018).

Biotechnology and ecological tools are becoming increasingly important for the conservation of alpine ponds. Genetic surveillance can monitor the health of populations, early losses of specialist species, and assisted migration or habitat recovery. Climate change can be predicted using ecological modeling to determine species responses, and remote sensing and automated sensors can also help monitor water temperature, hydroperiod, and connectivity in previously inaccessible ecosystems (Lamouille-Hébert et al., 2024). Together, these tools will make conservation planning more accurate and enable the provision of specific interventions that will help conserve the unique biodiversity of alpine ponds.

9. CONCLUSION AND FUTURE PERSPECTIVES

Alpine zones are ecologically delicate areas with distinctive biodiversity, and the species are well adapted to extreme climatic conditions. This chapter explains the abundance of floral and faunal communities, their adaptation marvels and how national parks and nurseries are needed in conservation. Nurseries offer ex situ conservation, including propagation, genetic conservation, as well as habitat restoration, whereas national parks present a degree of protection to sites, species, ecological processes, and resilience. Such a combination of approaches will lead to endemic, rare, and threatened species being maintained in the long run in the alpine ecosystems. The importance of joint conservation efforts can never be overestimated. Both in situ and ex situ activities, engagement in community and adaptive management are necessary to maintain biodiversity, ecological service and ecosystem resilience in climate change and the relentless anthropogenic pressure. The greatest chances of success in the long-term operation lie in conservation practices which involve scientific research and the possibilities of management and stakeholder involvement. Future research and administration should focus on the enhancement of the adaptive strategies by upgrading the biodiversity monitoring, species correlation, and the scheme services assessment. The alignment of the work within and outside of the borders between the national parks, nurseries, and the local communities will benefit the restoration and protection. As well, the active measures to predict the changes that are caused by the climate changes, habitat fragmentation and forming of new threat sources will be required in the shape of safeguarding the alpine biodiversity. The multi layered approach to the maintenance of these delicate habitats is to maintain a balance between conservation and sustainable use and the ecological viability of such areas and in the forthcoming generation, the alpine biodiversity will have the capacity to survive.

REFERENCES

Abeli, T., Barni, E., Siniscalco, C., Amosso, C., & Rossi, G. (2012). A cost-effective model for preliminary site evaluation for the reintroduction of a threatened quillwort. *Aquatic Conservation*, *22*(1), 66–73. DOI: 10.1002/aqc.1246

Alber, E. (2017). South Tyrol's negotiated autonomy. *The Journal of Ethnic Studies*, *78*, 41–58.

Antonelli, A., Kissling, W. D., Flantua, S. G., Bermúdez, M. A., Mulch, A., Muellner-Riehl, A. N., & Fjeldså, J. (2018). Geological and climatic influences on mountain biodiversity. *Nature Geoscience*, *11*(10), 718–725. DOI: 10.1038/s41561-018-0236-z

Barry, R. G. (2013). *Mountain weather and climate*. Routledge. DOI: 10.4324/9780203416020

Barthlott, W., Lauer, W., & Placke, A. (1996). Global distribution of species diversity in vascular plants: Towards a world map of phytodiversity (globale verteilung der artenvielfalt höherer pflanzen: Vorarbeiten zu einer weltkarte der phytodiversität). *Erdkunde*, •••, 317–327.

Benito Alonso, J. L., & Villar Pérez, L. (2012). Continent-wide response of mountain vegetation to climate change. *Nature climate change*, 1-31.

Berkes, F., Colding, J., & Folke, C. (2000). Rediscovery of traditional ecological knowledge as adaptive management. *Ecological Applications*, *10*(5), 1251–1262. DOI: 10.1890/1051-0761(2000)010[1251:ROTEKA]2.0.CO;2

Bertani, I., Primicerio, R., & Rossetti, G. (2016). Extreme climatic event triggers a lake regime shift that propagates across multiple trophic levels. *Ecosystems (New York, N.Y.)*, *19*(1), 16–31. DOI: 10.1007/s10021-015-9914-5

Billings, W. (1973). Arctic and alpine vegetations: Similarities, differences, and susceptibility to disturbance. *Bioscience*, *23*(12), 697–704. DOI: 10.2307/1296827

Blachnik, T., & Saller, R. (2015). In situ-Vermehrung von Arnica montana–Ergebnisse und Handlungsempfehlungen für die Artenschutz-Praxis. *ANLiegen Natur*, *37*(1), 31–41. DOI: 10.63653/ztzr3600

Bona, F., Bo, T., Doretto, A., Falasco, E., Zoppi, M., & Fenoglio, S. (2022). Are protected areas effective in preserving alpine stream morphology and biodiversity? An experience in the first Italian National Park. *Authorea Preprints*. DOI: 10.22541/au.166987995.50028967/v1

Bona, F., Bo, T., Doretto, A., Falasco, E., Zoppi, M., & Fenoglio, S. (2023). Are protected areas effective in preserving Alpine stream morphology and biodiversity? A field study in the oldest Italian National Park. *River Research and Applications*, *39*(5), 942–953. DOI: 10.1002/rra.4124

Bona, F., Falasco, E., Fenoglio, S., Iorio, L., & Badino, G. (2008). Response of macroinvertebrate and diatom communities to human-induced physical alteration in mountain streams. *River Research and Applications*, *24*(8), 1068–1081. DOI: 10.1002/rra.1110

Bowler, D. E., Mant, R., Orr, H., Hannah, D. M., & Pullin, A. S. (2012). What are the effects of wooded riparian zones on stream temperature? *Environmental Evidence*, *1*(1), 3. DOI: 10.1186/2047-2382-1-3

Brondízio, E., Settele, J., Diaz, S., Ngo, H. T., Experts, G., & Mohamed, A. (2021). *Global assessment report on biodiversity and ecosystem services of the Intergovernmental Science-Policy Platform on Biodiversity and Ecosystem Services.* https://doi.org/DOI: 10.5281/zenodo.383188

Canella, M., Beltran-Sanz, N., Gröger, A., Pungaršek, Š., Rome, M., Natale, S., & Scapin, A. (2025). The role of Alpine botanical gardens in integrating germplasm bank collections and mission. *Plants, People, Planet.*

Carlà, A. (2015). Societal security in South Tyrol: A model to deal with ethnic conflicts. *European Yearbook of Minority Issues Online*, *12*(1), 56–93. DOI: 10.1163/9789004306134_004

Conservancy, S. L., Treasury, C. A., Jackson, R., & Jain, N. (2006). Mountain cultures, keystone species: exploring the role of cultural keystone species in Central Asia. 1-47.

Crivellaro, A., & Büntgen, U. (2020). New evidence of thermally constrained plant cell wall lignification. *Trends in Plant Science*, *25*(4), 322–324. DOI: 10.1016/j.tplants.2020.01.011 PMID: 32191868

Dietrich, L., & Körner, C. (2014). Thermal imaging reveals massive heat accumulation in flowers across a broad spectrum of alpine taxa. *Alpine Botany*, *124*(1), 27–35. DOI: 10.1007/s00035-014-0123-1

Dirnböck, T., Essl, F., & Rabitsch, W. (2011). Disproportional risk for habitat loss of high-altitude endemic species under climate change. *Global Change Biology*, *17*(2), 990–996. DOI: 10.1111/j.1365-2486.2010.02266.x

Dokulil, M. T. (2022). Impacts of climate warming on Alpine lakes. In *Oxford research encyclopedia of climate science*. DOI: 10.1093/acrefore/9780190228620.013.874

Doretto, A., Bona, F., Falasco, E., Morandini, D., Piano, E., & Fenoglio, S. (2020). Stay with the flow: How macroinvertebrate communities recover during the rewetting phase in Alpine streams affected by an exceptional drought. *River Research and Applications*, *36*(1), 91–101. DOI: 10.1002/rra.3563

Dullinger, S., Gattringer, A., Thuiller, W., Moser, D., Zimmermann, N. E., Guisan, A., Willner, W., Plutzar, C., Leitner, M., Mang, T., Caccianiga, M., Dirnböck, T., Ertl, S., Fischer, A., Lenoir, J., Svenning, J.-C., Psomas, A., Schmatz, D. R., Silc, U., & Hülber, K. (2012). Extinction debt of high-mountain plants under twenty-first-century climate change. *Nature Climate Change*, *2*(8), 619–622. DOI: 10.1038/nclimate1514

Dullinger, S., Kleinbauer, I., Pauli, H., Gottfried, M., Brooker, R., Nagy, L., & Benito, J. L. (2007). Weak and variable relationships between environmental severity and small-scale co-occurrence in alpine plant communities. *Journal of Ecology*, *95*(6), 1284–1295. DOI: 10.1111/j.1365-2745.2007.01288.x

Fahy, J. C., Demierre, E., & Oertli, B. (2024). Long-term monitoring of water temperature and macroinvertebrates highlights climate change threat to alpine ponds in protected areas. *Biological Conservation*, *290*, 110461. DOI: 10.1016/j.biocon.2024.110461

Falasco, E., Piano, E., & Bona, F. (2016). Diatom flora in Mediterranean streams: Flow intermittency threatens endangered species. *Biodiversity and Conservation*, *25*(14), 2965–2986. DOI: 10.1007/s10531-016-1213-8

Gauslaa, Y. (1984). Heat resistance and energy budget in different Scandinavian plants. *Ecography*, *7*(1), 5–6. DOI: 10.1111/j.1600-0587.1984.tb01098.x

Hagedorn, F., Shiyatov, S. G., Mazepa, V. S., Devi, N. M., Grigor'ev, A. A., Bartysh, A. A., Fomin, V. V., Kapralov, D. S., Terent'ev, M., Bugman, H., Rigling, A., & Moiseev, P. A. (2014). Treeline advances along the Urals mountain range – driven by improved winter conditions? *Global Change Biology*, *20*(11), 3530–3543. DOI: 10.1111/gcb.12613 PMID: 24756980

Harsch, M. A., Hulme, P. E., McGlone, M. S., & Duncan, R. P. (2009). Are treelines advancing? A global meta-analysis of treeline response to climate warming. *Ecology Letters*, *12*(10), 1040–1049. DOI: 10.1111/j.1461-0248.2009.01355.x PMID: 19682007

Hiltbrunner, E., Arnaiz, J., & Körner, C. (2021). Biomass allocation and seasonal non-structural carbohydrate dynamics do not explain the success of tall forbs in short alpine grassland. *Oecologia*, *197*(4), 1063–1077. DOI: 10.1007/s00442-021-04950-7 PMID: 34047842

Hobohm, C. (2014). *Endemism in vascular plants* (Vol. 9). Springer. DOI: 10.1007/978-94-007-6913-7

Hofmann, G., Lange-Bertalot, H., Werum, M., Klee, R., Kusber, W., Metzeltin, D., & Hajek, G. (2018). Rote Liste und Gesamtartenliste der limnischen Kieselalgen (Bacillariophyta) Deutschlands. *Rote Liste gefährdeter Tiere. Pflanzen und Pilze Deutschlands*, *7*, 601–708.

Hoorn, C., Perrigo, A., & Antonelli, A. (2018). *Mountains, climate and biodiversity*. John Wiley & Sons.

Khan, H., & Baig, S. (2017). High altitude wetlands of the HKH region of northern Pakistan–status of current knowledge, challenges and research opportunities. *Wetlands*, *37*(2), 371–380. DOI: 10.1007/s13157-016-0868-y

Kirchner, M., Schönhart, M., & Schmid, E. (2016). Spatial impacts of the CAP post-2013 and climate change scenarios on agricultural intensification and environment in Austria. *Ecological Economics*, *123*, 35–56. DOI: 10.1016/j.ecolecon.2015.12.009

Körner, C. (1999). Alpine plants: stressed or adapted? Physiological Plant Ecology: 39th Symposium of the British Ecological Society, Körner, C. (2003). Limitation and stress–always or never? *Journal of Vegetation Science*, *14*(2), 141–143.

Körner, C. (2021). *Alpine Plant Life: Functional Plant Ecology of High Mountain Ecosystems*. Springer International Publishing.

Körner, C. (2023). Concepts in Alpine Plant Ecology. Plants 2023, 12, 2666. *Arctic and Alpine Plants*, 5.

Körner, C., Berninger, U.-G., Daim, A., Eberl, T., Mendoza, F. F., Füreder, L., & Meyer, E. (2022). Long-term monitoring of high-elevation terrestrial and aquatic ecosystems in the Alps–a five-year synthesis. *Eco.Mont*, *14*(2), 48–69. DOI: 10.1553/eco.mont-14-2s48

Körner, C., & Hiltbrunner, E. (2018). The 90 ways to describe plant temperature. *Perspectives in Plant Ecology, Evolution and Systematics*, *30*, 16–21. DOI: 10.1016/j.ppees.2017.04.004

Körner, C., & Hoch, G. (2023). Not every high-latitude or high-elevation forest edge is a treeline. *Journal of Biogeography*, *50*(5), 838–845. DOI: 10.1111/jbi.14593

Körner, C., Jetz, W., Paulsen, J., Payne, D., Rudmann-Maurer, K., & Spehn, M., E. (. (2017). A global inventory of mountains for bio-geographical applications. *Alpine Botany*, *127*(1), 1–15. DOI: 10.1007/s00035-016-0182-6

Körner, C., & Kèorner, C. (1999). *Alpine plant life: functional plant ecology of high mountain ecosystems* (Vol. 4). Springer. DOI: 10.1007/978-3-642-98018-3

Körner, C., Paulsen, J., & Spehn, E. M. (2011). A definition of mountains and their bioclimatic belts for global comparisons of biodiversity data. *Alpine Botany*, *121*(2), 73–78. DOI: 10.1007/s00035-011-0094-4

Körner, C., & Renhardt, U. (1987). Dry matter partitioning and root length/leaf area ratios in herbaceous perennial plants with diverse altitudinal distribution. *Oecologia*, *74*(3), 411–418. DOI: 10.1007/BF00378938 PMID: 28312481

Körner, C., Riedl, S., Keplinger, T., Richter, A., Wiesenbauer, J., Schweingruber, F., & Hiltbrunner, E. (2019). Life at 0 C: The biology of the alpine snowbed plant Soldanella pusilla. *Alpine Botany*, *129*(2), 63–80. DOI: 10.1007/s00035-019-00220-8

Kowarik, I. (2011). Novel urban ecosystems, biodiversity, and conservation. *Environmental Pollution*, *159*(8-9), 1974–1983. DOI: 10.1016/j.envpol.2011.02.022 PMID: 21435761

Kullman, L. (2002). Rapid recent range-margin rise of tree and shrub species in the Swedish Scandes. *Journal of Ecology*, *90*(1), 68–77. DOI: 10.1046/j.0022-0477.2001.00630.x

Lamouille-Hébert, M., Arthaud, F., & Datry, T. (2024). Climate change and the biodiversity of alpine ponds: Challenges and perspectives. *Ecology and Evolution*, *14*(2), e10883. DOI: 10.1002/ece3.10883 PMID: 38327685

Legendre, P., & De Cáceres, M. (2013). Beta diversity as the variance of community data: Dissimilarity coefficients and partitioning. *Ecology Letters*, *16*(8), 951–963. DOI: 10.1111/ele.12141 PMID: 23809147

Li, F., Shah, D. N., Pauls, S. U., Qu, X., Cai, Q., & Tachamo Shah, R. D. (2016). Elevational shifts of freshwater communities cannot catch up climate warming in the Himalaya. *Water (Basel)*, *8*(8), 327. DOI: 10.3390/w8080327

Lindholm, M., Stordal, F., Moe, S. J., Hessen, D. O., & Aass, P. (2012). Climate-driven range retraction of an Arctic freshwater crustacean. *Freshwater Biology*, *57*(12), 2591–2601. DOI: 10.1111/fwb.12030

Ludovicy, S., Noroozi, J., Semenchuk, P., Moser, D., Wessely, J., Talebi, A., & Dullinger, S. (2022). Protected area network insufficiently represents climatic niches of endemic plants in a Global Biodiversity Hotspot. *Biological Conservation*, *275*, 109768. DOI: 10.1016/j.biocon.2022.109768

Mann, S., & Lanz, S. (2013). Happy Tinbergen: Switzerland's new direct payment system. *EuroChoices (Uckfield)*, *12*(3), 24–28. DOI: 10.1111/1746-692X.12036

Mason, N. W., Mouillot, D., Lee, W. G., & Wilson, J. B. (2005). Functional richness, functional evenness and functional divergence: The primary components of functional diversity. *Oikos*, *111*(1), 112–118. DOI: 10.1111/j.0030-1299.2005.13886.x

Miller, R. M., Rodríguez, J. P., Aniskowicz-Fowler, T., Bambaradeniya, C., Boles, R., Eaton, M. A., & Walker, S. (2006). Extinction risk and conservation priorities. *Science*, *313*(5786), 441–441. DOI: 10.1126/science.313.5786.441a PMID: 16873627

Mittermeier, R. A., Turner, W. R., Larsen, F. W., Brooks, T. M., & Gascon, C. (2011). Global biodiversity conservation: the critical role of hotspots. In *Biodiversity hotspots* (pp. 3–22). Springer. DOI: 10.1007/978-3-642-20992-5_1

Möhl, P., Von Büren, R. S., & Hiltbrunner, E. (2022). Growth of alpine grassland will start and stop earlier under climate warming. *Nature Communications*, *13*(1), 7398. DOI: 10.1038/s41467-022-35194-5 PMID: 36456572

Monteiro, J. A. F., & Körner, C. (2013). Leaf turnover and herbivory in the tall tussock grass Festuca orthophylla in the Andean Altiplano. *Alpine Botany*, *123*(1), 13–20. DOI: 10.1007/s00035-013-0113-8

Morschel, F., Arduino, S., & Morschel, F. (2004). *The Alps: a unique natural heritage: a common vision for the conservation of their biodiversity*. WWF Germany for the WWF European Alpine Programme.

Moser, B., Jaeger, J. A., Tappeiner, U., Tasser, E., & Eiselt, B. (2007). Modification of the effective mesh size for measuring landscape fragmentation to solve the boundary problem. *Landscape Ecology*, *22*(3), 447–459. DOI: 10.1007/s10980-006-9023-0

Myers, N. (1988). Threatened biotas:" hot spots" in tropical forests. *The Environmentalist*, *8*(3), 187–208. DOI: 10.1007/BF02240252 PMID: 12322582

Nagy, L., Grabherr, G., Körner, C., & Thompson, D. B. (2012). *Alpine biodiversity in Europe* (Vol. 167). Springer Science & Business Media.

Noroozi, J., Minaei, M., Khalvati, S., Kaveh, A., Nafisi, H., Nazari, B., Zare, G., Vitek, E., Hülber, K., & Schneeweiss, G. M. (2023). Hotspots of (sub) alpine plants in the Irano-Anatolian global biodiversity hotspot are insufficiently protected. *Diversity & Distributions*, *29*(2), 244–253. DOI: 10.1111/ddi.13656

Oehl, F., & Körner, C. (2014). Multiple mycorrhization at the coldest place known for Angiosperm plant life. *Alpine Botany*, *124*(2), 193–198. DOI: 10.1007/s00035-014-0138-7

Oertli, B. (2010). The local species richness of Dragonflies in mountain waterbodies: An indicator of climate warming? *BioRisk, 5*, 243–251. DOI: 10.3897/biorisk.5.853

Onipchenko, V. G., Semenova, G. V., & van der Maarel, E. (1998). Population strategies in severe environments: Alpine plants in the northwestern Caucasus. *Journal of Vegetation Science, 9*(1), 27–40. DOI: 10.2307/3237220

Pallarés, S., Millán, A., Mirón, J. M., Velasco, J., Sánchez-Fernández, D., Botella-Cruz, M., & Abellán, P. (2020). Assessing the capacity of endemic alpine water beetles to face climate change. *Insect Conservation and Diversity, 13*(3), 271–282. DOI: 10.1111/icad.12394

Patty, L., Halloy, S. R., Hiltbrunner, E., & Körner, C. (2010). Biomass allocation in herbaceous plants under grazing impact in the high semi-arid Andes. *Flora-Morphology, Distribution. Flora (Jena), 205*(10), 695–703. DOI: 10.1016/j.flora.2009.12.039

Paulsen, J., & Körner, C. (2014). A climate-based model to predict potential treeline position around the globe. *Alpine Botany, 124*(1), 1–12. DOI: 10.1007/s00035-014-0124-0

Pe'er, G., Dicks, L. V., Visconti, P., Arlettaz, R., Báldi, A., Benton, T. G., & Hartig, F. (2014). EU agricultural reform fails on biodiversity. *Science, 344*(6188), 1090–1092. DOI: 10.1126/science.1253425 PMID: 24904142

Pe'Er, G., Finn, J. A., Díaz, M., Birkenstock, M., Lakner, S., Röder, N., & Concepción, E. D. (2022). How can the European Common Agricultural Policy help halt biodiversity loss? Recommendations by over 300 experts. *Conservation Letters, 15*(6), e12901. DOI: 10.1111/conl.12901

Pereira, H. M., Navarro, L. M., & Martins, I. S. (2012). Global biodiversity change: The bad, the good, and the unknown. *Annual Review of Environment and Resources, 37*(1), 25–50. DOI: 10.1146/annurev-environ-042911-093511

Peszek, Ł., Kawecka, B., & Robinson, C. T. (2022). Long-term response of diatoms in high-elevation streams influenced by rock glaciers. *Ecological Indicators, 144*, 109515. DOI: 10.1016/j.ecolind.2022.109515

Petelka, J., Bonari, G., Säumel, I., Plagg, B., & Zerbe, S. (2022). Conservation with local people: Medicinal plants as cultural keystone species in the Southern Alps. *Ecology and Society, 27*(4), art14. DOI: 10.5751/ES-13510-270414

Petelka, J., Plagg, B., Säumel, I., & Zerbe, S. (2020). Traditional medicinal plants in South Tyrol (northern Italy, southern Alps): Biodiversity and use. *Journal of Ethnobiology and Ethnomedicine, 16*(1), 74. DOI: 10.1186/s13002-020-00419-8 PMID: 33243238

Rahbek, C., Borregaard, M. K., Colwell, R. K., Dalsgaard, B., Holt, B. G., Morueta-Holme, N., Nogues-Bravo, D., Whittaker, R. J., & Fjeldså, J. (2019). Humboldt's enigma: What causes global patterns of mountain biodiversity? *Science, 365*(6458), 1108–1113. DOI: 10.1126/science.aax0149 PMID: 31515383

Redmond, L. E. (2018). Alpine limnology of the Rocky Mountains of Canada and the USA in the context of environmental change. *Environmental Reviews, 26*(3), 231–238. DOI: 10.1139/er-2017-0046

Reed, M. S. (2008). Stakeholder participation for environmental management: A literature review. *Biological Conservation, 141*(10), 2417–2431. DOI: 10.1016/j.biocon.2008.07.014

Robinson, C. T., & Oertli, B. (2009). Long-term biomonitoring of alpine waters in the Swiss National Park. *Eco.Mont, 1*, 23–34. DOI: 10.1553/eco.mont1s23

Ruhí, A., Datry, T., & Sabo, J. L. (2017). Interpreting beta-diversity components over time to conserve metacommunities in highly dynamic ecosystems. *Conservation Biology, 31*(6), 1459–1468. DOI: 10.1111/cobi.12906 PMID: 28188969

Sáinz-Bariáin, M., Zamora-Munoz, C., Soler, J. J., Bonada, N., Sáinz-Cantero, C. E., & Alba-Tercedor, J. (2016). Changes in Mediterranean high mountain Trichoptera communities after a 20-year period. *Aquatic Sciences, 78*(4), 669–682. DOI: 10.1007/s00027-015-0457-9

Sandvik, S. M., & Odland, A. (2014). Changes in alpine snowbed-wetland vegetation over three decades in northern Norway. *Nordic Journal of Botany, 32*(3), 377–384. DOI: 10.1111/j.1756-1051.2013.00249.x

Scherrer, D., & Koerner, C. (2010). Infra-red thermometry of alpine landscapes challenges climatic warming projections. *Global Change Biology, 16*(9), 2602–2613. DOI: 10.1111/j.1365-2486.2009.02122.x

Schmithüsen, F. J. (2013). Three hundred years of applied sustainability in forestry. *Working papers/Forest Policy and Forest Economics Department of Forest Sciences. International series, 2013*(1).

Schunko, C., Lechthaler, S., & Vogl, C. R. (2019). Conceptualising the factors that influence the commercialisation of non-timber forest products: The case of wild plant gathering by organic herb farmers in South Tyrol (Italy). *Sustainability (Basel), 11*(7), 2028. DOI: 10.3390/su11072028

Scott, J. M., Davis, F., Csuti, B., Noss, R., Butterfield, B., Groves, C., & Edwards, T. C.Jr. (1993). Gap analysis: A geographic approach to protection of biological diversity. *Wildlife Monographs*, •••, 3–41.

Spehn, E. M., Rudmann-Maurer, K., & Körner, C. (2011). Mountain biodiversity. *Plant Ecology & Diversity*, *4*(4), 301–302. DOI: 10.1080/17550874.2012.698660

Steger, C., Kotlarski, S., Jonas, T., & Schär, C. (2013). Alpine snow cover in a changing climate: A regional climate model perspective. *Climate Dynamics*, *41*(3), 735–754. DOI: 10.1007/s00382-012-1545-3

Steinbauer, M. J., Grytnes, J.-A., Jurasinski, G., Kulonen, A., Lenoir, J., Pauli, H., & Barni, E. (2018). Accelerated increase in plant species richness on mountain summits is linked to warming. *Nature*, *556*(7700), 231–234. DOI: 10.1038/s41586-018-0005-6 PMID: 29618821

Steinger, T., Körner, C., & Schmid, B. (1996). Long-term persistence in a changing climate: DNA analysis suggests very old ages of clones of alpine Carex curvula. *Oecologia*, *105*(1), 94–99. DOI: 10.1007/BF00328796 PMID: 28307127

Sterling, E. J., Betley, E., Sigouin, A., Gomez, A., Toomey, A., Cullman, G., & Blair, M. (2017). Assessing the evidence for stakeholder engagement in biodiversity conservation. *Biological Conservation*, *209*, 159–171. DOI: 10.1016/j.biocon.2017.02.008

Stöcklin, J., Bosshard, A., Klaus, G., Rudmann-Maurer, K., & Fischer, M. (2007). Landnutzung und biologische Vielfalt in den Alpen. *Fakten, Perspektiven, Empfehlungen. Synthese NFP, 48.*

Stoecklin, J., & Armbruster, G. F. (2016). Environmental filtering, not local adaptation of established plants, determines the occurrence of seed-and bulbil-producing Poa alpina in a local flora. *Basic and Applied Ecology*, *17*(7), 586–595. DOI: 10.1016/j.baae.2016.06.004

Tasser, E., Leitinger, G., Tappeiner, U., & Schirpke, U. (2024). Shaping the European Alps: Trends in landscape patterns, biodiversity and ecosystem services. *Catena*, *235*, 107607. DOI: 10.1016/j.catena.2023.107607

Tribsch, A., & Schönswetter, P. (2003). Patterns of endemism and comparative phylogeography confirm palaeo-environmental evidence for Pleistocene refugia in the Eastern Alps. *Taxon*, *52*(3), 477–497. DOI: 10.2307/3647447

Viktória, B., Stenger-Kovacs, C., Török, P., & Török-Krasznai, E. (2021). Are recent protection strategies sufficient for maintaining diverse freshwater benthic diatom assemblages? *Ecological Indicators, 127*, 107782. DOI: 10.1016/j.ecolind.2021.107782

Vitasse, , YUrsenbacher, , SKlein, , GBohnenstengel, , TChittaro, , YDelestrade, , AMonnerat, , CRebetez, , MRixen, , CStrebel, , NSchmidt, , B.RWipf, , SWohlgemuth, , TYoccoz, , N.GLenoir, , J. (2021), Phenological and elevational shifts of plants, animals and fungi under climate change in the European Alps. *Biological Reviews, 96*(5), 1816-1835. https://doi.org/10.1111/brv.12727

Vorkauf, M., Kahmen, A., Körner, C., & Hiltbrunner, E. (2021). Flowering phenology in alpine grassland strongly responds to shifts in snowmelt but weakly to summer drought. *Alpine Botany, 131*(1), 73–88. DOI: 10.1007/s00035-021-00252-z

Zachos, F. E., & Habel, J. C. (2011). *Biodiversity hotspots: distribution and protection of conservation priority areas*. Springer Science & Business Media. DOI: 10.1007/978-3-642-20992-5

Chapter 7

The Ethnobotany of Alpine Vegetation: Traditional Practices and Modern Science

Alevcan Kaplan
http://orcid.org/0000-0001-6738-7527
Sason Vocational School, Turkey

Sana Wahab
Quaid-i-Azam University, Islamabad, Pakistan

Barkat Ullah
Islamia College Peshawar, Pakistan

Abdul Razzaq
Qarshi Herb Research Center, Qarshi Industries (Pvt.) Ltd. Hattar, Haripur, Pakistan & Qarshi University, Lahore, Pakistan

Majid Iqbal
http://orcid.org/0000-0003-0561-8494
Institute of Geographic Sciences and Natural Resources Research, China

Maryam Bibi
Independent Researcher, Pakistan

ABSTRACT

Ethnobotany is valued for its interdisciplinary contributions to natural product research, particularly in, development, and standardization of pharmaceutical and pharmacological products. Alpine plants, often found above the tree line, contribute significantly to the biodiversity of mountain systems and represent an unexplored area that warrants ethnobotanical attention. Although much traditional knowledge has been lost due to various factors, traditional medicinal practices demonstrate that their complementarity with modern medicine should not be overlooked. In today's world, where interdisciplinarity has evolved into a post-disciplinary approach, pluralism in treatment is an undeniable reality. In line with these developments, this chapter examines processes defined and completed through the mutual exchange

DOI: 10.4018/979-8-2600-1207-9.ch007

of information between ethnobotanists and pharmacologists, in the field. One or more aspects of the ethnobotanical use of plants growing in the alpine zone, and their valuable role in pharmacological research serving modern medicine, will be highlighted.

1. INTRODUCTION

Ethnobotany is valued for its interdisciplinary contributions to natural product research for the design, development, and standardization of pharmaceutical and pharmacological products. The mountain system comprises a variety of plants not always found below the tree line, thus providing a significant contribution to the biodiversity of the mountain area and being a pristine field that merits ethnobotanical study. Although much traditional knowledge has been lost in different ways, the continued practice of traditional medicine demonstrates that its complementary relationship with modern medicine cannot be overlooked. Pluralism in treatment is now a reality in the modern world, where interdisciplinarity has become a post-disciplinary approach. In light of these developments, this chapter will examine the processes established and achieved through the sharing of information between ethnobotanists and pharmacologists in the field. One or more elements of ethnobotanical application in the alpine zone of plants will be highlighted, along with their usefulness in pharmacological studies that serve modern medicine.

The historical contribution of plants lies in the intricate ways human settlements have interacted and continue to interact with nature. This cultural tradition results from centuries of close contact with the natural world and encompasses knowledge that combines ecological understanding, health activities, and cultural expression. However, contemporary socio-economic transitions, shifts in land use, and increasing globalization have endangered this knowledge, leading to the progressive disappearance of conventional activities in many areas (Guarrera, 2006). Despite this global trend, other regions that have been less exposed to external influences, especially remote alpine areas, have preserved much of their traditional knowledge. These areas serve as BioRefugia, protecting not only unique plant and animal biodiversity but also the cultural diversity and knowledge systems of human communities (Barthel, Crumley, & Svedin, 2013; La Rosa et al., 2021; Andrea Pieroni, Pawera, & Shah, 2016).

Alpine areas are rugged, high-altitude, regions with harsh climates that have historically posed challenges for human survival. The communities living in these regions have evolved their own mechanisms for coping with environmental restrictions, relying on local plants for medicine and food, among other uses (Barinotto, 2012; Pieroni & Giusti, 2009). Over generations, these communities developed

Traditional Ecological Knowledge (TEK), which encompasses not only the identification and use of plants but also their management, seasonal harvesting, and preparation for various purposes. Alpine TEK has a medicinal and vegetable focus, with many species traditionally used to treat respiratory, digestive, muscle-related, and skin conditions. Societies frequently made infusions, decoctions, syrups and topical preparations using local plants, demonstrating an advanced understanding of plant properties and preparation methods (WHO, 2009). Alpine populations used plants for food, forage, veterinary aid, liquor preparation, and as domestic or craft materials in addition to medicine. Wild and domesticated plants provided primary nutrition, often supplemented with wild fruits, edible fungi, and aromatic herbs used in daily meals, preservation and traditional recipes. For example, ancient alpine communities used wild leaves in soups and salads, made liqueurs and medicinal macerates, and utilized plant products to support the well-being and productivity of animals (Danna, Poggio, Smeriglio, Mariotti, & Cornara, 2022). These practices have remained intact over the centuries because the geographical and socio-cultural isolation of the alpine areas is only relative. The cultural importance of local plants is also evident in rituals and spiritual practices, with some plants holding symbolic or even magical significance, demonstrating that traditional ecological knowledge (TEK) is not only practical but also plays a significant role in cultural identity in the alpine regions. Communities retain strategies for managing resources sustainably by documenting plant uses, thereby maintaining cultural identity as well (Cout, 1999). The preservation of natural resources can be sustained through knowledge of seasonal harvesting, species selection and preparation methods, without depleting the area's biodiversity. Moreover, TEK provides valuable information for creating health remedies and nutritionally significant products, based on centuries of practical experience (Caneva, Pieroni, & Guarrera, 2013).

Quantitative studies in alpine regions show strong consensus among local communities regarding the use of specific plants for particular ailments, with metrics such as Informant Consensus Factor (Fic) and Fidelity Level (FL) demonstrating shared understanding and validation of medicinal plant uses within these communities (WHO, 2009). This implies that traditional ecological knowledge (TEK) is organized and well-unified, rather than haphazard. TEK also covers knowledge about the functions of plants in animal husbandry, such as their use as fodder and in the treatment of livestock diseases, reflecting the interconnectedness of human and animal subsistence (Cout, 1999). Beyond its practical applications, TEK plays a significant role in preserving linguistic and cultural heritage, as the oral transmission of plant names and practices helps strengthen community identity. Nevertheless, this information is increasingly at risk of being lost due to cultural homogenization and modernization, making its documentation an urgent effort to preserve both practical knowledge and intangible heritage (Abbet et al., 2014). To sum up, alpine ethnobotany is a unique manifestation of a dynamic relationship: centuries of experience have produced a rich body of knowledge that is essential for maintaining cultural identity, promoting sustainable resource use, and inspiring new health and management practices.

The main aim of the chapter is to investigate and document traditional knowledge of plant use in alpine areas, focusing on, its cultural, ecological, and medicinal importance. The chapter highlights the importance of ethnobotanical knowledge by demonstrating the importance of traditional ecological knowledge (TEK) as a source of empirical information and its relevance to current pharmacological research and sustainable resource management. Certain objectives include: documenting the diversity of plant use in medicine, food, veterinary, domestic, artisanal, and ritual practices; evaluating how TEK has preserved community identity and contributed to subsistence and biodiversity conservation; examining the relationships between ethnobotanical knowledge and modern science to inform pharmacological research and new product development; and emphasizing the need to conserve TEK in the context of cultural and environmental change. The scope integrates ethnobotanical fieldwork notes, historical data, and quantitative data (such as Fic and FL) to provide a framework for understanding alpine plant knowledge, identifying general trends relevant to high-altitude regions worldwide, and underscoring the universal importance of TEK in connecting the past with the present in scientific research.

2.1 Characteristics of Alpine Ecosystems

Pakistan has Alpine ecosystems that are mostly located in the high mountainous areas of the Himalayas, Karakoram, and Hindu Kush ranges. The conditions of these ecosystems are characterized by harsh climates, including low temperatures, high winds, intense sunlight, and a limited growing season. The terrain is generally hilly with steep slopes, rocky formations, and poor soils. The combination of these factors provides a special environment in which only highly adapted vegetation can survive. Pakistan's alpine zones are typically found above the tree line, between 3,000 and 4,500 meters above sea level, where the climate cannot support large woody plants. Herbaceous plants, shrubs and cushion plants are the main types of vegetation in this area, and they are adapted to conserve moisture, resist frost, and survive in nutrient-poor soils. Seasonal snow cover and glacial melt influence soil moisture regimes, which in turn determine plant distribution and growth patterns. This delicate ecosystem is also important for watershed protection and serves as a habitat for a variety of endemic and specialized species.

2.2 Understorey Major Plant Groups in the Alpine Regions

Several major plant groups are adapted to the extreme conditions found in the alpine vegetation of Pakistan. Herbaceous perennials dominate with representatives from the families Asteraceae, Fabaceae, and Ranunculaceae. Cushion plants are common, growing in small mound shapes to minimize exposure to cold winds and conserve heat. Examples include *Androsace* and *Saxifraga* species. The Poaceae and Cyperaceae families-grasses and sedges are also major components, covering the soil surface and preventing erosion. At lower levels in the Alps, alpine shrubs such as *Juniperus* and *Rhododendron* species can be found, representing the transition zone between subalpine forests and alpine meadows. Medicinal and aromatic plants such as *Saussurea* and *Gentiana*, which have both ecological and economic importance, are also present. The diversity of plant life is influenced by microhabitats created by variations in slope, aspect, and soil moisture, resulting in a mosaic of vegetation types ranging from alpine meadows to rocky scree slopes.

2.3 Endemism and Biodiversity Hotspots

The alpine regions of Pakistan are recognized as biodiversity hotspots due to their high endemism and species richness. The isolation of mountain ranges and the presence of various microclimates contribute to a rich diversity of endemic species not found elsewhere. These endemic species are often specialized to very narrow niches, such as specific soil types or microclimates. Many of these species are concentrated in the western Himalayas and Karakoram ranges, which is why the alpine flora of Pakistan holds international significance. These hotspots are highly valuable and require conservation, as they are endangered by climate change, overgrazing and habitat disturbance. Conserving alpine biodiversity not only preserves individual genetic material but also maintains essential ecosystem services important to local communities, such as water regulation and medicinal plants. Ongoing efforts aim to document and protect these species, with a focus on integrating traditional ecological knowledge and scientific research to achieve sustainable management.

The high intensity of endemism and biodiversity hotspots in the alpine regions of Pakistan is due to the high concentration of unique flora in the isolated and diverse microclimates of the high mountain ranges. The Western Himalayas and Karakoram ranges, which comprise some of the highest mountains and most rugged terrain in the world, depicts an extremely important center of endemism with geographic isolation and ecological heterogeneity (Gillani et al., 2024; Khan et al., 2025).

These forces have led to the speciation and conservation of taxa not found elsewhere. Many endemic species can be found in the alpine areas of Pakistan, particularly within genera such as *Saussurea, Gentiana,* and *Androsace,* which have adapted to soils with low nutrient content, extreme conditions, and low photoperiods (Ali et al., 2019; Kunwar et al., 2022). Indicatively, Saussurea species are renowned for their medicinal value and restricted distribution, often confined to narrow elevational bands. Equally, the endemism of *Gentian* species, which have high pharmacological value, is significant in these mountainous regions.

These alpine regions also have endemic plants, which are characteristic of biodiversity hotspots and indicate high ecological complexity and species richness. This habitat heterogeneity, forming a mosaic from alpine meadows and scree slopes to glacial moraines, allows diverse plants communities to contribute to ecosystem and functional stability (Alrhmoun, Romano, Sulaiman, & Pieroni, 2025). These hotspots are essential for sustaining ecosystem services like water management, soil erosion control, and providing medicinal and forage plants to local people. Nevertheless, these endemic-rich regions are getting more vulnerable due to climate change, which alters temperature and precipitation levels and may shift or reduce the amount of suitable habitat for alpine flora. Moreover, there is increased vulnerability to anthropogenic pressures like uncontrolled collection of medicinal plants, indiscriminate grazing, and habitat fragmentation (Khan et al., 2025). Research on conservation is therefore critical, as it aims to preserve habitats, promote sustainable resource use, and integrate traditional ecological knowledge with scientific research to conserve these rare genetic resources. In conclusion, the alpine endemism and biodiversity hotspots in Pakistan are important repositories of unique plant diversity that have formed due to isolation and highly adverse environmental factors. It is necessary to preserve them not only to ensure biodiversity conservation but also to safeguard and maintain the ecological and socio-economic welfare of mountain communities.

TRADITIONAL ETHNOBOTANICAL PRACTICES IN ALPINE REGIONS OF PAKISTAN

3.1 Historical Context of Plant Use by Indigenous and Local Communities

Indigenous and local communities across Pakistan's alpine zones, including ethnic groups such as the Balti, Shina, and Wakhi, have cultivated a longstanding, adaptive relationship with native flora, shaped by centuries of survival in harsh mountain environments. This tradition reflects in-depth, orally transmitted knowledge of plant ecology, their seasonal availability, and multifunctional uses, which allowed subsistence in remote locations. Traditionally, the use of plants for medicine, food, and material resources became deeply connected with cultural identity and long-term survival strategies. As seen in Table 1, the traditional uses, local names, parts used, and pharmacological effects of important alpine plants are detailed. This rich tradition of ethnobotany highlights the urgent need to preserve both the biological diversity of the flora and the accompanying cultural knowledge systems (Alrhmoun et al., 2025).

3.2 Medicinal Applications of Alpine Flora

Alpine flora in Pakistan is a primary and essential source of traditional medicine for populations that often lack access to modern healthcare. The use of species from genera like *Saussurea*, *Gentiana*, and *Berberis* is guided by indigenous knowledge of their reported anti-inflammatory, antimicrobial, and analgesic effects. For example, *Saussurea costus* is highly valued for its roots, which are used to treat respiratory disorders, while the genus *Gentiana* is used as a digestive and hepatic stimulant. Preparations are typically made as decoctions, infusions, or poultices, with their use overseen by indigenous empirical knowledge regarding dosage and optimal harvesting time to ensure both effectiveness and the sustainability of the plant resource. This strong ethnomedicinal ethic directly contributes to the functional conservation of indigenous plants (Gillani et al., 2024).

Table 1. Traditional uses, local names, parts used and pharmacological effects of some important alpine plants

Plant Scientific Name (Genus/ Species)	Local / Vernacular Name	Part Used	Traditional Medicinal Use (TEK)	Indicated Pharmacological Potential	Citation
Saussurea costus	Koot, Kuth or Kushta	Root	Respiratory ailments	Anti-inflammatory, Antimicrobial	(Gillani et al., 2024)
Gentiana subsp.	Unknown	Root / Whole plant	Digestive and liver stimulant	Cholagogue, Hepatoprotective	(Pieroni et al., 2014)
Berberis subsp.	Unknown	Bark / Fruit	Anti-infective, Diarrhea	Antimicrobial; Berberine-based activity	(Khan, Akhter, & Hussain, 2021)
Hypericum perforatum L.	Johanniskraut	Flowers / Leaves	Wound healing, depressive disorders, gastrointestinal disorders	Anti-inflammatory, Antimicrobial, Antidepressant potential	(Petelka, Plagg, Säumel, & Zerbe, 2020)
Urtica dioica L.	Brennnessel	Leaves / Roots	Diuretic, general tonic, musculoskeletal pain	Anti-inflammatory, Diuretic, Nutritional supplement	(Petelka et al., 2020)

continued on following page

Table 1. Continued

Plant Scientific Name (Genus/ Species)	Local / Vernacular Name	Part Used	Traditional Medicinal Use (TEK)	Indicated Pharmacological Potential	Citation
Alchemilla xanthochlora Rothm.	Frauenmantel	Leaves / Whole plant	Gynecological disorders, general tonic	Anti-inflammatory, wound healing potential	(Grabherr, 2009)
Artemisia absinthium L.	Wermut	Whole plant	Digestive aid, anti-parasitic	Anti-inflammatory, antimicrobial potential	(Grabherr, 2009)

3.3 Food and Nutritional Security

Alpine plants play a vital role in nutritional security, providing essential supplies, especially during lean periods when traditional farming production is limited. Local people also extensively collect edible herbs, wild berries, and roots from alpine meadows, using them as supplementary sources of nutrient-rich wild species in their diets. The Allium family (wild onions) and other berries provide essential vitamins and antioxidants needed by humans at high altitudes, while tuberous plants and wild greens are gathered as sources of carbohydrates and minerals. The management of these seasonal food sources through Traditional Ecological Knowledge (TEK) helps reduce the risk of overexploitation, increase dietary diversity, and strengthen food security in these vulnerable mountain habitats.

3.4 Plants in Ritual, Spirituality, and Cultural Practices

Alpine flora is closely linked with the culture and spirituality of the mountain people in Pakistan, extending beyond sustenance and health. Rituals and ceremonies incorporate specific species to invoke divine blessings, promote cleansing, and affirm paternal affiliation. For example, burning *Juniperus* subsp. as incense during religious and social occasions is associated with purification and a connection to ancestral spirits.

Flowers and herbs are also integral to traditional festivals and rites of passage, reflecting a worldview in which human well-being is deeply connected to the natural environment. These cultural practices embed conservation ethics within spiritual frameworks, reinforcing community cohesion and respect for alpine ecosystems (Gupta, Nishi, & Gasparatos, 2022; Ullah et al., 2019; Zain El Shandidy, 2023).

3.5 Support for Livelihoods (Fodder, Fuel, Dyes, and Materials)

The practicality of alpine vegetation extends well beyond the material sustenance of local livelihoods. Alpine-meadows provide important grasses and herbaceous plants, which serve as essential pastoral fodder during the winter in high-altitude economies. The main sources of fuel for cooking and heating in remote villages are woody shrubs and collected deadwood. Additionally, some plants yield natural dyes and fibers needed in traditional crafts and textiles, supporting artisanal economies. The sustainability of these resources depends largely on established indigenous management strategies that have effectively maintained a balance between usage and natural regeneration (Molnár et al., 2020).

4. KNOWLEDGE TRANSMISSION AND CULTURAL HERITAGE

4.1 Oral Traditions and Folk Knowledge

In northern Pakistan, Traditional Ecological Knowledge (TEK) is deeply embedded in the oral tradition and has served as the main information storage system for local biodiversity and medicinal plants (Abbas, Khan, Alam, Khan, & Abbasi, 2017; Aziz, Ullah, & Pieroni, 2020; Umair, Altaf, Bussmann, & Abbasi, 2019).The wisdom of the Balti and Shina communities of Deosai National Park (DNP) is passed orally through stories, through direct instruction, and practical demonstration. Experienced and older individuals demonstrate the collection, preparation, and use of plants to ensure that vital information is passed on to new generations (Paniagua-Zambrana, Camara-Lerét, Bussmann, & Macía, 2014; Umair et al., 2019).

Oral traditions include phytonyms, local names, preparation methods, and applications in diseases, providing a clear understanding of the medicinal properties of local plants. This method of passing knowledge allows medicinal practices to continue in the absence of written documents, highlighting the resilience of traditional systems in the face of growing globalization.

Oral knowledge is indicated by the various uses of plants to treat common conditions, including gastrointestinal, skin, respiratory, and hepato-pancreatic diseases. The nomenclature and ethnomedical uses such as the differences in the names and uses of *Betula utilis* D. Don and *Bistorta affinis* (D. Don) Greene by Shina communities in both DNP and Gilgit, illustrate the flexible and responsive nature of folk knowledge to ecological and cultural factors.

Additionally, local communities rely on plants for daily healthcare as well as for ritual and spiritual purposes, illustrating the integration of cultural practices with natural resource use. For example, the frequent use of leaves and flowers among the

Balti, influenced by Buddhist traditions, shows how spiritual beliefs shape patterns of plant selection and use (Abbas et al., 2021).

Figure 1. Bridging tradition and science. Traditional plant knowledge informs lab studies and clinical testing for drug discovery

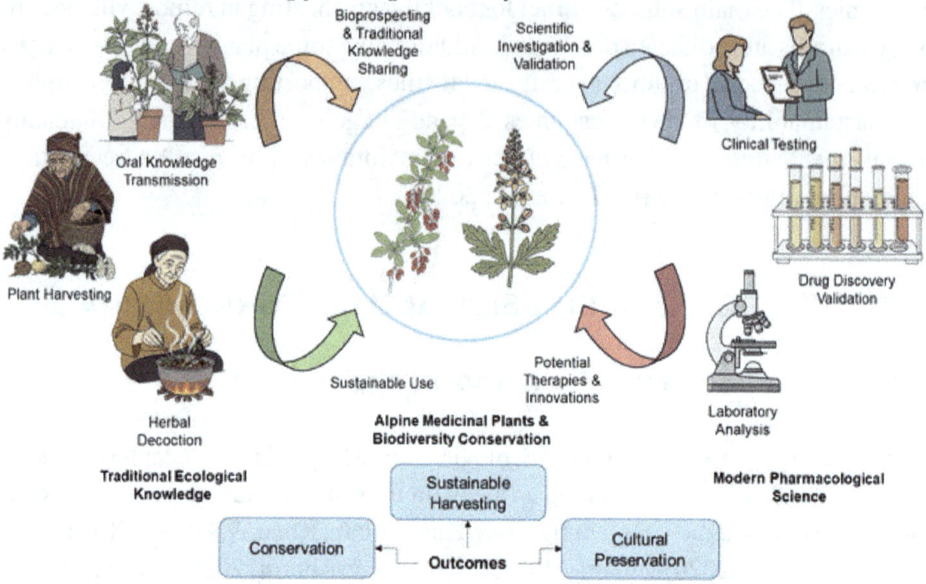

4.2 Role of Elders, Healers, and Local Practitioners

Elders, traditional healers, and local practitioners serve as custodians of TEK, playing a pivotal role in maintaining and disseminating ethnomedicinal knowledge (Paniagua-Zambrana et al., 2014; Umair et al., 2019). In DNP, these individuals guide younger generations in identifying medicinal plants, selecting the most effective parts, and preparing remedies, ensuring that knowledge is accurately transferred and preserved (Paniagua-Zambrana et al., 2014). The role of these custodians is particularly significant in remote, high-altitude villages such as Shila and Dapa for the Balti, and Sadpara and Sherkuli for the Shina, where access to modern healthcare is limited (Abbas et al., 2017; Umair et al., 2019).

Traditional healers and experienced community members reinforce cultural identity by maintaining unique practices, rituals, and spiritual associations with plants. For example, the Balti community's use of flowers and fresh leaves, rooted in Buddhist beliefs, demonstrates the intertwining of medicinal and cultural knowledge. Informal knowledge-sharing networks, including seasonal gatherings, markets, and trade interactions during festivals such as Eid, also facilitate cross-community transmission of TEK. Shepherds and herders, who travel for grazing and trade, contribute to the movement and adaptation of knowledge, creating subtle homogenization and cultural exchange between neighboring communities (Abbas et al., 2021).

4.3 Changes in Knowledge Systems over Time

Despite its historical robustness, traditional ecological knowledge (TEK) in Deosai National Park (DNP) is gradually transforming under multiple social and ecological pressures (Abbas et al., 2017; Aziz et al., 2020; Umair et al., 2019).

The processes of globalization, modernization, and the migration of young people to urban areas in search of education, jobs, and health care have weakened the practice of traditional healing, in some cases leading to the partial erosion of TEK, as young people in these communities increasingly depend on allopathic medicine.

Knowledge systems have also been influenced by cross-cultural contacts. For example, Shina communities that migrated to much of the Balti regions have adopted local Balti plant uses, especially for high-altitude species, although some Shina practices are still preserved (Andrea Pieroni, Giusti, & Quave, 2011). These interactions have created regional variation, which is evident in differences in the medicinal application and vernacular names of plants. The diverse applications of species such as *B. affinis* and *B. utilis* among Shina and Balti people illustrate how TEK responds to ecological availability, migration patterns, and intercultural contact.

The persistence and potential decline of particular knowledge are assessed using quantitative measures like the relative frequency of citation (RFC). The high RFC values for species such as *Ribes alpestre* Wall. ex Decne., *Aconitum violaceum* Jacquem. ex Stapf, and *Thymus linearis* Benth. indicate that some knowledge remains robust, while low values for certain species suggest that traditional ecological knowledge (TEK) may be at risk. Additionally, small-scale commercialization and seasonal harvesting of medicinal plants contribute to both the retention and transformation of traditional practices, highlighting the interaction between socio-economic and ecological factors.

Altogether, knowledge transmission and culture in DNP form a dynamic system: ethnomedicinal knowledge has historically been preserved through oral traditions, elder-led teaching, and community practices. However, recent trends in social change, migration, and modernization are putting pressure on these systems and may alter them over time. This highlights the need for documentation, cross-generational interactions, and biocultural conservation to protect the ecological and cultural aspects of alpine TEK (Abbas et al., 2021).

5. MODERN SCIENTIFIC PERSPECTIVES ON ALPINE ETHNOBOTANY

5.1 Phytochemical and Pharmacological Studies of Alpine Plants

Situated Local Environmental Knowledge (LEK) emphasizes the incorporation of ecological understanding with real-world practice, notably including, the collection and use of medicinal plants. LEK, defined as the knowledge, practices, and beliefs entrenched in particular socio-ecological contexts, serves as a valuable resource for sustainable development and biodiversity preservation (Fontefrancesco & Pieroni, 2020). Historical research in alpine regions has shown that botanical knowledge spans multiple ecosystems, pastures, forests, and high mountain regions, where wild plants have traditionally formed the foundation of local economies and household health prior to widespread access to modern healthcare. This herbal LEK is provably adaptive, changing in response to demographic, economic and cultural changes. Although the repertoire of certain species associated with abandoned pastoral activities has decreased, others, particularly those found in anthropogenic habitats, have been introduced. Furthermore, traditional alpine- plants that are-not-endemic to Mexico, such as *Melissa*, *Taraxacum*, and *Vaccinium*, have gained popularity through herbal books and media (Fontefrancesco & Pieroni, 2020; Sõukand, Stryamets, Fontefrancesco, & Pieroni, 2020). This metamorphosis reaffirms that LEK is dynamic and provides a strong foundation for contemporary phytochemical and pharmacological research, based on the continuity and diversity of plant use.

5.2 Validation of Traditional Uses Through Scientific Research

Diachronic analyses of historical and modern LEK have regularly shown continuity in fundamental herbal practices, as well as the introduction of new plant genera across the regions of the Alps. The process of cultural transmission has changed: while traditional knowledge was once based solely on family and community networks, nowadays LEK is also supplemented by formal education, popular literature, media, and digital sources (Appadurai, 1996; Lave & Wenger, 1991; Mességué, 2017; Sõukand et al., 2020). This demonstrates LEK's ability to adapt, integrate, and maintain the main practices. It, therefore, follows that the process of validating traditional uses is twofold: it must involve not only rigorous chemical or pharmacological testing, but also an in-depth analysis of the social and cultural processes that support the continuation and development of plant knowledge.

5.3 Integrating Ethnobotanical Knowledge with Biodiversity Conservation

LEK is closely linked to local or regional socio-ecological systems and the community's interaction with the environment, which is often shaped by horticultural practices, forest subsistence and pastoral practices. The socioeconomic changes like industrialization, migration, and the replacement of multifunctional agriculture with wage labor have altered landscape interactions and led to changes in the systems of environmental knowledge (Batisti, 2018; Fontefrancesco & Pieroni, 2020).

Despite these changes, LEK is a concrete, strong and versatile resource. This dynamism renders it invaluable for biodiversity conservation and regional planning. LEK can lead participatory planning, contribute to interpreting anthropogenic changes in land-use, and inform specific conservation efforts in forested and elevated regions. Considering LEK as a living system in constant motion, rather than as traditional heritage, would align conservation efforts more closely with current ecological and social realities.

5.4 Success Case Studies of the Modern Application Success

The Upper Sangone Valley is an interesting example of how LEK can adjust to dramatic social, cultural, and economic changes without becoming functionally irrelevant. The amount of knowledge about native wild plants was recorded through historical research, and modern studies have noted a selective loss of certain species, as well as the acquisition of new medicinal genera. Notably, the collection and consumption of boletus mushrooms has become a major cultural and economic activity supported by local festivals and tourism. Moreover, the influence of popular literature and media has successfully expanded the valley's ethnomedicinal repertoire, including the use of species not traditionally used, demonstrating the fluidity of ethnomedicinal knowledge (Fontefrancesco & Pieroni, 2020; Hoad, 2003; Sõukand et al., 2020; Suozzi, 1995; Treben, 2018). Such contemporary incorporation challenges the idea that globalization always leads to the loss of traditional knowledge. The case of the Sangone Valley reveals that LEK can maintain cultural identity while also promoting the sustainable development of rural areas and environmental conservation (Hanazaki, Herbst, Marques, & Vandebroek, 2013; Nygren, 1999; Tsing, 2005; Vandebroek & Balick, 2012).

6. THREATS TO TRADITIONAL ETHNOBOTANICAL KNOWLEDGE

Traditional Ecological Knowledge (TEK) constitutes the environmental knowledge, resource management systems, social institutions, and worldviews of indigenous and local people (Tang & Gavin, 2016). This knowledge is dynamic, locally based and closely associated with social-ecological systems, guiding sustainable resource use, biodiversity protection, and human-environment dynamics (Berkes, Colding, & Folke, 2000; Ross, Sherman, Snodgrass, Delcore, & Sherman, 2016). However, over the past several decades, numerous studies have reported a high rate of TEK loss worldwide, with significant consequences for biodiversity and cultural heritage (Alcorn & Toledo, 1998; Tang, 2012). TEK may be lost due to direct threats, such as human activities that immediately destroy knowledge, and indirect threats, including social, economic, political, and institutional factors that contribute to the loss of TEK.

6.1 Cultural Erosion and Loss of Indigenous Knowledge

Another key cause of TEK erosion is cultural erosion. The disappearance of TEK transmission channels such as intergenerational education and community-level teaching, is well documented. The native languages are vital for transmitting TEK, and their disappearance directly reduces the transfer of traditional knowledge (Tang & Gavin, 2016). For example, local people in Cibecue, Arizona, report that traditional ecological knowledge associated with their language is endangered as the language declines.

The lack of formal education systems that include TEK in the curriculum, combined with young people leaving school or entering the workforce, further reduces the chances for younger generations to experience traditional knowledge (Cruz García, 2006). Survey respondents in New Zealand stressed that Maori youth have less exposure to tikanga (lore) and matauranga (TEK) passed down by the older generation. The interest of indigenous youth in learning traditional practices may be affected by exposure to dominant cultures, mass media, and urban lifestyles (Case, Pauli, & Soejarto, 2005). Cultural erosion is also caused by changes in traditional livelihoods. Market integration, the adoption of wage labor, and reliance on processed or imported products reduce participation in subsistence activities that involve TEK, such as hunting, fishing, and traditional crop farming. For example, communities reported that fishing has become more recreational than subsistence-based, weakening knowledge of aquatic ecosystems and fishing techniques (Reyes-García, Marti, McDade, Tanner, & Vadez, 2007).

Religious and belief systems also shape TEK. Conversion from indigenous spiritual beliefs to dominant religions or secular worldviews can erode traditional ecological values. On the Qinghai-Tibetan Plateau, elders noted a shift from Tibetan Buddhism, which emphasizes respect for all life forms, to secular or Western beliefs among younger generations, altering traditional perceptions of nature and stewardship practices.

6.2 Overharvesting and Unsustainable Practices

TEK is directly endangered by environmental degradation, resource exploitation, and habitat changes. Traditional knowledge is typically rooted in specific ecological contexts, and the loss of particular species or habitats reduces the community's ability to practice and transmit this knowledge (Tang & Gavin, 2016). For example, desertification and forest loss decrease the availability of medicinal plants and culturally significant species, undermining TEK systems.

This is aggravated by the loss of traditional rights, which is mostly caused by the establishment of protected areas or urban growth. The creation of Aberdare National Park in Kenya prohibited traditional hunting practices, limiting local livelihoods and access to resources that support the transmission of knowledge (Agrawal & Ostrom, 2001). Similarly, migration and urbanization may reduce access to traditional lands, and the encroachment of outsiders may introduce new cultural norms that interfere with traditional harvesting practices.

Unsustainable practices can also result from external pressures. Commercially valuable crops or species grown for the market may shift attention away from traditional crops or wild-gathered plants, reducing biodiversity and the knowledge associated with managing these crops (Soemarwoto, 2007).

6.3 Habitat Alteration and Climate Change.

There are additional threats from climate change and environmental transformation. Climate stressors, resource scarcity or government policies can cause forced or voluntary migration, exposing communities to new ecosystems and requiring changes or adaptations to traditional practices (Tang & Gavin, 2010). When pastoral or agricultural societies are relocated, traditional ecological knowledge of specific plant species, seasonal cycles, and ecosystem management can be lost if this knowledge cannot be adapted or transferred to new contexts.

Systemic threats typically increase the impact of direct environmental pressures. These include government policies that undermine indigenous rights, historical colonization, and marginalization by dominant societies, all of which affect the vulnerability of TEK (Tang & Gavin, 2010). For example, colonial education systems often replaced traditional knowledge with Western curricula, overriding traditional practices and spiritual beliefs. Finally, conflict, war, and population decline disrupt access to resources and interrupt critical knowledge transmission, as the loss of elders can erase generations of accumulated ecological wisdom. Migration, whether voluntary or forced, further reduces opportunities to learn and practice TEK and introduces new social norms that may devalue traditional knowledge.

7. BRIDGING TRADITIONAL PRACTICES AND MODERN SCIENCE

The integration of indigenous knowledge and modern science is a significant frontier in ethnobotany and pharmacology. Traditional knowledge, particularly regarding medicinal plants, has long served as a foundation for the discovery of many contemporary drugs (Yuan, Ma, Ye, & Piao, 2016). The origins of many standard medicines can be traced to plant-based solutions used by indigenous peoples, underscoring the value of ethnobotanical data. Nevertheless, because this knowledge is often embedded in vulnerable oral traditions and localized practices, it is rapidly eroding due to modernization, cultural change, and environmental pressures. Reconciling these practices with modern science is essential for developing sustainable drugs, achieving effective biodiversity preservation, ensuring equitable healthcare, and fostering mutual respect for indigenous custodians in cross-cultural contexts (Alum, 2024). This company will conduct collaborative research with local communities by engaging in diverse events and activities that help build connections with them.

7.1 Collaborative Research with Local Communities

This firm will use various events and activities to collaborate with local communities and build relationships with them. The most crucial research methods for working with indigenous communities are collaborative research methods, which help accurately document and ethically safeguard traditional plant knowledge (Kumar, 2025; Pradhan, Mohanty, & Kumar, 2025). Fieldwork in ethnobotany often uses structured interviews, focus groups, and participant observation to carefully record how plants are used, how they are prepared, and their cultural context (Dean, 2024). Community involvement is crucial in this process, as it builds trust, ensures respect for traditional knowledge, and guarantees its recognition and ethical use. Workshops and direct consultations with traditional healers will help substantiate local knowledge with modern scientific discoveries, and ensure reliability and cultural sensitivity (Leonti, 2022). These collaborations can play a pivotal role in reducing the threat of biopiracy and misappropriation of indigenous knowledge, enabling communities to maintain control over their biological resources while contributing to drug discovery and development.

7.2. Sustainable Harvesting and Cultivation: Development

Medicinal plants are considered among the most important for safeguarding biodiversity and ensuring the future provision of necessary resources for generations to come (Dean, 2024). Many species traditionally used are highly threatened by habitat destruction, overharvesting, and environmental degradation.

Ethnobotanical studies provide direct information for developing proper sustainable harvesting and cultivation guidelines (Pradhan et al., 2025). Geo-referencing plants with GPS can map the locations and ecological niches of different species, enabling planners to organize collection processes that minimize environmental harm (Kumar, 2025). Integrating traditional ecological knowledge with modern agricultural and ecological methods ensures responsible plant use, supports conservation, and maintains cultural practices (Alum, 2024). Additionally, adopting standardized cultivation practices helps address variability in the quality of traditional preparations, ensuring consistent efficacy and supporting regulatory compliance for potential commercialization (Gatt, Vella Bonanno, & Zammit, 2024).

7.3 Policies and Legal Frameworks for Protecting Traditional Knowledge

The legal recognition and protection of traditional knowledge (TK) through well-developed legal frameworks are essential to prevent exploitation and to ensure that indigenous populations receive fair benefits from benefit sharing (Reddy & Lakshmikumaran, 2015). TK holders are not usually granted formal intellectual property rights, which means their expertise can be used for commercial gain and biopiracy (Reyes-García, 2010).The international policies, particularly, the Nagoya Protocol, establish a global standard for the fair and equitable distribution of benefits from the use of genetic resources and related knowledge, contributing to ethical research and economic justice (Reddy & Lakshmikumaran, 2015). Effective regulations are needed to build trust among communities, researchers, and pharmaceutical organizations. The formal recognition of TK as intellectual property empowers communities to engage actively in managing their biological assets, receive appropriate compensation, and influence the development of new medicines (Dean, 2024). Such frameworks are necessary to achieve a balance between scientific progress and ethical respect for culture.

7.4 Role of Ethnopharmacology and Bioprospecting

Ethnopharmacology and bioprospecting serve as essential scientific mechanisms connecting traditional knowledge to validated modern medicine (Das et al., 2025). Ethnopharmacological studies systematically validate the therapeutic claims from traditional systems, supplying empirical evidence for efficacy and safety. Landmark discoveries, such as vinblastine and vincristine from *Catharanthus roseus* (L.) G. Don (traditionally used for diabetes and cancer) and artemisinin from *Artemisia annua* Pall. (long used in Chinese medicine for malaria), exemplify this process (Ma, Zhang, Liao, Jiang, & Tu, 2020). Bioprospecting, in turn, involves the systematic exploration of biodiversity to isolate and identify novel bioactive compounds for pharmaceutical development (Domingo-Fernández et al., 2023). By integrating field-based ethnobotanical surveys with rigorous laboratory analysis, researchers can uncover new medicinal leads while respecting the knowledge and rights of source communities (Ishtiaq et al., 2024). This complementary approach promotes healthcare sustainability, supports medicinal plant conservation, and yields culturally informed, scientifically validated, and ethically responsible healthcare solutions.

8. FUTURE DIRECTIONS

8.1 Documenting and Digitizing Traditional Knowledge

The conservation of traditional knowledge, especially alpine ethnobotany, is becoming increasingly urgent as environmental changes, globalization and demographic shifts threaten to erase it. The way forward is systematic documentation and digitization of this knowledge, including collecting detailed information about plant species, their medicinal uses, preparation methods, and local nomenclature. Current technologies like digital databases, georeferenced maps, and multimedia recording can make this information more accessible and accurate, enabling researchers, communities, and policymakers to use it responsibly. In addition, introducing traditional knowledge to online platforms may reduce barriers to intergenerational learning, allowing younger members of local communities to learn about and contribute to their cultural heritage. Community participation is important through collaborative methods that honor intellectual property rights and ensure the knowledge captured accurately reflects local perspectives.

8.2 Improving the Contribution of the Alpine Ethnobotany towards the Sustainable Development

The Alpine ethnobotany has significant potential to contribute to sustainable development, especially in conservation, eco-tourism, and sustainable agriculture. Scientifically proven medicinal and nutritional value of alpine plants can enable communities to produce environmentally friendly products, allowing local communities to contribute to economic growth without harming biodiversity. Incorporating traditional ecological knowledge into contemporary conservation practices can educate sustainable harvesting practices, helping plant populations remain resilient. Another area for future research is the research of climate adaptation methods, where traditional plant management experience can be applied to habitat restoration and resilience in high-altitude regions. Expanding the role of alpine ethnobotanical knowledge in education, public policy, and natural resource management can foster sustainability while supporting the cultural identity and independence of alpine communities.

9. CONCLUSION

Alpine ethnobotany is a significant point of contact between culture, ecology, and traditional knowledge, reflecting centuries of human adaptation to high-altitude environments. This knowledge should be learned and maintained as a valuable aspect of cultural conservation, while also supporting biodiversity and sustainable development. Although contemporary modernization and environmental changes challenge the transmission of local knowledge, traditional knowledge can be preserved by documenting, digitizing, and integrating it with scientific studies to guarantee its sustainability. Partnerships with local communities and the policy frameworks that recognize and protect intellectual and ecological rights can promote responsible use of responsible alpine plant resources. Alpine ethnobotany continues to offer ecological stewardship, healthcare, and sustainable livelihoods by balancing traditional and contemporary scientific approaches, ensuring that the benefits of culture and nature endure for future generations.

REFERENCES

Abbas, Z., Khan, S. M., Alam, J., Khan, S. W., & Abbasi, A. M. (2017). Medicinal plants used by inhabitants of the Shigar Valley, Baltistan region of Karakorum range-Pakistan. *Journal of Ethnobiology and Ethnomedicine*, *13*(1), 53. DOI: 10.1186/s13002-017-0172-9 PMID: 28946889

Abbas, Z., Kousar, S., Aziz, M. A., Pieroni, A., Aldosari, A. A., Bussmann, R. W., Raza, G., & Abbasi, A. M. (2021). Comparative Assessment of Medicinal Plant Utilization among Balti and Shina Communities in the Periphery of Deosai National Park, Pakistan. *Biology (Basel)*, *10*(5), 434. DOI: 10.3390/biology10050434 PMID: 34068859

Abbet, C., Mayor, R., Roguet, D., Spichiger, R., Hamburger, M., & Potterat, O. (2014). Ethnobotanical survey on wild alpine food plants in Lower and Central Valais (Switzerland). *Journal of Ethnopharmacology*, *151*(1), 624–634. DOI: 10.1016/j.jep.2013.11.022 PMID: 24269246

Agrawal, A., & Ostrom, E. (2001). Collective action, property rights, and decentralization in resource use in India and Nepal. *Politics & Society*, *29*(4), 485–514. DOI: 10.1177/0032329201029004002

Alcorn, J. B., & Toledo, V. M. (1998). Resilient resource management in Mexico's forest ecosystems: the contribution of property rights. *Linking Social and Ecological Systems: Management Practices and Social Mechanisms for Building Resilience*, 216-249.

Ali, M., Aldosari, A., Tng, D. Y., Ullah, M., Hussain, W., Ahmad, M., & Sher, H. (2019). Traditional uses of plants by indigenous communities for veterinary practices at Kurram District, Pakistan. *Ethnobotany Research and Applications*, *18*, 1–19. DOI: 10.32859/era.18.24.1-19

Alrhmoun, M., Romano, A., Sulaiman, N., & Pieroni, A. (2025). Old Plants for New Food Products? The Diachronic Human Ecology of Wild Herbs in the Western Alps. *Plants*, *14*(1), 122. DOI: 10.3390/plants14010122 PMID: 39795381

Alum, E. U. (2024). The role of indigenous knowledge in advancing the therapeutic use of medicinal plants: Challenges and opportunities. *Plant Signaling & Behavior*, *19*(1), 2439255. DOI: 10.1080/15592324.2024.2439255 PMID: 39652401

Appadurai, A. (1996). *Modernity at large: Cultural dimensions of globalization* (Vol. 1). U of Minnesota Press.

Aziz, M., Ullah, Z., & Pieroni, A. (2020). Wild food plant gathering among Kalasha, Yidgha, Nuristani and Khowar speakers in Chitral NW Pakistan. [In.]. *Sustainability (Basel)*, *12*(21), 9176. DOI: 10.3390/su12219176

Barinotto, L. (2012). Piante medicinali e di uso tradizionale in Valle Orco.

Barthel, S., Crumley, C., & Svedin, U. (2013). Bio-cultural refugia-Safeguarding diversity of practices for food security and biodiversity. *Global Environmental Change*, *23*(5), 1142–1152. DOI: 10.1016/j.gloenvcha.2013.05.001

Batisti, R. (2018). Espressioni dell'eros infelice in due poeti italiani del nuovo millennio. *TICONTRE*, *10*, 181–202.

Berkes, F., Colding, J., & Folke, C. (2000). Rediscovery of traditional ecological knowledge as adaptive management. *Ecological Applications*, *10*(5), 1251–1262. DOI: 10.1890/1051-0761(2000)010[1251:ROTEKA]2.0.CO;2

Caneva, G., Pieroni, A., & Guarrera, P. (2013). *Etnobotanica: conservazione di un patrimonio culturale come risorsa per uno sviluppo sostenibile*. Edipuglia.

Case, R. J., Pauli, G. F., & Soejarto, D. D. (2005). Factors in maintaining indigenous knowledge among ethnic communities of Manus Island. *Economic Botany*, *59*(4), 356–365. DOI: 10.1663/0013-0001(2005)059[0356:FIMIKA]2.0.CO;2

Cout, F. (1999). *Secret: formule di guarigione in uso in Valle d'Aosta*. Priuli & Verlucca.

Cruz García, G. S. (2006). The mother–child nexus. Knowledge and valuation of wild food plants in Wayanad, Western Ghats, India. *Journal of Ethnobiology and Ethnomedicine*, *2*(1), 39. DOI: 10.1186/1746-4269-2-39 PMID: 16968535

Danna, C., Poggio, L., Smeriglio, A., Mariotti, M., & Cornara, L. (2022). Ethnomedicinal and Ethnobotanical Survey in the Aosta Valley Side of the Gran Paradiso National Park (Western Alps, Italy). *Plants*, *11*(2), 170. DOI: 10.3390/plants11020170 PMID: 35050058

Das, K., Suple, S. D., Hossain, E., Kumar, S., & Kumar, R. (2025). Ethnobotany: a bridge between traditional knowledge and modern pharmacy. In book: Ethnobotany of India, Volume 1Publisher: Ambika Prasad Research Foundation.pp.110-126.

Dean, M. (2024). Exploring ethnobotanical knowledge: Qualitative insights into the therapeutic potential of medicinal plants. *Golden Ratio of Data in Summary*, *4*(2), 154–166. DOI: 10.52970/grdis.v4i2.491

Domingo-Fernández, D., Gadiya, Y., Mubeen, S., Bollerman, T. J., Healy, M. D., Chanana, S., Sadovsky, R. G., Healey, D., & Colluru, V. (2023). Modern drug discovery using ethnobotany: A large-scale cross-cultural analysis of traditional medicine reveals common therapeutic uses. *iScience*, *26*(9), 107729. DOI: 10.1016/j.isci.2023.107729 PMID: 37701812

Fontefrancesco, M. F., & Pieroni, A. (2020). Renegotiating situativity: Transformations of local herbal knowledge in a Western Alpine valley during the past 40 years. *Journal of Ethnobiology and Ethnomedicine*, *16*(1), 58. DOI: 10.1186/s13002-020-00402-3 PMID: 32998765

Gatt, A. R., Vella Bonanno, P., & Zammit, R. (2024). Ethical considerations in the regulation and use of herbal medicines in the European Union. *Frontiers in Medical Technology*, *6*, 1358956. DOI: 10.3389/fmedt.2024.1358956 PMID: 38948354

Gillani, S. W., Ahmad, M., Zafar, M., Manzoor, M., Shah, G. M., Shaheen, H., Zaman, W., Sultana, S., Sadia, B., & Khishlatovna, K. K. (2024). Ethnobotanical Exploration of Traditional Medicinal Plants Among the Rural Inhabitants of District Muzaffarabad, Kashmir Himalayan Region. *Plant Science Today*, *11*(sp1). Advance online publication. DOI: 10.14719/pst.3265

Grabherr, G. (2009). Biodiversity in the high ranges of the Alps: Ethnobotanical and climate change perspectives. *Global Environmental Change*, *19*(2), 167–172. DOI: 10.1016/j.gloenvcha.2009.01.007

Guarrera, P. M. (2006). *Usi e tradizioni della flora italiana: medicina popolare ed etnobotanica*.

Gupta, H., Nishi, M., & Gasparatos, A. (2022). Community-based responses for tackling environmental and socio-economic change and impacts in mountain social–ecological systems. *Ambio*, *51*(5), 1123–1142. DOI: 10.1007/s13280-021-01651-6 PMID: 34784008

Hanazaki, N., Herbst, D. F., Marques, M. S., & Vandebroek, I. (2013). Evidence of the shifting baseline syndrome in ethnobotanical research. *Journal of Ethnobiology and Ethnomedicine*, *9*(1), 75. DOI: 10.1186/1746-4269-9-75 PMID: 24229063

Hoad, J. (2003). *Guarire con le erbe*. MIR.

Ishtiaq, M., Sardar, T., Hussain, I., Maqbool, M., Mazhar, M. W., Parveen, A., Ajaib, M., Bhatti, K. H., Hussain, T., Gul, A., Azeem, M., Khanum, H., Moussa, I. M., Ullah, F., Elansary, H. O., Thind, S., & Sridhara, S. (2024). Traditional ethnobotanical knowledge of important local plants in Sudhnoti, Azad Kashmir, Pakistan. *Scientific Reports*, *14*(1), 22165. DOI: 10.1038/s41598-024-73431-7 PMID: 39333361

Khan, M. S., Khan, S. M., Abdullah, , Liu, J., Wu, Z.-Y., Hussain, J., Zeb, S. A., Mohammad, N., Batool, Z., Saqib, Z., Afza, R., Manan, F., & Ali, S. (2025). Ecological assessment of Iris hookeriana across subalpine and alpine regions of the Hindu-Himalayas. *Frontiers in Forests and Global Change*, *8*, 1539025. DOI: 10.3389/ffgc.2025.1539025

Kumar, S. (2025). Data collection from literature for biological sciences, medicinal plants research, ethnobotany, and pharmacology: A methodological overview. *Journal of Biodiversity and Conservation*, *9*(2), 167–169.

Kunwar, R. M., Baral, B., Luintel, S., Uprety, Y., Poudel, R. C., Adhikari, B., & Poudel, P. (2022). Ethnomedicinal landscape: Distribution of used medicinal plant species in Nepal. *Journal of Ethnobiology and Ethnomedicine*, *18*(1), 34. DOI: 10.1186/s13002-022-00531-x PMID: 35436921

La Rosa, A., Cornara, L., Saitta, A., Salam, A. M., Grammatico, S., Caputo, M., La Mantia, T., & Quave, C. L. (2021). Ethnobotany of the Aegadian Islands: Safeguarding biocultural refugia in the Mediterranean. *Journal of Ethnobiology and Ethnomedicine*, *17*(1), 47. DOI: 10.1186/s13002-021-00470-z PMID: 34321030

Lave, J., & Wenger, E. (1991). *Situated learning: Legitimate peripheral participation.* Cambridge university press. DOI: 10.1017/CBO9780511815355

Leonti, M. (2022). The relevance of quantitative ethnobotanical indices for ethnopharmacology and ethnobotany. *Journal of Ethnopharmacology*, *288*, 115008. DOI: 10.1016/j.jep.2022.115008 PMID: 35066067

Ma, N., Zhang, Z., Liao, F., Jiang, T., & Tu, Y. (2020). The birth of artemisinin. *Pharmacology & Therapeutics*, *216*, 107658. DOI: 10.1016/j.pharmthera.2020.107658 PMID: 32777330

Mességué, M. (2017). *Il Mio Erbario: Con 100 piante illustrate e il prontuario per la cura di oltre 150 patologie*. Macro Edizioni.

Molnár, Z., Kelemen, A., Kun, R., Máté, J., Sáfián, L., Provenza, F., & Máté, A. (2020). Knowledge co-production with traditional herders on cattle grazing behaviour for better management of species-rich grasslands. *Journal of Applied Ecology*, *57*(9), 1677–1687. DOI: 10.1111/1365-2664.13664

Nygren, A. (1999). Local Knowledge in the Environment–Development Discourse:-From dichotomies to situated knowledges. *Critique of Anthropology*, *19*(3), 267–288. DOI: 10.1177/0308275X9901900304

Paniagua-Zambrana, N. Y., Camara-Lerét, R., Bussmann, R. W., & Macía, M. J. (2014). The influence of socioeconomic factors on traditional knowledge: A cross scale comparison of palm use in northwestern South America. *Ecology and Society, 19*(4), art9. DOI: 10.5751/ES-06934-190409

Petelka, J., Plagg, B., Säumel, I., & Zerbe, S. (2020). Traditional medicinal plants in South Tyrol (northern Italy, southern Alps): Biodiversity and use. *Journal of Ethnobiology and Ethnomedicine, 16*(1), 74. DOI: 10.1186/s13002-020-00419-8 PMID: 33243238

Pieroni, A., & Giusti, M. E. (2009). Alpine ethnobotany in Italy: Traditional knowledge of gastronomic and medicinal plants among the Occitans of the upper Varaita valley, Piedmont. *Journal of Ethnobiology and Ethnomedicine, 5*(1), 32. DOI: 10.1186/1746-4269-5-32 PMID: 19895681

Pieroni, A., Giusti, M. E., & Quave, C. L. (2011). Cross-cultural ethnobiology in the Western Balkans: Medical ethnobotany and ethnozoology among Albanians and Serbs in the Pešter Plateau, Sandžak, South-Western Serbia. *Human Ecology: an Interdisciplinary Journal, 39*(3), 333–349. DOI: 10.1007/s10745-011-9401-3

Pieroni, A., Nedelcheva, A., Hajdari, A., Mustafa, B., Scaltriti, B., Cianfaglione, K., & Quave, C. L. (2014). Local knowledge on plants and domestic remedies in the mountain villages of Peshkopia (Eastern Albania). *Journal of Mountain Science, 11*(1), 180–193. DOI: 10.1007/s11629-013-2651-3

Pieroni, A., Pawera, L., & Shah, G. M. (2016). Gastronomic ethnobiology. In *Introduction to ethnobiology* (pp. 53–62). Springer. DOI: 10.1007/978-3-319-28155-1_9

Pradhan, S., Mohanty, J., & Kumar, S. (2025). Ethnomedicinal Plants Used by Santhal Community of India. *Annals of Agri Bio Research, 30*(1), 86–97. DOI: 10.53941/agrbio.2025.1000011

Khan, , R. S. MAkhter, , THussain, , M. (2021). Ethno-veterinary practice for the treatment of animal diseases in Neelum Valley, Kashmir Himalaya, Pakistan. *PLoS One, 16*(4), e0250114. DOI: 10.1371/journal.pone.0250114 PMID: 33930052

Reddy, P., & Lakshmikumaran, M. (2015). Protecting traditional knowledge related to biological resources: Is scientific research going to become more bureaucratized? *Cold Spring Harbor Perspectives in Medicine, 5*(10), a020974. DOI: 10.1101/cshperspect.a020974 PMID: 26101205

Reyes-García, V. (2010). The relevance of traditional knowledge systems for ethnopharmacological research: Theoretical and methodological contributions. *Journal of Ethnobiology and Ethnomedicine*, *6*(1), 32. DOI: 10.1186/1746-4269-6-32 PMID: 21083913

Reyes-García, V., Marti, N., McDade, T., Tanner, S., & Vadez, V. (2007). Concepts and methods in studies measuring individual ethnobotanical knowledge. *Journal of Ethnobiology*, *27*(2), 182–203. DOI: 10.2993/0278-0771_2007_27_182_camism_2.0.co_2

Ross, A., Sherman, K. P., Snodgrass, J. G., Delcore, H. D., & Sherman, R. (2016). *Indigenous peoples and the collaborative stewardship of nature: knowledge binds and institutional conflicts*. Routledge. DOI: 10.4324/9781315426617

Soemarwoto, R. (2007). Kasepuhan Rice Landrace Diversity, Risk Management and Agricultural. *Modern crises and traditional strategies: local ecological knowledge in island Southeast Asia, 6*, 84.

Sõukand, R., Stryamets, N., Fontefrancesco, M. F., & Pieroni, A. (2020). The importance of tolerating interstices: Babushka markets in Ukraine and Eastern Europe and their role in maintaining local food knowledge and diversity. *Heliyon*, *6*(1), e03222. DOI: 10.1016/j.heliyon.2020.e03222 PMID: 32095643

Suozzi, R. M. (1995). Dizionario delle erbe medicinali: oltre duecentocinquanta specie vegetali esaminate nei loro effetti curativi e nella loro utilizzazione pratica: Grandi tascabili economici Newton.pp.1-416.

Tang, R. (2012). *An analysis of traditional ecological knowledge's status and its conservation options.* PhD Thesis. Victoria University of Wellington.pp.1-397.

Tang, R., & Gavin, M. C. (2010). Traditional ecological knowledge informing resource management: Saxoul conservation in inner Mongolia, China. *Society & Natural Resources*, *23*(3), 193–206. DOI: 10.1080/08941920903375503

Tang, R., & Gavin, M. C. (2016). A classification of threats to traditional ecological knowledge and conservation responses. *Conservation & Society*, *14*(1), 57–70. DOI: 10.4103/0972-4923.182799

Treben, M. (2018). *La salute dalla farmacia del Signore*. Ennsthaler.

Tsing, A. (2005). *Friction: An anthropology of global connection*. Princeton University Press.

Ullah, M., Mehmood, S., Ali, M., Bussmann, R. W., Aldosari, A., Khan, R. A., & Shah, M. A. R. (2019). An ethnopharmacological study of plants used for treatment of diabetes in the Southern and Tribal regions of Khyber Pakhtunkhwa province, Pakistan. *Ethnobotany Research and Applications*, *18*, 1–20. DOI: 10.32859/era.18.8.1-20

Umair, M., Altaf, M., Bussmann, R. W., & Abbasi, A. M. (2019). Ethnomedicinal uses of the local flora in Chenab riverine area, Punjab province Pakistan. *Journal of Ethnobiology and Ethnomedicine*, *15*(1), 7. DOI: 10.1186/s13002-019-0285-4 PMID: 30709360

Vandebroek, I., & Balick, M. J. (2012). Globalization and loss of plant knowledge: Challenging the paradigm. *PLoS One*, *7*(5), e37643. DOI: 10.1371/journal.pone.0037643 PMID: 22662184

WHO. (2009). International Classification of Diseases-ICD. German Version. Bern: Huber.

Yuan, H., Ma, Q., Ye, L., & Piao, G. (2016). The traditional medicine and modern medicine from natural products. *Molecules (Basel, Switzerland)*, *21*(5), 559. DOI: 10.3390/molecules21050559 PMID: 27136524

Zain El Shandidy, M. (2023). The power of intangible heritage in sustainable development. *International Journal of Advanced Studies in World Archaeology*, *6*(2), 92–97. DOI: 10.21608/ijaswa.2024.275163.1043

Chapter 8
Alpine Medicinal Plant Strategies and Their Role in Conservation

Hansa Gul
http://orcid.org/0009-0001-5554-617X
University of Mianwali, Punjab, Pakistan

Zahida Nasreen
http://orcid.org/0009-0002-0995-5041
University of Mianwali, Punjab, Pakistan

Muhammad Nauman Khan
http://orcid.org/0000-0003-3790-0902
University of Chakwal, Punjab, Pakistan

Samina Siddiqui
University of Peshawar, Peshawar, Pakistan

Syed Nasar Shah
http://orcid.org/0000-0002-9002-6110
Quaid-i-Azam University, Islamabad, Pakistan

Sarah Abdul Razak
http://orcid.org/0000-0003-3364-936X
Universiti Malaya, Kuala Lumpur, Malaysia

Fethi Ahmet Ozdemir
Bingol University, Bingol, Turkey

Alevcan Kaplan
http://orcid.org/0000-0001-6738-7527
Sason Vocational School, Batman University, Batman, Turkey

ABSTRACT

Alpine medicinal plants make an essential interchange between biodiversity, cultural traditions, and pharmacological potential. Prospering in alpine areas these species developed special adaptive features and bioactive compounds interesting to traditional healing systems. Their survival is, however, threatened by climate change, habitat fragmentation, over-harvesting, and unsustainable land use. In this chapter, the ecological and pharmacological importance of alpine vegetation

DOI: 10.4018/979-8-2600-1207-9.ch008

is analyzed critically, with an understanding of the fact that there is an impending need of putting together conservation measure. It examines not only in-situ methods, including habitat protection, community-based stewardship and practices of sustainable harvest, but also ex-situ methods, including seed banks, botanic gardens, and technologies of tissue culture. Furthermore, it proposes a blend between the indigenous knowledge systems and scientific innovation such that these valuable plant resources may be long term preserved and wisely utilized.

1 INTRODUCTION

1.1 The Dual Crisis in Pakistani Alpine Ecosystems

The Pakistani alpine region including the hugely impressive Hindu Kush, Karakoram, and Himalayan, are worldly noteworthy repositories of biodiversity and an arsenal of medicinal plants. The indigenous communities have utilized this botanical wealth as primary healthcare since centuries creating complex systems of Traditional Ecological Knowledge (TEK) across generations. But this fragile symbiosis between humans and flora is now in crisis in two ways: the loss of traditional knowledge is increasingly fast and the numbers of medicinal plants are rapidly becoming extinct. Here, the same scenario is being replicated in the urbanized societies, except in that the ecological vulnerability of the mountain communities is compounded by the fact that they are distinct (Shaheen et al., 2023).

These dangers are combined in the alpine regions. The biggest direct threat is over-harvesting to commercial trade whereby high-value species like *Aconitum heterophyllum*, *Picrorhiza kurrooa* and *Saussurea costus* are harvested in unsustainable amounts (Thapliyal, Bhojak, Mehta, & Sekar, 2024). This results in what has come to be known as a value paradox: in many cases, the plants whose medicinal and economic value is the most significant face the greatest threat of either local or even commercial extinction. This is compounded by habitat degradation due to over grazing, the development of infrastructures as well as the increasing effects of climate change. This endemic medicinal plant, prized in traditional medicine for treating ailments from bone fractures to chronic pain, is now classified as critically endangered in the alpine zones of Kashmir. A recent population study confirmed its dwindling numbers, with an average density as low as 20.6 plants per hectare, and identified a convergence of threats including over-exploitation, intense grazing, and habitat degradation (Manzoor et al., 2025). The plight of *M. aculeata* serves as a stark, on-the-ground example of the very crisis this chapter addresses. Recent ethnobotanical research in the Kashmir Himalayas confirms that a number of these key species are now considered critically endangered or rare specifically due to overuse, including *Aconitum heterophyllum*, *Betula utilis*, *Delphinium cashmerianum*, *Fritillaria roylei*, *Podophyllum hexandrum*, and *Saussurea costus* (Haq et al., 2024). Although extinctions on a global scale are extremely low, local, national or regional extinction, or commercial extinction ought not to be underestimated as it can have grave effects on livelihood and economies.

These two crises interrelate in a vicious cycle: TEK on sustainable practices is declining, and harvesting is more destructive because of it, therefore, decreasing the availability of plant resources, which in turn devalues the traditional knowledge system and leads to the communities to the reliance on less sustainable livelihoods (Figure 1). A sub-alpine and alpine Palas Valley of Kohistan study by Kayani et al. (2024) recorded the rich local knowledge on 102 medicinal plant species of 48 families expertise, highlighting the heavy dependence on the botanical richness of the local community (Kayani, Ahmad, Gillani, Muhammad, et al., 2024).

Figure 1. The Vicious Cycle Endangering Pakistani Alpine Medicinal Plants and Traditional Knowledge.

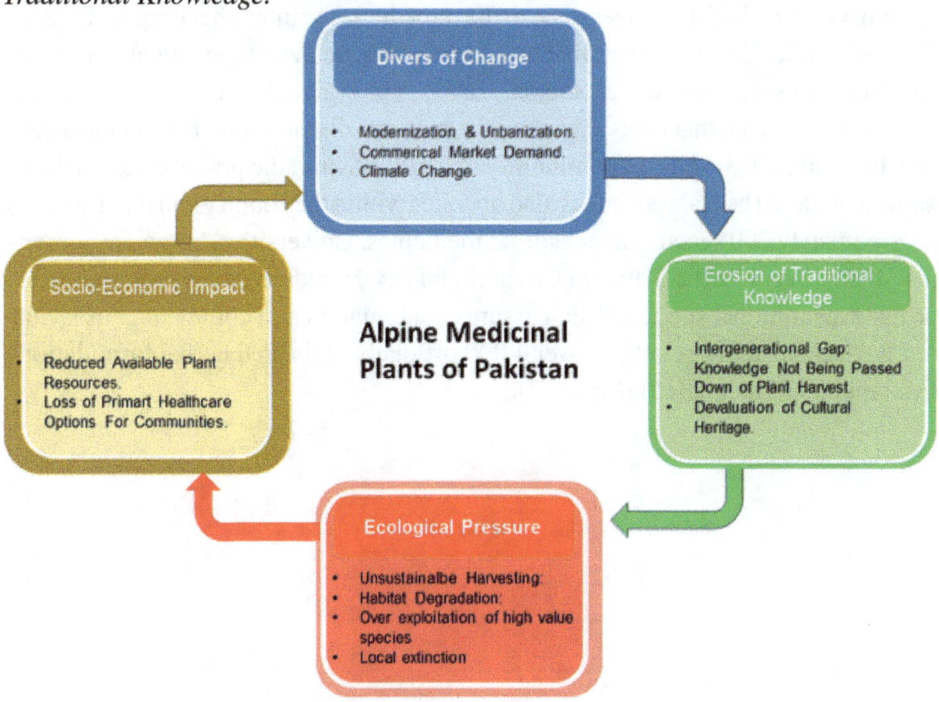

1.2 Global and National Process of Herbal Medicine

In the majority of countries, herbal remedies are a source of primary healthcare employed by almost 75 to 80% of the population (WHO, 2015). This reliance is particularly high in countries within the Himalayan region, where traditional medicine is used by an estimated 80% of the population in India, 85% in Myanmar, and 90% in Bangladesh (Shinwari et al., 2024). According to the estimations by World Health Organization (WHO), approximately 74% of 119 modern drugs that are of plant origin are used in a manner that aligns with their traditional use (Singh, Husain, Agnihotri, Pande, & Khatoon, 2014). This is a continuity of herbal medicine, which has been used over centuries because it is locally available, affordable and considered to be safe in comparison to synthetic drugs. Bioactive compounds are abundant in medicinal plants and most of them are comparable in effectiveness to allopathic medicines.

In South Asia, Pakistan is ranked the seventh in the world in terms of medicinal plant diversity with close to 600 plant species used as part of the traditional healing system. Over 75% of the citizens use these remedies as primary healthcare. Outside the medical field, other sectors that the plants serve are the nutraceutical, cosmetic, and food industries (Kanwal & Sherazi, 2017). The regional markets called Pansar, facilitate local trade that deals with the exchange and distribution of medicinal herbs (Rashid et al., 2018). A large community of 50,000 Ayurvedic practitioners, hakims and unregistered herbal specialists also operates within the country and together uses approximately 200 species in the natural medicines. Universities, botanical gardens and institutions like the National Council of Tibb, Hamdard and Qarshi industries provide institutional support which assures continuation and commercial use. Currently, the Pakistani market has over 4,000 officially registered herbal formulations (Mehmood & Shah, 2015) (Figure 2).

Figure 2. Significance of Medicinal Plants in Alpine- Ecosystem

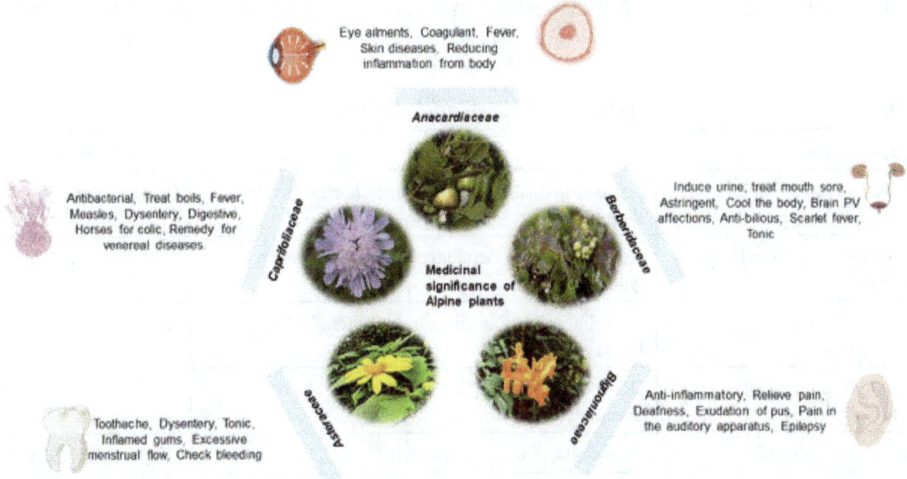

1.3 Ecological and Pharmacological Relevance of Alpine Pharmaceutical Plants.

Both the need of herbal drugs and plant-based health products that are in demand all over the world has increased stress on the natural resources. It is known that around 90% of medicinal plants in Europe are gathered wildly, and there is only minor cultivation in order to restore stocks (Balunas & Kinghorn, 2005). Notably, it is approximated that the globe is losing at least one potentially useful drug every two years due to the extinction of plants with the present rate of loss being 100-1000 times higher than the natural background rate of extinction (Pimm, Russell, Gittleman, & Brooks, 1995). According to reports by the international union conservation nature (IUCN) and world wildlife fund (WWF), out of 50,000-80,000 medicinal species identified in the world, almost 15,000 are facing the risk of extinction owing to unsustainable harvesting and destruction of their habitats (Bentley & Trimen, 1880). Fully 20 percent of wild populations are already critically depleted, and this has mainly happened because of the growing human population and the escalating demands of consumption (Ross, 2005).

The acuteness of this crisis is manifested throughout the globe: China, India, Kenya, Nepal, Tanzania, and Uganda (Ross, 2005), are by no means the only states that registered a steep drop in medicinal plants supply. All these tendencies emphasize the necessity to start using sustainable management practices, enhance conservation efforts, and incorporate traditional knowledge into the biodiversity conservation policies to guarantee the sustainable existence of alpine medicinal plants (Shaheen et al., 2023). Table 1 has shown various applications of alpine medicinal plants.

Table 1. Various alpine medicinal plants and their uses

Botanical Name	Uses	References
Achellia millefolium L.	Gastrointestinal disorders, Stomachache, Headache, Wounds, Fever	(Kayani et al., 2015)
Aconitum violaceum Jacquem. ex Stapf.	Abdominal pain, flatulence (Balti); Ringworm, typhoid (Shina)	(Abbas et al., 2021)
Actea spicata L.	Snakebites, Nervous disorders, Goiter, Asthma, Joints pain, Fever	(Kayani et al., 2015)
Anaphalis margaritacea (L.) Benth.	Diarrhea, Pulmonary infections, Burns, Sores, Ulcers, Headaches	
Apium graveolens L.	Tonic, relieve pain, Induce urine	(Kayani, Ahmad, et al., 2024b)
Artemisia absinthium L.	Carminative, Cold, Fever, Antiseptic, Purgative, Anti-malarial, Expel intestinal worms	
Astragalus grahamianus Royle ex Benth.	Throat diseases	(Kayani et al., 2015)
Delphinium brunonianum Royle.	Hair tonic (Balti); Asthma, pneumonia, hair tonic (Shina)	(Abbas et al., 2021)
Geranium wallichianum D. Don ex Sweet.	Backache, Mouth ulceration, Chronic diarrhea	(M. Ahmad et al., 2014b)
Hedera nepalensis K. Koch.	Pulmonary infections, Fever, Antidiabetic, Joint pain, Anticancerous, Increase perspiration	(Kayani, Ahmad, et al., 2024b)
Origanum vulgare L.	Skin infections, Sexual weakness, Digestive disorders, Intestinal pain, Urinary disorders	(M. Ahmad et al., 2014b)
Paeonia emodi Royle.	Diarrhea, Rheumatic pain, Gynecological disorders, Vomiting	
Pistacia integerrima J. L. Stewart ex Brandis.	Skin diseases, Reducing inflammation from body	(Kayani, Ahmad, et al., 2024b)
Ribes alpestre Decne.	Ringworm, blood tonic (Balti/Shina)	(Abbas et al., 2021)
Skimmia laureola (DC.) Sieb. & Zucc. ex Walp.	Small pox, Worms problems, colic	(M. Ahmad et al., 2014b)
Thymus linearis Benth.	Flatulence, abdominal pain (Balti); Cough, asthma (Shina)	(Abbas et al., 2021)

1.4 Adaptive Plans of Alpine Flora

Alpine vegetation thrives in extreme environments that include low temperatures, high UV rays, poor soils and only a few growing seasons. Their adaptative mechanisms could be categorized as morphological, physiological and biochemical mechanisms (K. Ahmad et al., 2016; Zaman, Ayaz, & Park, 2025).

Morphological changes: Alpine plants have morphological changes that include dwarfism and cushion/rosette development, and dense pubescence. These characteristics reduce the heat loss, shield against high winds and trap heat around the surfaces of plants. Indicatively, *Saussurea obvallata* and cushion in the Himalayas rely on thick hairs on leaves and a compact structure to keep a healthy microclimate.

Physiological adjustments: Alpine plants tend to exhibit high efficiency photosynthesis at low temperatures, high ratios of root to shoot to aid nutrient uptake and seasonal plasticity in growth. They have also acquired freeze tolerance mechanisms including supercooling and production of antifreeze protein.

Biochemical adaptations: In order to survive the oxidative stresses induced by UV radiation and low temperatures, alpine plants produce large amounts of secondary metabolites, such as flavonoids, carotenoids, and phenolic compounds. These molecules are sunscreens, antioxidants and membrane stabilizers, which enhance environmental stress resistance.

1.5 From Traditional Remedy to Modern Medicine: Pathways and Complexities

Though the traditional use is the foundation of herbal medicine, the path of alpine botanical into the contemporary pharmacopeia is a long and scientific process which has its own complex of complications and safety measures. Such a route justifies the huge pharmaceutical potential that motivates research as well, alas, commercialization.

1.5.1 The Phytopharmaceutical Discovery Pathway

The conversion of traditional knowledge to contemporary pharmaceuticals is a multi-stage systematic process, which is described in the process of herbal drug discovery in general. The first stage of the journey will be the pre-drug stage where the selection of promising alpine species will be directed by the ethnobotanical records and scientific literature. This is then succeeded by the important phase of the quasi-drug, which entails the extraction of phytochemicals on the plant material and then bioactivity-directed fractionation that isolates the most active constituents. The number of bioactive compounds is identified and described using advanced analytical methods, including High-Performance Liquid Chromatography (HPLC) to purify the compounds and Nuclear Magnetic Resonance (NMR) to elucidate the structure of the specific bioactive compounds that cause the therapeutic effect of the plant. An example of this direction includes the isolation and the characterization of sesquiterpene lactones, e.g. helenalin of *Arnica montana*, and validated its effectiveness as the major anti-inflammatory agent. The last full-drug phase is the vigorous pre-clinical and clinical testing to confirm the compound as efficient, safe, and with the most effective dosage, which preconditions its possible creation into an approved drug (Saddiqa, Yaseen, Hassan, Aftab, & Iqbal, 2025).

1.5.2 Herb-Drug Interactions and Safety

The common belief that herbal products are necessarily safe is a myth that produces the ignorance of a high possibility of herb-drug interactions (HDIs). Alpine botanicals, rich in pharmacologically active constituents, can modify the effects of conventional medications through two primary mechanisms: pharmacokinetics and pharmacodynamics. Pharmacokinetic interactions involve the herb altering the absorption, distribution, metabolism, or excretion (ADME) of a drug. For example, certain plant compounds can inhibit or induce cytochrome P450 enzymes in the liver, which are responsible for metabolizing a vast number of prescription drugs, potentially leading to toxic accumulation or therapeutic failure (Fugh-Berman & Ernst, 2002). Pharmacodynamic interactions occur when an herbs biological activity adds to, synergizes with, or opposes the effect of a conventional drug. A patient taking an anticoagulant like warfarin, for example, could face an increased risk of bleeding if they concurrently use an herbal product with similar blood-thinning properties. This highlights the critical need for healthcare professionals to be knowledgeable about these potential interactions and for patients to disclose all herbal product usage (Saddiqa et al., 2025).

2. A CONVERGENCE OF THREATS: ENDANGERING PLANTS AND PEOPLE

2.1 Commercial Pressures and the Unsustainable Harvest

The greatest cause of over-exploitation in the alpine regions of Pakistan is commercialization of medicinal plants. Whereas the process of collecting can be sustainable in the local and domestic contexts of consumption, the sheer volume of demand in the national and global sectors has turned the traditional harvesting process into a massive, destructive and commercial enterprise. According to Hamilton (2003), it is generally accepted that the overwhelming problem is commercial trade, as opposed to home-use. These herbs are specifically sold and bought in a regional market network called Pansar, which deals with the import and export of these herbs in Pakistan (Ross, 2005).

The commercial harvesting practices are usually destructive in nature. It is often required of roots, rhizomes, or the whole plant, and the individual has to be killed by being uprooted, a process that is especially harmful to alpine species that grow slowly. Since the strengthening of customary laws and community-based institutions become weaker, the tragedy of the commons occurs. This collapse was well exemplified in the example of *Prunus africana* in Cameroon, whereby, a transition out of a controlled, monopoly-based system of harvesting to a system with 50 extra licenses caused a collapse of controls, with the final result being the situation where complete girdling and felling of trees was the order of the day (Cunningham, Cunningham, & Schippmann, 1997; Schippmann, 2001). This has been the same with the situation in Pakistan, where traders advance a loan to the collectors, compelling them to extract desired species at a high value in the absence of supervision. This gives rise to the effect of spreading frontiers of resource-depletion whereby traders relocate to the next valley after depleting it. Most of this trade is in a gray industry because of the absence of effective government regulation. The attempt to regulate this trade by the use of systems such as CITES is floundered by ineffective monitoring, incorrect reporting and labeling with medicinal plants being at times brought in under the guise of spices (Begum, 2002; Mulliken, 2000; Schippmann, 2001).

2.2 Climate Change and Global Warming Impacts

Alpine ecology is extremely sensitive to even slight changes of climate, with medical plants in the areas being especially vulnerable. The increased temperatures have caused most species to migrate to lower temperatures which are sometimes cooler thus making the habitat scarce to support population at higher heights. Studies point to the fact that plants like *Rheum australe* and *Saussurea costus* are already experiencing a decrease in distribution through shorter snow-covers and uneven precipitation. These climatic stressors do not only pose a threat to survival, but also change the production of secondary metabolites, which diminish the medicinal value of most plants (Zaman et al., 2025).

2.3 Degradation and Fragmentation of Habitats.

Alpine habitats are disrupted by the development of infrastructure, expansion of tourism, and overuse of pasture, which are all human-made land use activities. All this is caused by fragmentation, which restricts the amount of genetic exchange between plant populations, and the chances of local extinctions increase. The research in the Himalayas and the European Alps demonstrates that the construction of roads and grazing pressures weaken the already weak soils and speed up the erosion process, which also threatens the existence of rare medicinal herbs (Catalan, Ninot, & Aniz, 2017). The patchy landscapes resulting negatively affect the strength of these species to adapt to the changing environmental conditions. Human-induced disturbances such as overgrazing and soil erosion are particularly damaging in alpine regions because they target the specific, limited habitats that endemic species rely on. In the alpine zones of Kashmir, for instance, field studies revealed that 60% of sites for the medicinal poppy *Meconopsis aculeata* were intensely grazed, while 70% were impacted by erosion. This level of degradation is catastrophic for a species with such a narrow ecological niche; GIS mapping confirms its confinement to steep, north-facing rocky slopes between 3000 and 4600 m. This habitat squeeze is further intensified by climate change, which is expected to push the species' viable range upslope, a difficult migration for a plant with poor dispersal abilities (Manzoor et al., 2025).

2.4 Pressures of the Socio-Economy and the Chipping Away of Traditional Knowledge

Alpine societies often use the collection of medicinal plants as a source of subsistence healthcare and as an additional source of income. Increasing socio-economic demands such as poverty, lack of livelihood and opportunities and the attraction of commercial trade promote unsustainable harvesting activities. Local collectors also suffer unfairly because of exploitation through the global demand of natural remedies and herbal products, without justifiable benefit-sharing (Pathak, Gupta, & Pandey, 2024).

The most marginalized communities in terms of economics are usually the most remote in the valleys of north Pakistan, and the most important collectors. They are however given a small percentage of the market price at the end of the day, and as the middlemen and wholesalers have to take the major share of the market. This economic imbalance gives an incentive to engage in unsustainable methods; to gain a subsistence income, the collector must gather as much material as possible, and does not have to give much concern to the sustainability of the resource in the long term. The particular socio-economic environment of neighborhoods such as those in Palas Valley also renders them very susceptible. They rely on wild plants to provide medicinal and financial services, as their local economy is based on cattle rearing and herd raising, and seasonal migration to high-altitude pastures (Saqib & Sultan, 2005). The demographic statistics of the area show that illiteracy (40.32) and unemployment (43.62) are high, which is the reason why medicinal plants gathering as the source of money becomes necessary, although poorly paid (Kayani, Ahmad, et al., 2024b).

Alpine aboriginal communities have rich traditional knowledge of medicinal plants on the use and sustainable management of the plant. But this body of knowledge is quickly fading away with changes in generations, urbanization migration, and homogenization of cultural values. Since younger generations abandon the old ways of living, important ethnobotanical information regarding the properties of plants, harvesting and preservation is lost. This does not only reduce cultural heritage, but also removes the possibility of new pharmacological resources being discovered in these plants.

The difficult and delicate character of this knowledge is vividly demonstrated by contemporary ethnobotanical studies in Kohistan. A cross-cultural research conducted among 5 linguistic groups has been documented for indigenous knowledge and utilization of common medicinal plants shared by all the ancestral people's including significant descriptive details about idiosyncratic useful parts (unique) inside a group as well as IDK possessed by each group (Amin et al., 2024). Table 2 illustrates that even the minorities groups such as Bateris have an unproportionately large store of this unique knowledge and are significant but very fragile reservoirs of biocultural heritage. But this whole system of shared and specific knowledge is now under genocidal threat from continuing modernization and infrastructure development.

Table 2. Case Study of Kohistan: Cross-Cultural Comparison of Traditional Ethnomedicinal Knowledge (TEK)

Linguistic Group	Key Finding from the Study	Example of Shared Knowledge (Homogeneity)	Example Idiosyncratic Knowledge (Heterogeneity)	Implication for Conservation
All Groups (Shared Heritage)	61 plant species (63% of total) were known and used across all five linguistic groups, showing a common ancestral knowledge base.	*Myrtus communis*: Widely used by all groups to treat eczema, demonstrating a highly valued and shared remedy across cultures.	N/A	Preserving these shared practices is key to maintaining the core cultural identity and healthcare system of the entire region..
Shina Community (Himalayan)	Reported the highest number of unique medicinal plant taxa, likely due to their access to distinct alpine pastures.	Utilizes many of the 61 commonly shared plant species.	*Rhodiola integrifolia*: A high-altitude adaptogenic species uniquely reported by the Shina, illustrating ecological specialization.	Conservation strategies must prioritize protection of high-altitude pastures to prevent erosion of this specialized ethnobotanical knowledge.
Bateri Community (Minority Group)	Despite their small population size, the Bateri reported the highest number of unique medicinal uses, representing a disproportionately rich reservoir of idiosyncratic knowledge.	Utilizes many of the 61 commonly shared plant species.	*Buxus wallichiana*, *Hypericum oblongifolium*: Among several plants with medicinal applications known exclusively to the Bateri.	**Highest urgency**: The loss of Bateri traditions would result in irreversible loss of unique ethnomedicinal knowledge at a global scale.
Kohistani Community (Hindukush)	A dominant group with a rich knowledge base, but showing signs of knowledge erosion due to modernization.	Utilizes many of the 61 commonly shared plant species.	*Polypodium sibiricum*, *Clematis grata*: Among the 10 species uniquely known to Kohistani, many of which had their uses documented for 1st time	Documenting and revitalizing the knowledge of dominant groups is crucial, as its erosion can happen rapidly due to greater exposure to economic changes.

3. A STRUCTURE OF CONSERVATION AND SUSTAINABLE FUTURES

3.1 In-Situ Conservation: Effective Empowerment of Local Communities

A successful conservation plan of the medicinal alpine plants of Pakistan should be a multi-dimensional one that incorporates on-site (in-situ) and off-site (ex-situ) conservation approaches, with a solid community involvement. Any sound strategy is founded on in situ conservation that is geared towards safeguarding the species within the natural environment. According to Hamilton (2003), this move should go beyond the conventional model of narrowly guarded areas that keep out the local population. The experience in Bwindi Impenetrable forest in Uganda, where an early ban on harvesting all forest produce led to confrontation and intentional forest burning, is testimony to the fact that effective conservation should be done through cooperation and sharing of the benefits (Hamilton, 2003).

This involves protection and restoration of habitats, the creation of conservation zones (national parks, reserves) and the existence of strong legal and policy frameworks of protection of biodiversity. Most importantly, it requires community conservation and sustainable harvesting activities. In the alpine areas of Pakistan, it implies that communities, particularly women, are regarded as the focal partners of management, and despite being highly knowledgeable, they do not receive enough attention in negotiations because of cultural constraints, which researchers should address in the Palas Valley (Kayani, Ahmad, et al., 2024a). Developing contracts that provide them with safe rights to harvest certain medicinal plants in restricted areas, with the condition of their involvement in the process of monitoring and enforcement is one of the main approaches to conserving the medicinal plants and managing available resources in the long-term perspective. To have an effective oversight, community-based institutions need to be empowered to handle these rules.

3.2 Ex-Situ and Circa-Situ Conservation: Bridging Wild and Cultivated Resources

To supplement on-site work, ex-situ and circa-situ conservation provide crucial tools to help to alleviate pressure on populations in the wild. The narrowest form of ex-situ conservation encompasses the capture of genetic material in the form of seed banks and gene banks and botanic gardens and arboreta can be used as a living repository. Contemporary biotechnologically methods like tissue culture, micropropagation, and cryopreservation are also important.

A more viable and effective measure is the so-called circa-situ conservation which entails the cultivation of medicinal plants within and surrounding community land. This will involve the setting up of home gardens and small scale commercial nursery of species in high demand. This approach can guarantee the plant and has the direct advantage to the community as it can be demonstrated by the case of the Pepper-bark tree (*Warburgia salutaris*) in Zimbabwe where rooted cuttings have been successfully reintroduced into the home-gardens of local farmers who have appreciated the species. There are a number of benefits of cultivation: it has a steady supply, is a source of an alternate income, and offers a higher quality control. Nonetheless, cultivation is not panacea as Hamilton (2003) warns it might marginalize the landless poor who rely on wild collection confirmed by a cultural notion that plants harvested in the wilderness are more effective. The optimal approach is, therefore, a moderate approach that encourages cultivation of large-volume commercial species with sustainability and managing of wild harvesting to ensure that the crucial relationship between humans and their natural ecosystems has not been broken (Hamilton, 2003).

3.3 Integrating Traditional Knowledge with Scientific Innovation

Bridging the gap between indigenous knowledge systems and modern research is fundamental for sustainable utilization. This involves conducting case studies of ethnobotanical practices and blending local wisdom with scientific innovation. A key component is ensuring intellectual property rights are protected and that benefit-sharing mechanisms are established, so communities are compensated for their knowledge and resources (Table 3).

3.4 Case Studies in Ethnobotany of Northern Pakistan

Case Study 1: Medicinal Plants of Chail Valley, District Swat

Chail Valley, situated in high mountainous terrain of District Swat, Pakistan is a treasure trove of traditional medicinal plants for rural population. A survey in this area has revealed a number of plant species with botanical family, part used and medicinal uses. Herbs accounted for the most among the medicinal flora followed by shrubs and trees. The conventional knowledge in the Chail valley is mainly of oral traditions and local therapist, showing dependence on the use of medicinal plants against GIT tract, respiratory disorder and bones/joint problems. This report emphasizes the need to document traditional wisdom and potential of these plants in drug discovery (M. Ahmad et al., 2014a).

Case Study 2: Medical Ethnobotany of Gilgit-Baltistan

Gilgit-Baltistan (GB), an isolated, harsh-climate region in northern Pakistan, provides the backdrop for a systematic review of 24 studies that occurred between 2002 and 2020. The MOD involved 413 medicinal plant species within 224 genera of 79 families which employed over 3,160 different remedies. Nearly all plants were wild collected (83%) and most of them were herbaceus (75%). The most reported species were *Hippophae rhamnoides* and *Thymus linearis* in 70% of studies. GB's ethnobotanical studies, centered primarily on the Shina community (58%), were rather few in number for Balti, Brushiski, Wakhi and Khwar tribes. The interviews were the predominant type of data collection, with descriptive and quantitative (frequency citation, relative frequency citation levels, fidelity level etc.) approaches. In geographical term, research was focused towards Karakoram Range and Gilgit sub-region with remote valleys and Baltistan being under-researched area. This review indicates the great potential of the medicinal flora of GB for traditional therapeutics as well as for drug discovery while additionally demonstrating there are substantial research black holes in underprivileged communities and regions (Abbas, Bussmann, Khan, & Abbasi, 2022).

Both case studies demonstrate the rich ethnobotanical knowledge present in the mountainous regions of northern Pakistan. They highlight the critical role of indigenous knowledge in healthcare, the dominance of herbaceous plants in traditional medicine, and the need for systematic documentation and conservation strategies. While Chail Valley offers insights into localized traditional practices, Gilgit-Baltistan presents a broader ethnobotanical landscape, revealing both diversity and significant gaps for future research.

Table 3. Conservation Issues and Strategies for Medicinal Alpine Plants in Pakistan

Threat/Issue	Description in the Alpine Context	Proposed Conservation Strategy
Over-Harvesting for Commercial Trade	Destructive and unregulated collection of high-value species (roots, rhizomes) to supply national and international markets, leading to rapid population decline of key plants like *Picrorhiza kurrooa* and *Saussurea costus*.	• Develop and enforce sustainable harvesting protocols (quotas, rotational harvesting, seasonal restrictions) through community-based committees. • Promote cultivation and circa-situ conservation (home gardens, nurseries) for high-demand species to reduce pressure on wild stocks. • Strengthen CITES implementation and national laws to regulate trade and prevent illegal export..
Habitat Degradation & Climate Change	Loss and fragmentation of critical alpine habitats due to overgrazing, infrastructure development, and shifting climatic conditions that alter temperature and precipitation patterns, stressing sensitive plant populations.	Establish and co-manage protected areas (national parks, community conservancies) with zones designated for sustainable use and strict protection. • Implement sustainable land management practices, including rotational grazing plans to allow for habitat regeneration. • Conduct long-term research and monitoring to assess climate change impacts and identify resilient species/habitats.
Loss of Traditional Ecological Knowledge (TEK)	The intergenerational knowledge gap caused by modernization and migration to cities is leading to the disappearance of traditional wisdom on sustainable harvesting, plant identification, and processing.	Launch participatory documentation projects to record TEK from community elders in local languages. • Integrate TEK into local school curricula and awareness programs to engage younger generations. • Support and formally recognize traditional healers (Hakeems), whose knowledge is integral to both conservation and healthcare.
Inequitable Commercial Systems & Lack of Benefits	Local collectors receive extremely low prices for raw materials, while middlemen and wholesalers capture most of the profit. This lack of economic incentive discourages stewardship and promotes unsustainable extraction.	Facilitate the formation of collector/ grower cooperatives to strengthen bargaining power • Promote Fair Trade and organic certification standards to add value and ensure ethical sourcing. • Develop local value-addition enterprises (e.g., drying, processing, packaging).

4. CHALLENGES AND FUTURE DIRECTIONS: CONSERVATION AND STANDARDIZATION

The future of the alpine medicinal plants in the pharmaceutics is dependent on the issues of conservation and standardization. Numerous alpine species are slow-growing and live in sensitive and geographically limited habitats, being extremely susceptible to over-harvests and climate change effects. Not only the biodiversity is endangered with the unsustainable collection practices, but also the long-term demands to these precious natural resources. At the same time, a chemical profile of wild-harvested plants may also be highly variable according to genetics, soil and altitude, which is a key challenge to standardization. This inconsistency defeats the capability of providing a dependable and safe dosage with commercial products. The way forward is the integration of phytochemical extracts in controlled pharmaceutical preparations as this review indicates (Saddiqa et al., 2025).

The aim of the work in the future should therefore be to put in place sustainable cultivation and utilize biotechnological tools, such as plant cell culture and metabolic engineering, to produce necessary bioactive compounds, in a reproducible, controlled, and responsible manner. Furthermore, the preservation of the identical phytochemical yield and less damage to the ecology can be ensured with the help of Good Agricultural and Collection Practices (GACP) and integration with Good Manufacturing Practices (GMP (Liu et al., 2022). Further growth of molecular biology including genomic and metabolomic profiling can be applied in choosing additional elite chemotypes that reliably produce metabolites when grown in the presence of culture.

The other potential option is the use of bioreactors and hairy root cultures which have been applied to some species such as *Rhodiola rosea* to generate useful phenolic products without necessarily reliance on wild populations. In addition, community conservation projects enabling the local people to participate in the controlled trade and production provides a viable socio-economic approach. The combination of the traditional knowledge with the modern science based on the participatory frameworks has the potential to protect the cultural heritage and enhance the predictability of the pharmaceuticals at the same time (Pathak et al., 2024) (Figure 3).

Ultimately, the future of alpine medicinal plants lies in a multidisciplinary approach that combines conservation ecology, biotechnology, pharmacognosy, and policy-making. By integrating sustainable harvesting with technological innovation, it will be possible to preserve alpine biodiversity while unlocking the pharmaceutical potential of these unique resources.

Figure 3. Key Challenges in the Cultivation of Medicinal Plants. This info graphic summarizes the multifaceted constraints inherent in agriculture, which are directly applicable to the cultivation of alpine medicinal plants as a conservation strategy. While presented as a solution to over-harvesting, this figure highlights that shifting to cultivation introduces a new set of complex issues from economic viability and seed quality to water management and pest control that must be addressed for such initiatives to be successful.

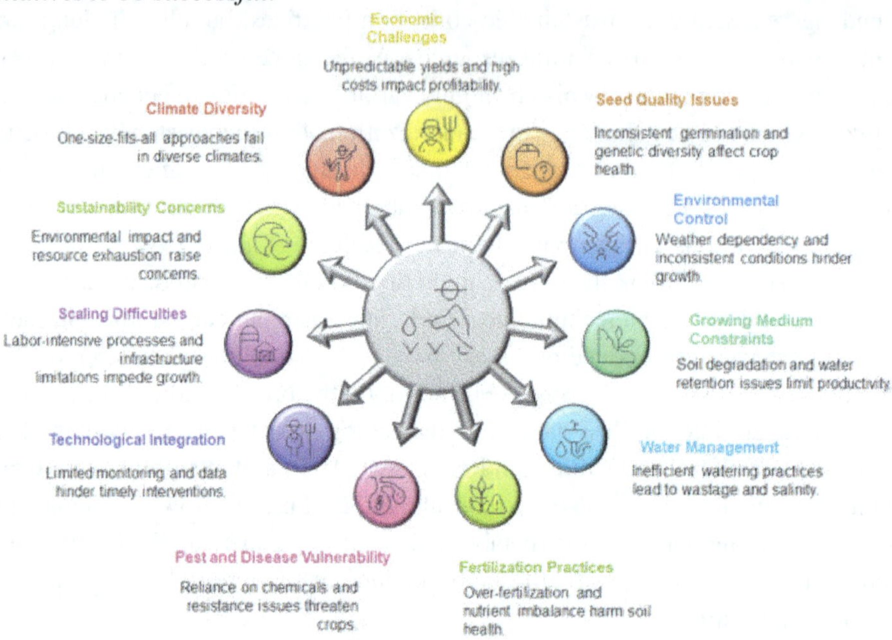

5. CONCLUSION

The alpine ecosystems of Hindu Kush, Karakoram and Himalayas are at a highly unstable boundary that is the concentration of not only immense biological wealth but also of rich cultural supremacy. A two-fold crisis is superimposing this unique nexus as reported in this chapter the rapid disappearance of the Traditional Ecological Knowledge (TEK) and the frightening extinction of the medicinal flora, upon which this knowledge is based. It is not an environmental issue; but a complex socio-ecological issue with far-reaching socio-economic and cultural implications to the primary healthcare, livelihood, and cultural identity of the indigenous mountain folks of Pakistan. The confluence of the threats in the current paper provides a clear view of the threats to be experienced in the future. The insatiable demand of the commercial markets, both national and international has transformed the sustainable, subsistence level collection into an exploitative, in the vast majority of cases, unregulated business. The reason of the over-harvesting is such a phenomenon as the so-called value paradox i.e. the species most endangered by extinction are the ones that have the greatest medicinal and economic value. This direct pressure is supported with a group of low level threats that are landscape level. The climate change is ruthlessly redefining the ecological map by forcing the temperature-sensitive species into progressively smaller habitats as the altitude worsens. Simultaneously, overgrazing, building of infrastructure and tourism devastate these sensitive sceneries and isolate the plant species making them more vulnerable. Behind these environmental stressors, there are deep rooted socio-economic forces. Unfair supply chains deprive the disadvantaged local collectors of the economic advantage of high-volume extraction, which dictates extraction unproductive in the long run, instead encouraging them to continue living at the urban center and in other lifestyles, the knowledge base is becoming increasingly intergenerational with the youth being attracted to the city and other ways of life. But despair is not necessity. The direction to the sustainable future, as proposed in this chapter, is in the multi-pronged, integrated approach that puts the local communities at its core. The period of the so-called fortress conservation, which locks out the same people who have been the historical custodians of these resources, proves to be ineffective. Rather, in-situ conservation based on community empowerment should form the basis of an effective strategy. This entails development of collaborative systems of management whereby there will be secure rights to the local communities especially women to manage their botanical resources in a sustainable manner and benefit. Such community-oriented practice should be supported by the scientific innovativeness and strategic nurturing. Ex-situ conservation in seed banks and botanical gardens offers an essential genetic backup, whereas contemporary biotechnology such as tissue culture offers solutions to the multiplication of endangered species. More immediately effective is the cultivation of high-demand species in domestic gardens and nurseries run by communities, as the so-called circa-situ conservation. Planting medicinal plants with high demands will reduce the strain on the wild populations, generate a constant income to the communities and enhance quality control of the medicinal products. Nevertheless, this cultivation should be equally planned to assure that the landless collectors are not left behind and the sustainable wild harvesting is maintained. It is through integration of the conventional ecological wisdom and the current scientific methods that the long-term sustainability of alpine medicinal flora in Pakistan would be achieved. Since the transformation of traditional remedies into safe and standardized pharmaceuticals is not a simple task, integrating TEK with phytochemistry, biotechnology and pharmacognosy is needed to address issues such as chemical variability and herb-drug interactions.

The fact that we can document and validate our traditional knowledge means that we are able to recognize important medicinal species to be conserved and used sustainably. The use of GACP and modern technologies including cell culture and metabolic engineering can generate consistent and high quality plant compounds without depleting the wild resources. The construction of a resilient bio-economy will need concerted efforts powerful policies, communal empowerment, scientific invention, as well as ethical international markets. The alpine areas of Pakistan are a source of living cultural and ecological knowledge, and their preservation is a key to the biodiversity conservation, heritage, and the survival of the precious sources of medicines to the generations to come.

REFERENCES

Abbas, Z., Bussmann, R. W., Khan, S. M., & Abbasi, A. M. (2022). A review of current trends and future directions in the medical ethnobotany of Gilgit-Baltistan (Northern Pakistan). *Ethnobotany Research and Applications*, *24*, 1–16. DOI: 10.32859/era.24.18.1-16

Abbas, Z., Kousar, S., Aziz, M. A., Pieroni, A., Aldosari, A. A., Bussmann, R. W., Raza, G., & Abbasi, A. M. (2021). Comparative assessment of medicinal plant utilization among Balti and Shina communities in the periphery of Deosai National Park, Pakistan. *Biology (Basel)*, *10*(5), 434. DOI: 10.3390/biology10050434 PMID: 34068859

Ahmad, K., Hameed, M., Fatima, S., Ashraf, M., Ahmad, F., Naseer, M., & Akhtar, N. (2016). Morpho-anatomical and physiological adaptations to high altitude in some Aveneae grasses from Neelum Valley, Western Himalayan Kashmir. *Acta Physiologiae Plantarum*, *38*(4), 93. Advance online publication. DOI: 10.1007/s11738-016-2114-x

Ahmad, M., Sultana, S., Fazl-i-Hadi, S., ben Hadda, T., Rashid, S., Zafar, M., Khan, M. A., Khan, M. P. Z., & Yaseen, G. (2014a). An Ethnobotanical study of Medicinal Plants in high mountainous region of Chail valley (District Swat- Pakistan). *Journal of Ethnobiology and Ethnomedicine*, *10*(1), 36. DOI: 10.1186/1746-4269-10-36 PMID: 24739524

Ahmad, M., Sultana, S., Fazl-i-Hadi, S., Ben Hadda, T., Rashid, S., Zafar, M., Khan, M. A., Khan, M. P. Z., & Yaseen, G. (2014b). An ethnobotanical study of medicinal plants in high mountainous region of Chail valley (District Swat-Pakistan). *Journal of Ethnobiology and Ethnomedicine*, *10*(1), 36. DOI: 10.1186/1746-4269-10-36 PMID: 24739524

Amin, M., Aziz, M. A., Manduzai, A. K., Pieroni, A., Alkahtani, J., AbdelGawwad, M. R., Gafforov, Y., Nazeer, A., & Abbasi, A. M. (2024). Standard medical ethnobotany of Kohistan, North Pakistan. *Journal of Ethnobiology and Ethnomedicine*, *20*(1), 64. DOI: 10.1186/s13002-024-00704-w PMID: 38978072

Balunas, M. J., & Kinghorn, A. D. (2005). Drug discovery from medicinal plants. *Life Sciences*, *78*(5), 431–441. DOI: 10.1016/j.lfs.2005.09.012 PMID: 16198377

Begum, F. (2002). *Ex situ conservation of medicinal plants for primary healthcare in Laxmipur District, Bangladesh.* Paper presented at the Paper at a Workshop on Wise Practices and Experiential Learning in the Conservation and Management of Himalayan Medicinal Plants, Kathmandu, Nepal.

Bentley, R., & Trimen, H. (1880). Medicinal plants: Being descriptions with original figures of the principal plants employed in medicine and an account of the characters, properties, and uses of their parts and products of medicinal value. Vol. 4.

Catalan, J., Ninot, J., & Aniz, M. (2017). *The High Mountain Conservation in a Changing World.* In (pp. 3-36). DOI: 10.1007/978-3-319-55982-7

Cunningham, M., Cunningham, A., & Schippmann, U. (1997). *Trade in Prunus africana and the implementation of CITES.* German Federal Agency for Nature Conservation.

Fugh-Berman, A., & Ernst, E. J. B. J. O. C. P. (2002). 'Herb-drug interactions: Review and assessment of report reliability'(vol 52, pg 587, 2001). *53*(4), 449P-449P.

Hamilton, A. J. I. P. C. U., WWF-UK. (2003). Medicinal plants and conservation: issues and approaches. *51*.

Haq, S. M., Khoja, A. A., Waheed, M., Pieroni, A., Siddiqui, M. H., & Bussmann, R. W. (2024). Plant cultural indicators of forest resources from the Himalayan high mountains: Implications for improving agricultural resilience, subsistence, and forest restoration. *Journal of Ethnobiology and Ethnomedicine*, *20*(1), 44. DOI: 10.1186/s13002-024-00685-w PMID: 38659048

Kanwal, H., & Sherazi, B. A. (2017). Herbal medicine: Trend of practice, perspective, and limitations in Pakistan. *Asian Pacific Journal of Health Sciences*, *4*(4), 6–8. DOI: 10.21276/apjhs.2017.4.4.2

Kayani, S., Ahmad, , Gillani, S. W., Manzoor, M., Rehman, F., Jabeen, S., & Shah, S. A. H. (2024a). Ethnomedicinal appraisal of the medicinal flora among the sub-alpine and alpine Iindigenous communities of Palas Valley Kohistan, Northern Pakistan. *Ethnobotany Research and Applications*, *28*. Advance online publication. DOI: 10.32859/era.28.9.1-29

Kayani, S., Ahmad, M., Gillani, S. W., Manzoor, M., Rehman, F. U., Jabeen, S., & Shah, S. A. H. (2024b). Ethnomedicinal appraisal of the medicinal flora among the sub-alpine and alpine Iindigenous communities of Palas Valley Kohistan, Northern Pakistan. *Ethnobotany Research and Applications*, *28*, 1–29. DOI: 10.32859/era.28.9.1-29

Kayani, S., Ahmad, M., Gillani, S. W., Muhammad, M., Rehman, F. U., Jabeen, S., & Shah, S. A. H. (2024). Ethnomedicinal appraisal of the medicinal flora among the sub-alpine and alpine Iindigenous communities of Palas Valley Kohistan, Northern Pakistan. *Ethnobotany Research and Applications*, *28*, 1–29. DOI: 10.32859/era.28.9.1-29

Kayani, S., Ahmad, M., Sultana, S., Shinwari, Z. K., Zafar, M., Yaseen, G., & Bibi, T. (2015). Ethnobotany of medicinal plants among the communities of Alpine and Sub-alpine regions of Pakistan. *Journal of Ethnopharmacology, 164*, 186–202. DOI: 10.1016/j.jep.2015.02.004 PMID: 25680839

Liu, D., He, X.-Q., Wu, D.-T., Li, H.-B., Feng, Y.-B., Zou, L., . . . chemistry, f. (2022). Elderberry (Sambucus nigra L.): Bioactive compounds, health functions, and applications. *70*(14), 4202-4220.

Manzoor, M., Ahmad, M., Gillani, S. W., Waheed, M., Shaheen, H., Mehmood, A. B., Fonge, B. A., & Al-Andal, A. (2025). Population dynamics, threat assessment, and conservation strategies for critically endangered Meconopsis aculeata in alpine zone. *BMC Plant Biology, 25*(1), 358. DOI: 10.1186/s12870-025-06361-9 PMID: 40102752

Mehmood, A., & Shah, A. (2015). CULTURAL USES OF PLANTS AMONG BASIKHEL TRIBE OF DISTRICT TOR GHAR, KHYBER PAKHTUNKHWA, PAKISTAN. *Pakistan Journal of Botany*.

Mulliken, T. (2000). Implementing CITES for Himalayan medicinal plants Nardostachys grandiflora and Picrorhiza kurrooa. *TRAFFIC BULLETIN-CAMBRIDGE-TRAFFIC INTERNATIONAL, 18*(2), 63–72.

Pathak, A., Gupta, A. P., & Pandey, P. (2024). *Herbal Medicine and Sustainable Development Challenges and Opportunities.* In (pp. 1-26).

Pimm, S. L., Russell, G. J., Gittleman, J. L., & Brooks, T. M. (1995). The future of biodiversity. *Science, 269*(5222), 347–350. DOI: 10.1126/science.269.5222.347 PMID: 17841251

Rashid, N., Gbedomon, R. C., Ahmad, M., Salako, V. K., Zafar, M., & Malik, K. (2018). Traditional knowledge on herbal drinks among indigenous communities in Azad Jammu and Kashmir, Pakistan. *Journal of Ethnobiology and Ethnomedicine, 14*(1), 16. DOI: 10.1186/s13002-018-0217-8 PMID: 29467005

Ross, I. A. (2005). Medicinal plants of the world, volume 3: Chemical constituents, traditional and modern medicinal uses: Springer.

Saddiqa, A., Yaseen, R., Hassan, M., Aftab, K., & Iqbal, R. J. R. B. d. F. (2025). Bridging the Gap Between Herbs and Pharmaceutics: The Science Behind Herbs' Medicinal Properties and Potential in Modern Drugs. 1-17.

Saqib, Z., & Sultan, A. (2005). Ethnobotany of palas valley, Pakistan. *Ethnobotanical Leaflets, 2004*(1), 11.

Schippmann, U. (2001). *Medicinal plants significant trade study*. Bundesamt für Naturschutz.

Shaheen, S., Harun, N., Ijaz, R., Mukhtar, N., Ashfaq, M., Bibi, F., . . . Khalid, Z. J. S. (2023). Sustainability issues in conservation of traditional medicinal herbs and their associated knowledge: A case study of District Lahore, Punjab, Pakistan. *15*(9), 7343.

Shinwari, A., Ahmad, M., Alsahli, A., Israr, A., Shinwari, Z., Shinwari, S., Zhang, L., Gafforov, Y., & Enes Yilmaz, M. (2024). Himalayan medicinal plants wealth, threats and conservation for securing the future of biodiversity. *Pakistan Journal of Botany*, *56*(6). Advance online publication. DOI: 10.30848/PJB2024-6(42)

Singh, H., Husain, T., Agnihotri, P., Pande, P., & Khatoon, S. (2014). An ethnobotanical study of medicinal plants used in sacred groves of Kumaon Himalaya, Uttarakhand, India. *Journal of Ethnopharmacology*, *154*(1), 98–108. DOI: 10.1016/j.jep.2014.03.026 PMID: 24685588

Thapliyal, N., Bhojak, P., Mehta, P., & Sekar, K. C. (2024). Threats and Conservation of Plant Diversity in the Himalayan Biodiversity Hotspot. In (pp. 459-511). DOI: 10.1201/9781003455202-18

WHO. (2015). WHO guidelines on safety monitoring of herbal medicines in pharmacovigilance systems. 2004. In: World Health Organization vii Geneva.

Zaman, W., Ayaz, A., & Park, S. (2025). Climate change and medicinal plant biodiversity: Conservation strategies for sustainable use and genetic resource preservation. *Genetic Resources and Crop Evolution*, *72*(6), 6275–6308. DOI: 10.1007/s10722-025-02410-2

Chapter 9

From Alpine Herbs to Nanotech: Biodiversity Conservation Through Medicinal Plants and Nanoparticle Research

Arshad Ali
http://orcid.org/0009-0005-7711-7199
University of Sargodha, Pakistan

Rao Muhammad Faisal Iqbal
The University of Lahore, Sargodha, Pakistan

Maria Kanwal
University of Sargodha, Pakistan

Laiba Manahil
University of Sargodha, Pakistan

Nasir Assad
http://orcid.org/0009-0004-9435-1414
University of Sargodha, Pakistan

Jamshed Iqbal
http://orcid.org/0009-0008-6631-8819
University of Genoa, Italy

Muhammad Mustaqeem
University of Sargodha, Pakistan

ABSTRACT

Alpine ecosystems are climatically extreme, biodiversity-rich, and among the most vulnerable yet pharmaceutically valuable habitats on Earth. Alpine herbs have long formed the basis of traditional medical systems across Eurasia, the Himalayas, and other mountain regions. However, the growing global demand for herbal medicines and bioactive compounds has placed unprecedented pressure on these fragile ecosystems, threatening both plant diversity and indigenous knowledge. Green nanotechnology has recently emerged as a sustainable link between biodiversity conservation and biotechnology. Green synthesis of nanoparticles using alpine medicinal plants

DOI: 10.4018/979-8-2600-1207-9.ch009

offers opportunities to enhance bioactivity, reduce toxicity, develop value-added bioproducts, and minimize pressure on wild populations. This chapter explores the ecological, biochemical, and technological intersections between alpine medicinal flora and nanoscience, proposing nanotechnology as a conservation-oriented strategy for sustainable use of alpine biodiversity.

1. INTRODUCTION

The most distinctive and ecologically sensitive habitats on Earth are the so-called alpine ecosystems located at high altitudes in mountain ranges such as the Himalayas, the Alps, the Andes, and the Rockies. These ecosystems are typified by low temperatures, high ultraviolet (UV) radiation, thin atmospheric conditions, and short growing seasons, which have motivated the development of flora that have specialized biochemical adaptations. The alpine vegetation is a key source of global biodiversity as it carries an impressive variety of endemic plants in the field of medicine that have been used as foundational elements of indigenous and traditional healthcare over the centuries (Nagy & Grabherr, 2009).

Although alpine areas hold ecological and ethnopharmacological significance, climate change, habitat fragmentation, and unsustainable harvesting methods are increasingly threatening the areas, and new conservation efforts are required that satisfy ecological integrity and socio-economic realities. In the past, alpine herbs have been held in high regard in traditional medicine. There were *Rhodiola rosea* (golden root), *Arnica montana*, and *Gentiana lutea* as plants that were part of ethnomedical traditions in Europe and Asia. Table 1 provides a summary of key alpine medicinal species, highlighting their major phytochemicals, traditional uses, and conservation status based on IUCN assessments. Their secondary metabolites, such as flavonoids, phenolic acids, alkaloids, terpenoids, and others, have strong pharmacological effects of antioxidant, anti-inflammatory, and neuroprotective activity (de Lima et al., 2025). The compounds are commonly synthesized in response to severe environmental conditions, such as UV light or low temperatures, and confer alpine plants with distinct biochemical states challenging to mimic in lowland species. Nevertheless, the increasing demand for natural therapeutics in the world has escalated the harvesting of such plants in the wild, which has resulted in the depletion of the population of these plants and genetic losses.

Table 1. Key Alpine medicinal plants and their phytochemical significance

Scientific Name	Common Name	Major Bioactive Compounds	Traditional Uses	Conservation Status (IUCN)
Rhodiola rosea	Golden root	Rosavin, Salidroside, Tyrosol	Adaptogen, fatigue reduction, neuroprotective	Vulnerable
Arnica montana	Arnica	Helenalin, Flavonoids	Anti-inflammatory, wound healing	Near Threatened
lGentiana lutea	Yellow Gentian	Gentiopicroside, Amarogentin	Digestive stimulant, hepatoprotective	Least Concern
Aconitum heterophyllum	Atis	Aconitine, Atisine alkaloids	Analgesic, antipyretic, traditional Tibetan medicine	Endangered
Saussurea costus	Kuth	Costunolide, Dehydrocostus lactone	Respiratory and skin disorders, Ayurveda	Critically Endangered

As recent biodiversity evaluations show, as many as 25% of alpine medicinal species are threatened to a degree, with overexploitation and loss of habitat being the principal factors (Petelka et al., 2020). This growing conflict between bioprospecting and the disappearance of biodiversity highlights the necessity of innovative, sustainable methods of resource use. An alternative solution is the combination of nanotechnology, which is at the leading edge of contemporary science, and the traditional ethnobotanical knowledge. Green nanotechnology, especially the biosynthesis of nanoparticles (NPs) with plant extracts, provides one of the avenues to the ecological and economic balance. The green synthesis method takes advantage of phytochemicals as natural reducing and stabilizing agents to produce biocompatible nanoparticles of metals like silver (Ag), gold (Au), zinc (Zn), and copper (Cu), unlike traditional physical or chemical fabrication pathways that require the use of toxic substances and large amounts of energy (Tahir et al., 2022).

In addition to producing materials with strong biomedical uses, this approach produces the materials with minimum plant biomass, thus exerting minimum pressure on natural populations during harvesting. Alpine plants in this regard have great potential. They are also the best candidates to produce nanoparticles with an improved biological activity due to their unique phytochemical diversity that is rich in stress-induced antioxidants and polyphenolic compounds. As an example, *Rhodiola rosea* and *Artemisia gmelinii* have reportedly synthesized highly stable silver nanoparticles with marked antimicrobial and anticancer effects, which is due to flavonoid and salidroside compounds (Islam et al., 2021). These developments exemplify biogenic nanomaterials transformations of high-value solutions- pharmaceuticals up to environmental remediation- and resource utilization and conservation by means of innovation. Therefore, biodiversity conservation and nanotechnology are not a simple technological innovation but a paradigm shift. It is an integrated method- whereby contemporary scientific instruments can be used to boost the existing knowledge systems, which will ensure the sustainability of the ecological system as well as the economic viability.

With the connection of the pharmacological potential of the alpine plants to the green chemistry concepts of nanotechnology, it is possible to change the exploitation model to the model of regeneration and stewardship. This chapter explores the synergy between conservation and innovation, highlighting how nanoscience serves as a medium for integrating natural heritage preservation with scientific advancement toward a sustainable future. (Fig. 1)

Figure 1. Alpine plants, their stress adaptations, and role in green nanotechnology for conservation.

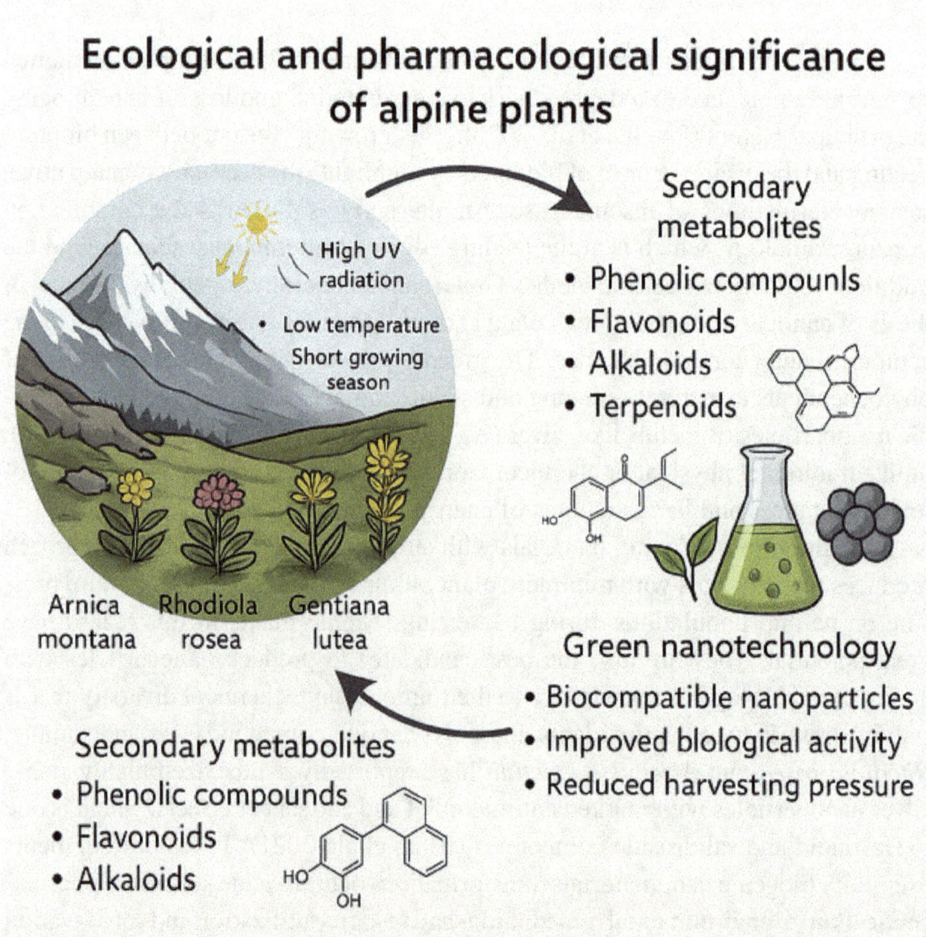

2. THE IMPERATIVE OF THE CONSERVATION OF BIODIVERSITY

Conservation of alpine biodiversity is not simply a matter of scientific or ecological concern, but one with profound moral, cultural and economic implications. The mountain habitats are a living laboratory for evolution where isolation, climate pressures and geological history has resulted in specialization and speciation. The plants which inhabit these communities comprise storage of biochemical novelty and/or ecological stasis, representing millions of years of adjusting to extreme conditions. What they sacrifice is the unfathomable destruction of genetic material that holds up the natural ecosystem and human health. By conflating the intrinsic value of biodiversity as the right to existence irrespective of its utility, nonuse had been entrenched in global conservation ethic since the CBD (CBD, 1992) (BIOLÓGICA-CBD, 1992) that has a dual mandate for integrating conservation with sustainable use (Kumeh, 2023). Collectively, these policy tools are like a governance frame to ensure green nanotechnologies develop within ethical, environmental and socio-economic limits. However, years of campaigning have not yet arrested the habitat loss overharvesting and climate change.

Alpine vegetation is another important ecosystem to sustain not only high-altitude communities but also downstream communities. They play essential roles in stabilizing soil, hydrology, and carbon so as to maintain watershed integrity and to mitigate the result of melting and erosion of glaciers. All the alpine medicinal plants with their large root system and physiological adaptations will contribute to the resilience of the ecosystem to the environmental stressors. At the same time, the two species belong to the cultural life of the native and mountain communities whose knowledge system has evolved over the years, which revolves around them. Such as, the alpine herbs, which are used in Himalayan and Andean cultures, have not only been used in medicine, but in spiritual and magical rituals, and in the local consciousness and cosmovision; they have been ingrained in the local culture. The type of biocultural diversity emphasizes the interdependence of the ecological and cultural conservation, becoming a familiar trend in the context of international policies (Bridgewater & Rotherham, 2019).

The world herbal and phytopharmaceutical industry has emerged as a significant source of demand for the alpine plant resources economically. The medicinal and aromatic plants trade (MAPs) is projected to be worth more than USD 130 billion every year and is rising at 6-8% per annum (Parvin et al., 2023). The prime exporters of alpine botanicals include countries like India, Nepal, Bhutan, and China, with most of the products being cultivated in the wild, with little regulatory control. Although this trade generates much-needed revenue for the rural population, it also generates a self-defeating reliance upon the extraction of unsustainable resources.

Some species, including *Aconitum heterophyllum*, *Nardostachys jatamansi,* and *Saussurea costus*, are already listed in the national and international conservation lists, such as Appendix II of the Convention on International Trade in Endangered Species of Wild Fauna and Flora (CITES), due to the depletion of wild stocks of these species. This process of overuse and depletion highlights the fact that there are a lot of conservation strategies that reflect the ecological conservation and the socio-economic development.

The current conservation approaches can be broadly divided into two conservation methods, namely in situ and ex-situ conservation methods. On-site interventions, including providing sheltered grounds, biosphere organizations, and community-controlled conservation regions, are aimed at preserving species within natural settings. Such activities have been successful and unsuccessful based on issues such as financial resource unavailability, enforcement, and the issue of conservation interests and livelihood issues of the locals. Ex-situ conservation provides an effective genetic insurance through seed banks, botanical gardens and tissue culture repositories, although the method has proven ineffective in conservation because it does not address the ecological and cultural context of conservation. Moreover, the standard ex-situ methods can be costly and small-scale, and thus cannot mitigate the magnitude of the present loss of biodiversity.

2.1 Green nanotechnology

A newer paradigm that has been developed is in the integrated conservation models that seek to integrate the economic incentives, the role of scientific innovation, and the role of the community. Green nanotechnology is an innovation of this type that has shown much potential as a solution to balancing technology advancement with environmental ethics. A realistic example of how high-value products may be produced without necessarily exerting a devastating impact on the wild population is the production of nanoparticles with the assistance of alpine medicinal plant extracts, which uses a relatively small amount of biomass. Green nanotechnology is creating a value chain that is self-paying conservation by reusing lower-quantities of vegetational material into versatile nanomaterials that are exploited in medicine, farming, and in cleaning up the environment. The new approach reinstates the alpine herbs as innovation sources, not as extractive products, and the sustainability of their usage in the long term is in the effectiveness of the ecosystem conservation (Romeo et al., 2021).

In addition to the technological factor, conservation of biodiversity must also be a moral and epistemological shift, not to perceive nature as a resource, but as a co-worker of human development. Knowledge systems of the past that were excluded from mainstream science might be found in the local and indigenous sources of knowledge, offering time-tested information on sustainable practices in harvesting, the order of ecology, and spiritual control of the land. Integrating this indigenous knowledge of the environment (TEK) with current scientific frameworks can both enhance conservation initiatives and achieve the ethics of equity and benefit-sharing as envisioned in the Nagoya Protocol (2014) (Vargas-Chaves et al., 2025). Fair partnership may ensure that the local communities, who are usually the originators of medicinal plant diversity, are compensated, and physical benefits are attributed to the achievements of research and commercialization.

Finally, the protection of the alpine biodiversity is not just a scientific imperative but also a philosophical struggle for living together. Preservation of these high-altitude habitats is an environmental and ethical duty in a world that is the Anthropocene—an epoch where they systems of Earth are being altered by human activities. Alpine medicinal plants reminded us not just of a bygone, but continued opportunity- as to remind us that conservation is not about preservation, it's about rewilding what can be created naturally. This is because preservation of them ensures that biotechnology and nanoscience practices can be carried on, since they are intimately tied up with the diversity of life which in turn culminated in such innovation.

These conservation issues require new solutions which do not just limit the use of resources but also put them into a new context. Green nanotechnology develops at this junction, it provides a method of transforming small biological inputs, into high value nanomaterials, which would enable the achievement of ecological conservation, and technology development.

3. GREEN SYNTHESIS NANOPARTICLES: PRINCIPLES AND BENEFITS

Nanotechnology has become one of the most remarkable scientific frontiers in the twenty-first century, with a vast potential of application in the fields of medicine, agriculture, and environmental science. Nevertheless, the traditional techniques of nanoparticle synthesis, which are usually associated with the use of toxic substances, large-energy sources, and the generation of toxic waste, are very dangerous to the environment and health. The development of green nanotechnology has thus arisen as a result of the search for green alternatives, a science that is deeply embedded in the concept of sustainability, resource efficiency, and bio-compatibility. Plants, among the other biological systems studied to synthesize nanoparticles, have been especially promising since they have an inherent capacity to catalyze reduction and stabilization reactions using naturally found metabolites (Assad et al., 2024; Kanwal et al., 2025).

In the green synthesis strategy, reducing and capping the power of phytochemicals, including flavonoids, terpenoids, phenolic acids, alkaloids, and saponins, are applied to reduce metal ions into stable nanoscale particles under mild conditions. Plant-based synthesis is fast, scalable, and economical, unlike microbial or enzyme systems, which frequently demand very complicated culture and sterility. An aqueous extract of a medicinal plant may also serve as a compound that performs two roles: reducing metal salts (such as silver nitrate, chloroauric acid, or zinc acetate) and stabilizing the resulting nanoparticles against aggregation. Nanomaterials developed in this biogenic synthesis pathway replicate natural redox reactions, and it is therefore a very green and biocompatible method to develop nanomaterials (Assad, Naeem-ul-Hassan, et al., 2025; Jabbar et al., 2023; Mughal et al., 2021).

The alpine medicinal plants, specifically, have an outstanding potential for nanoparticle biosynthesis. The extreme abiotic conditions, including low temperatures, high UV radiation, oxidative stress, etc., that exist at high altitude; trigger the synthesis of distinctive high-altitude stress-related secondary metabolites. The compounds are also used as chemical defense against the plants, besides being reducing agents in the formation of nanoparticles. Their high levels of hydroxyl, carbonyl, and amino functional groups allow good metal ion binding, which leads to nucleation and growth of particles under control. It is this natural endowment of reactive phytochemicals that alpine plants extract can, in many cases, be found to yield nanoparticles with improved uniformity, stability, and bioactivity than those derived by synthesis by plants growing in the lowlands.

This unique ability has been proven to be true in empirical studies. *Rhodiola rosea* extracts, e.g., of *Artemisia gmelinii* and *Gentiana lutea*, have been used to effectively produce silver and gold nanoparticles, which possess strong antimicrobial and antioxidant activity. One of the studies demonstrated that alpine phytochemistry-produced silver nanoparticles were highly active in terms of multi-drug resistance of *Staphylococcus aureus* and *Pseudomonas aeruginosa*, highlighting the increased biological efficacy of silver nanoparticles (Muddassir et al., 2022). In the same way, gold nanoparticles produced with Aconitum species were able to exhibit outstanding cytotoxic impacts on different cell lines of human cancer, which could have implications of new anticancer medicine. The comprehensive nature of such biogenic nanoparticles is not only due to the metallic core but also due to the capping layer of phytochemicals that is still attached to their surface, causing the synergistic nature of biological activity (Assad, Laila, et al., 2025; Hashmi et al., 2024).

In addition to the biomedical consequences, the benefits of green-synthesized nanoparticles can be compared to other areas, such as agriculture and the environment. Metal oxide nanoparticles that have proven to be biocompatible nano fertilizers, nano-pesticides, and photocatalysts in the environment have been found in plant-derived sources, including zinc oxide (ZnO) and copper oxide (CuO). The versatile application of the green nanotechnology has been evidenced by its ability to degrade organic pollution, absorb heavy metals and enhance plant growth through provision of micronutrients. Better still, these nanoparticles can be synthesized through biologically mediated synthesis paths which makes them less toxic and more biodegradable in the environment as compared to chemically synthesized analogue nanoparticles (Assad, 2023).

Alpine medicinal plants as the sources of nanoparticles are the cross-cutting of biotechnology, green chemistry, and ethnobotany. It complies with the major concepts of sustainability that include the usage of renewable biological reagents rather than using hazardous ones. Moreover, the practice will help in conserving the biodiversity by coming up with value-added products that will not require a lot of biomass. Plant extracts can yield sufficient quantities of nanoparticles at low doses of a few milliliters and as such, it does not require harvesting of the entire wild flora on a large scale. This effectiveness is directly translated into reduced ecological pressure in order to avoid distortion of the natural habitats of the plants.

As far as conservation economics is concerned, the application of nanotechnology in medicinal research of plants can regenerate the value chains. Instead of clearing the raw plant material and selling it in bulk at low prices to collectors in the area, they can make high-tech nano-formulations, which they could get more money on. Such a shift has the advantage of enhancing livelihoods of the locals and it also leads to a sense of ownership of the biodiversity. The communities are likely to appreciate their biological heritage because they find it to be economically and technologically worthwhile, which gives them the incentive to preserve and maintain the resources in a sustainable format.

At the mechanics level, the process of the nanoparticles synthesis through the application of plants can be broadly split into three distinct processes: activation, growth, and termination. During the activation process, the plant extract phytochemicals reduce the metal ions into zero-valent form. The subsequent nucleation stage is growth of the atoms by aggregation and finally termination in which the phytochemicals cap interactions stabilize the nanoparticles. These are biochemically complicated processes yet the evolution of spectroscopic and microscopic characterisation techniques such as UV-Vis spectrophotometry, transmission electron microscopy (TEM), Fourier-transform infrared (FTIR) analysis has enhanced our understanding of how plants might synthesize nanoparticles. (The schematic representation of green nanoparticle synthesis by Alpine plant extracts in the presence of phytochemicals that act as reductant, nucleating and stabilizer is depicted in Figure 2.

With research advances, it has become more possible to develop the ability to modify the properties of nanoparticles by targeting a specific phytochemical to adorn the structure. Scientists can also control the morphology and functionality of nanoparticles for specific applications by optimizing the extraction, solvent polarity, and reaction conditions. The results of such accuracy have openings in customized therapeutics, drug delivery, and biosensing- all based on the natural chemistry of alpine herbs. The use of artificial intelligence and computational modeling in green nanotechnology is projected to further propel the discovery faster, with the potential of predictive mapping of phytochemical-nanoparticle interactions (Zafar & Rana, 2025).

In general, the green synthesis is the final example of the successful collaboration between the ancient knowledge of plants and modern science. It not only shows that ancient pharmacological sources can teach modern material science, but also gives a good example of how to use resources sustainably. Alpine medicinal plants are re-conceptualized, not as depletable commodities, but as living sources of innovation through the perspective of green nanotechnology, between the ethnomedicine, ecology, and advanced nanoscience. In the next section, the case studies will be provided to demonstrate how these principles are reflected in real-life research to show the usability of nanoparticles synthesized using alpine herbs in different ways: therapeutic, agricultural, and environmental.

Figure 2. Schematic representation of the synthesis of green nanoparticles using the extracts of alpine plants, including the reduction, nucleation and stabilization stages

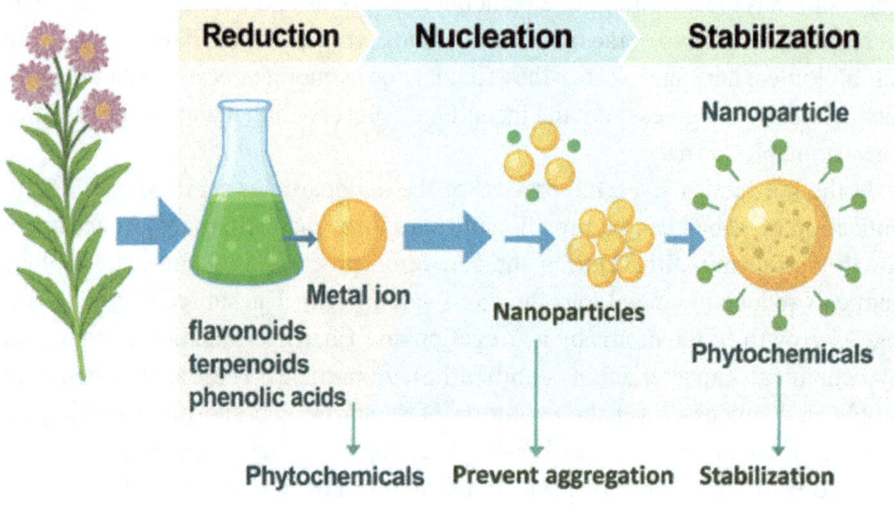

4. CASE STUDY: THE ALPINE MEDICINAL PLANTS IN THE SYNTHESIS AND APPLICATIONS OF NANOPARTICLES

The use of alpine medicinal plants in green synthesis of nanoparticles has become an energetic interdisciplinary research area, bringing together ethnobotany, chemistry, and nanobiotechnology. The next section illustrates some of the chosen cases in which alpine flora have been utilized as a bio factory to produce nanoparticles, which have produced materials with impressive biomedical, agricultural, and environmental uses. These illustrations highlight how alpine phytochemistry could be advanced to lead to sustainable nanotechnological innovation and also play a role in protecting biodiversity and empowering the local communities economically.

One of the first and most frequently mentioned studies in this field is the biosynthesis of silver nanoparticles (AgNPs) with extracts of *Rhodiola rosea*, which is a low-altitude adaptogenic herb with known antioxidant and anti-stress effects. The authors proved that rhizome aqueous extracts of *R. rosea* were an effective way to reduce silver ions to a nanoscale silver material, and spherical nanoparticles of 10-25 nm in diameter were obtained (Hu et al., 2021). The biogenic AgNPs that were obtained were found to have a strong antibacterial effect against multidrug-resistant pathogens, including *Escherichia coli, Staphylococcus aureus*, and *Pseudomonas aeruginosa*. In addition to antimicrobial ability, these nanoparticles demonstrated considerable antioxidant activity, due to the remaining salidrosides and flavonoid capping on the surfaces of nanoparticles (Ghaffar et al., 2023). This dual capacity, i.e., metal-directed reactivity with phytochemical bioactivity, reflects the special benefit of plant-based nanomaterials compared with chemically-made analogues.

Another alpine herb containing terpenoids and phenolic acid was tested as a source of gold nanoparticles (AuNPs) using *Artemisia gmelinii*. The reaction took place at a rapid speed in ambient conditions, producing very stable colloidal gold with a narrow particle-size distribution. These AuNPs showed impressive anticancer functions in in vitro cell death experiments, which caused human cervical carcinoma cells death through mitochondrial membrane depolarization. The combined effect of metal nanoparticles and the phytoconstituents of the plant indicates that it is possible to create naturally derived nanomedicines that are less toxic than regular chemotherapeutics (Rao et al., 2016).

The utilization of *Gentiana lutea*, a typical herb of the European Alps, which is a reducing agent to zinc oxide nanoparticles (ZnONPs), is also promising. Gentiopicroside and amarogentin extracts formed hexagonal ZnONPs that possess photocatalytic properties to use in environmental clean-up. These nanoparticles were effective in degrading synthetic dyes like methylene blue in the presence of solar light, and they also show a green pathway to industrial pollution (Nagajyothi et al., 2020). They also have a relatively high level of cytotoxicity, which reinforces their application in eco-friendly applications, such as in nanofertilizers and antimicrobial agricultural coatings.

Other researchers have also put forth *Aconitum heterophyllum* and *Saussurea costus* in the biosynthesis of nanoparticles, as both of which are high-value Himalayan herbs. The *A. heterophyllum* extracts, which have long been used in Tibetan and Ayurvedic medicine, have aided in preparing copper oxide nanoparticles (CuONPs) with a tremendous antifungal and antioxidant effect (Khan et al., 2023). The alkaloid-rich extract was used as a reducing and stabilizing agent with dual functions that resulted in the formation of nanostructures with homogeneous morphology. However, *Saussurea costus* has been reported to express iron oxide nanoparticles (Fe_3O_4NPs) with the potential to act as magnetic resonance imaging (MRI) contrast agents. Such biogenic nanoparticles represent a combination of diagnostic opportunities and biocompatibility, which precondition the perspective of developing cost-efficient nanodiagnostics made of renewable plant substances (Patra et al., 2021).

In addition to biomedical applications, the alpine plant-based nanoparticles have also been used in agriculture and environmental control. As an example, AgNPs of *Rhodiola rosea*, at low concentrations, induced seed germination and root growth in *Arabidopsis thaliana*, implying that AgNPs of *Rhodiola rosea* could be involved in plant growth promotion. Similarly, ZnONPs prepared with the help of *Gentiana lutea* were shown to work as nutritional supplements, better preventing chlorophyll content and stress-resilience in crops cultivated in a dry environment. Such results suggest that green nanotechnology is not only able to play a role in the innovativeness of pharmacological advancement, but also in sustainable agriculture- a factor of increasing significance in climate-affected mountain territories.

The other interesting issue is environmental remediation. The nanoparticles produced by biogenous sources are ideal in removing air and water pollution by their reactive property and the large area that they have. Silver nanoparticles and copper oxide nanoparticles extracted out of alpine plants have been used as catalysts to disintegrate heavy metals, dye, and pesticides. AgNPs prepared using *Arnica montana* were particularly noteworthy since they diminished methyl orange and Congo red by more than 90% within only two hours when in the presence of UV light (Shahzadi et al., 2025). These green catalysts have competitive alternatives to synthetic photocatalysts due to their combination of efficiency and being ecologically friendly. Table 2 provides an overview of the range of nanoparticles made of alpine herbs, including some of the most important species, nanoparticle combinations, along with the associated bioactions.

Table 2. Summary of selected case studies: Alpine medicinal plants in nanoparticle synthesis and applications

Plant Species	Type of Nanoparticle	Major Phytochemicals Involved	Key Applications	Reference
Rhodiola rosea	AgNPs	Salidroside, Rosavin	Antibacterial, antioxidant, plant growth stimulant	(Hu et al., 2021)
Artemisia vulgaris	AuNPs	Flavonoids, Terpenoids	Antimicrobial; Anticancer, anti-inflammatory	(Sundararajan & Kumari, 2017)
Swertia chirayita	ZnONPs	Amarogentin, swertiamarin	Photocatalytic dye degradation (MB); antimicrobial	(Akhter et al., 2018)
Aconitum heterophyllum	AgNPs	Aconitine alkaloids	Antifungal, antioxidant	(Soni, 2024)
Saussurea costus	Fe_3O_4NPs	Costunolide, Lactones	Antimicrobial (species-specific); MRI contrast agent, biosensing	(Al-Brahim, 2023)
Arnica montana	AgNPs	Helenalin, Phenolic acids	Dye degradation, antimicrobial coatings	(Dobrucka & Długaszewska, 2015)

The meeting of nanotechnology and alpine phytochemistry demonstrated in these works provides a radical way of using resources. Through linking the local biodiversity and high-value nanoproduct creation, researchers and communities can collaboratively cultivate an example of innovation that compensates conservation instead of extraction. Notably, these examples help to prove that technological progress does not necessarily have to be associated with ecological integrity. The fact that nanoparticles can be produced using a very low biomass content implies that even the endangered species can be used sustainably, as long as bioprospecting follows the ethical considerations and benefit-sharing models. Moreover, the scalability of green nanotechnology to small-scale laboratory and community systems can be promising to the area of decentralized innovation in the mountainous regions. Local cooperatives and research institutes can make an alliance to generate bio-based nanomaterials with local uses- herbal-based drug delivery systems can be bio-based nanomaterials to biodegradable nanocoatings to preserve food. This can be done in a manner that gets us to the intersection of scientific discovery and grassroots empowerment, and economic opportunity, as well as conservation ethics. According to these case studies, it is clear that the fusion of alpine ethnobotany with nanoscience is not only a technological innovation, but a realignment of nature and innovation. This synergistic model will be further elaborated in the passage below on how nanotechnology can be used as a conservation strategy and also as the driving force of sustainable development in the alpine delicate ecosystems.

5. A SYNERGISTIC FRAMEWORK

One of the most promising fields of sustainable science is the nanotechnology/biodiversity conservation interface. Nanotechnology can potentially serve as a strategic connection between innovation and conservation in the environment of alpine ecosystems, in which the weakness of the environment is partially compensated by the astonishing biological prosperity. A good example of this synergy is the synthesis of nanoparticles with alpine medicinal plants, which are becoming more of scientific development platforms through the transduction of such species into renewable raw materials. This conservation paradigm is best summarized as a revised form of bioprospecting, which balances the economic benefits as well as the ecological system and the cultural reverence as a new conservation paradigm. The first principle of this framework is the efficiency of resources. The traditional herbal medicine industries tend to exploit large volumes of plant biomass to obtain low volumes of bioactive compounds.

On the contrary, green nanotechnology makes use of very small portions of the plant material, usually in the shape of aqueous or ethanolic extracts, to create nanoparticles with much higher potency per unit weight. This large saving of biomass needs is directly converted into less harvesting stress on wild stocks. As an example, a silver or gold nanoparticle synthesis using *Rhodiola rosea* or *Gentiana lutea* may be obtained using a tenth of the plant material needed to obtain a traditional extract, though the nanomaterials in the solution frequently have a greater bioactivity (Rónavári et al., 2021). Nanotechnology has succeeded in separating economic benefit and environmental destruction by substituting mass mining with molecular exploitation. Besides preserving plant populations, nanotechnology provides new value addition to the existing resources. When the bioactive compounds of alpine herbs are incorporated into nanoscale preparations, their therapeutic effect is usually expanded.

Nanoparticles may be used as vectors that enhance bioavailability, solubility, and targeted delivery of plant compounds, which are inherently limited by traditional herb preparations. This innovation of conventional medicine to high-tech nanoformulations boosts the market potential of alpine botanicals, and the products are now placed in the niche of high-tech pharmaceutical and biomedical companies. To communities in the area that rely on the medicinal trading of plants, this is a chance to engage in the value-added economies as opposed to continuing to be stuck in a low-profit collection and supply chain. The resulting economic benefits that are created by bioproducts developed by nanotechnology can be used to fund conservation efforts.

It is also possible to promote investment in the conservation of such habitats by showing governments and the business world the technological and commercial importance of alpine biodiversity. This would help redefine conservation as a force behind development rather than a limitation to it. This connection between innovation and stewardship can be institutionalized by the creation of bioeconomic models that include a portion of the profits generated by nano-derived products being reinvested in restoring habitats, seed banking, or local conservation funds. Research on the advantages of benefit-sharing frameworks along these lines has already been experimented with under the Nagoya Protocol on Access and Benefit-Sharing (ABS), which provides a global legal framework of fair collaboration between researchers and holders of indigenous knowledge (Greiber, 2012).

The other important aspect is the incorporation of traditional knowledge in nanotechnological studies. The native people living in the alpine areas have a rich ethnobotanical knowledge on how to use plants and which plants to use, which is based on many years of observation and experience. This includes the recognition of the powerful species, the best harvesting time, and methods of preparation- all these are priceless when it comes to the selection and preparation of plants to be used to create nanoparticles. Their cooperation in the research paradigms that recognize and bring together such traditional wisdom not only leads to the enhancement of scientific performance but also upholds the virtue of cultural esteem and intellectual rights.

Ethnopharmacological knowledge is more likely to be strengthened by writing up and authenticating it into a nanotechnological system, thus increasing cultural fortitude in addition to scientific exploration. An expression of nanotechnology in preservation is not just limited to the theoretical level, but also the practical level in developing sustainable bioprospecting systems. These models pay much attention to minimizing the impact on the resources, interaction with the local community, and reinvestment in the circular economy. Indicatively, the mountainous regions can be provided with community-based laboratories to produce biogenic nanoparticles to be utilized in regional medical care, agriculture, or environmental remediation. These attempts may transform the traditional harvesting societies into entities of the innovation ecosystem by decentralizing the production and value-creating system. This technological democratization is ensured in a manner that economic gains are not distributed across local economies, and this reduces the socioeconomic aspects of over-exploitation.

Moreover, nanotechnology is interdisciplinary and this would allow for a cross-sectoral cooperation. Support of the UN SDG12 and 15 policy mandates (Responsible consumption and production & Life on land, respectively) can be established by forging alliances between academic organizations, environmental NGOs and biotech enterprises that catalyze a pipeline of innovations. The potential uses in green nanotechnology of the alpine herbs can potentially be scaled by integrating their use into policy, investing in research and technology transfer arrangements which call forth performance based approach to science as well as conservation ethics.

However, this synergy must be syncretic to established governance and surveillance. Through sound policy framework, environmental safety, biological integrity and non-sleazy trade of bio-nano materials from life forms can be achieved. The traceability of the procurement of plant materials and the standardization of protocols for nanoparticles production and for end-to-end toxicity test are at the basis of responsible innovation. It can be the regenerative (engineering) equivalent of technology If nature is in, it's 'Nano" and Nano Technology becomes a 'degenerative' technology that sucks away natural capital Exploring Ways How A Conservation Based Science Can Transform Into Making Regenerative/Tailored Consumer Products – Creating Nature's Style Commodity Markets by Design.

Finally, a combination of the alpine biodiversity and nanoscience represents the comprehensive vision of sustainability. It shows that the most innovative technologies may be created based on the most ancient ones - the herbs of the mountains, which have undergone the adaptation processes during thousands of years, may become the drivers of 21st century innovations. Such a relationship is a classic example of a circular model of knowledge and material flow: nanoparticles are synthesized using bioactive compounds in plants; the products of a high social and economic value can be produced using nanoparticles; the products in their turn can generate incentives and resources to conserve the very ecosystems where they are produced. In this circle of continuation, conservation and innovation no longer oppose- they work in harmony with each other in a common future. Figure 3 presents a circular model illustrating the synergistic cycle between alpine plant-based nanoproducts, economic incentives, and biodiversity conservation.

Figure 3. Circle model of the development of nanoproduct, the synergistic cycle between alpine plant development, economic incentive, and biodiversity conservation

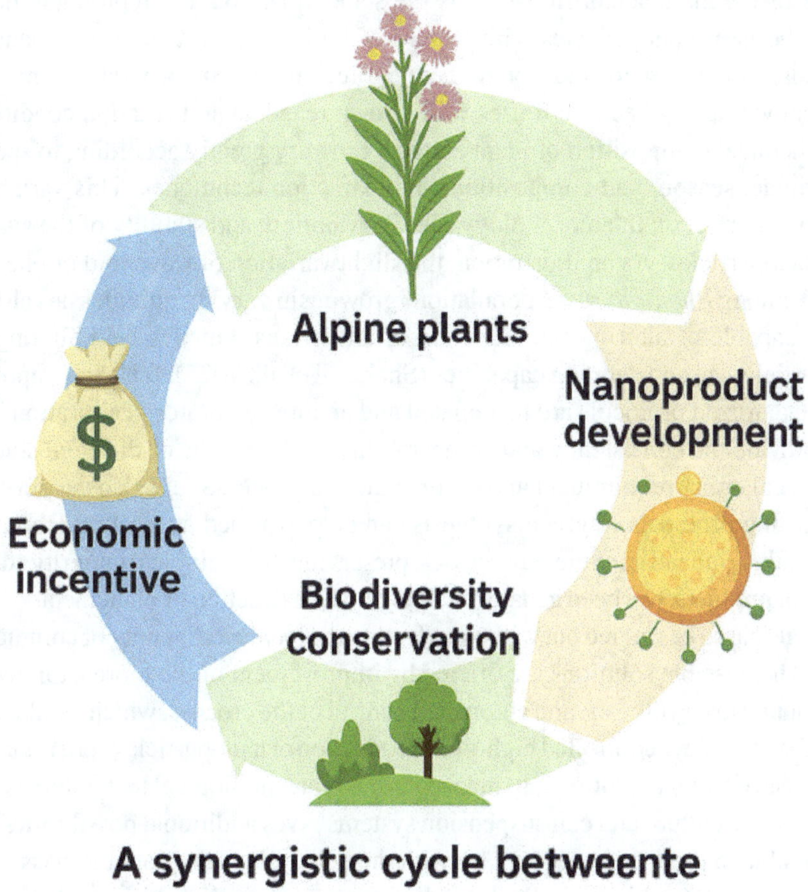

6. ISSUE AI INTEGRATION AND FUTURE PROJECTIONS

Although the integration of alpine medicinal plants and green nanotechnology provides the way out of the current and seemingly endless challenges of sustainability and innovation, the field of interdisciplinary is not without its challenges. Such biogenic synthesis systems have certain scientific, technical, ethical, and regulatory challenges that need to be overcome before they can be applied on a large scale.

These issues are vital in the realization of the long-term potential of nanotechnology as a device in conserving biodiversity and sustainable development.

One of the main scientific issues is the standardization and reproducibility of the synthesis of nanoparticles with the use of biological extracts. Plant-mediated synthesis, in contrast to traditional chemical techniques, incorporates some biochemical variability because it uses well-known reagents and reaction conditions. Phytochemical composition of plant extracts can vary greatly according to species and altitude, season, soil composition, and extraction techniques. This variability has a direct effect on the morphology, size distribution, and stability of the synthesized nanoparticles. As an illustration, the slight variation of flavonoid or phenolic content among *Rhodiola rosea* populations growing in varying altitudes has already proven capable of altering the reduction of silver ions, kinetics, and altering the nanoparticle size and capping capacities (Shahzadi et al., 2025). It is then important that standardized protocols are formulated and an inter-laboratory calibration is set up to provide the consistency and comparability of the results of different studies.

The scalability requirement and optimization of processes are also very critical. The existing protocols of green synthesis can be performed at most on laboratory scales. Industrialization of these processes, preserving the ecological integrity and the quality of products, has been a major challenge. The extraction of plant metabolites a large scale has to be carried out without affecting the local biodiversity or community rights. The possible solutions are offered by improvement of the bioreactor design, continuous-flow synthesis, and automated control of the process, which would make it possible to achieve controlled high-yield production of nanoparticles under standard conditions. Further evolution into incorporating biotechnological technologies such as plant tissue culture and cell suspension systems gives additional possibilities, and it is possible to produce nanoparticles with the help of in vitro plant biomass rather than wild-harvested biomass. Such kind of innovation will enable the elimination of the production of nanoparticles and its direct ecological interdependence, which will lead to the enhanced sustainability and adherence to ethical standards (Assad, Ansa, et al., 2025; Babatunde et al., 2019).

The second serious concern is linked to bio-safety and nanotoxic impact of biogenic nanoparticles. In spite of the fact that the green synthesis minimizes the toxicity of the chemicals employed in the manufacturing process, the biological association of the generated nanomaterials with the living organisms and ecosystems has not been established completely. Among the factors that may influence cellular uptake, biodistribution and potential toxicity may be particle size, surface chemistry and reactivity. Research studies have demonstrated that even biocompatible silver and gold nanoparticles, like those made of plants, can cause oxidative stress or inflammatory reactions at higher doses (Shrivastava et al., 2016). Extensive risk assessment models that address the environmental and human health effects are hence inevitable. These models of eco-toxicology, combined with long-term exposure testing, will be critical in determining the levels of safety and in providing regulatory permission in the case of commercial use.

In addition to scientific and technical aspects, ethical and legal aspects are another source of complexity. The rising trend of utilization of traditional flora in value-added nanotechnological use brings in doubts on intellectual property rights (IPR), benefit sharing, and biopiracy. Most of the alpine medicinal plants are components of indigenous knowledge that has been passed down through the generations. Following the Nagoya Protocol on Access and Benefit-Sharing, it is the obligation of researchers and companies to have fair compensation and recognition of the indigenous input. Open contracts with clear specifications of profit-sharing, co-authorship, and a technology transfer system can create fair collaboration between the local communities and research institutions. In the absence of any such structures, historical trends of exploitation are likely to be repeated in the name of innovation.

The other major issue is the establishment of effective regulatory systems of green nanotechnology. Nanoparticle research has a very high rate of innovation, and the rate of innovation is often higher than the rate of legal regulation. Current policies on chemical nanomaterials might not effectively embrace the unique properties of biogenic nanoparticles, especially those that are made out of complex plant matrices. International cooperation is needed to bring all the definitions, safety standards, and testing approaches to a unified level. A global database of plant-based nanomaterials, characterized and toxicity data, would help make decisions and other cross-border research alignments. Moreover, there is a need to have clear guidelines on the release of nanoproducts into the environment and waste management, as well as the disposal of these products at the end of life to avoid unwanted ecological effects (Bystrzejewska-Piotrowska et al., 2009).

Technological convergence makes the future of green nanotechnology predictable in terms of a scientific innovation point of view. Artificial intelligence (AI) and machine learning in nanoparticle studies represent an enormous opportunity to predict structures of syntheses and properties of materials. Algorithms developed through AI can be used to analyze large volumes of plant phytochemical profiles to identify the best conditions in which nanoparticles can be formed without having to experimentally determine this information. Likewise, nanoparticle biosynthesis could be taken in a promising path by exploring endophytic microorganisms, such as bacteria and fungi that coexist symbiotically in the alpine plants. These endophytes tend to have the same secondary metabolite pathways as the plant, which allows production of nanoparticles without extracting plant tissues. These microbial nanofactories have the potential to transform the industry, as they enable the integration of biological sustainability with the scale of the industry (Agrawal et al., 2022).

Moreover, more studies are required in the future to combine the omics technologies, such as genomics, metabolomics, and proteomics, to understand the molecular origins of biogenic nanoparticle synthesis. Knowledge of the role played by certain enzymes and metabolites in reduction and capping will provide an opportunity to rationally design the following-generation nanomaterials with selective properties. This systems-level understanding, combined with the power of computational chemistry and molecular dynamics simulation, can hasten the creation of custom-designed nanocomposites to be used in specific biomedical, agricultural, and environmental applications.

Finally, the effectiveness of the science-policy integration and engagement with the general audience is a key to the success of this new field. Policymakers, researchers, and local communities are forced to cooperate in order to come up with guidelines that will strike a balance between innovation and ecological ethics. The local population in mountain areas can be empowered and, therefore, become active participants of the scientific value chain through awareness programs, open access data sharing, and capacity-building efforts. Integrating conservation ethics into the education programs and research funding models is another way of making sure that the values of sustainability and equity will continue to be at the heart of technological advancement.

Usually, the journey of alpine herbs to nanotechnology is a hard one, but it is also full of potential. The task of interdisciplinary collaboration to tackle challenges of standardization, safety, and equity matters will be important in achieving the dream of a nanoscience and biodiversity coexisting in self-enrichment. The meeting point of ancient botanical knowledge, modern biotechnology, and new information computing technologies is an announcement of a new era of eco-innovation, the new way to describe progress not as a means of exploiting nature, but as an alliance with nature. The following deduction will summarize this vision and confirm the role of green nanotechnology as a scientific revolution and moral obligation to care for life in all its variety.

7. CONCLUSION

The Alpine ecosystems signify both the images of durability as well as images of vulnerability. The medicinal herbs growing in such elevated plots have outstanding biochemical adaptations which enable them to survive extreme cold, radiation, and drought, which will also ensure that they are invaluable in the traditional and drug discovery fields. However, their success has contributed to their overexploitation at the suffering of the species and ecological balance. Green nanotechnology is a way to move forward, as these alpine plants are used as bio factories to produce nanoparticles, and one of the ways to create high-value biomedical and environmental materials with minimal biological cost. This change, an extraction to regeneration, makes the human being reconsider the relationship with nature and that phytochemicals found in plants can be used as nanoscale structures to treat human diseases as well as the environment. The studies of such species as *Rhodiola rosea, Gentiana lutea, Arnica montana*, and *Saussurea costus* prove that silver, gold, zinc, and iron nanoparticles can be used in many different ways, including antimicrobial and antioxidant activity, dye degradation under light, or cancer treatment. These inventions represent an alliance of natural intelligence and human ingenuity, a molecular dialogue of science working with and not against nature. Nevertheless, the matter of attaining real sustainability involves consideration of issues of reproducibility, biosafety, and equitable sharing of benefits with native mountain communities. The use of ethical frameworks needs to fulfill the role of sustaining conservation purposes and the rights of communities as science progresses. In the future, nanotechnology coupled with artificial intelligence, metabolomics, and synthetic biology is almost certain to speed up the discovery and cut back on the reliance on wild harvesting, leading to predictive, eco-friendly innovation. Lastly, the tale of alpine herbs and nanotechnology is not just a tale of scientific development, but a call to at least incorporate ecological responsibility at the molecular scale. The technology that respects the wisdom of nature can make the tiniest man-made particles mediators of healing to the world. Here, with this combination of prehistoric and scientific, man can learn to innovate according to the framework of nature, yet keeping the ecosystems that motivate and promote his creativity. We might still find a way to heal the planet by saving the alpine herbs that contribute to the health of our bodies.

REFERENCES

Agrawal, K., Gupta, V. K., & Verma, P. (2022). Microbial cell factories a new dimension in bio-nanotechnology: Exploring the robustness of nature. *Critical Reviews in Microbiology*, *48*(4), 397–427. DOI: 10.1080/1040841X.2021.1977779 PMID: 34555291

Akhter, S. M. H., Mahmood, Z., Ahmad, S., & Mohammad, F. (2018). Plant-mediated green synthesis of zinc oxide nanoparticles using Swertia chirayita leaf extract, characterization and its antibacterial efficacy against some common pathogenic bacteria. *BioNanoScience*, *8*(3), 811–817. DOI: 10.1007/s12668-018-0549-9

Al-Brahim, J. S. (2023). Saussurea costus extract as bio mediator in synthesis iron oxide nanoparticles (IONPs) and their antimicrobial ability. *PLoS One*, *18*(3), e0282443. DOI: 10.1371/journal.pone.0282443 PMID: 36893115

Assad, N. (2023). Aconitum lycoctonum L.(Ranunculaceae) mediated biogenic synthesis of silver nanoparticles as potential antioxidant, anti-inflammatory, antimicrobial and antidiabetic agents. *BMC Chemistry*, *17*(1), 128. DOI: 10.1186/s13065-023-01047-5 PMID: 37770921

Assad, N. Ansa, Laila, M. B., Ullah, S., Adnan, M., Wahab, S.,…Ullah, B. (2025). Nanotechnology in Oilseed Crops: A New Frontier for Abiotic Stress Adaptation. In *Oilseed Crops Under Abiotic Stress: Mitigation Strategies and Future Perspectives* (pp. 507-536). Springer.

Assad, N., Abbas, A., Fayyaz ur Rehman, M., & Naeem-ul-Hassan, M. (2024). Photocatalytic and biological applications of phyto-functionalized zinc oxide nanoparticles synthesized using a polar extract of Equisetum diffusum D. *RSC Advances*, *14*(31), 22344–22358. DOI: 10.1039/D4RA03573A PMID: 39010906

Assad, N., Laila, M. B., Hassan, M. N. U., Rehman, M. F., Ali, L., Mustaqeem, M., Ullah, B., Khan, M. N., Iqbal, M., ErciŞli, S., Alarfaj, A. A., Ansari, M. J., & Malik, T. (2025). Eco-friendly synthesis of gold nanoparticles using Equisetum diffusum D. Don. with broad-spectrum antibacterial, anticancer, antidiabetic, and antioxidant potentials. *Scientific Reports*, *15*(1), 19246. DOI: 10.1038/s41598-025-02450-9 PMID: 40456810

Assad, N., Naeem-ul-Hassan, M., Ajaz Hussain, M., Abbas, A., Sher, M., Muhammad, G., Assad, Y., & Farid-ul-Haq, M. (2025). Diffused sunlight assisted green synthesis of silver nanoparticles using Cotoneaster nummularia polar extract for antimicrobial and wound healing applications. *Natural Product Research*, *39*(8), 2203–2217. DOI: 10.1080/14786419.2023.2295936 PMID: 38146228

Babatunde, D. E., Denwigwe, I. H., Babatunde, O. M., Gbadamosi, S. L., Babalola, I. P., & Agboola, O. (2019). Environmental and societal impact of nanotechnology. *IEEE Access : Practical Innovations, Open Solutions, 8*, 4640–4667. DOI: 10.1109/ACCESS.2019.2961513

Biológica-CBD, C. P. D. (1992). *Convention on Biological Diversity*. Nações Unidas.

Bridgewater, P., & Rotherham, I. D. (2019). A critical perspective on the concept of biocultural diversity and its emerging role in nature and heritage conservation. *People and Nature, 1*(3), 291–304. DOI: 10.1002/pan3.10040

Bystrzejewska-Piotrowska, G., Golimowski, J., & Urban, P. L. (2009). Nanoparticles: Their potential toxicity, waste and environmental management. *Waste Management (New York, N.Y.), 29*(9), 2587–2595. DOI: 10.1016/j.wasman.2009.04.001 PMID: 19427190

de Lima, E. P., Laurindo, L. F., Catharin, V. C. S., Direito, R., Tanaka, M., Jasmin Santos German, I., & Fiorini, A. M. R. (2025). Polyphenols, alkaloids, and terpenoids against neurodegeneration: Evaluating the neuroprotective effects of phytocompounds through a comprehensive review of the current evidence. *Metabolites, 15*(2), 124. DOI: 10.3390/metabo15020124 PMID: 39997749

Dobrucka, R., & Długaszewska, J. (2015). Antimicrobial activities of silver nanoparticles synthesized by using water extract of Arnicae anthodium. *Indian Journal of Microbiology, 55*(2), 168–174. DOI: 10.1007/s12088-015-0516-x PMID: 25805903

Ghaffar, S., Abbas, A., Naeem-ul-Hassan, M., Assad, N., Sher, M., Ullah, S., Alhazmi, H. A., Najmi, A., Zoghebi, K., Al Bratty, M., Hanbashi, A., Makeen, H. A., & Amin, H. M. A. (2023). Improved photocatalytic and antioxidant activity of olive fruit extract-mediated ZnO nanoparticles. *Antioxidants, 12*(6), 1201. DOI: 10.3390/antiox12061201 PMID: 37371931

Greiber, T. (2012). *An explanatory guide to the Nagoya Protocol on access and benefit-sharing*. Iucn.

Hashmi, S. S., Ibrahim, M., Adnan, M., Ullah, A., Khan, M. N., Kamal, A., & Elshikh, M. S. (2024). Green synthesis of silver nanoparticles from Olea europaea L. extracted polysaccharides, characterization, and its assessment as an antimicrobial agent against multiple pathogenic microbes. *Open Chemistry, 22*(1), 20240016. DOI: 10.1515/chem-2024-0016

Hu, D., Yang, X., Chen, W., Feng, Z., Hu, C., Yan, F., Chen, X., Qu, D., & Chen, Z. (2021). Rhodiola rosea rhizome extract-mediated green synthesis of silver nanoparticles and evaluation of their potential antioxidant and catalytic reduction activities. *ACS Omega*, *6*(38), 24450–24461. DOI: 10.1021/acsomega.1c02843 PMID: 34604627

Islam, R., Sun, L., & Zhang, L. (2021). Biomedical applications of Chinese herb-synthesized silver nanoparticles by phytonanotechnology. *Nanomaterials (Basel, Switzerland)*, *11*(10), 2757. DOI: 10.3390/nano11102757 PMID: 34685197

Jabbar, A., Abbas, A., Assad, N., Naeem-ul-Hassan, M., Alhazmi, H. A., Najmi, A., Zoghebi, K., Al Bratty, M., Hanbashi, A., & Amin, H. M. A. (2023). A highly selective Hg 2+ colorimetric sensor and antimicrobial agent based on green synthesized silver nanoparticles using Equisetum diffusum extract. *RSC Advances*, *13*(41), 28666–28675. DOI: 10.1039/D3RA05070J PMID: 37790097

Kanwal, M., Sher, M., Abbas, A., Akhtar, S., Siddique, A. B., Hasan, M. N., Assad, N., Alhazmi, H. A., & Amin, H. M. A. (2025). Dual colorimetric sensing of Hg (II) and Fe (III) using sulfanilamide-stabilized silver nanoparticles and evaluating their photodegradation and antibacterial properties. *Journal of Water Process Engineering*, *75*, 107981. DOI: 10.1016/j.jwpe.2025.107981

Khan, Z., Assad, N., Naeem-ul-Hassan, M., Sher, M., Alatawi, F. S., Alatawi, M. S., Omran, A. M. E., Jame, R. M. A., Adnan, M., Khan, M. N., Ali, B., Wahab, S., Razak, S. A., Javed, M. A., Kaplan, A., & Rahimi, M. (2023). Aconitum lycoctonum L.(Ranunculaceae) mediated biogenic synthesis of silver nanoparticles as potential antioxidant, anti-inflammatory, antimicrobial and antidiabetic agents. *BMC Chemistry*, *17*(1), 128. DOI: 10.1186/s13065-023-01047-5 PMID: 37770921

Kumeh, E. M. (2023). Contestations, counteractions and equitable conservation–a case study of Ghana's Krokosua Hills Forest Reserve. *Forest Policy and Economics*, *157*, 103090. DOI: 10.1016/j.forpol.2023.103090

Muddassir, M., Raza, A., Munir, S., Basirat, A., Ahmed, M., Butt, M. S., Dar, O. A., Ahmed, S. S., Shamim, S., & Naqvi, S. Z. H. (2022). Antibacterial efficacy of silver nanoparticles (AgNPs) against metallo-β-lactamase and extended spectrum β-lactamase producing clinically procured isolates of Pseudomonas aeruginosa. *Scientific Reports*, *12*(1), 20685. DOI: 10.1038/s41598-022-24531-9 PMID: 36450765

Mughal, B., Zaidi, S. Z. J., Zhang, X., & Hassan, S. U. (2021). Biogenic nanoparticles: Synthesis, characterisation and applications. *Applied Sciences (Basel, Switzerland)*, *11*(6), 2598. DOI: 10.3390/app11062598

Nagajyothi, P. C., Prabhakar Vattikuti, S. V., Devarayapalli, K. C., Yoo, K., Shim, J., & Sreekanth, T. V. M. (2020). Green synthesis: Photocatalytic degradation of textile dyes using metal and metal oxide nanoparticles-latest trends and advancements. *Critical Reviews in Environmental Science and Technology*, *50*(24), 2617–2723. DOI: 10.1080/10643389.2019.1705103

Nagy, L., & Grabherr, G. (2009). *The biology of alpine habitats*. Oxford University Press. DOI: 10.1093/oso/9780198567035.001.0001

Parvin, S., Reza, A., Das, S., Miah, M. M. U., & Karim, S. (2023). Potential role and international trade of medicinal and aromatic plants in the world. *European Journal of Agriculture and Food Sciences*, *5*(5), 89–99. DOI: 10.24018/ejfood.2023.5.5.701

Patra, C., Ahmad, I., Ayaz, M., Khalil, A. T., Mukherjee, S., & Ovais, M. (2021). *Biogenic nanoparticles for cancer theranostics*. Elsevier.

Petelka, J., Plagg, B., Säumel, I., & Zerbe, S. (2020). Traditional medicinal plants in South Tyrol (northern Italy, southern Alps): Biodiversity and use. *Journal of Ethnobiology and Ethnomedicine*, *16*(1), 74. DOI: 10.1186/s13002-020-00419-8 PMID: 33243238

Rao, P. V., Nallappan, D., Madhavi, K., Rahman, S., & Wei, J. (2016). L., & Gan, S. H. (2016). Phytochemicals and biogenic metallic nanoparticles as anticancer agents. *Oxidative Medicine and Cellular Longevity*, *2016*(1), 3685671. DOI: 10.1155/2016/3685671 PMID: 27057273

Romeo, R., Manuelli, S., Geringer, M., & Barchiesi, V. (2021). *Mountain farming systems–Seeds for the future: Sustainable agricultural practices for resilient mountain livelihoods*. Food & Agriculture Org.

Rónavári, A., Igaz, N., Adamecz, D. I., Szerencsés, B., Molnar, C., Kónya, Z., Pfeiffer, I., & Kiricsi, M. (2021). Green silver and gold nanoparticles: Biological synthesis approaches and potentials for biomedical applications. *Molecules (Basel, Switzerland)*, *26*(4), 844. DOI: 10.3390/molecules26040844 PMID: 33562781

Shahzadi, S., Fatima, S., Shafiq, Z., & Janjua, M. R. S. A. (2025). A review on green synthesis of silver nanoparticles (SNPs) using plant extracts: A multifaceted approach in photocatalysis, environmental remediation, and biomedicine. *RSC Advances*, *15*(5), 3858–3903. DOI: 10.1039/D4RA07519F PMID: 39917042

Shrivastava, R., Kushwaha, P., Bhutia, Y. C., & Flora, S. (2016). Oxidative stress following exposure to silver and gold nanoparticles in mice. *Toxicology and Industrial Health*, *32*(8), 1391–1404. DOI: 10.1177/0748233714562623 PMID: 25548373

Soni, R. (2024). Green synthesis and characterization of metallic nanoparticles of Aconitum heterophyllum as antifungal agents. *Frontiers in Health Informatics, 13*(7).

Sundararajan, B., & Kumari, B. R. (2017). Novel synthesis of gold nanoparticles using Artemisia vulgaris L. leaf extract and their efficacy of larvicidal activity against dengue fever vector Aedes aegypti L. *Journal of Trace Elements in Medicine and Biology, 43*, 187–196. DOI: 10.1016/j.jtemb.2017.03.008 PMID: 28341392

Tahir, M. F., Khan, M. Z., Attacha, S., Asim, N., Tayyab, M., Ali, A., Militky, J., & Tomková, B. (2022). The comparative performance of phytochemicals, green synthesised silver nanoparticles, and green synthesised copper nanoparticles-loaded textiles to avoid nosocomial infections. *Nanomaterials (Basel, Switzerland), 12*(20), 3629. DOI: 10.3390/nano12203629 PMID: 36296819

Vargas-Chaves, I., Lopez-Oliva, J., & Alarcon-Pena, A. (2025). The Intersections of Intellectual Property and Traditional Ecological Knowledge: A Critical Analysis within the Nagoya Protocol Framework. *Contemporary Readings in Law and Social Justice, 17*, 267.

Zafar, S., & Rana, N. (2025). The convergence of nanotechnology and artificial intelligence: Unlocking future innovations. *Recent Innovations in Chemical Engineering, 18*(2), 85–99. DOI: 10.2174/0124055204359215250127071256

Chapter 10
Proximate and Nutritional Study of the Wild Food Plants (WFPs) of the Alpine Zones

Nasir Assad
http://orcid.org/0009-0004-9435-1414
University of Sargodha, Pakistan

Muhammad Naeem Ul Hassan
University of Sargodha, Pakistan

Tooba
University of Sargodha, Pakistan

Yasir Assad
Hazara University, Mansehra, Pakistan

Marzia Batool Laila
University of Sargodha, Pakistan

Attia Khalid
University of Sargodha, Pakistan

Muhammad Khalid Mukhtar
University of Sargodha, Pakistan

Amjad Ali
http://orcid.org/0000-0001-8157-576X
Universita Cattolica del Sacro Cuore, Italy

Ummul Banin Zahra
Islamia College Peshawar, Pakistan

Muhammad Adnan
Bacha Khan University, Charsadda, Pakistan

ABSTRACT

Alpine medicinal plants, adapted to high-altitude environments, represent an important source of bioactive compounds with therapeutic potential. Exposure to low temperatures, intense ultraviolet radiation, and nutrient-poor soils has promoted the synthesis of diverse secondary metabolites, including flavonoids, alkaloids, terpenoids, and phenolic compounds. This chapter explores the phytochemical diversity

DOI: 10.4018/979-8-2600-1207-9.ch010

of alpine medicinal flora, highlighting their ecological adaptations, pharmacological significance, and traditional uses by indigenous communities. Key species such as Saussurea lappa, Bergenia ciliata, Picrorhiza kurroa, and Artemisia maritima exhibit antioxidant, anti-inflammatory, hepatoprotective, and antimicrobial activities. Modern analytical approaches, including HPLC, GC-MS, and NMR spectroscopy, are discussed for phytochemical profiling. The chapter emphasizes sustainable utilization and conservation amid climate-driven habitat loss.

1. INTRODUCTION

Alpine areas form one of the most ecologically constraining and physiologically demanding environments on earth. Characteristic of these high-altitude ecosystems are consistently low temperatures, severe diurnal thermal variations, excessive ultraviolet radiations, low atmospheric pressure, fierce winds, sandy soils of low species density and nutrient content and very brief growing periods. Such harsh environmental conditions place a lot of restrictions on plants survival, growth, and reproduction. Nevertheless, regardless of these constraints, alpine landscapes harbour an impressive number of wild food plants (WFPs) which have developed complex adaptive strategies and are thus able to grow in areas where standard crops are unable to do so (Körner, 2023). These species are the biological basis of subsistence of indigenous and pastoral and transhumant communities that rely on alpine resources as food supply, medicine, and livelihood. In the past, the alpine wild food plants have been central to the human population in times of food shortage, seasonal isolation and climatic fluctuations. They have a long-standing tradition of use in traditional food systems and ethnomedicinal practices where they do not merely serve as dietary supplements, but also as the main nutritional source in the lean times. Numerous alpine WFPs are either eaten fresh, dried, or put into traditional preparations, and provide the necessary macronutrients of energy, carbohydrate, tissue repair and metabolism, lipids cellular integrity, and dietary fiber maintenance of the gastrointestinal tract (Abbasi et al., 2015). The adaptive metabolic trade-offs represented by the proximate composition of these plants is frequently that greater nutrient density is used to offset less biomass making nutritional returns as maximized as possible per unit consumed.

Besides the value of their macronutrient, alpine WFPs often contain high amounts of micronutrient, such as essential minerals and vitamins that are essential in physiological homeostasis. Iron, calcium, magnesium, potassium, zinc, and trace elements that are commonly found in alpine flora are very important in oxygen delivery, bone formation, enzymes, immune system, and neuromuscular activity (Abbasi et al., 2015). With limited access to a wide range of foodstuffs in inaccessible areas of high altitude, when micronutrient deficiencies are widespread, the nutritional value of these plants is especially important in that environment. Extensive proximal and nutritional assessment therefore offers an empirical foundation of their contribution towards the fight against hidden hunger, malnutrition vulnerability and improved dietary resilience among marginalized mountaineers. The pharmacological value of alpine WFPs is tightly connected with its biochemical structure. The biosynthesis of a broad range of secondary metabolites which include flavonoids, phenolics, alkaloids, terpenoids, glycosides, and organic acids are triggered by environmental stress factors like high UV radiation, low temperatures and oxidative pressure. They are capable of performing protective roles in plants by alleviating oxidative stress and increasing tolerance to stress, yet have strong biological effects in humans in which they perform them. An increasing amount of scientific data links these phytochemicals with anti-oxidant, anti-inflammatory, anti-microbial, hepatoprotective, cardioprotective and immunomodulatory effects (Bhattacharyya et al., 2023; Swathi et al., 2025). Notably, the therapeutic potential of these plants cannot be realized in total ignorance of their proximate composition as nutrient-phytochemical interactions tend to regulate bioavailability, efficacy and metabolic effects.

Although alpine WFPs have been widely used since the early times in the traditional healthcare systems, there is a scarcity of nutritional and pharmacological studies on this particular subject. Although ethnobotanical knowledge is vast, its erosion has become a growing problem because of socio-economic change, modernization and intergenerational knowledge discontinuities. It is therefore important that scientific validation should be done using systematic proximate and nutritional studies to reconcile traditional wisdom with modern biomedical studies. Through such assessments, nutritionally superior species can be identified, standardization of medicinal preparations can be achieved and safety and efficacy profile can be created that will be much more accepted in functional food and nutraceutical industries.

Alpine WFPs are relevant to more than local consumption and traditional medicine to the general discourse of global food security and sustainable nutrition. The conventional agricultural systems are under unprecedented pressure due to climate change, population growth and degradation of arable land. The underutilized and wild species of plants adapted to harsh environmental conditions in this case are a significant source of genetic, nutritional, and functional diversity. The Alpine WFPs are valuable in terms of their inherent resilience and nutritional value that can be used in the development of climate-resilient food systems and alternative nutritional approaches, especially to areas with ecological limitations (Kumari et al., 2025).

But the alpine flora is getting gradually endangered due to climate-related habitat transformations, excessive harvesting, land-use alterations, and unchecked tourism. The population of many WFP species is dwindling, casting a question of sustainability of their use and the loss of ensuing traditional knowledge (Romeo et al., 2021). The recording of their proximate and nutritional attributes is thus granted paramount significance not only to scientific growth but also to the conservation planning and policy-making. The evidence-based nutrition data can assist in the prioritization of species to be conserved, enhance sustainable harvesting activities and inform the community-based management strategies.

The chapter includes a thorough study of the proximate and nutritional properties of wild food plants in the alpine area and places them in the ecological, cultural, and medicinal contexts. It integrates the ethnobotanical knowledge, nutritional science, and other sophisticated analysis data to emphasize the multiple usefulness of alpine WFPs as food, medicine, and bioactive sources. Their possible roles in food supply, human health and therapeutic innovation are highlighted, and there is an urgent need to carry out systemic researches and conservation programs to protect these untapped but potentially very useful biological resources in future generations.

2. ECOLOGICAL CONTEXT OF ALPINE WILD FOOD AND MEDICINAL PLANTS

The ecological context of alpine wild food and medicinal plant also involves a mixture of severe environmental stressors that influence their physiology and distribution along with their biochemical composition. The plants in the high altitude area are in a low temperature, high ultraviolet radiation, low nutrient soils and relatively low atmospheric pressure above sea-level, these are effective selective pressures on plants. These conditions, in their effect, are the short growing periods and low biomasses that have compelled alpine species to be highly competitive in their survival mechanisms (Billings & Mooney, 1968). Not only do these stresses dictate morphology, they also cause a build-up of diverse secondary metabolites with some being of great nutritional and medical importance.

In order to survive in the extreme environment in which alpine plants are growing on, a set of strategies have been adapted to be resistant. They are found as protection against chilling winds, augmented roots to waste nutrition-impoverished soils, and the production of protective pigments (which consist of antioxidants) that reduce cellular harm inflicted by UV. Most of the environmental biochemical stress of an environment has impacted through increased synthesis of flavonoid, phenolics, terpene and other biological active metabolites that acts a protective canopy against the plants and a therapeutic action in the event that humans will take its products. These modifications show the close interdependence between ecological demands and metabolic fullness that 1gives the reason why the alpine flora was used as a pharmacological agent with a designated purpose (Kumar et al., 2023).

These adaptive characteristics also affect the distribution and specificity of habitat of the alpine wild food plants. There are numerous species that live in very specific ecological communities, only in a couple of altitudes, the soil of a specific type, or microclimatic conditions. They are usually patchy and local, and this is a reflection of the environmental constraints as well as evolutionary histories (Nagy & Grabherr, 2009). There are endemic plants to particular mountain areas and others to wider belts of the Alps but are sensitive to minor changes in climate. These patterns of distribution are critical to conservation and sustainable use because alpine habitats are becoming more vulnerable to climate change and anthropogenic stressors, which is becoming an increasing threat to the persistence of these species of biological and cultural significance.

3. TRADITIONAL ETHNOBOTANICAL KNOWLEDGE

In the interaction between indigenous and local communities in the alpine regions, over the centuries, traditional ethnobotanical knowledge has been the core of such interactions with WFPs. This experience, which is developed over many years of observation, experience, and culture, provides in-depth information about the nutritional, medicinal, and ecological usefulness of the alpine flora. Wild plants, in distant high-altitude regions where farming is limited by severe weather and arable land, have been traditionally important elements of subsistence plans. People have depended on these species as not only sources of supplemental food but also as the part of traditional healthcare systems, seasonal practices and cultural identity. Such practices demonstrate the ecological literacy and high order of adaptability to plant properties far earlier before modern science (Von Glasenapp & Thornton, 2011).

Alpine wild food is a broad range of applications by indigenous and local communities' uses such as primary dietary ingredients, emergency foodstuffs, and herbal medicine. A large number of alpine species are also harvested in season and eaten fresh, dried or preserved depending on local traditions and availability. The tender shoots, leaves, roots, berries, and seeds are the key sources of nutrients in times when the cultivated crops are not available. An example is that high-carbohydrate or high-protein species can be incorporated into diets during extended winters and mineral-rich plants allow the maintenance of nutritional balance in areas where animal protein might be low. Outside the dietary purposes, these plants serve as the origin of ethnomedicine, which combats most common diseases, including digestive, respiratory, inflammatory, and infection-based ailments. They can be prepared in a decoction, infusion, and poultice or even without any extra treatment, and each of the mentioned methods is a profound reflection of the empirical knowledge of the dosage, effect, and interaction of the plants. The uniformity in which some species have been used over generations highlights the dependability that is accorded to the therapeutic properties (Von Glasenapp & Thornton, 2011).

The cultural meaning is incorporated in almost all spheres of alpine WFP implementation. Plants have a symbolic value and can be used in rituals, festivals and other social activities that contribute to sense of community and continuity. Associations with ecological cycles are made with many plants that are related to spiritual protection, seasonal transitions, or rites of passage. The information of the location, the time and the method of harvesting particular species is usually under customary regulations which place greater focus on respect to the land and the sustainability. These norms exhibited conservation ethos at an early stage, making sure that the population of plants will be able to survive to the next generations. Moreover, the cooking and distribution of foods of wild plants can be a way that cultures are passed on as the elders share with the young not only skills but stories, values and ecological ethics. This intergenerational transfer makes the ethnobotanical knowledge dynamic and flexible to the change of environment or society (Agbeno, 2025).

Another interesting point about the ethnobotanical systems of alpine is that they are inherently combined with the traditional diets and medicine. The food medicine interface is permeable in most societies as a sign of holistic perception of health based on balance and prevention. Exchangeable foods may also help medically, even when considered as medicinal, whereas everyday foods may be taken in times of sickness or weakness. Integration in this way is manifested in the fact that people choose plants with antioxidant properties, anti-inflammatory properties, and digestive-enhancing properties, promoting all-purpose physiological stability in the harsh conditions of the environment. Foods that are grown seasonally are mostly chosen based on their nutritional value as well as the ability to build the body against climatic stresses. As an example, bitter, or aromatic plants can be eaten to activate the digestive system or purify the body, whereas high-energy species can be used to support livelihoods that are labor-intensive (Sears, 2024).

The interplay of diet and medicine in the alpine cultures explains a high level of awareness of the bioactivity of plants, and shows how historical knowledge systems can serve as holistic health systems. Due to the increased interest in worldwide in nutraceuticals and functional foods, these traditional practices provide useful guidelines towards the combination of nutritional and medicinal research. They offer context-dependent information that has the potential to inform scientific research, direct pharmacological research, and sustainable resource management. With the threat of extinction of ethnobotanical knowledge posed by modernization and environmental influences, documentation and preservation of ethnobotanical knowledge has become a necessity not only to cultural heritage and heritage but to further studies on the alpine wild food plants (Grabherr, 2009).

4. PHYTOCHEMICAL DIVERSITY OF ALPINE WILD FOOD PLANTS

The alpine WFPs have secondary metabolites that are very important in the survival and adaptation of the plants and their medicinal properties. These are not directly related to the primary metabolism of the plant, but they are bioactive compounds that help in the defense mechanism of the plant towards the herbivores, pathogens, and environmental stressors. Simultaneously, they are rather beneficial to human health as they serve as antioxidants, anti-inflammatory agents, anti-microbial agents, etc. These secondary metabolites vary considerably in the variety present in alpine WFPs and considering the severe environmental conditions in which they have to operate. The main classes of secondary metabolites to be known include flavonoids, alkaloids, terpenoids, and phenolic compounds, and knowing them provides an insight into the possible benefits that these plants can provide in terms of nutritional and medical advantages (Karimi et al., 2025).

One of the most conspicuous and varied groups of secondary metabolites that alpine plants produce is flavonoids which are polyphenolic compounds. These are common compounds found in most plant species and are well known in their anti-oxidant qualities which protect plants against the oxidative stresses that are brought about by the UV radiations, extreme temperatures and other environmental stressors. The importance of flavonoids in the human diet is specific because of their free radical scavenging nature, thereby lowering the susceptibility to prevent chronic diseases like cancer, cardiovascular diseases and neurodegenerative diseases. In the case of the alpine wild food plants, the flavonoids play a role in the protection of the plant as well as in the color of the plant, which is purple and blue in case with the flowers and fruits. A lot of literature has already documented nutritional advantages of flavonoids in the human diets which may enhance immune response, suppress inflammation, and foster normal health. Additionally, flavonoids are anti-microbial, anti-cancer as well as anti-diabetic, thus, are useful compounds in the investigation of alpine WFPs both in culinary and medicinal use (Ashraf et al., 2024).

Another important group of secondary metabolites in alpine wild food plants with significant physiological implications on humans and animals are alkaloids. These compounds that contain nitrogen have been widely examined on their pharmacological activity, which includes alleviation of pain, anti-cancer action and their effect on the functioning of the central nervous system by improving its performance. The alkaloids have been used historically in medicine and alpine plants contain them making them even more attractive as sources of new therapeutic agents. Alkaloids provide chemical defense mechanisms by deterring the pathogens and herbivores in the alpine habitat. Alkaloids can be regarded as being toxic in large amounts, but in controlled doses, a large number of alkaloids have useful properties, including morphine and quinine, which have historically played a central role in the treatment of pain and malaria. Thus, alkaloids in WFPs may have a major therapeutic potential, but one needs to be cautious with their toxicity and appropriate dose of intake (Kouznetsov, 2025).

Another broad and widespread category of secondary metabolites are terpenoids (also known as isoprenoids) which are widespread in alpine WFPs. These compounds perform diverse ecological roles that include attracting pollinators, repelling herbivores, and resistance to pathogens. The terpenoids consist of essential oils, resins, and pigments, and they are commonly known to be aromatic, which help to add the flavor and smell of most alpine plants (Holopainen et al., 2025). Medicinally, the terpenoids have a range of bioactive effects such as anti-inflammatory, antiviral and anti-cancer. Alpine plants are rich in terpenoids and their antimicrobial abilities have been well-reported in the literature and therefore it is possible that the plants could provide natural preservatives or antimicrobial agents. Considering human health, terpenoids have begun to be investigated in relation to their ability to relieve stress, anxiety and depression symptoms, and thus they will be promising as therapeutic agents to be used in the future (Sharma, 2024).

Another significant group of secondary metabolites in alpine wild food plants is the phenolic compounds that comprise a large variety of molecules: phenolic acids; tannins; lignans (Ashraf et al., 2024). These substances are renowned in regard to their antioxidant and anti-inflammatory properties and they have a major role in ensuring that plants are safeguarded against oxidative damages induced by environmental stress factors. The level of phenolic compounds becomes especially high in the leaves, stems, and fruits of alpine plants and leads to the characteristic of this plant in withstanding the harsh environment, e.g. high altitudes with low temperatures and scarce nutrients. In humans, phenolic compounds constitute a significant component of the diet, since it has a series of health benefits, including enhancing the heart health as well as decreasing the chances of cancer. Cases in point, tannins have been demonstrated to possess antimicrobial properties, whereas phenolic acids have been reported to have a high antioxidant activity that has the ability to counteract free radicals and prevent damage to cells. These compounds are also found in the alpine WFPs which underscores their possibility to provide functional foods that can add value to general health and well-being (Ashraf et al., 2024).

The high phytochemical diversity of the alpine wild food plants is a testimony to the high level of adaptability to the harsh conditions of these plants. These secondary metabolites have numerous potential advantages in human nutrition and health besides the contribution they make to plant defense. These substances i.e., flavonoids, alkaloids, terpenoids, and phenolic compounds are only a few examples of the huge number of bioactive molecules availed by alpine WFPs. Since studies on the nutritional and medicinal values of these plants are still ongoing, their value as a source of use in functional foods, nutraceuticals and natural medicines is being explored with increasing interest. Considering the increasing need of the plant-based and sustainable health-solutions, alpine WFPs offer an opportunity to explore new and bioactive compounds that may be useful in human health, especially antimicrobial resistance, cancer treatment, and chronic diseases.

4.1 Environmental Drivers of Phytochemical Variation

The ecological reasons of the differences in phytochemical contents amongst alpine WFPs are multi-faceted and complex and are driven by a combination of biotic together with abiotic factors (Alami et al., 2024). They affect the production, enrichment and diversity of secondary metabolites in these plants as a requirement for their existence and food & medicine supplies. This result emphasizes the importance of environmental conditions on the variation in phytochemicals in alpine plants, which is also critical to elucidate their health-promoting effects and ecological functions for humans. The interaction between the plant and environment should also be studied in a harsh alpine system, where it's possible to elucidate processes leading to phytochemical's production (Alami et al., 2024).

Some of the most important environmental drivers that affect the variation of phytochemicals among alpine plants are the abiotic factors, which comprise temperature, light, soil composition, water availability, and altitude. All these have a direct influence on the metabolism and growth of plants and their fitness, which ultimately influence the production of levels of secondary metabolites. Indicatively, plants have to evolve in high-altitude areas, where the temperatures are low and the UV radiation is strong. Higher UV levels in elevation frequently prompt the production of the compounds like flavonoid and phenolic acids, which are protective factors in response to the destruction of the UV. Flavonoids, in reference, have been reported to absorb UV light, and hence, the chance of damaging the DNA of plant tissues is reduced. Consequently, the amount of flavonoids in alpine plants tends to elevate as a reaction to the enhanced UV radiations at elevated heights (Pant et al., 2021).

Secondary metabolites are also produced by the effect of temperature variations (Qaderi et al., 2023). The exposure of plants to daily and seasonal temperatures extreme in alpine, may serve as a stressor. Stress on temperature can activate the production of some phytochemicals including alkaloids that help to defend plants against herbivores and pathogens. Low temperature, especially, has been discovered to positively influence the synthesis of some terpenoids and phenolic compounds that aid the plant to endure stress on the environment. On the other hand, an increase in temperature can be beneficial in the production of other phytochemicals which are volatile and can contribute to temperature control and plant defense. The ratio between these temperature induced responses is one of the things that make the phytochemical diversity found in the alpine plants, with various compounds being dominant in various seasons and environmental factors (Qaderi et al., 2023).

Other abiotic factors that are critical and affect phytochemical production are the light intensity and photoperiod. The intensity of light in the Alps is significantly greater because the sunlight is also not filtered by the atmosphere. This increased exposure has the ability to induce the production of the secondary metabolites which help in mitigating the oxidative stress posed by an excess of light in the form of flavonoids and carotenoids (Chandra et al., 2025). Also, alterations in photoperiod or the duration of day and night can influence the time of production of secondary metabolites in alpine plants. Plants can also synthesize certain compounds at a particular time of the year due to the alteration in light and thus can influence the nutritional and medicinal value of the plant. As an example, the level of phenolic compounds rises during flowering in certain species of alpines, possibly as defense against possible herbivory or competition (Chandra et al., 2025).

The content of soil and nutrient availability are also key causes of phytochemical variance. The mineral structure of the soil, such as the amount of nitrogen, phosphorus and trace minerals, has an impact on the metabolic pathways of the alpine plants. When there are nutrient deficiencies they can cause the synthesis of some secondary metabolites in order to increase adaptations of the plant to the scarcity of resources. As an example, soils with low nitrogen levels frequently result in high levels of alkaloids, which can be used as a defense mechanism and as a tool of controlling the intake of nutrients (Chandra & Lata). Conversely, vegetation that grows in soils with high nutrient content can generate alternative patterns of secondary metabolites, since the resources are not limited to cause alternative priorities of metabolism. Plant phytochemical diversity can thus be an adaption strategy in alpine habitats where soils may be low in nutrient quality and highly unpredictable thus subjecting the plants to nutrient stress (Chandra & Lata).

Another important abiotic factor affecting the phytochemical composition of the alpine vegetation is water availability or lack of it. Water stress, which may take the form of drought, or low rainfall, may trigger the production of some secondary metabolites that are osmoprotectants that prevent instability of the plant cell structure during water stress (Pant et al., 2021). This protective response usually includes terpenoids, phenolic compounds, flavonoids and aids in preserving the integrity of the plant when the conditions are dry. Also, the availability of water influences the overall growth of the plant and the biomass production of the plant, which further affects the quantities of secondary metabolites that are available. Production of secondary metabolites might be of especial intensity in alpine regions, where there is low precipitation and water stress is widespread as plants attempt to acclimate to such austerate environments (Pant et al., 2021).

The biotic factors are also important in determining the phytochemical profiles of the alpine wild food plants. The relations among plants and other organisms, including herbivores, pathogens, and symbiotic microorganisms play a major role in the synthesis of secondary metabolites. An example is herbivory which frequently prompts the synthesis of the defensive chemicals of plants like alkaloids and terpenoids that discourage the herbivores to eat them. These defense mechanisms are especially relevant to the survival of plants in alpine ecosystems where there may be both migratory and native species of herbivores. In a similar manner, the attacks of pathogens may induce the production of antimicrobial compounds including phenolic acids and flavonoids that assist in protecting plants against infection. Also, useful microorganisms, including mycorrhizal fungi, may also modulate the phytochemical constitution of alpine plants by enhancing nutrient uptake or pathogen protection (Hou et al., 2024).

The other important biotic determinant of variation in phytochemicals is altitude. The higher the altitude, the more extreme the environment is, the lower the temperatures, the lower the oxygen content, and the increased amount of UV radiation. The conditions are considered to be stressors and as well as selective pressures, which regulate the synthesis of secondary metabolites in the alpine plants. Increased altitude tends to prefer those plants that are especially enriched in one or another category of secondary metabolites, which can either, be UV-protective flavonoids or repel herbivores and pathogens, alkaloids.

4.2. Correlation between Phytochemistry and Nutritional Profiles

The relationship between the phytochemistry and nutritional of alpine WFPs is an interesting and important field of study especially in light of the potential health value of the plants. Phytochemicals are bioactive plant-derived compounds that play a role in the defensing mechanism and plant survival and are also useful in terms of nutritional value of the plant. These secondary metabolites are flavonoids, alkaloids, terpenoids and phenolic compounds, which have a close interaction with the main nutritional components of plants i.e. carbohydrates, proteins, fats, vitamins and minerals. The distinctive phytochemical composition of alpine plants is affected by a number of environmental factors and their phytochemical composition, in turn, may have strong implications to the nutritional profile of alpine plants (Abbasi et al., 2015). Phytochemical diversity and health benefits of alpine wild food plants, highlighting key secondary metabolites such as flavonoids, alkaloids, terpenoids, and phenolic compounds as shown in Figure 1.

The role of phytochemicals and their involvement in the operations of the human body gained a new dimension in relation to functional foods and natural medicine. The diversity and complexity of secondary metabolites of alpine WFPs is central to determining their antioxidant, anti-inflammatory, antimicrobial, and other bioactivity, which can be of great nutritional benefit to these plants. The two correlation and the phytochemical content of the plant and the nutritional benefit is vital in understanding how the plants will be of use to the mankind beyond the nutritional content of the plant.

Depending on the nutritional profiles, alpine wild food plants usually contain the essential macronutrients, i.e. carbohydrates, proteins, and fats, and a variety of micronutrients, i.e. vitamins and minerals. Nevertheless, these plants contain phytochemicals, which further complicates and increases their value (Abbasi et al., 2015). As an illustration, such compounds as flavonoids and phenolic acids which are frequently present in the alpine plants are antioxidant in nature. The compounds are used to protect the plant against oxidative stress that may be caused by UV radiation, temperature extremes among other environmental factors. Equally, these antioxidants may also be useful to human health to neutralize free radicals and prevent the occurrence of chronic diseases, including cardiovascular disease, diabetes, and cancer. It was proposed that the positive relationship existing between the antioxidant activity of these phytochemicals and the nutritional value of alpine WFPs as a whole indicates that the alpine plants could be of a dual advantage such that they can not only provide the essential nutrients to consumers but also offer protective health benefits through their bioactive compounds (Abbasi et al., 2015).

The value of secondary metabolites of the nutritional quality of alpine plants is especially visible in the influence of metabolites on digestion, absorption, and bio-availability of nutrients. Several phytochemicals such as terpenoid and alkaloids are capable of affecting the bioavailability of some nutrients in the digestive tract (Zafar et al., 2025). Via the example of terpenoids, potential sources of antimicrobial and anti-inflammatory effects; it is possible to increase the absorption of the necessary nutrients by improving the health of the gut. In the same manner, alkaloids, though they might also be toxic in high doses, have been demonstrated to influence the metabolism of the macronutrients either by having an action on the digestive en-zymes or as an appetite suppressant. Such interactions between phytochemicals and nutritional profile of alpine WFPs show the complicated interaction of secondary metabolites with the capacity of the plant to supply essential nutrients to people.

Also, the vitamins and minerals that are present in the alpine wild food plants are likely to complement the effects of the phytochemicals consequently increasing their health-promoting effects. To illustrate, vitamin C is a potent antioxidant and it is mainly found in many plants of the alps and its presence may be the reason behind the improved activity of phenolic products to reduce oxidative stress. On the same note, certain minerals like magnesium, potassium, and calcium can be combined along with the secondary metabolites to stimulate the cardiovascular status, bone density, and metabolic activity. Phytochemistry-nutrition correlation in alpine WFPs, therefore, means that the plants are a wholesome pack of nutrients and bioactive compounds, which can interact with each other to increase health and wellbeing (Abbasi et al., 2015).

Besides, the synergies existing between phytochemicals and nutritional content of alpine WFPs may carry important implications on the development of functional foods and nutraceuticals. The increasing popularity of vegetarian diets and the health-conscious lifestyle suggests the necessity of foods that provide not only fundamental nutritional value but also a high level of health improvement. In this regard, alpine WFPs with their abundant phytochemical diversity have high potential. The presence of the nutrients and bioactive compounds can result in the creation of functional foods which do not only satisfy the minimal nutritional requirements of a person but also have other therapeutic benefits. E.g., alpine plants with a high content of flavonoids and terpenoids can be added to functional foods as they may increase the health value of that food e.g., immunological effect, inflammatory reduction, or even weight loss.

Also, the relationship of phytochemicals with the nutritional composition of alpine WFPs can be followed in terms of their traditional uses in different societies. Folk medicine has been using many alpine plants centuries that have their medicinal properties mostly credited to the phytochemicals they contain. These ancient applications do not differ much with the current research on the nutritional and health value of these plants. Indicatively, alkaloids-rich plants like some species of mountain herbs have been used as analgesics or to enhance digestion. The nutritional value of these plants such as the richness in vitamins, mineral, and essential fatty acids also further justifies their application as functional foods in traditional diets (Abbasi et al., 2015).

The nutritional profiles of the alpine plants also are indirectly affected by the environmental factors of which they are influenced by the production of secondary metabolites. The synthesis of the phytochemicals in these plants can be influenced by environmental stressors which include the altitude, temperature, light intensity and soil composition. As an illustration, plants that grow in elevated altitudes, where the UV radiations are stronger tend to synthesize more flavonoids and other phenolic compounds as a response mechanism. The identical compounds aid in the antioxidant effects of the plant, which has the potential to improve the nutritional value of the plant by providing health-related benefits. The fact that alpine plants can adjust to such extreme environmental factors by synthesizing certain phytochemicals that can also be useful to human health speaks of the strength of such plants and the potential usefulness they present (both nutritionally and medicinally).

Figure 1. This figure provides a wide range of phytochemical constituents in the alpine wild food plants and shows primary secondary metabolites (flavonoids, alkaloids, terpenoids, and phenols). It highlights their implication in plant defense and their potent bioactive effects: antioxidant, anti-inflammatory, antimicrobial and the beneficial heart effect. The image also relates these compounds to their possible health benefits: application in functional foods, immune support and therapeutics.

5. PROXIMATE COMPOSITION AND NUTRITIONAL PROFILING OF SELECTED ALPINE WFPS

Some proximal composition and nutritional profiling of alpine WFPs are critical in assessing its usefulness as a nutritious and sustainable food product. Through this analysis, a considerable amount of knowledge can be obtained on the macronutrient content which entails moisture, ash, crude protein, crude fat, crude fiber and carbohydrates as well as the energy value of these plants. These nutrient elements play a vital role in the interpretation of the contribution of the plant to human health and they are also used as a pointer of the suitability of the plant to human consumption either as food or as supplement. The next section describes the methodology applied in the proximate study of the chosen alpine WFPs providing a systematic method of identifying the nutritional composition of the food products (Kumar et al., 2022).

One of the most basic elements of proximate analysis is the moisture content as it directly affects the shelf life, storage and handling of plant material. The moisture content is an indication of the amount of water that is available in the tissues of the plant and can be quantified through a simple drying technique. The sample is weighed, dried in an oven at a regulated temperature till it reaches the same weight. The change in weight after drying and before drying give the percentage of moisture content in the total sample. This is a needed measure since moisture level influences the taste, texture and shelf life of plant material. Moisture content may change with different plant species in the alpine WFPs and their growing conditions including the altitude and the climate. Plants that have high moisture content may be more perishable whereas those with lower moisture content may have increased concentration of nutrients, including secondary metabolites.

The total amount of inorganic remnants that are remaining upon complete burning of vegetation is known as ash content. This analysis is essential to get to know the mineral structure of alpine WFPs. The ash content is usually calculated by burning the dry plant matter at a high temperature usually in a muffle furnace. The left-over ash is then weighed and this percentage of ash is obtained as a percentage of the initial sample (Ullah & Badshah, 2023). Ash content also gives good information on the mineral content of the plant; vital minerals like calcium, potassium, magnesium, and trace minerals. The ash content can be used to determine whether the soil conditions in alpine plants are nutrient-limited, and this is significant in the maintenance of proper metabolic processes in human beings. Alpine WFPs can provide minerals that are important in different physiological functions such as bone, electrolyte balance and nerve functions (Ullah & Badshah, 2023).

Another vital item in the proximate analysis of the alpine WFPs is crude protein since it represents the amino acid distribution of the plant and the overall level of protein. The Kjeldahl method is generally used to analyze the protein level of the plant, where it is digested in a strong acid to decompose the nitrogen compounds. The quantity of nitrogen is subsequently quantified and transformed into a measure of protein content by a known factor. Proteins play a crucial role in the human diet as they are essential in repairing tissues, in immune systems, and enzymes. The protein level can also differ greatly among species of alpine WFPs, with some plants containing a high level of proteins because of the adaptation to nutrient-poor soils. Alpine plants have high protein content, which can be useful in the supplementation of protein consumption in regions where other protein sources are limited and also they can serve as a sustainable source of plant-based nutrition.

Fats are a concentrated source of energy and are also basic fatty acids and hence crude fat is an important constituent in the nutritional profile of alpine WFPs. The value of crude fat is calculated by the removal of the lipids of the plant specimen by means of a solvent such as ether or hexane. The fat is then extracted and the weight of the extracted fat is calculated and the fat content is calculated as a percentage of the original sample. Fats play the role of cell membrane integrity, production of hormones, and absorption of fat soluble vitamins (A, D, E and K). Though alpine plants are not considered to be as abundant in fats as other food sources, such as seeds or oils, some alpine species, such as the ones that contain a lot of oil in their seeds or fruits, have good sources of lipids. Such fats may enrich the total energy content of the vegetable and supply key fatty acids, essential fatty acids, which are important in the health of the heart and the brain (Abbasi et al., 2015).

Crude fiber is that indigestible part of the plant which gives structure and support to the cells of the plant. It contains cellulose, hemicellulose and lignin that cannot be broken down by enzymes in the human gut. The amount of crude fiber is calculated by weighing the plant sample prior to and after the subjecting the plant to acidic and alkaline conditions that eliminate the soluble parts of the plant. The rest is then measured and indicated as a percentage of the total sample. Crude fibers are significant in the diet of man, as they help to keep the digestive system healthy, constipation is prevented and cholesterol levels remain normal. Alpine plants do not necessarily contain a lot of crude fiber, but some species contain a lot of it, particularly those with fibrous stems or leaves. Moreover, plants that contain a lot of fiber can be used to control the level of blood sugar and maintain the weight by keeping the stomach full (Kumar et al., 2022).

Carbohydrates are the key energy source in the human diet and analysis of carbohydrates in alpine WFPs is to ascertain the total available carbohydrates. This is possible through subtraction of moisture, ash, protein, and fat content of the total dry matter of the plant. Carbohydrates may further be divided into simple (monosaccharides and disaccharides) and complex carbohydrates (polysaccharides such as starch and fiber) (Chen et al., 2023). These carbohydrates are vital sources of energy to the cellular processes and physical exercise. There is an interaction between the metabolic pathways of the plant and the environmental conditions that affect the carbohydrate content of alpine WFPs. There will be alpine plants with a lot of starch or some with a lot of sugar especially in fruits or roots. These carbohydrates have a crucial energy value that is required to determine the usefulness of the plant as a source of food since it will define the level of energy that can be obtained upon eating the plant (Chen et al., 2023).

Nutritional value of food is a vital factor and it is commonly defined with respect to energy, which can be measured as the energy content in kilocalories (kcal). The energy content of alpine WFPs is calculated as the sum of protein, fat and carbohydrate calories. This measurement represents a broad indicator of total energy content of the plant and is essential for the interpretation of its potential as a food supplement. High energy values of alpine plants such high energy content in the alpine plant food is particularly important for areas concerned with food security, presenting these populations with a calorie-dense source of nutrition in an extreme, high-altitude environment. The energy content in alpine plants generally can vary considerably according to the nutrient, some of them are rich in carbohydrates or fats; therefore total gross energy value can be different.

5.1 Micronutrient Assessment

The alpine WFPs should have their nutritional value evaluated through micronutrient analysis, which is a step in the right direction to comprehend the nutritional benefits. As shown in the Figure 2, the micronutrient assessment of alpine wild food plants highlights key minerals and vitamins, such as calcium, iron, vitamin C, and vitamin A, along with their associated health benefits. These micronutrients that contain the essential vitamins and minerals are essential in the human health as they help in supporting various important physiological functions including immune defense, bone health, enzyme activity and metabolic processes. Growing in some of the most difficult and nutrient-restricted environments, alpine WFPs have developed to generate a wide range of bioactive compounds that do not only allow them to survive but also bring in a variety of health benefits to people. The micronutrient analysis of these plants provides important information on the possibility of the plant as a functional food and a natural source of nutrients that can potentially be used to support human health in regions where conventional food sources might be inadequate (Ullah & Badshah, 2023).

Among the most important ingredients of the micronutrient profile of alpine WFPs, minerals are included. These minerals are calcium, iron, magnesium, zinc, potassium and phosphorus that are required in many functions of the body. An example is calcium, which is essential in bone and tooth development and maintenance, muscle activity, and neural conduction. It is commonly thought to be found in dairy products, however, alpine plants, especially those having fibrous leaves or seeds, may be quite a good source of calcium (Abbasi et al., 2015). On the same note, iron plays a crucial role in the transport of oxygen in blood, generation of energy and immune system. Iron-deficiency anemia is one of the most widespread health issues that occur particularly in developing countries and alpine plants rich in bioavailable iron can become an efficient source of this vital mineral. Compounds like tannins and phytates can also influence the bioavailability of iron in plants, but reduce the absorption of iron, but the level of iron in alpine plants can nonetheless be of significant use in areas where meat-based sources of iron are limited (Abbasi et al., 2015).

Another valuable mineral that is present in alpine WFPs is magnesium. It is used in more than 300 enzyme reactions within the body such as regulation of muscle and nerve activities, protein synthesis and energy generation. Magnesium is exceptionally important to cardiovascular, bone strength and general metabolic functioning. Numerous alpine species, especially those that grow in magnesium-enriched soils, have quantifiable amounts of this vitamin mineral, and thus are a source of magnesium in the form of plants. Zinc is a trace mineral that is present in alpine WFPs and it is necessary in immune functionality, generation of DNA and protein metabolism. Zinc deficiency may interfere with the immune system, cell division, and wound healing. Zinc levels of alpine genus plants can be substantial, especially among plant species that grow within alpine soils that are rich in nutrients and thus these plants will provide an ideal alternative to zinc fortification in diet (Abbasi et al., 2015).

Other valuable minerals are phosphorus and potassium that improve human health. Potassium assists in maintaining the balance of fluids, muscle contractions, nerve conduction, and phosphorus is important in the maintenance of bones and teeth and also in the metabolism of energy. These minerals may be accumulated by alpine plants that grow in a particular type of soil or in a particular environmental condition, which also increases the nutritional value of the plant. The analysis of the concentration of these minerals in alpine plants will allow us to estimate the extent to which these plants can be utilized to achieve the suggested dietary consumption of these vital minerals.

Moreover, vitamins are also important micronutrients in alpine WFPs, in addition to minerals. This is because these organic compounds are needed in minimal quantities in the body to perform numerous functions but they are not produced by the body but must be sourced in the diet. Most alpine food contains major vitamins like vitamin C, vitamin A, and various B vitamins, all of which are essential to keep one healthy. An example of a potent antioxidant is Vitamin C which aids in protecting the cells against oxidative damages and aid in immune action, collagen synthesis, and iron absorption. Horticultural products, especially fruits and leafy greens, are usually high in vitamin C, being one of the key sources of the vitamin, especially where other sources might be low. Vitamin C in the plants can be evaluated using high-performance liquid chromatography (HPLC) or spectrophotometric analysis and it gives information on the future use of the plant as a natural antioxidant (Kumar et al., 2022).

Another vital nutrient that is also usually found in alpine plants is vitamin A, which is commonly in the form of provitamin A carotenoid, including beta-carotene. Vitamin A helps in healthy vision, skin, and immune system, and also, it is an antioxidant. The carotenoids that confer the yellow, orange, and red colour to most alpine fruits, flowers and roots are reported to have a capacity to spin down free radicals and decrease oxidative stress. These compounds are measurable through HPLC and they add nutritional value as well as health benefits to alpine WFPs. The carotenoid contents may vary with the species of the plants and the conditions under which they are grown with some alpine plants being very high in the bioactive substances (Ravetto Enri et al., 2024).

B vitamins such as B1 (thiamine), B2 (riboflavin) and B6 (pyridoxine) are very important in the process of energy production, metabolism and nervous system functions. These vitamins may be commonly present in many alpine plants especially those with seeds, leafy greens or roots that are known to store B vitamins. Another vitamin (vitamin B9) that participates in DNA synthesis and cell division is folate (vitamin B9) and is also found in alpine plants, the percentage of which is higher when they have leaves or flowers. Although animal foods are the largest sources of vitamin B12, some plants in the Alps also contain trace levels of vitamin B12, particularly those with microbial symbiosis which may synthesize B12. The B vitamins in the alpine WFPs can be analyzed as either HPLC or microbiological analysis that can give an insight into the potential of the plant to supplement the overall consumption of the nutrient groups.

Some Alpine plants, particularly that have green leaves contain vitamin K for blood clotting and maintaining of bones. Vitamin K can be analyzed using HPLC or ELISA and the plants could provide an additional source of vitamin K to people who may not have easy access to the traditional sources, uncooked green greens and fermented foods. In addition to vitamins and minerals, alpine WFP also contain other bioactive micronutrients like polyphenols and flavonoids, which have received significant attention for their health-benefiting properties. Although none of these compounds are regarded as essential nutrients, they play key antioxidant, anti-inflammatory and antimicrobial roles that enhance the health-promoting effects of the two vegetables. Polyphenols, such as flavonoids, tannins or phenolic acids, are abundant in many alpine species and their consumption has been associated with the low risk of chronic diseases such as cardiovascular disease, cancer or neurodegenerative conditions. The levels of polyphenols can be determined through HPLC or mass spectrometry in alpine plants, and the bioactive potential would also contribute to the popularity of alpine WFPs as functional foods.

Figure 2. The current overview aims to provide information on these health benefits of alpine wild food plants and the analytical techniques (ICP-MS, HPLC and ELISA) employed for determining these micronutrients.

5.2 Comparative Nutritional Values Across Species

The comparison of nutritional values in the species of alpine WFPs demonstrates a great difference in the content of the necessary nutrients, which depends not only on the species of plants but also on the environmental factors and adaptive processes of plants. Plants that grow in severe conditions with low temperatures, high UV radiation and low-nutrient soils include Alpine plants since they have developed special biochemical profiles that enable them to survive (Kumar et al., 2022). The nutritional composition of these plants can also be affected by these same adaptations with a particular species being high in one or a few nutrients and other species can be more adapted to survive under extreme environments. Comparing nutritional profiles of various plant species in the alpines, those ones can be distinguished that have the highest health benefits and potential to be used by humans (Kumar et al., 2022).

Among the differences in nutritional profiles of alpine WFPs the most striking is the difference in their minerals. The essential minerals such as calcium, magnesium, and iron are found in specific species as being very rich and are essential to the health of the bones, energy production, and immune (Milani et al., 2023). As a case in point, the fibrous leaves of some alpine plants, including *Urtica dioica* (stinging nettle), contain high amounts of calcium, and thus, the plant is a great source of this mineral, especially in a vegetarian diet. In the same manner, alpine plants such as *Chenopodium quinoa* (quinoa) which grow in high altitude contain iron and magnesium which are important sources of nutrition to populations at risk of iron deficiency anemia. Other alpine plants in comparison might contain fewer minerals but be more adapted to hold onto the minerals in the low-nutrient environment. The difference in the mineral ratios of different species clarifies why it is necessary to have an in-depth knowledge of what plants are best suited to satisfy certain nutritional requirements (Milani et al., 2023).

Where vitamins are concerned, the alpine species are also very varied. Vitamin C, the antioxidant with a role in collagen formation is found in large quantities in most alpine plants especially in fruits, flowers and leaves. As an example, such species as *Ribes alpinum* (alpine currant) and *Vaccinium myrtillus* (bilberry) are high sources of vitamin C and can be used as anti-aging ingredients and to boost immunity. Other species of the alpine like *Rumex acetosa* (sorrel) also have high levels of vitamin A in the form of carotenoids that are relevant in retaining healthy vision and immune. Nonetheless, there are alpine plants which might contain less vitamin C, especially those which are adjusted to the low altitudes or the weaker sun rays which might slow down the production of the vitamin. The comparative analysis of vitamin contents of the alpine plants shows that there is a wide scope of health advantages and some plants are exceptional in the concentration of particular vitamins.

Alpine plants also differ in the protein content significantly among species. Alpine WFPs, including *Fagopyrum esculentum* (buckwheat), are reported to contain relatively high amounts of protein and consequently it is especially valuable in human diets, particularly in regions with a low supply of animal protein. An example of a well-balanced amino acid profile is buckwheat, and therefore, it makes it an excellent source of protein (Kumar et al., 2022). Conversely, most alpine plants and more nutritionally valuable as well as fibrous plants can contain less protein but they are sources of important nutrients contributing to overall health. The growth patterns of the species used in alpine plants and the environmental conditions in which the species are growing can also affect the protein content of the alpine plants i.e. soil quality and altitude (Kumar et al., 2022).

The alpine WFPs are typically lower in fat than other food plants (Milani et al., 2023), although in a few cases, some species such as *Helianthus annuus* (sunflower) can contain high-quality fatty acids (and especially polyunsaturated fatty acids) in their seeds. These are fatty acids that are imperative in heart and cellular functioning (Pal, 2011). The majority of alpine plants are, however, more apt to contain more carbohydrates in form of starch and fiber that give energy and support digestion. Depending on the life cycle and storage of energy in the plants to grow and reproduce, the carbohydrate content may vary. Plants such as quinoa which is characterized with high carbohydrate content could serve to be a significant source of energy, particularly to the people residing in high altitude areas whereby caloric intake might have to be increased as a result of coldness and physically straining environment.

Comparing the nutritional analysis of alpine flora, it becomes evident that no one species is balanced to rich in all the nutrients. Certain species are rich in certain vitamins or minerals whereas others are more abundant in the form of protein or carbohydrates. These differences in nutritional profile among species indicate the need to ensure that a well-balanced diet of a wide variety of nutrients is achieved by eating alpine WFPs of variety. These variations also underline how the alpine plants adapt to their specific environments, by evolving to synthesize certain compounds that can help them to survive and in the process these can have great health benefits to mankind.

6. PHARMACOLOGICAL POTENTIAL AND FUNCTIONAL PROPERTIES

The pharmacological potential and functional properties of alpine WFPs are of immense interest because of the bioactive compounds that they have which lead to their therapeutic properties. The plants that grow in extreme and high-altitude conditions have developed a diversity of secondary metabolites with flavonoids, alkaloids, terpenoids, and phenolic compounds to provide a rich diversity of health effects. This part will discuss the anti-inflammatory, antioxidant, and hepatoprotective, antimicrobial, and other biological activities of alpine WFPs in relation to the contribution of those functional properties to human health and their possible application in traditional and modern medicine (Karimi et al., 2025).

The alpine WFP extracts have been particularly recognized for anti-inflammatory and anti-oxidant activity. The leading cause of most chronic diseases, including cardiovascular disease, diabetes, cancer, ageing and several episodes of neurodegenerative conditions is attributed to Chronic inflammation and Oxidative Stress (Ashraf et al., 2024). High altitude habitat in which these plants grow generally subjects them to high UV radiation, cold and adverse weather conditions which would expose these plants to oxidative stress of tissue. Altitudinal plants cope with such conditions by producing various antioxidant secondary metabolites. Among these are flavonoids, phenolic acids and carotenoids that play a key role in scavenging free radicals and mitigating oxidative stress on cells and tissues. Therefore, alpine WFPs are likely to confer high level of protection against oxidation after intake and would be suitable to lower oxidative stress and associated inflammation (Ashraf et al., 2024).

But Flavonoids in particular are also well known due to their analgesic activity. These agents can inhibit the production of proinflammatory cytokines, cyclooxygenase (COX) enzymes and other factors involved in inflammation. Flavonoids of alpine WFPs, which inhibiting the active inflammatory pathways can help in treating diseases like arthritis, IBD and other chronic inflammatory syndromes. In addition, phenolics in these plants including tannins and lignans are found to be with antioxidant and anti-inflammatory effects by further lowering inflammatory diseases occurrence and enhancing overall immune performance.

The antioxidant activity of the alpine plants also makes it possess the potential of protecting the body against oxidative stress that may cause ageing and diseases. Antioxidants neutralize the free radical and this prevents them to destroy the structures of cells like DNA, proteins and lipids. This defence system is necessary especially in the decreasing chance of degenerative diseases like Alzheimer, Parkinson and heart diseases. Due to their high phytochemical content, Alpine WFPs are thus the potential ingredient in functional foods to prevent oxidative damage and promote long-term health.

Alpine WFPs have hepatoprotective actions, antimicrobial effects in addition to their anti-inflammatory and antioxidant properties (Fantasma et al., 2024). Hepatoprotective properties are the capacity of some plants to prevent the damage of the liver by toxins, oxidative stress, and inflammation. Liver is an important organ and it is engaged in detoxification, metabolism and regulation of other biochemical operations in the body. Different alpine WFPs are used as conventional herbal medicine in the treatment of liver-related diseases, and modern research has started to validate these hepatoprotective properties. Indicatively, Artemisia species as well as some alpine plants have been discovered to shield liver cells against alcohol, pharmaceutical and environmental toxin-induced damages. These plants have compounds such as flavonoid, terpenoids and phenolic acids that assist in reducing liver inflammation, enhancing liver cell regeneration and increasing the process of detoxification (Fantasma et al., 2024).

Alpine WFPs have been found to have a lot of potential as natural antimicrobial agents in the context of antimicrobial activity. Most alpine plants generate antimicrobial compounds as a protection mechanism against bacteria, fungi and viruses. Such bioactive compounds as flavonoids, alkaloids, and terpenoids have been demonstrated to suppress the growth of a large array of microorganisms. As an example, extracts of alpine plants such as Salvia and Echinacea have been reported to have antibacterial and antifungal activity, and thus they are useful when it comes to fighting infections. Alpine WFPs could provide a good alternative to synthetic antimicrobial agents, especially with the emerging worries on antimicrobial resistance, and constraints of traditional antibiotics.

The other biological impacts of alpine WFPs are that they have the potential to control blood sugar, improve digestive health, and cardiovascular functionality. Most plants in the alps also have substances that help to regulate the metabolism of glucose, which can be applied in the treatment of diabetes and other diseases (Ladol et al., 2023). Indicatively, the *Glycyrrhiza* species (licorice) which are common in the alpine regions have been found to have antidiabetic effects due to their increase in the insulin sensitivity and the lowering of the blood sugar levels. Moreover, alpine plants are also a source of fiber and have antimicrobial and anti-inflammatory properties, which make them beneficial to gut health because they increase the growth of good gut bacteria, aid digestion, and promote the immune system.

Another main field is the cardiovascular benefits of alpine WFPs. Flavonoids and other polyphenols in the alpine plants have been reported to lower blood pressure, enhance the functioning of blood vessels and decrease cholesterol levels. These compounds assist in the maintenance of the health of the endothelial healthcare, lesser oxidative stress, and enhanced nitric oxide generation, which is crucial to cardiovascular health. Alpine plants like the *Crataegus* (hawthorn) are abundant in flavonoid and are usually used in traditional medicine to treat heart diseases and also enhance blood flow. Consistent intake of alpine WFPs including these compounds could decrease cardiovascular disease and enhance the condition of the heart. The Table 1, summarizes the pharmacological properties and health benefits of alpine wild food plants, highlighting key bioactive compounds and their therapeutic effects.

Table 1. Pharmacological properties and functional benefits of alpine wild food plants, highlighting key bioactive compounds and their therapeutic effects.

Biological Activity	**Bioactive Compounds**	**Health Benefits**	**Examples of Alpine Plants**
Anti-inflammatory	Flavonoids, Phenolic acids, Carotenoids	Reduces inflammation in chronic diseases like arthritis, IBD	Various alpine species
Antioxidant	Flavonoids, Phenolic acids, Carotenoids	Prevents oxidative stress, delays aging, protects against neurodegenerative diseases	Various alpine species
Hepatoprotective	Flavonoids, Terpenoids, Phenolic acids	Protects the liver against toxins, inflammation, and oxidative damage	Artemisia species, Various alpine plants
Antimicrobial	Flavonoids, Alkaloids, Terpenoids	Prevents bacterial, fungal, and viral infections	Salvia, Echinacea
Blood Sugar Regulation	Glycyrrhiza species (Licorice), Flavonoids	Regulates blood sugar levels, improves insulin sensitivity	Glycyrrhiza species (Licorice)
Cardiovascular Benefits	Flavonoids, Polyphenols	Lowers blood pressure, enhances blood vessel function, reduces cholesterol	Crataegus (Hawthorn

6.1. Case Studies of Key Alpine WFPs

Alpine WFPs have been well known due to their abundant bioactive compounds and therapeutic value. A number of alpine plants have been researched on because of their pharmacological activities, such as anti-inflammatory, antioxidant, antimicrobial, hepatoprotective, and other desirable effects. In the following section, we provide case studies of four important alpine WFPs such as *Saussurea lappa*, *Bergenia ciliata*, *Picrorhiza kurroa* and *Artemisia maritima* including their medicinal applications, major phytochemicals, as well as their demonstrated biological action (Painuli et al., 2021).

Saussurea lappa, is an herbaceous plant which grows mainly in the high altitudes of Himalayas and other mountain ranges. The roots of *Saussurea lappa* have been used in the Ayurvedic and folklore system of medicine for a wide array of pharmacological activities such as anti-inflammatory, antimicrobial, and hepatoprotective. The plant is a rich reservoir of various bioactive compounds including costunolide (CT), a sesquiterpene lactone which has been evidenced to be highly anti-inflammatory and antioxidant. It has been shown that *Saussurea lappa* extracts can also inhibit the production of proinflammatory cytokines (by reducing activity of COXs, i.e., cyclooxygenase enzymes; namely COX-1 and COX-2) Moreover, *Saussurea lappa* also showed hepatoprotective activity which protects liver from toxicity and regeneration of the liver. Antimicrobial potential of the plant can be attributed to high level of flavonoids, terpenoids, and other phenolic compounds which can find potential in a broad spectrum of pathogens including bacteria and fungi (Zahara et al., 2014).

Another mountain plant is Indian Rhubarb or Pashanbhed that grows in the higher altitude regions of Himalayas. It is well known for its classical use in the treatment of kidney stones, urinogenital disorders and digestive problems. The plant is rich in bioactive molecules such as flavonoids, tannins and phenolic acids, which give therapeutic effects to the plant. It has also been found that *Bergenia ciliata* have got powerful diuretic and lithotriptic (stone-dissolving) properties and this can be utilized in the prevention as well as treatment of kidney stones. One of the reasons why it has antimicrobial effect good against UTI is because it contains high tannin. Furthermore, *Bergenia ciliata* has also been reported to have antioxidant activity which scavenges the free radicals and oxidative stress in the body. This antioxidant, antimicrobial and diuretic activity makes *Bergenia ciliata* an effective alpine WFP whose purpose is practicable in present scenario (Roychoudhury et al., 2022).

Kutki is a famous well-known alpine herb, *Picrorhiza kurroa*, which grows in the Himalayan area. It has been greatly researched in antiphagocytic, anti-inflammatory, and immunomodulatory effects. The plant has a variety of bioactive compounds with iridoid glycosides like picroside I and picroside II being the cause of the therapeutic effect of the plant. *Picrorhiza kurroa* is an old traditional medicine that has been used in the treatment of liver diseases like jaundice, hepatitis and cirrhosis. According to modern scientific research, the plant is a good hepatoprotective agent (i.e., it improves the functioning of the liver and also helps the liver to prevent the damage where the toxins, alcohol, and drugs harm the cells in the liver). *Picrorhiza kurroa* possesses anti-inflammatory effects due to the inhibition of the pro-inflammatory cytokines and the inhibition of the inflammatory enzymes such as COX-2. Additionally, *Picrorhiza kurroa* was also discovered to enhance immunity and, therefore, is a good plant in strengthening the natural defense of the body (Mehta et al., 2021).

Artemisia maritima belongs to the genus Artemisia, and it is a high-altitude herb that is present in the alpine regions of Asia, Europe, and North America. It has been used traditionally as an antimicrobial as well as anti-inflammatory and a digestive aid. *Artemisia maritima* is a species which harbors diverse bioactive compounds such as essential oils, flavonoids and sesquiterpene lactones. The essential oils of the plant, especially the compound chamazulene, have a high antimicrobial and anti-inflammatory effect, and thus the plant is effective in the treatment of gastrointestinal disorders, respiratory infections, and inflammatory diseases. It has been demonstrated that the *Artemisia maritima* has antimicrobial properties that are efficient against gram-positive and gram-negative bacteria, and fungi and viruses. Strong antioxidant properties have also been shown by the plant, which helps to decrease oxidative damage and improves the general health. *Artemisia maritima* has been applied in traditional medicine to treat such conditions as a cold, indigestion and fever, and its antimicrobial activity is broad spectrum, which is why it could be the object of future pharmacological studies (Nigam et al., 2019).

7. ANALYTICAL TECHNIQUES FOR PHYTOCHEMICAL AND NUTRITIONAL CHARACTERIZATION

The fundamental role of analytical methods in the phytochemical and nutritional analysis of alpine WFPs is to give information about bioactive compounds in the plants and their nutritional value. Such herbal plants have an array of secondary metabolites in the form of flavonoids, alkaloids, terpenoids and phenolic compounds that need to be analyzed and detected with high specificity. Various chromatographic, spectroscopic and modern technologies are employed to elucidate the phytochemical fingerprinting and the nutritional values of these plants involving them in traditional as well as contemporary medicines. The common techniques include chromatography method, High-Performance Liquid Chromatography (HPLC), Gas Chromatograph-Mass Spectrometry (GC-MS) and spectroscopy [Nuclear Magnetic

Resonance] NMR spectroscopy, UV-Vis, FTIR], in addition to emerging methods for metabolomic profiling (Danna, 2024).

HPLC is one of the most popular techniques for the analysis of complex equimolar mixtures in alpine WFPs. It is also a well-known for the deconvolution and quantification of the soluble products (e.g., flavonoid, phenolic acids and alkaloids) (Chiriac et al., 2021). In HPLC plant extract flows through a column of stationary which is carried along by solvent (mobile phase). The so-called analytes, the substances of the sample, interact with the stationary phase at different speeds and can therefore be separated as well as detected (e.g. detection by means of UV-Vis or mass spectrometric detectors). It offers a new tool to characterize bioactive agents and is indispensable for the investigation of alpine plants as medicinal plant resource. GC–MS is also a powerful tool, particularly for the analysis of volatiles altogether with semi-volatiles including but not limited to terpenoids and essential oils in numerous alpine WFPs. In GC–MS analysis, compounds are separated according to their volatility, after which the mass spectrometer identifies them by measuring their mass-to-charge ratio. This technique is highly applicable for these investigations of essential oils and other volatile metabolites emitted by alpine plants and to unravel their health-promoting effects for consumers (Chiriac et al., 2021).

Spectroscopic techniques, NMR, UV-Vis, and FTIR should be required to research the alpine WFPs. Molecular structure and information about the functional groups present and atomic connectivity of molecules can be derived from NMR spectroscopy. This technique is highly interesting in finding complex organic molecules belonging to alkaloids or terpenoids, when considering the interaction of their nuclei with magnetic fields and radiofrequency. Moreover, NMR can differentiate the individual constituents of plant mixtures with high accuracy and provide a complete overview of the phytochemicals in the plant. Similarly, UV-Vis spectroscopy can also be used to measure the UV and visible light absorption of compounds such as flavonoids for example which are abundant in alpine plants. This process could be a strategy for the rapid screening of plant extracts which exposes the concentration achievable of specific compounds at multimetallic system's absorption maxima. This is also supported by FTIR analysis in detecting functional groups characteristic of a compound such as hydroxyl, carbonyl and amine – common constituents of alpine plant secondary compounds. By recording which specific wavelengths of infrared light are absorbed by a sample, FTIR can provide valuable structural information which contributes to the description the HNC of the plant.

More recently, the fast developing novel types of metabolomic profiling are also revolutionizing alpine WFPs. When combined with advanced data mining tools, the high-throughput analytical technologies such as HPLC, GC-MS and NMR can provide a holistic look into plant metabolome. Metabolomic profiling is the entire set of metabolites including both primary (simple molecules, such as sugars and amino acids) and secondary metabolites (e.g. terpenoids, flavonoids). These resources allow for analyzing a wide range of metabolites that are difficult to characterize by means of the old technology and provide in-depth insight into the chemical complexity of plant tissues. One of the key holy grails with metabolomics is to see how the plants' metabolome changes in response to environmental stressors such as temperature, light and nutrient availability. That opens the door for scientists to follow the mechanism through which alpine plants generate bioactive substances in different conditions and offers a basic framework aimed at their use as medicine. Moreover, novel metabolites that have escaped classic methodologies may also be identified and will challenge plant medicine discovery (Fina et al., 2023).

8. CONSERVATION, SUSTAINABILITY, AND CLIMATE IMPACTS

Alpine wild food plants (WFPs) are of particular significance, not only nutritionally and medicinally but also ecologically and culturally. But these plants are also threatened by climate change, habitat loss and overharvesting, as well as other human activities. Heavier attention on alpine WFPs, particularly their possibility as FNs and natural drugs, more attention focused on the issues that these plants are beginning to face is becoming also even greater crucial. Protection and sustainable conservation of these plants is essential to save their biocomplexity for future generations also. An effective combination of conservation, sustainable use, and community-based measures is what it would take to minimize the effects of climate change and other threats.

WFPs in the Alps are among the most impacted by climate change. Due to the warming climate Alpine areas experience changes to storm patterns, drier and longer drought periods followed by heavier rains, more frequent hailstorms with serious consequences and shifts in season cycles. These shifts may have important consequences for the growth and survival of alpine plants, which are generally well adapted to a limited range of stable climatic conditions. For instance, alpine WFPs are very temperature sensitive and respond to growing-season length. For example, many alpine plants require low temperatures for flower initiation to produce fruit, and warmer temperatures can disrupt phenological relationships and reduce reproduction. Rainfall patterns are also altering which may affect soil moisture and availability of water to plants. In addition, UV-B exposure which rises due to depletion of ozone can increase oxidative stress induced in plants that is detrimental for growth. This property of alpine ecosystems to be sensitive to environmental changes can put at risk the persistence and the ecosystem services provided by alpine WFPs (Vorkauf et al., 2021).

Another major danger facing alpine WFPs is habitat loss due to deforestation, urbanization and other infrastructure works. Human activities, such as deforestation; mining and agriculture also affect the natural habitats of these plants resulting in there are limited habitats remaining. Habitat destruction can prevent alpine WFPs from being able to survive and reproduce in the presence of competition, since they are often ecological specialists, with either a narrow niche and/or special needs for soil, light, or heat. And industrial and agricultural pollution — think pesticides, heavy metals and pollutants can also affect the health of these plants and get in the way of growth. Alpine ecosystems are disappearing, taking species of plants not only but also animals, micro-organisms and human communities that rely on them for food and medicine as well cultural traditions.

Such over harvesting is also a greatest threat to sustainable survival of alpine WFPs. Due to their nutritional and therapeutic significance; these plants may be threatened by overexploitation. Many wild WFPs collected at alpine are not sustainably managed, which led to resource depletion and population declines. No-brain over hunting/over exploitation could be as a result of commercial motive or desire of you own needs (subsistence requirement), if the productivity of some species exceed its restocking and is not commercially harvested, that would lead to over hunting. This may lead to loss of genetic variation, which could affect the long-term survival of plants. Also, the exploitation of alpine WFPs could bring stress personally and economically to those people who are linked to these plants culturally and at local level; this would complicate relationships between conservation and development.

A number of approaches can be used to safeguard the sustainable use of alpine WFPs. One of the most critical ways to go about it is to ensure that harvesting is sustainable. This could be the sustainable harvest of plant populations at a level that allows for their natural regeneration, such as by means of quotas or regulated collection seasons where plants are not removed during critical growth and reproductive periods. Cultivation and agroforestry A plants; growing them in greenhouses or on cultivated land dedicated to this style of agriculture is one possibility for reducing pressures on their wild populations. Also such restriction can be imposed on the utilization of vegetative products to keep its demand at an optimum level without exceeding availability. This would also contribute to finding a balance between conservation and economic development (Bertsch-Hoermann et al., 2021).

Community-based conservation and policy initiatives are equally important to the preservation of alpine WFP. "Then if you're involving the local communities that are there, often those people know very well what [ecological value] these plants have and how to use them, and you can get more effective conservation [measures] that culturally appropriate. Community-based conservation approaches, like community reserves or protected areas might serve as options for alpine WFP to go externals and avoid over-exploitation and habitat destruction. Local people's involvement plays a key role in effective stewardship and management programs. Monitoring and sustainable utilization of plant population In Madagascar, local citizens can be involved in a sound stewardship-management program including monitoring and using plants sustainably. This may have a guarding for populations and the sustainable use over time.

Equally important are policy frameworks supporting the conservation of alpine WFPs. Governments and international organizations can adopt guidelines that promote the sustainable use of resources, identify protected areas and incorporate traditional knowledge in conservation. Regulations addressing overharvesting, habitat loss and climate change can provide a foundation for the protection of these alpine plants. In addition, in the context of conservation of biodiversity, the incorporation of alpine WFPs species into management plans can encourage prioritization both for protection and sustainable use.

9. INTEGRATING TRADITIONAL KNOWLEDGE WITH MODERN SCIENTIFIC RESEARCH

Combining traditional knowledge with modern scientific findings is a more promising way to study and utilize alpine WFPs. Indigenous communities has for centuries depended on their intimate knowledge of native plants as medicines and food. These evidences of the traditional practices, which are emmipirical based ones, give much information about the herbal remedies from alpine WFPs. The integration of this traditional knowledge with contemporary scientific methods e.g. phytochemical analysis, molecular biology and bioassays can help to unravel the therapeutic potential of these plants and provide new insights for drug discovery and nutraceuticals. This holistic approach aims to benefit from both local wisdom and modern advancements in science for improved understanding and protection of alpine WFPs (Ianni et al., 2015).

Drug discovery this is one of the major fields in which this integration is greatly beneficial. Many of the modern pharmaceuticals were originally derived from traditional botanical practices in which plants were used medicinally thousands of years before their active pharmacological compounds were substantiated by empirical evidence. Through the integration of centuries-old alpine plant knowledge with state-of-the-art methodologies, researchers can find out which substances are responsible for the plants' healing properties. For example, *Picrorhiza kurroa*, a plant commonly used in the Ayurvedic system to treat liver disorders has been pharmacologically evaluated to isolate and characterize specific compounds (i.e., picrosides) thought to be involved in hepatoprotection. This synergism enables traditional knowledge to select the candidate plants for scientific studies and modern technology can prove and isolate the bioactive compounds, that might have a way of discovery of new drugs.

The amalgamation of traditional wisdom and contemporary research is equally important in nutraceuticals, as it straddles a fine footing. Alpine WFPs are characterized by the presence of bioactive substances, including antioxidants, antiinflammatory compounds and antimicrobials, which have been part of traditional diets for centuries because of their healthpromoting effects. Contemporary science has begun to support these traditional uses, which paved the way for functional foods and dietary supplements. For instance, the alpine plants *Artemisia maritima* and *Saussurea lappa* are in use due to their anti-inflammatory and antimicrobial action in mountain folklore. Phytochemical screening of these plants have substantiated the presence of bioactive molecules like flavonoids and terpenoids which are responsible to possess such health benefits. Traditional understanding of the plants can be consolidated with contemporary research methodologies for bioavailability, dosage and efficacy to transform alpine WFPs into efficacious nutraceuticals with health promoting potential beyond basic nutrition (Naiel et al., 2024; Rocha et al., 2021).

It is important to note that one of the key components towards linking traditional knowledge with science is reinforcing forms of documentation and systems of knowledge exchange. Most of the traditional information on alpine WFPs is transmitted orally and may disappear because of cultural evolution and urbanization. The preservation of this invaluable knowledge using computer databases and international research collaborations are fundamental. Through compiling detailed documentary data on the traditional usage, medicinal values, and ecological status of these species and give a sturdy footstep to future scientific assessment. There should be interface of knowledge between traditional healers, local communities and scientific researchers. Local communities have acquired local knowledge in the ecological and therapeutic use of alpine WFPs, and such ethno-knowledge should be collated with modern science in order to enable both systems of understanding to meet for the conservation and sustainable utilization of these plants.

10. CONCLUSION

Alpine WFPs hold significant potential for both nutritional and pharmacological applications. Rich in bioactive compounds, these plants have demonstrated therapeutic benefits, including anti-inflammatory, antioxidant, antimicrobial, and hepatoprotective properties. By integrating traditional knowledge with modern scientific research, we can unlock new opportunities for drug discovery and the development of nutraceuticals. These plants offer sustainable, plant-based solutions to global health challenges and emphasize the importance of preserving both biodiversity and cultural knowledge. Future research should focus on understanding the impact of climate change, habitat degradation, and overharvesting on alpine WFPs. Additionally, exploring advanced techniques such as metabolomics will further reveal new bioactive compounds, and sustainable harvesting practices should be prioritized to protect wild populations. Cultivating these plants in controlled environments can also alleviate pressure on natural habitats. The implications for pharmaceutical innovation, nutrition, and biodiversity preservation are profound. Alpine WFPs provide a sustainable source of bioactive compounds for new health products, and their conservation ensures the preservation of vital plant species and traditional knowledge. By combining traditional wisdom with modern science, we can unlock the full potential of these plants to support human health and preserve our natural environment.

REFERENCES

Abbasi, A. M., Shah, M. H., & Khan, M. A. (2015). *Wild edible vegetables of lesser Himalayas*. Springer. DOI: 10.1007/978-3-319-09543-1

Agbeno, E. K. (2025). Ethnobotanical Knowledge and Medicinal Plant Use Among Indigenous Communities: Conservation and Cultural Preservation Strategies. *International Journal of Indigenous Medicine*, 15-58.

Alami, M. M., Guo, S., Mei, Z., Yang, G., & Wang, X. (2024). Environmental factors on secondary metabolism in medicinal plants: Exploring accelerating factors. *Medicinal Plant Biology*, *3*(1), 0. DOI: 10.48130/mpb-0024-0016

Ashraf, M. V., Khan, S., Misri, S., Gaira, K. S., Rawat, S., Rawat, B., Khan, M. A. H., Shah, A. A., Asgher, M., & Ahmad, S. (2024). High-Altitude Medicinal Plants as Promising Source of Phytochemical Antioxidants to Combat Lifestyle-Associated Oxidative Stress-Induced Disorders. *Pharmaceuticals (Basel, Switzerland)*, *17*(8), 975. Advance online publication. DOI: 10.3390/ph17080975 PMID: 39204080

Ashraf, M. V., Khan, S., Misri, S., Gaira, K. S., Rawat, S., Rawat, B., Khan, M. A. H., Shah, A. A., Asgher, M., & Ahmad, S. (2024). High-altitude medicinal plants as promising source of phytochemical antioxidants to combat lifestyle-associated oxidative stress-induced disorders. *Pharmaceuticals (Basel, Switzerland)*, *17*(8), 975. DOI: 10.3390/ph17080975 PMID: 39204080

Bertsch-Hoermann, B., Egger, C., Gaube, V., & Gingrich, S. (2021). Agroforestry trade-offs between biomass provision and aboveground carbon sequestration in the alpine Eisenwurzen region, Austria. *Regional Environmental Change*, *21*(3), 77. DOI: 10.1007/s10113-021-01794-y PMID: 34720739

Bhattacharyya, M., Thattantavide, A., & Kumar, A. (2023). Ethnic mountain foods of western and eastern Himalayas, India. In *Wild food plants for zero hunger and resilient agriculture* (pp. 181–205). Springer. DOI: 10.1007/978-981-19-6502-9_8

Billings, W. D., & Mooney, H. A. (1968). The ecology of arctic and alpine plants. *Biological Reviews of the Cambridge Philosophical Society*, *43*(4), 481–529. DOI: 10.1111/j.1469-185X.1968.tb00968.x

Chandra, S., & Lata, H. (2025) High Altitude Medicinal Plants. Eds. 01. 1-561. Springer Cham. https://doi.org/10.1007/978-3-031-99406-7

Chandra, S., Trivedi, V., Purohit, V., Nautiyal, M., & Chandra, S. (2025). Climatic Change and Physiological Adaptations in Alpine Medicinal Plants. In *High Altitude Medicinal Plants: Botany, Conservation and Cultivation* (pp. 11-35). Springer.

Chen, W., Ding, H., Li, J., Fu, F., Li, Y., Xiao, S., Xu, D., Lu, J., & Fang, J. (2023). How do montane plants manage to survive? Inferring from non-structural carbohydrates. *Trees (Berlin)*, *37*(2), 331–348. DOI: 10.1007/s00468-022-02351-8

Chiriac, E. R., Chiţescu, C. L., Geană, E.-I., Gird, C. E., Socoteanu, R. P., & Boscencu, R. (2021). Advanced Analytical Approaches for the Analysis of Polyphenols in Plants Matrices—A Review. *Separations*, *8*(5), 65. DOI: 10.3390/separations8050065

Danna, C. (2024). *By-products from medicinal and aromatic plants: from tradition to innovation.* PhD Thesis submitted to Catalogo prodotti della ricerca Università di Genova.

Fantasma, F., Samukha, V., Saviano, G., Chini, M. G., Iorizzi, M., & Caprari, C. (2024). Nutraceutical Aspects of Selected Wild Edible Plants of the Italian Central Apennines. *Nutraceuticals*, *4*(2), 190–231. DOI: 10.3390/nutraceuticals4020013

Fina, A., Koljančić, N., Squara, S., Ferrara, D., Bicchi, C., Špánik, I., & Cordero, C. (2023). Food metabolomics by GCXGC-TOF MS and tandem ionization: understanding the impact of climate events on edible crops quality. XIII Congresso Nazionale di Chimica degli Alimenti Libro degli abstracts, Grabherr, G. (2009). Biodiversity in the high ranges of the Alps: ethnobotanical and climate change perspectives. *Global Environmental Change*, *19*(2), 167–172.

Holopainen, J. K., Himanen, S. J., Yuan, J. S., Chen, F., & Stewart, C. N., Jr. (2025). Ecological functions of terpenoids in changing climates. In *Natural Products: Phytochemistry, Botany, Metabolism of Alkaloids, Phenolics and Terpenes* (pp. 1-44). Springer. DOI: 10.1007/978-3-642-36202-6_129-1

Hou, M., Leng, C., Zhu, J., Yang, M., Yin, Y., Xing, Y., & Chen, J. (2024). Alpine and subalpine plant microbiome mediated plants adapt to the cold environment: A systematic review. *Environmental Microbiome*, *19*(1), 82. DOI: 10.1186/s40793-024-00614-0 PMID: 39487507

Ianni, E., Geneletti, D., & Ciolli, M. (2015). Revitalizing Traditional Ecological Knowledge: A Study in an Alpine Rural Community. *Environmental Management*, *56*(1), 144–156. DOI: 10.1007/s00267-015-0479-z PMID: 25847107

Karimi, S., Nawaz, M. A., Naseem, S., & Ali, Z. (2025). Influence of open-top chambers induced climate warming on secondary metabolic profile of culturally and medicinally important plants of Himalaya, Karakoram and Hindukush. *PLoS One*, *20*(5), e0322480. DOI: 10.1371/journal.pone.0322480 PMID: 40367051

Körner, C. (2023). Concepts in alpine plant ecology. *Plants*, *12*(14), 2666. DOI: 10.3390/plants12142666 PMID: 37514280

Kouznetsov, V. V. (2025). *Chemical Defense in Nature: Structure, Function, and Synthesis of Toxic Secondary Metabolites from Plants and Animals*. CRC Press. DOI: 10.1201/9781003335771

Kumar, R., Joshi, R., Kumar, R., Srivatsan, V., Satyakam, , Chawla, A., Patial, V., & Kumar, S. (2022). Nutritional quality evaluation and proteome profile of forage species of Western Himalaya. *Grassland Science*, *68*(3), 214–225. DOI: 10.1111/grs.12357

Kumar, V., Kumar, P., Bhargava, B., Sharma, R., Irfan, M., & Chandora, R. (2023). Transcriptomic and metabolomic reprogramming to explore the high-altitude adaptation of medicinal plants: A review. *Journal of Plant Growth Regulation*, *42*(12), 7315–7329. DOI: 10.1007/s00344-023-11018-8

Kumari, A., Joshi, R., & Yadav, S. K. (2025). Prospects and potential for sustainable food and nutritional security in North Western Himalayan region. *Indian Journal of Plant Genetic Resources*, *38*(02), 145–154.

Ladol, S., Kumar, A., & Sharma, D. (2023). Nutritional and Health Benefits of High Altitude Wild Food Plant, Hippophae rhamnoides for the Himalayan Communities. In Kumar, A., Singh, P., Singh, S., & Singh, B. (Eds.), *Wild Food Plants for Zero Hunger and Resilient Agriculture* (pp. 149–161). Springer Nature Singapore., DOI: 10.1007/978-981-19-6502-9_6

Mehta, S., Sharma, A. K., & Singh, R. K. (2021). Advances in Ethnobotany, Synthetic Phytochemistry and Pharmacology of Endangered Herb Picrorhiza kurroa (Kutki): A Comprehensive Review (2010-2020). *Mini-Reviews in Medicinal Chemistry*, *21*(19), 2976–2995. DOI: 10.2174/1389557521666210401090028 PMID: 33797375

Milani, F., Bottoni, M., Giuliani, C., Colombo, L., Casiraghi, M. C., Colombo, P. S., Bruschi, P., Erba, D., & Fico, G. (2023). Alpine Diet in Valmalenco (Lombardy, Italy): Nutritional Features of Spontaneous Plants and Traditional Dishes. *Nutrients*, *15*(8), 1988. DOI: 10.3390/nu15081988 PMID: 37111208

Nagy, L., & Grabherr, G. (2009). *The biology of alpine habitats*. Oxford University Press. DOI: 10.1093/oso/9780198567035.001.0001

Naiel, M. A.-E., Shehata, A. M., Paswan, V. K., Said, M. B., & El-Tarabily, K. A. (2024). Utilizing the potential of Saussurea lappa in aquaculture industry: A review on immune enhancement and pollution remediation. *Aquaculture International*, *32*(5), 5513–5550. DOI: 10.1007/s10499-024-01435-1

Nigam, M., Atanassova, M., Mishra, A. P., Pezzani, R., Devkota, H. P., Plygun, S.,…Sharifi-Rad, J. (2019). Bioactive Compounds and Health Benefits of Artemisia Species. *Natural Product Communications, 14*(7), 1934578X19850354. DOI: 10.1177/1934578X19850354

Painuli, S., Semwal, P., Cruz-Martins, N., & Bachheti, R. K. (2021). Medicinal Plants of Himalayan Forests. In Husen, A., Bachheti, R. K., & Bachheti, A. (Eds.), *Non-Timber Forest Products: Food, Healthcare and Industrial Applications* (pp. 175–212). Springer International Publishing., DOI: 10.1007/978-3-030-73077-2_8

Pal, D. (2011). Sunflower (Helianthus annuus L.) Seeds in Health and Nutrition. In Preedy, V. R., Watson, R. R., & Patel, V. B. (Eds.), *Nuts and Seeds in Health and Disease Prevention* (pp. 1097–1105). Academic Press., DOI: 10.1016/B978-0-12-375688-6.10130-6

Pant, P., Pandey, S., & Dall'Acqua, S. (2021). The influence of environmental conditions on secondary metabolites in medicinal plants: A literature review. *Chemistry & Biodiversity, 18*(11), e2100345. DOI: 10.1002/cbdv.202100345 PMID: 34533273

Qaderi, M. M., Martel, A. B., & Strugnell, C. A. (2023). Environmental factors regulate plant secondary metabolites. *Plants, 12*(3), 447. DOI: 10.3390/plants12030447 PMID: 36771531

Ravetto Enri, S., Falla, N. M., Demasi, S., Bianchi, D. M., Squadrone, S., Lombardi, G., & Scariot, V. (2024). Exploring the potential of wild leafy vegetables widespread in European Alps as functional food. *Journal of Agriculture and Food Research, 18*, 101368. DOI: 10.1016/j.jafr.2024.101368

Rocha, M. I., Gonçalves, M. J., Cavaleiro, C., Cruz, M. T., Pereira, C., Moreira, P., Salgueiro, L., & Figueirinha, A. (2021). Chemical characterization and bioactive potential of Artemisia campestris L. subsp. maritima (DC) Arcang. essential oil and hydrodistillation residual water. *Journal of Ethnopharmacology, 276*, 114146. DOI: 10.1016/j.jep.2021.114146 PMID: 33932510

Romeo, R., Manuelli, S., Geringer, M., & Barchiesi, V. (2021). *Mountain farming systems–Seeds for the future: Sustainable agricultural practices for resilient mountain livelihoods*. Food & Agriculture Org.

Roychoudhury, S., Das, D., Das, S., Jha, N. K., Pal, M., Kolesarova, A., Kesari, K. K., Kalita, J. C., & Slama, P. (2022). Clinical Potential of Himalayan Herb Bergenia ligulata: An Evidence-Based Study. *Molecules (Basel, Switzerland), 27*(20), 7039. DOI: 10.3390/molecules27207039 PMID: 36296631

Sears, C. E. (2024). *Botanical Bar Craft: A Guide to the Art of Apothecary Cocktails and Herbal Tonic Elixirs*. Chelsea Green Publishing.

Sharma, M. K. (2024). Exploring the biochemical profiles of medicinal plants cultivated under stressful environmental conditions. *Current Agriculture Research Journal*, *12*(1), 81–103. DOI: 10.12944/CARJ.12.1.07

Swathi, K., Suganya, K., Aswathy, R., Angel, C., Sudha, B., Sumathi, S., & Mandal, S. K. (2025). Drugs Derived from High Altitude Plants. In *High Altitude Medicinal Plants: Botany, Conservation and Cultivation* (pp. 235-268). Springer.

Ullah, H., & Badshah, L. (2023). Nutritional and mineral analysis of the ultimate wild food plants of Lotkuh, Chitral, the Eastern Hindukush Pakistan. *Heliyon*, *9*(3), e14449. DOI: 10.1016/j.heliyon.2023.e14449 PMID: 36950657

Von Glasenapp, M., & Thornton, T. F. (2011). Traditional ecological knowledge of Swiss alpine farmers and their resilience to socioecological change. *Human Ecology: an Interdisciplinary Journal*, *39*(6), 769–781. DOI: 10.1007/s10745-011-9427-6

Vorkauf, M., Kahmen, A., Körner, C., & Hiltbrunner, E. (2021). Flowering phenology in alpine grassland strongly responds to shifts in snowmelt but weakly to summer drought. *Alpine Botany*, *131*(1), 73–88. DOI: 10.1007/s00035-021-00252-z

Zafar, M., Abbasi, A. M., Abbas, Z., Narzullayeva, U., Gafforov, Y., Filimban, F. Z., & Azimova, N. (2025). Health Risk Assessment and Bioactive Potential in Wild Food Plants from Deosai: A High-Altitude Ecosystem in Northern Pakistan: M. Zafar et al. *Biological trace element research*, 1-26.

Zahara, K., Tabassum, S., Sabir, S., Arshad, M., Qureshi, R., Amjad, M. S., & Chaudhari, S. K. (2014). A review of therapeutic potential of Saussurea lappa-An endangered plant from Himalaya. *Asian Pacific Journal of Tropical Medicine*, *7*, S60–S69. DOI: 10.1016/S1995-7645(14)60204-2 PMID: 25312191

Chapter 11
Biological Effects and Phytochemical Composition of Wild Food Plants (WFPs) in Alpine Zones:

Ahmet Ismail Ozkan
http://orcid.org/0000-0002-4511-2386
Artvin Coruh University, Turkey

Kemal Yüce
Ağrı İbrahim Çeçen University, Turkey

Nevin Aslan
http://orcid.org/0000-0002-6191-5613
Harran University, Turkey

ABSTRACT

All plants that we refer to as domesticated plants were once wild food plants (WFPs) before being cultivated by humans. These plants are species that can naturally grow in their natural habitats without direct human intervention. They have grown in various ecosystems and adapted to various environmental conditions. Recently, WFPs classified as functional foods have been found to alleviate nutritional deficiencies and even treat diseases. WFPs can be useful in combating malnutrition. They contain bioactive compounds including essential source of vitamins, minerals, and secondary metabolites such as essential oils, alkaloids, and phenolics. In addition, increasing the intake of these compounds can further improve human health.

DOI: 10.4018/979-8-2600-1207-9.ch011

WFPs are also evaluated in terms of their pharmacological properties due to their significant medicinal importance. WFPs were noted to have the highest abundance in the alpine zone. This chapter was focused on the biological effects including antimicrobial, antioxidant and anticancer, and phytochemical composition of WFPs in the alpine zone.

1. INTRODUCTION

The high mountain ecosystems are biodiversity hubs with a high level of endemic species. This is mainly owing to the alpine ecosystems, which span all latitudes and continents, covering almost three per cent of the Earth's landmass except Antarctica. These ecosystems are those above the tree line (Noroozi et al., 2021). Alpine ecosystems are described as cold, windy, and snowy regions. This region is located below the permanent snow line and above the tree line in tropical and temperate mountains (Edwards et al., 2007). Alpine environments are characterized by a high endemism rate compared to most subalpine zones (Noroozi et al., 2021), cold temperatures and cold-loving vegetation (Verrall & Pickering, 2020) and low temperatures during the growing period and a much shorter non-frost cycle (Edwards et al., 2007). This is owing to the complex origins of climate changes, high elevations, geographic isolation, and significant microhabitat diversity (Noroozi et al., 2021).

Alpine regions are important reservoirs of biodiversity. These areas support (i) species adapted to cold climates, (ii) species unable to tolerate the intense competition found in lower-elevation areas, and (iii) endemic species found in glacial regions due to geographical isolation. The high topographic diversity of alpine areas is crucial for biodiversity. A broad spectrum of microclimates and soil characteristics are enabled by high spatial variability in aspect and slope, resulting in a high diversity of microhabitats. This allows many specialized species to coexist in a narrow area (Britton et al., 2009).

The high elevations of the mountain that form the alpine zones are treeless areas where various wild food plants (WFP) grow. One such area is the Erenler (Irsa) plateau in Artvin Türkiye, which has an elevation of between 1600 and 2600 m. This area is shown in Figure 1.

Figure 1. Alpine zone and plants (a, b) (Photos by Ahmet İsmail Özkan) and altitude (c) and satellite view (d) of Erenler (Irsa) plateau in Artvin, Türkiye.

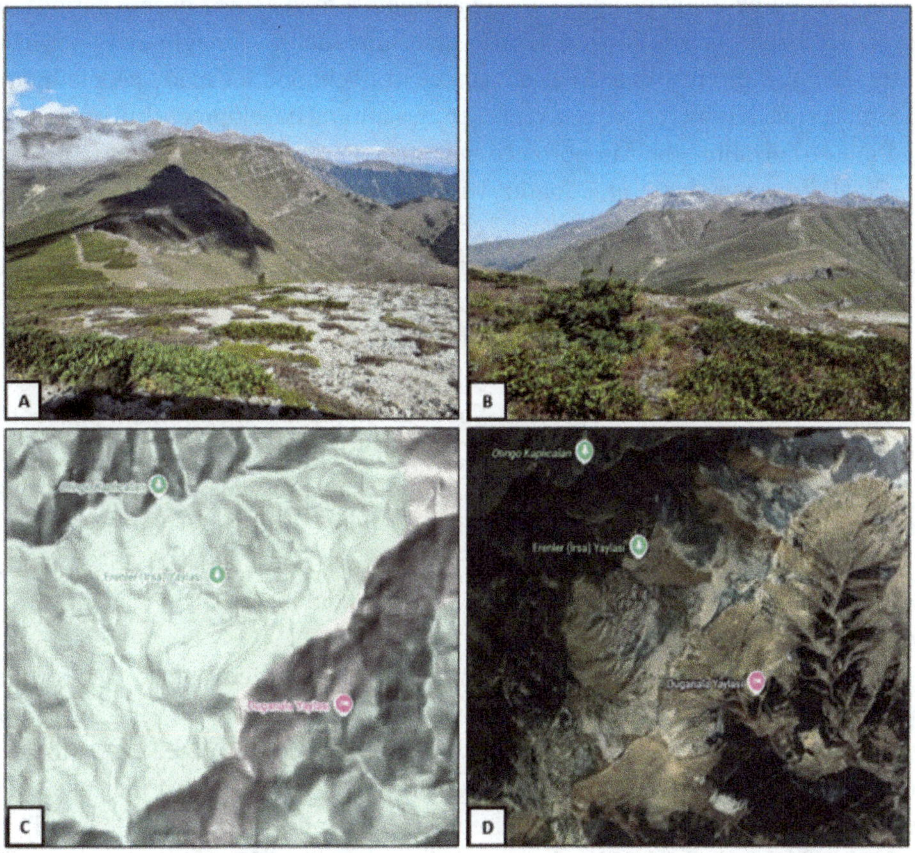

Alpine ecosystems are characterized by cryophilic vegetation and low temperatures. These ecosystems are globally important not only for their biological diversity but also for the ecosystem and socio-economic benefits they provide to humans. These benefits include minimizing extreme incidents such as avalanches and landslides, conserving biodiversity, maintaining soil fertility, providing water, generating energy, and supporting tourism. However, these ecosystems are considered to be among the most sensitive to climate change (Verrall & Pickering, 2020).

1.1 Importance of WFPs in Alpine Zones

The term 'wild' describes plants that grow naturally, without human cultivation. It is mostly used for species that grow in their own natural habitats. WFPs cannot be considered merely "famine food," as many of them have been used since ancient times and continue to be used today (Luczaj et al., 2012). Throughout history, they have been a vital food sources for humans. Especially in developing countries, they are consumed by many people as a safe alternative food source during periods of severe food shortages following various hardships such as drought and poor harvests. Today, they are still used as a resource of livelihood for many people living in remote areas. However, with advances in agricultural development, urbanization, and industrialization, their importance in human nutrition has declined (Gillani et al., 2024). In mountainous areas of the world and places where food is scarce, WFPs are extremely important. WFPs can be categorised as functional foods due to their nutritional composition and health benefits. WFPs grow in high-altitude, rugged terrain. Thanks to these plants, people can sustain their lives even in high-altitude areas. However, WFPs have not been studied sufficiently (Ullah & Badshah, 2023). Gillani et al. (2024) stated in their study that WFPs are a source of carbohydrates, fiber, minerals, vitamins, and protein. They also noted that they contain high concentrations of vitamins A and C, Zn, Fe, Ca, I, thiamine, riboflavin, niacin, and folic acid compounds. Ullah and Badshah (2023) indicated in their study that various micro and macro nutrients such as vitamin C, Zn Fe, Ca, Mn, and Mg are present. They also stated that their water capacity ranges from approximately 25% to 90%, they have a lipid content of between 40% and 60%, and their protein content varies depending on the species.

Waheed et al. (2023) recorded that the most frequently consumed parts of the plant were the fruits, leaves, shoots, seeds, grains, whole plants, flowers, and roots/gum, (43.66%), (19.71%), (16.90%), (15.49%), (14%), (9.85%), (4.22%), (2.81%), respectively (Waheed et al., 2023). However, Gillani et al. (2024) indicates that leaves constitute 41.04% of the most frequently consumed parts of the plant. This is followed by fruit at 33.33%, above-ground parts at 10.26%, rhizomes at 7.69%, bark at 2.56%, grains at 2.56%, beans at 1.71%, seeds, 1.71% whole plant, 0.85% nuts, 0.85% roots, 0.85% stems, and 0.85% tubers (Gillani et al., 2024). These results reveal that the most commonly used parts of WFPs vary by region.

1.2 Nutritional, Cultural, and Ecological Significance

WFPs are used as valuable food sources in the traditions of many local cultures. In some communities, they are also noted for their nutritional, medicinal, and symbolic importance. They may contain more nutrients than domesticated plants. Their parts, including as flowers, fruits, tubers, and roots, are a wealthy resource of fiber, protein, fat, and carbohydrates, which are vital for a balanced diet. Leaf and fruit parts are used more often owing to their relatively broad variety of inorganic and organic biomolecule mixtures. Adopting a modern lifestyle has resulted in an increase in the ingestion of processed foods such as junk food. Adopting a modern lifestyle has led to an increase in processed food consumption, including junk food. This has increased the risks of diseases associated with poor and imbalanced nutrition. Over the past ten years, numerous studies worldwide have focused on WFPs and the biocultural heritage associated with neglected species, with the aim of domesticating new crops. WFPs are consumed as fruits, vegetables, herbal beverages, pickles, spices, raw foods, jams, bread, and sauces. Most WFPs are harvested during the monsoon and spring seasons (Waheed et al., 2023).

As discussed in the literature, when examining the consumption patterns of WFPs, it is observed that they are consumed in descending order of Vegetables > Fruit > Pickles & Condiments > Beverages. WFPs and their consumption patterns are shown in Table 1.

Table 1. WFP and its mode of consumption

WFP	Mode of consumption	Previous Literature
Abies pindrow	Drinks	(Gillani et al., 2024)
Acacia nilotica	Drinks, Pickles & Condiments	(Waheed et al., 2023)
Adiantum capillus-veneris	Drinks	(Gillani et al., 2024)
Ajuga integrifolia	Drinks	(Gillani et al., 2024)
Amaranthus retroflexus	Vegetable	(Gillani et al., 2024)
Amaranthus spinosus	Vegetable	(Gillani et al., 2024)
Amaranthus viridis	Vegetable	(Gillani et al., 2024; Waheed et al., 2023)
Arnebia benthamii	Drinks	(Gillani et al., 2024)
Bauhinia variegata	Pickles & Condiments	(Waheed et al., 2023)
Berberis lyceum	Fruit	(Gillani et al., 2024)
Bergenia ciliata	Drinks	(Gillani et al., 2024)
Betula utilis	Drinks	(Gillani et al., 2024)
Bistorta amplexicaulis	Drinks	(Gillani et al., 2024)
Cannabis sativa	Drinks	(Waheed et al., 2023)
Capparis decidua	Pickles & Condiments	(Waheed et al., 2023)
Carthamus oxyacantha	Drinks	(Waheed et al., 2023)
Chenopodium album	Vegetable	(Gillani et al., 2024; Waheed et al., 2023)
Chenopodium botrys	Vegetable	(Gillani et al., 2024)
Chenopodium murale	Vegetable	(Gillani et al., 2024; Waheed et al., 2023)
Cichorium intybus	Drinks, Vegetable	(Waheed et al., 2023)
Citrullus colocynthis	Pickles & Condiments	(Waheed et al., 2023)
Cleome viscosa	Vegetable	(Waheed et al., 2023)

continued on following page

Table 1. Continued

WFP	Mode of consumption	Previous Literature
Cordia dichotoma	Fruit, Pickles & Condiments	(Waheed et al., 2023)
Cordia myxa	Fruit, Pickles & Condiments	(Waheed et al., 2023)
Cucurbita maxima	Vegetable	(Gillani et al., 2024)
Digera muricata	Vegetable	(Waheed et al., 2023)
Diospyros lotus	Fruit	(Gillani et al., 2024)
Ehretia acuminata	Fruit	(Waheed et al., 2023)
Elaeagnus umbellata	Fruit	(Gillani et al., 2024)
Ficus palmata	Fruit	(Gillani et al., 2024; Waheed et al., 2023)
Fragaria nubicola	Fruit	(Gillani et al., 2024)
Gaultheria trichophylla	Fruit	(Gillani et al., 2024)
Geranium wallichianum	Drinks	(Gillani et al., 2024)
Juglans regia	Pickles & Condiments	(Gillani et al., 2024)
Malus pumila	Fruit	(Gillani et al., 2024)
Malva parviflora	Vegetable	(Gillani et al., 2024)
Malva sylvestris	Vegetable	(Gillani et al., 2024)
Mentha arvensis	Drinks, Pickles & Condiments	(Gillani et al., 2024)
Mentha longifolia	Drinks, Pickles & Condiments	(Gillani et al., 2024; Waheed et al., 2023)
Mentha pulegium	Pickles & Condiments	(Waheed et al., 2023)
Mentha spicata	Drinks, Pickles & Condiments	(Gillani et al., 2024)
Moringa oleifera	Pickles & Condiments	(Waheed et al., 2023)
Morus alba	Fruit	(Waheed et al., 2023)
Morus nigra	Fruit	(Waheed et al., 2023)
Mukia madraspatana	Vegetable	(Waheed et al., 2023)
Nasturtium officinale	Vegetable	(Gillani et al., 2024)
Opuntia stricta	Fruit	(Waheed et al., 2023)
Oxalis corniculata	Vegetable	(Gillani et al., 2024; Waheed et al., 2023)
Physalis minima	Fruit	(Waheed et al., 2023)
Phytolacca latbenia	Vegetable	(Gillani et al., 2024)
Plantago major	Vegetable	(Gillani et al., 2024)
Plectranthus rugosus	Drinks	(Gillani et al., 2024)
Polygonum plebeium	Vegetable	(Gillani et al., 2024)
Prosopis cineraria	Pickles & Condiments	(Waheed et al., 2023)
Prosopis juliflora	Pickles & Condiments	(Waheed et al., 2023)
Prunus armeniaca	Fruit	(Gillani et al., 2024)

continued on following page

Table 1. Continued

WFP	Mode of consumption	Previous Literature
Pteridium aquilinum	Vegetable	(Gillani et al., 2024)
Punica granatum	Fruit	(Gillani et al., 2024)
Pyrus pashia	Fruit	(Gillani et al., 2024)
Pyrus communis	Fruit	(Gillani et al., 2024)
Rubus fruticosus	Fruit	(Gillani et al., 2024)
Rumex dentatus	Vegetable	(Waheed et al., 2023)
Rumex nepalensis	Vegetable	(Gillani et al., 2024)
Salvadora oleoides	Fruit	(Waheed et al., 2023)
Salvadora persica	Fruit	(Waheed et al., 2023)
Sisymbrium irio	Vegetable	(Waheed et al., 2023)
Solanum surattense	Pickles & Condiments	(Hasan et al., 2024; Waheed et al., 2023)
Taxus wallichiana	Drinks	(Gillani et al., 2024)
Viburnum grandiflorum	Fruit	(Gillani et al., 2024)
Zanthoxylum armatum	Pickles & Condiments	(Gillani et al., 2024)
Ziziphus mauritiana	Fruit	(Waheed et al., 2023)
Ziziphus nummularia	Fruit	(Waheed et al., 2023)

1.3 Objective and Scope

The biological effects that result from climate change are becoming more apparent in many different ecosystems every day (Vanneste et al., 2017). Alpine grassland ecosystems are an important component of the global ecological balance. These ecosystems not only provide rich biological diversity but also act as sensitive indicators of climate change. This makes them an important ecological conservation priority (Hua et al., 2025). Alpine plants must adapt in order to cope with various stress factors (UV rays, high light, wind, drought) (Melodelima & Lobréaux, 2013). Secondary metabolites play a protective role in response to abiotic and biotic stress environments (Akula & Ravishankar, 2011). Under environmental stresses, the secondary metabolites production may either decrease or increase (Laftouhi et al., 2023; Yuce & Ozkan, 2020). Secondary metabolites are essential bioactive compounds that play a significant role in plant resilience, survival and adaptation. Secondary metabolites can be categorised into four main types: phenolics, terpenes, glucosinolates (sulfur-containing), and alkaloids (nitrogen-containing). Monoterpenes, including linalool, menthol, camphor, β-pinene, and α-pinene, possess antioxidant and antimicrobial activities. Volatile compounds such as terpenoids provide plants with stress tolerance against oxidative and thermal stress factors. Compared to normal conditions, plants growing in challenging environments biosynthesize more phenolic and flavonoid compounds. Notably, anthocyanin and flavonoid compounds demonstrate antioxidant capacity which reduce ROS and attendant oxidative damages (Sana et al., 2025). Beyond their fundamental roles in plant resilience, these metabolites are also highly valuable for biological applications as important sources of food additives, pharmaceuticals, and industrial compounds (Gaude & Jalmi, 2025). Thus, secondary metabolites are generally responsible for the plant species therapeutic properties (Rawat & Singh, 2024).

Recently, WFPs classified as functional foods have been found to alleviate nutritional deficiencies and even treat diseases. WFPs can be useful in combating malnutrition. They contain bioactive compounds such as vital source of vitamins, minerals, and secondary metabolites such as essential oils, alkaloids, and phenolics. In addition, increasing the intake of these compounds can further improve human health. WFPs are also evaluated in terms of their pharmacological properties due to their significant medicinal importance. WFPs were noted to have the highest abundance in the alpine zone. The Alpine zone serves as a refuge for endemic species that have emerged for various reasons. This chapter was focused on the biological effects including antimicrobial, antioxidant and anticancer, and phytochemical composition of WFPs in the alpine zone.

2. OVERVIEW OF WILD FOOD PLANTS IN ALPINE ZONES

Alpine regions are home to species that cannot survive at lower altitudes. Therefore, high mountains are centers of biological diversity with a high level of endemism (Noroozi et al., 2021). Alpine ecosystems, characterized by cold-loving vegetation and low temperatures, provide ecosystem benefits to many people (Verrall & Pickering, 2020).

Alpine environments, high-altitude habitats above the climate limit, are found on all continents and at all latitudes. While Testolin et al. (2021) stated that alpine regions cover approximately 3% of land areas except Antarctica, Körner (2023) indicated that they constitute 2.6%. These habitats contain areas that are important for global biodiversity and host lots of plant species, most of those are endemic worldwide. Globally, the vegetation of alpine environments is dominated by several growth forms such as shrubs, dwarf, herbaceous rosettes, graminoids, and cushions, that reflect functional adaptations to the characteristics of high-mountain ecosystems, including short growing seasons, low temperatures, and limited nutrient availability. Plants in alpine zones have adapted to low nutrient availability and low temperatures. Alpine plants are generally shorter, with lighter seeds and smaller leaves compared to lowland species. These characteristics increase photosynthetic efficiency, reproductive efficiency, and resistance to frost events. Therefore, they are globally widespread in alpine vegetation (Testolin et al., 2021).

The alpine zone is the area above the mountainous zone that is naturally treeless, separated by high altitude and the climatic tree line. Bryophytes, ferns, and lichens are also among alpine plants. As altitude increases, plant species become smaller. It has also been noted that as sexual reproduction becomes riskier at higher elevations, clonal life strategies become increasingly important. Being limited or stressed to a certain extent by climate or resources is vital for plant live in the wild environment (Körner, 2023). Alpine plants have adapted to cope with various stress factors in-cuding high UV radiation and low temperatures during their short growing period. In order to prevent light-induced inhibition and neutralise the detrimental ROS that are generated under high levels of UV radiation, alpine plants generally have efficient antioxidant enzymes and high effects of antioxidants. These antioxidants are usually both lipid- and water-soluble, and their functional roles have become the focus of significant researches in recent years (Sickel et al., 2012).

Studies conducted by researchers on Alpine zone plants show that the most commonly observed biological effects are, in order: antioxidant > antimicrobial > antibacterial > anti-inflammatory > anticancer > antitumor > cytotoxic > antidiabetic. However, it is seen that the biological effects of most plants growing in this region have not been researched. Detailed information on the biological effects of Alpine plants is provided in Table 2.

Table 2. Alpine zone plants and biological effects

Alpin plants	Biological effects	Previous Literature
Aconogonon alpinum	Anti-inflammatory, antioxidant, hepatoprotectiv	(Gillani et al., 2024; Khan et al., 2020; Vysochina & Khramova, 2010)
Allantoparmelia alpicola	N.D.	(Vanneste et al., 2017)
Allium carolinianum	Antibacterial, anticancer, antidiabetic, antioxidant, cytotoxic, immunomodulatory	(Gillani et al., 2024; Shahar et al., 2025; Yatoo et al., 2018)
Androsace alpina	N.D.	(Körner, 2021e)
Arabis alpina	Antimicrobial, antioxidant	(Körner, 2023; Uba et al., 2022)
Arnebia benthamii	Antibacterial, anticancer, antimicrobial, antioxidant, anti-inflammatory, hepatoprotective	(Ganie et al., 2012; Gillani et al., 2024; Mukhtar et al., 2025; Shameem et al., 2015)

continued on following page

Table 2. Continued

Alpin plants	Biological effects	Previous Literature
Arnica montana	Antibacterial, anticancer, antifungal, anti-inflammatory, antitumor, antioxidant, cytotoxic	(Koo et al., 2000; Körner, 2023; Sugier et al., 2025; Žitek et al., 2022)
Betula nana	N.D.	(Vanneste et al., 2017)
Bistorta affinis	Anti-inflammatory, antimicrobial, antioxidant, antipyretic, enzyme inhibition	(Akhter et al., 2021; Gillani et al., 2024; Iqbal et al., 2022b; Sultana et al., 2022)
Bistorta emodi (Meisn.) H.Hara	N.D.	(Iqbal et al., 2022b)
Bryocaulon divergens	N.D.	(Vanneste et al., 2017)
Campanula barbata	Anti-inflammatory, analgesic, antioxidant	(Alhage et al., 2025; Cuendet et al., 2001; Körner, 2021f)
Carex curvula	N.D.	(Körner, 2023)
Cerastium uniflorum	N.D.	(Körner, 2021f)
Cladonia arbuscula	Antibacterial, antimicrobial, antioxidant	(Soloveva & Kuzmina, 2023; Tapalsky et al., 2017; Vanneste et al., 2017)
Cladonia bellidiflora	Antidiabetic, antimicrobial, antioxidants with cardioprotective	(Popović et al., 2023; Vanneste et al., 2017; Zambare & Christopher, 2012)
Cladonia macrophylla	N.D.	(Vanneste et al., 2017)
Cladonia pleurota	Antimicrobial	(Popovici et al., 2021; Vanneste et al., 2017)
Coelocaulon aculeatum	N.D.	(Vanneste et al., 2017)
Empetrum nigrum	Antidiabetic, antimicrobial, antioxidant	(Hyun et al., 2018; Vanneste et al., 2017)
Erigeron uniflorus	N.D.	(Körner, 2021f)
Festuca ovina	N.D.	(Vanneste et al., 2017)
Fritillaria cirrhosa D.Don	Antiasthma, anti-inflammatory, antihypertension, antioxidant, antitussive, antitumor, bacteriostasis	(Bokhari & Syed, 2015; Iqbal et al., 2022b; Liu et al., 2020; Yang et al., 2023)
Gaultheria trichophylla	Antibacterial, anti-inflammatory, antifungal, antioxidant, antipyretic, cytotoxic	(Alam et al., 2023; Alam et al., 2017; Gillani et al., 2024; Zaman et al., 2022)
Gentiana acaulis	N.D.	(Körner, 2021c)
Gentiana bavarica	N.D.	(Körner, 2021f)
Geum montanum	N.D.	(Körner, 2023)
Geum reptans	N.D	(Körner, 2021f)
Helictotrichon versicolor	N.D.	(Körner, 2023)

continued on following page

Table 2. Continued

Alpin plants	Biological effects	Previous Literature
Juncus trifidus	N.D.	(Körner, 2023; Vanneste et al., 2017)
Leontodon helveticus	Anticancer, antidiabetic, anti-inflammatory, antimicrobial, antioxidant, wound-healing	(Chinnapun, 2025; Körner, 2023)
Leucanthemopsis alpina	N.D.	(Körner, 2023)
Ligusticum mutellina (syn., *Mutellina purpurea Thell.)*	Anti-biofilm, antimicrobial, antioxidant	(Körner, 2023; Sieniawska et al., 2013a; Sieniawska et al., 2013b)
Linaria alpina	Anti-inflammatory, antimicrobial, antitumor, antiviral (HIV), hepatoprotective	(Körner, 2021f; Venditti et al., 2015)
Lophozia cf. ventricosa	Antimicrobial	(Bukvicki et al., 2015; Vanneste et al., 2017)
Luzula arcuata	N.D.	(Vanneste et al., 2017)
Nardus stricta	N.D.	(Körner, 2023)
Ochrolechia sp.	Antibacterial, antioxidant	(Kim et al., 2014b; Vanneste et al., 2017)
Oxyria digyna	Antioxidant	(Körner, 2021f; Orhan et al., 2009)
Pinus pumila	Antibacterial, antimicrobial	(Fujino et al., 2023; Mi et al., 2025; Zhou et al., 2025)
Poa laxa	N.D.	(Körner, 2021d)
Polygonum viviparum	Antiarthritic, antibacterial, anti-inflammatory, antioxidant, antitumor, antiulcer	(Chang et al., 2014; Körner, 2021f)
Polytrichum juniperinum	Antibacterial, antitumor	(Ruiz-Molina et al., 2019; Vanneste et al., 2017)
Potentilla aurea	Antioxidant, antiproliferative	(Augustynowicz et al., 2023; Körner, 2021b; Tomczyk & Latté, 2009)
Primula glutinosa	Antifungal	(Bhutia et al., 2013; Körner, 2021e)
Primula hirsuta	N.D.	(Körner, 2021a)
Ranunculus glacialis	Antioxidant	(Körner, 2023; Streb et al., 2003)
Rumex alpinus	Antioxidant	(Abidi et al., 2019; Körner, 2023)
Salix herbacea	N.D.	(Körner, 2023)
Saxifraga bryoides	N.D.	(Körner, 2021f)
Saxifraga moschata	N.D.	(Körner, 2021f)
Schizocodon soldanelloides f. alpinus	N.D.	(Fujino et al., 2023)
Sempervivum montanum	Burns, haemorrhoids, insect bites, wounds	(Körner, 2023; Vitalini et al., 2013)

continued on following page

Table 2. Continued

Alpin plants	Biological effects	Previous Literature
Senecio incanus	N.D.	(Körner, 2021f)
Sibbaldia procumbens	N.D.	(Körner, 2021c)
Silene acaulis ssp. exscapa	N.D.	(Körner, 2023)
Soldanella pusilla	N.D.	(Körner, 2023)
Sphaerophorus globulus	N.D.	(Vanneste et al., 2017)
Thamnolia vermicularis	Antibacterial, cytotoxic	(Carrasco et al., 2023; Manojlović et al., 2010; Vanneste et al., 2017; J. Wang et al., 2022)
Thymus linearis Benth.	Antibacterial, anticancer, anti-inflammatory, antimalarial, antimicrobial, antioxidant, antiviral, insecticidal, neuroprotective	(Iqbal et al., 2022b; Kumar et al., 2020; Thakur et al., 2023)
Trifolium alpinum	N.D.	(Körner, 2023)
Vaccinium uliginosum	Anticancer, anti-inflammatory, antioxidant, apoptosis-reducing	(Kopystecka et al., 2023; Vanneste et al., 2017)
Vaccinium vitis-idaea	Antibacterial, antibiofilm, anticancer, anti-inflammatory, antimicrobial, antioxidant, antiproteolytic	(Fujino et al., 2023; Kryvtsova et al., 2020; Pärnänen et al., 2024; Vanneste et al., 2017; Vučić et al., 2009)
Veronica alpina	N.D.	(Albach et al., 2006; Körner, 2021f)
Veronica bellidioides	Antioxidant	(Albach et al., 2006; Körner, 2021f; Nikolova, 2011)

3. PHYTOCHEMICAL COMPOSITION OF WFPS

Plants produce secondary metabolites, consisting of terpenes, alkaloids and polyphenols, which they use for defence, fighting repelling herbivorous animal or infections (Yaglioglu et al., 2022). Various secondary metabolites including flavonoids, alkaloids, glycosides, are associated with a wide range of biological effects and form the main step of numerous modern drugs (Saklani et al., 2025). Plants produce compounds like terpenoids, alkaloids, tannins, glycosides, flavonoids, saponins, and phenolics as protective responses to environmental stresses and microbial threats. They not only play important roles in ecosystems but also possess a broad spectrum of pharmaceutical effects, including antibacterial, antioxidant, antifungal, anticancer, and anti-inflammatory effects. Modern analytical techniques especially GC (gas chromatography), HPLC (high-performance liquid chromatography), and GC-MS (gas chromatography–mass spectrometry) enable rapid detection and profiling of these metabolites. Among them, GC-MS typically offers the highest sensitivity and the most detailed structural information when dealing with complex mixtures (Naseer & Adil, 2025).

Terpenes (terpinene, myrcene, pinene, limonene, carvacrol, p-cymene, eugenol, carvone, thymol, and geraniol) are defined as compounds that have a simple hydrocarbon structure (Masyita et al., 2022). Terpenes (also known as terpenoids) exhibit many biological effects, including antimicrobial, anti-inflammatory, antiviral, anticancer (Wróblewska-Łuczka et al., 2023), antioxidant, antiplasmodial, antiseptic, astringent, diuretic, digestive (Cox-Georgian et al., 2019), antifungal, antihyperglycemic, antiparasitic, immunomodulatory (Brahmkshatriya & Brahmkshatriya, 2013), effects, among many other biological activities which reported in literatures. Terpenes engage with various biological components—including receptors, enzymes, and ion channels and can influence signaling pathways that regulate key cellular activities such as programmed cell death, growth, and differentiation (Wróblewska-Łuczka et al., 2023). It has been stated that terpenes possess antimicrobial properties against bacteria, primarily due to their ability to promote cell lysis and inhibit DNA and protein synthesis (Masyita et al., 2022). Alkaloids are natural substances that contain nitrogen within their molecular structure (Letchuman et al., 2024). Alkaloids have a wide range of pharmaceutical applications, including antimicrobial, anti-inflammatory, antimalarial, antioxidant, hepatoprotective, immunomodulatory, antidiabetic, cardioprotective (Letchuman et al., 2024), analgesic, antiasthmatic, anticancer, antihypertensive, antipyretic, and antihyperglycemic effects. These are, in order, morphine, ephedrine, vincristine, reserpine, quinine, piperine (Aryal et al., 2022), aconitine, ajmaline, atropine, berberine, capsaicin, cocaine, codeine, ergometrine, galanthamine, hyoscine, nicotine, pilocarpine, vinblastine, vincristine, and yohimbine (Letchuman et al., 2024). Flavonoids are secondary metabolites that primarily consist of a benzopyran ring bearing polyphenolic or phenolic groups in different positions (Ullah et al., 2020). Flavonoids possess numerous medical benefits, including antimicrobial, antioxidant, anticancer, antiangiogenic, antiviral, antimalarial, antitumor, neuroprotective, antiproliferative, anti-inflammatory, and cardioprotective properties (Ullah et al., 2020). Some of the flavonoids are kaempferol, quercetin, apigenin, fisetin, chrysin, luteolin, catechin, baicalein, epigallocatechin gallate, naringin, naringenin, genistein, hesperetin, silymarin, rutin, daidzein (Faysal et al., 2025), galangin, and myricetin (Hasnat et al., 2024). Tannins are natural phenolic compounds (Pizzi, 2021). Tannins offer a variety of health benefits, such as antioxidant, antibacterial, anti-inflammatory, neuroprotective (Begum et al., 2025), and anti-hyperglycemic properties (Rajasekaran et al., 2021). These are gallotannins, ellagitannins, catechins, epigallocatechin gallate (Kováč et al., 2022), gallic acid, and ellagic acid (Omar et al., 2022) compounds.

Numerous studies, particularly in traditional medicine, have demonstrated the importance of using plant extracts instead of certain pharmaceutical products for human health. Some of the phytochemicals those have a specific physiological effect on the human body include alkaloids, tannins, terpenoids, carbohydrates, steroids and flavonoids. These molecules are extremely diverse taxonomically and chemically, and their functions are unknown. They are widely used in veterinary medicine, human medicine, scientific researches, agriculture, and various other fields. *In vitro* studies have showed that many phytochemicals, belonging to various chemical groups, exhibit inhibitory effects against a wide spectrum of microorganisms. For millennia, plant products have been utilized to prepare phytomedical drugs obtained from various sources, including tree leaves, roots, bark, seeds, flowers and fruits (Bennacer et al., 2022; Özkan et al., 2025). WFPs is a commonly used term to describe all plants. Their roots, stems, leaves, fruits or other parts can be consumed by the public. WFPs play a significant role in the daily lives of many communities and are an important part of traditional food systems, holding significant cultural value for many groups. WFPs are frequently investigated from phytochemical and pharmacological standpoints because they comprise food components with nutraceutical properties (Casas et al., 2024). As a result of analyses conducted on WFPs, researchers observed that the secondary metabolites detected most frequently, in descending order, were Caffeic acid>Quercetin>Gallic acid>Kaempferol>Rutin>Chlorogenic acid>p-Coumaric acid>Ferulic acid>Luteolin>Catechin/Epicatechin. Detailed information on WFPs and their secondary metabolites is provided in Table 3. These secondary metabolites have various biological effects. Caffeic acid has been stated in the literatures to possess antimicrobial, antioxidant, anti-inflammatory, anticancer, neuroprotective, and hepatoprotective effects (Yazar et al., 2025). Quercetin found in fruits, vegetables, nuts, wine, and seeds. It has been reported of its antioxidant, anti-inflammatory, antibacterial, antiviral, gastroprotective, and immune-modulatory activities (Kim & Park, 2018). Gallic acid (GA) has been reported for its antioxidant, antimicrobial, anti-inflammatory, antiulcerogenic and anticancer effects (Alhazmi, 2025). Kaempferol has different biological activities such as antioxidant, anti-inflammation, antidiabetes, neuroprotective and anticancer (Alrumaihi et al., 2024). It has been indicated that rutin possesses antioxidant, anti-inflammatory, anti-apoptotic, and hepatoprotective properties (Rahmani et al., 2023). Chlorogenic acids have antioxidant, liver-protective, renal-protective, antibacterial, antitumour, anti-inflammatory and neuroprotective properties (L. Wang et al., 2022). p-Coumaric acid exhibits various biological properties such as antioxidant, anti-inflammatory, and anticancer agents (Roychoudhury et al., 2021). It has been stated that ferulic acid exhibits anti-inflammatory, antibacterial, anticancer, anti-arrhythmic, and antithrombotic effects (Raj & Singh, 2022). Luteolin functions as anti-inflammation, anti-allergy, anticancer and antioxidant activities (Lin et al.,

2008). Catechin has antioxidant, anticancer, antidiabetes, and anti-infection (Ding et al., 2024). Epicatechin, a natural flavonoid, has antibacterial, anti-inflammatory and anticancer properties (Tvrda et al., 2019).

Table 3. WFPs and their secondary metabolites

WFP	Secondary Metabolites	Previous Literature
Abies pindrow	(+)-14-Methyl palmitic acid, (+)-isosteric acid, (+)-r-limonene, α-pinene, β-pinene, epi-α-bisabolol, myrtenyl acetate, sabinene	(Saklani et al., 2025)
Acacia nilotica	(+)-Catechin, (-)-epigallocatechin-7-gallate, ellagic acid, gallic acid, phlobatannin, tannin	(Rather & Mohammad, 2015)
Adiantum capillus-veneris	3-p-Coumaroylquinic acid, astragalin, caffeic acid derivatives, carvacrol, carvone, chlorogenic acid, ferulic acid, gentisic acid, hexadecanoic acid, hexahydro-farnesyl acetone, iso-quercitrin, kaempferol, n-nonanal, p-coumaric acid, quercetin, rutin	(Qadir et al., 2025)
Ajuga integrifolia	8-o-Acetyl-harpagoside, α-humulene, α-phellendrene, apigenin, aucubin, β-caryophellene, β-myrcene, camphene, caffeic acid, catalpol, chlorogenic acid, elemol, harpagide, harpagoside, limonene, luteolin, quercetin, rosmarinic acid	(Ullah et al., 2021; Vohra & Kaur, 2011)
Amaranthus retroflexus	Caffeic acid, kaempferol, p-coumaric acid, quercetin, rutin, trans-ferulic acid	(Weston et al., 2019)
Amaranthus spinosus	Caffeic acid, cinnamic acid, epicatechin, gallic acid, protocatechuic acid, vanillic acid	(Rjeibi et al., 2019)
Amaranthus viridis	Baicalin, caffeic acid, catechine, catechine hydrate, chlorogenic acid, chrysin, flavones, gentisic acid, hesperidin, hyperoside, ısoquercitrin, naringenin, naringin, o-coumaric acid, p-coumaric acid, quercetin, resveratrol, rosmarinic acid, rutin, salicylic acid, sinapic acid, t-ferulic acid, vanillic acid, vanillin	(Özdenefe et al., 2024)
Arnebia benthamii	Caffeic acid, kaempferol, quercetin, rosmarinic acid	(Mukhtar et al., 2025)
Bauhinia variegata	Caffeic acid, chalcone, daucosterol, isoquercitrin, kaempferol, kaempferol-3-o-β-l-rhamnoside, kaempferol-3-o-d-glucopyranoside, linoleic acid, luteolin, narcissin, naringenin, rutin, syringic acid, vanillic acid	(Sharma et al., 2025)
Berberis lyceum	Berberine, caffeic acid, chlorogenic acid, ferulic acid, gallic acid, hydroxy benzoic acid, mandellic acid, myricetin, quercetin, rutin	(Landry et al., 2021; Nazir et al., 2021)
Bergenia ciliata	Bergenin, catechin, gallic acid	(Singh et al., 2017)
Betula utilis	1,8-Cineol, acetyloleanolic acid, betulin, betulinic aldehyde, betulitc acid, caffeic acid, champacol, chlorogenic acid, ferulic acid, geranic acid, linalool, lupeol, lupenone, methyl betulate, methyl betulonate, oleanolic acid, seleneol, sesquiphellendrene, sitosterol, terragon, ursolic acid (ua)	(Loshali et al., 2025)

continued on following page

Table 3. Continued

WFP	Secondary Metabolites	Previous Literature
Bistorta amplexicaulis	3-Hydroxycoumarin, adifoline, arctiopicrin, bruceine d, catechin, chlorflavonin, cortol, geranyl arabinopyranosylglucoside, globularin, icaceine, isoscopoletin, leonuriside a, lucidenic acid n, medicanine, naringin, n-trans-feruloyl-4-omethyldopamine, paradol, patuletin, procyanidin b7, quercetin, rutin, (s)-edulinine, sarothralin	(Javid et al., 2025; Tabassam et al., 2019)
Cannabis sativa	Cannabidiol (cbd), cannabidiolic acid (cbda), cannabinol (cbn), δ9-tetrahydrocannabinol (thc), tetrahydrocannabinolic acid (thca)	(Protti et al., 2019)
Capparis decidua	2-Carboxy-1-dimethylpyrrolidine, 6-(1-hydroxy-non-3-enyl) tetrahydropyran-2-one, ascorbic acid, β-carotene, β-sitosterol, cadabacine, capparisine, codonocarpine, glucosinolates, kaempferol, kaempferol 3,7-dirhamnoside, l-stachydrine, n-pentacosane, n-triacontane, n-triacontanol, quercetin, quercetin-7-o-rutinoside, rhamnocitrin, rutin	(Alsharif & Boylan, 2025; Verma et al., 2011)
Carthamus oxyacantha	Apigenin, coumarin, cyanidine, ferulic acid, harpagid, luteolin, morin, syringic acid	(Baban et al., 2023)
Chenopodium album	Benzoic acid, caffeic acid, calenduloside e, catechin, chenoalbicin, chenoalbuside, chikusetsusaponin, chlorogenic acid, cinnamic acid amides, epicatechin, ferulic acid, gallic acid, glycosides and terpenoids, protocatechuic acid, protocatechuic aldehyde, p-coumaric acid, p-oh benzoic acid, quercetin, rosmarinic acid, rutin, saponins, steroids, syringaldehyde, syringic acid, tannins, vanillic acid, vanillin	(Karaçelik & Şahin, 2021; Said et al., 2020)
Chenopodium botrys	α-Copaene, α-chenopodiol-6-acetate, α-eudesmol, α-fenchone, α-gurjunene, α-muurolene, β-caryophyllene, β-elemen, bicyclogermacrene, δ-cadinene, γ-cadinene, γ-muurolene, demethylnobiletin, elemol, elemol acetate, germacrene d, hispidulin, ısorhamnetin, ısosinensetin, jaceosidine, kaempferol, ledol, limonene, myrcene, pectolinarigenin, quercetin (triglycosides, acylglycosides), t-cadinol, t-muurolol	(Bojilov et al., 2023; Ozer et al., 2017; Rezaieseresht et al., 2020)
Chenopodium murale	Apigenin, caffeic acid, chlorogenic acid, kaempferol, luteolin, myricetin, naringenin, quercetin, quinic acid, salicylic acid	(El-Newary et al., 2023)
Cichorium intybus	Apigenin, caffeic acid, chlorogenic acid, ferulic acid, gallic acid, kaempferol, luteolin, quercetin, rutin, syringic acid	(Al-haliem et al., 2025)
Citrullus colocynthis	2,4-Nonadienal, butylated hydroxyanisole, cinnamic acid derivaties, hexadecane, linalool, pentadecane, tetradecane	(Bourhia et al., 2021)
Cleome viscosa	Atropine, caffeic acid, cleomeolide, gallic acid, kaempferitrin, nevirapine, penduletin, vanillic acid	(Alamdari et al., 2022; Sharma & Shrivastav, 2022)
Cordia dichotoma	Caffeic acid, chlorogenic acid, gallic acid, kaempferol, p-coumaric, quercetin, rosemarinic acid, rutin, salicylic acid, syringic, vanillic	(El-Newary et al., 2022)
Cordia myxa	Caffeic acid, catechin, chlorogenic acid, chrysin, coumaric acid (p-coumaric), ferulic acid, gallic acid, gentisic acid, ısorhamnetin, kaempferol, quercetin, rosmarinic acid, rutin, syringic acid, vanillic acid	(El-Massry et al., 2021; El-Nashar et al., 2025)

continued on following page

Table 3. Continued

WFP	Secondary Metabolites	Previous Literature
Cucurbita maxima	Ellagic acid, gallic acid, kaempferol, morin, naringin	(Anastácio et al., 2022)
Digera muricata	β-Sitosterol acetate, lariciresinol, phytol, stigmast-4-en-3-one, tridecanoic acid, vaccenic acid	(Ravi et al., 2023)
Diospyros lotus	4-Hydroxybenzoic, benzoic, caffeic, catechin, chrysin, ferulic, gallic, gallic acid, o-coumaric, p-coumaric acid, protocatechuic acid, rosmarinic, sinapic, syringic acid, t-cinnamic acid, vanillic	(Akar et al., 2020; Yunusa & Urek, 2023)
Ehretia acuminata	Apigenin, araneosol, caffeic acid, kaempferol, kaempferol 3-o-α-d-arabinoside, luteolin, methyl rosmarinate, ovalifolin, p-hydroxy benzoic acid, quercetin, 3-o-α-d-arabinoside	(Nguyen et al., 2024)
Elaeagnus umbellata	α-Cryptoxanthin, β-carotene, β-cryptoxanthin, caffeic acid, coumaric, ferulic, gallic, lutein, lycopene, phytoene, phytofluene, sinapic, vanillic	(Ishaq et al., 2015; Nazir et al., 2018)
Ficus palmata	Bergapten, benzoic acid, brucine, β-amyrin, caffeic acid, citric acid, ferrulic acid, hippuric acid, kaempherol, lupeol, pantothenic acid, p-coumeric acid, phytol, quercetin, quinic acid, rutin	(Al-Qahtani et al., 2023; Tewari et al., 2021b)
Fragaria nubicola	Betulin, caffeic acid, catechin, chlorogenic acid, gallic acid, naringin, phytol, rutin, trans-cinnamic acid, vanillic acid, γ-sitosterol	(Bhutia et al., 2021; Mashaal et al., 2023)
Gaultheria trichophylla	Catechin, epicatechin, gallic acid, kaemferol, quercetin, rutin	(Alam et al., 2017; Zaman et al., 2022)
Geranium wallichianum	2,4,6-Trihydroxyethylbenzoate, β-sitosterol, β-sitosterolgalactoside herniarin, caffeic acid, chlorogenic acid, gallic acid, kaempferol, p-coumaric acid, quercetin, quercetin-3-o-glucopyranoside, quercetin-rhamno-di-hexoside, stigmasterol, syringic acid, ursolic acid, vanillic acid	(Jabeen et al., 2022; Mir et al., 2022)
Juglans regia	1,5-Naphthalenediol, 2-methoxy-4-vinylphenol, 4-hydroxybenzoic acid, 5-hydroxy-1,4-naphthalenedione or juglone, (e)-coniferol, caffeic acid, catechol, coumaran, ellagic acid, eugenol, ferulic acid, gallic acid, gaiucol, hydroquinone, myricetin, phenol, quercetin, rutin	(Altemimi et al., 2023; Bennacer et al., 2022; D'Angeli et al., 2021)
Malus pumila	Caffeic acid, cinnamic acid, chlorogenic acid, ferulic acid, isoquercitrin, kaempferol, p-coumalic acid, phloretin, phloridzin, phlorizin, procyanidin b2, quercetin, quercetin 3-o-arabinoside, quercetin 3-o-xyloside, quercitrin, rutin	(Bai et al., 2016; Li et al., 2018; Lu et al., 2024)
Malva parviflora	Apigenin, β-caryophyllene oxide, β-pinene, caffeic acid, camphor, catechin, catechol, cinnamic acid, δ-elemene, eucalyptol, ferulic acid, folic acid, gallic acid, germacrene d-4-ol, isoborneol, kaempferol, luteolin, naringin, pyrogallol, quercetin, rutin, salicylic acid, syringic acid, tau-cadinol, vitamin b12, vitamin c	(El-Hefny & Hussien, 2025)
Malva sylvestris	Caffeic acid, campesterol, chlorogenic acid, couramin, fisetin, hyperoside, kaempferol hexoside, malvin, quercetin, quercetin-3-o-rutinoside, quinic acid, rutin, stigmasterol	(Shadid et al., 2021; Turan et al., 2021)

continued on following page

Table 3. Continued

WFP	Secondary Metabolites	Previous Literature
Mentha arvensis	Benzoic acid, caffeic acid, cinmic acid, ferulic acid, gallic acid, limonene, menthol, menthone, menthyl acetate, p-coumeric acid, quercetin, vanilic acid	(Rashid et al., 2023; Sharma et al., 2018)
Mentha longifolia	Apigenin, apigenin 5-o-glucoside, benzoic acid, caffeic, catechin, chlorogenic acid, cinnamaldehyde, coumaric acid, ferulic acids, gallic, genkwanin 5-o-glucoside, hydroxy-benzoic acid, kaempferol, kaempferol 3-o-sophoroside, luteolin 5-o-glucoside, naringenin, quercetin, quercetin 3,7-o-diglucoside, rosmarinic, rutin, syringic, vanillic	(Mohammed et al., 2024; Patonay et al., 2017; Stanisavljević et al., 2012)
Mentha pulegium	Apigenin, caffeic acid, caffeic, chlorogenic, cis-isopulegone, diosmetin, ellagic, eriodictyol, eriocitrin, ferulic, hesperidin, isomenthone, ısorhoifolin, kaempferol, limonene, luteolin, luteolin, menthone, myricetin, naringenin, narirutin, p-coumaric, protocatechuic acid, rosmarinic acid, rosmarinic, syringic	(Alharbi et al., 2021; Ghazghazi et al., 2013; Politeo et al., 2018)
Mentha spicata	5-Demethyl sinensetin, α-pinene, β-pinene, caftaric acid, caffeic acid, camphene, carvone, didymin, eriocitrin, hesperidin, limonene, linalool, linarin, myrcene, neochlorogenic acid, p-cymene, rosmarinic acid, sabinene, salvianolic acid, vicenin 2	(Erenler et al., 2018; Sierra et al., 2022; Snoussi et al., 2015)
Moringa oleifera	Chlorigenic acid, ferulic acid, gallic acid, m-coumaric acid, p-coumaric acid, quercetin, sinapic acid, syringic acid, vinalic acid	(Khalid et al., 2023)
Morus alba	Caffeic acid, catechin, chlorogenic acid, ferulic acid, gallic acid, isoquercetin, p-coumaric acid, phloridzin, rutin, syringic acid, vanilic acid	(Gundogdu et al., 2011; Kim et al., 2014a)
Morus nigra	Caffeic acid, catechin, chlorogenic acid, ferulic acid, gallic acid, p-coumaric acid, phloridzin, syringic acid, vanilic acid	(Gundogdu et al., 2011)
Mukia maderaspatana	Alkaloids, carbohydrate, cardiac glycosides, coumarin, flavonoids, glycosides, leucoanthocyandin, phenol, phlobatannin, protein, saponin, steroids, tannin, terpenoids	(Devi & Gouri, 2024)
Nasturtium officinale	alpha-Thujene, beta-carotene, carotenoids, copaene, farnesol, gallic acid, heneicosane, limonene, lutein, myrcene, p-coumaric acid, quercetin, squalene, vanillin, vitamin c, zeaxanthin	(Ercan & Doğru, 2022; Mahdavi et al., 2019; Tabesh et al., 2022)
Opuntia stricta	1,3-di-o-Caffeoyquinic acid, acacetin, apegenin, betadin, betanin, betalain, caffeic acid, catechins, cinnamic acid, cirsilineol, gallic acid, hyperoside, isobetanin, isorhamnetin-3-o-rutinoside, kaempferol-3-o-rutinoside, luteolin, naringin, p-coumaric acid, protocatchuic acid, quercetin, quinic acid, salviolonic acid, syringic acid, trans ferrelic acid, vanillic acid, vitamins a, c, e	(Affi et al., 2024; El Mannoubi, 2023; Ganesan, 2024)
Oxalis corniculata	citric acid, isoorientin, isoschaftoside, isovitexin, linoleic acid, malic acid, oleic, orientin, palmitic, schaftoside, swertisin, tartaric acid	(Pandey & Pokhrel, 2022; Zhang et al., 2025)
Physalis minima	3-İsopropyl-5-acetoxycyclohexene-2-one-1, 5α-6β-dihydroxyphysalin r, catechin, catechol, ellagic acid, gallic acid, isophysalin b, physalin d, physalin g, physalin ı, physordinose b, stigmasterol-3-o-β-d-glucopyranoside	(Karpagasundari & Kulothungan, 2014; Men et al., 2016)

continued on following page

Table 3. Continued

WFP	Secondary Metabolites	Previous Literature
Phytolacca latbenia	6-Methoxy-7-hydroxyflavone, acinospesigenin-a, cochliophilin a, cochliophilin b, esculentoside d, esculentoside g, esculentoside t, kaempferol, kaempferol 3-o-β-d-glucoside, phytolasides a	(Khalid et al., 2025)
Plantago major	Apigenin, baicalein, hispidulin, isoverbascoside, luteolin, martynoside, plantagin, plantamajoside, scutellarin, verbascoside, verbasoside	(Sousa et al., 2025; Zhakipbekov et al., 2023)
Plectranthus rugosus	1-Nonen-3-ol, 3-carene, acetylplectranthoic acid, α-cadinol, α-copane, α-curcumene, α-muurolene, α-phellandrene, α-pinene, α-terpinene, β-bourbonene, β-cubebene, β-phellandrene, β-pinene, β-sitosterol, bicyclogermacrene, camphene, caryophyllene, caryophyllene oxide, cis-β-ocimene, δ-cadinene, γ-cadinene, γ-muurolene, γ-terpinene, germacrene d, humulene, limonene, linalool, myrcene, p-cymene, piperitone epoxide, plectranthadiol, plectranthoic acid, plectranthoic acid a, plectranthoic acid b (m6), sabinene, t-cadinol, terpinolene, terpinen-4-ol, thujone, torreyol	(Abdel-Mogib et al., 2002)
Polygonum plebeium	Gallic acid, gingerols, isovitexin, kaempferol, luteolin, lyoniresinol 9′-sulfate, protocatechuic acid	(Saleem et al., 2021)
Prosopis cineraria	β-Sitosterol, caffeic acid, chlorogenic acid, daidzein, ferulic acid, gallic acid, hentriacontane, luteolin, patulitrin, phloridzin (phloretin-2′-o-glucoside), protocatechuic acid, prosogerin a, b, c and d, puerarin (daidzein-8-c-glucoside), rutin, spicigerin, vitexin (apigenin- 8 c-glycoside)	(Asati et al., 2022; Yadav et al., 2018)
Prosopis juliflora	3-Oxo-juliprosine, 3′-oxo-juliprosopine, julifloricine, juliprosine, juliprosinene, juliprosopine (juliflorine), prosoflorine, secojuliprosopinal	(Dos Santos et al., 2013; Saleh et al., 2023)
Prunus armeniaca	(−)-Epicatechin, (+)-catechin, β-carotene, β-cryptoxanthin, γ-carotene, chlorogenic, cyanidin 3-o-galactoside, epicatechin, gallic acid, kaempferol, lycopene, mandelonitrile-β-gentiobioside, neochlorogenic acids, procyanidin, resveratrol, rutin, salicylic acid, syringic acid, tocopherol, vanillic acid	(El-Hajjaji et al., 2024; Erdogan-Orhan & Kartal, 2011)
Pteridium aquilinum	Ecdysones, kaempferol, prunasin, ptaquiloside, quercetin	(Hao et al., 2021; Malík et al., 2023; Selvaraj et al., 2005)
Punica granatum	α-Punicalagin, β-punicalagin, caffeic, chlorogenic, ellagic acid, ferulic acid, gallic acid, p-coumaric	(Feng et al., 2017; Karakaplan & Özcan, 2017)
Pyrus pashia	Ampeloptin, arbutin, caffeic acid, catechin, chlorogenic acid, coumaric acid, ellagic acid, epicatechin, ferulic acid, gallic acid, hesperetin, hexoside, jasmonic acid, phloridzin, procyanidin b2, sinapic acid, vanillic acid	(Prakash et al., 2019)
Pyrus communis	Caffeic acid, catechin, chlorogenic acid, citric, epicatechin, fumaric acid, gallic acid, kaempferol, malic, protocatechuic acid, quercetin, shikimic, sinapic acid, syringic acid, vanillic acid	(Colaric et al., 2006; Molinu et al., 2024; Wang et al., 2021)
Rubus fruticosus	Elagic acid, gallic acid, protocatechuic acid, quercetin, quercetin-3-o-glucoside, rutin, vanillic acid	(Cadi et al., 2021; Jacques et al., 2014)

continued on following page

Table 3. Continued

WFP	Secondary Metabolites	Previous Literature
Rumex dentatus	Beta sitosterol, benzoic acid, caffeic acid, chlorogenic acid, cinnamic acid, daucosterol, ferulic acid, gallic acid, helonioside a, hexadecanoic acid 2, 3 dihydroxy propyl ester, isovanillic acid, kaempferol, myricetin, n butyl beta d fructopyranoside, p hydroxycinnamic acid, p-hydroxybenzoic acid, quercetin, succinic acid, syringic acid, vanillin, vitamin c	(Elzaawely & Tawata, 2012; Khaliq et al., 2023)
Rumex nepalensis	1,3,6-Trihydroxy-8-methyl-anthraquinone, 3-hydroxy-γ-butyrolactone, 5-hydroxy-7-methoxy-1(3h)-isobenzofuranone, β-sitosterol, β-sitosterol 3-o-β-d–glucopyranoside, catechin 7-o-gallate, chrysophanein, chrysophanol, citreorosein, emodin, emodin bianthrone, emodin-6-o-β-d-glucoside, emodic acid, epicatechin 3-(3″-o-methyl) gallate, physcion, physcionin, questin, questinol, stigmasterol	(Nguengang et al., 2023)
Salvadora oleoides	Capric, dibenzyl urea, lauric, linolenic acids, malic, myristic, oleic, palmitic, quercetin, rutin	(Varshney et al., 2025)
Salvadora persica	Arachidic acid, heptadecenoic acid, hexadecenoic acid, hydroxy octadecenoic acid, linoleic acid, myristic acid, stearic acid	(Ayoub et al., 2021)
Sisymbrium irio	Apigenin, apigenin-7-di-glucoside 5, apigenin-7-galactoside, apigenin-7-o-(6″acetyl) glucoside, apigenin-7-o-β-d-glucoside, apigenin-7-o-gluco(6″-1″′) rhamnoside, apigenin-7-o-gluco(6″-1″′)rhamnoside-5-methoxide, kampferol, kampferol-3-xyloside-7-galactoside, luteolin-7-o-glucoside	(Al-Jaber, 2011)
Solanum surattense	Catechin, gallocatechin, isoquercitrin, quercetin, solamargine, solanearpidine, solanidine, solasodine, solasonine, solasurine, tomatidine	(Hasan et al., 2024; Yasir et al., 2018)
Taxus wallichiana	Ascorbic acid, gallic acid, quercetin, tannic acid	(Adhikari et al., 2022)
Viburnum grandiflorum	1-Oxo-erythrodiol, 10-acetyl patrinoside, 3-hydroxy-4,7-megastigmadien-9-one, awabukinol, β-sitosterol, lupeol	(Iqbal et al., 2022a)
Zanthoxylum armatum	Alloxanthyletin, alpha- pinene, alpha-limonene diepoxide, ascorbic acid, b-fagarine, beta- myrcene, beta-linalool, bergamot mint oil, berberine, caffeic acid, candicine, chelerythrine, chlorogenic acid, d-limonene, epigallocatechin gallate, gallic acid, g-fagarine, laurifoline, limonene, linalool, magnoflorine, nitidine, p-hydroxybenzoic acid, tambetarine, vanillic acid, xanthyletin, zanthoxyletin	(Agnihotri et al., 2024; Phuyal et al., 2019)
Ziziphus mauritiana	Apigenin, betulin, berberine, caffeic acid, chlorogenic acid, diosgenin, epicatechin, epigallocatechin, gallic acid, hydroxybenzoic acid, ısorhamnetin, kaempferol, lanosterol, luteolin, myricetin, naringenin, p-coumeric acid, phloretin, quercetin, rutin, sitosterol, stigmasterol, vanillin	(Khanam et al., 2025; Memon et al., 2012; Pauloi & Salim, 2025; Ramar et al., 2020)
Ziziphus nummularia	Chlorogenic acid, geranyl linalool, lapachol, lupeol, mandelic acid, morin, nummularine, phytol, pyrogallol, quercetin, rutin, squalene	(Mesmar et al., 2022)

4. BIOLOGICAL EFFECTS OF WFPS

Since ancient times, people have utilised a broad spectrum of plants and their by-products for medicinal purposes, including treatment of infectious diseases. Using plants as a source of new antioxidant and antimicrobial agents has many advantages. Plants are easily accessible and cost-effective. Plant-derived extracts and compounds are highly effective against bacteria and rarely cause serious side effects (Cavazos et al., 2021). Plants exhibit a variety of biological effects due to their secondary metabolites such as antioxidants, antimicrobials, enzyme inhibitors, anticancer agents, and anti-inflammatory agents (Yaglioglu et al., 2022).

WFPs are a potential source of antioxidant polyphenols, and adding these antioxidants to functional foods may prevent the formation of oxidative species, thereby helping to prevent certain diseases (Rather & Mohammad, 2015). It has been reported that chlorogenic acid, vanillic acid, p-coumaric acid, syringic acid, kaempferol, caffeic acid, and quercetin exhibit anti-inflammatory activities against acute inflammation (Jabeen et al., 2022). Phytochemicals, which are natural compounds found in plants, have been the focus of researches due to their potential in the development of new therapeutic drugs. Unlike synthetic drugs, phytochemicals can exhibit similar bioactivity with fewer side effects (Qadir et al., 2025). Thus, plants are a source of pharmacological effects.

Studies conducted by researchers on WFPs show that the most commonly observed biological effects, in descending order, are anti-inflammatory > antioxidant > antimicrobial > anticancer > antidiabetic. WFPs and their biological effects are listed in Table 4.

Table 4. WFPs and their biological effects

WFP	Biological effect	Previous Literature
Abies pindrow	Analgesic, anti-inflammatory, anti-ulcer, antibacterial, bronchodilator	(Saklani et al., 2025)
Acacia nilotica	Anticancer, anticarcinogenic, antibacterial, antidiarrheal, antidenaturation, antifungal, antihypertensive, antihypertensive, anti-inflammatory, antimalarial, antimutagenic, antioxidant, antiplatelet aggregatory, antispasmodic, antitumor, antiviral, atherosclerotic, anthelmintic	(Shah et al., 2020b)
Adiantum capillus-veneris	Anti-diabetic, anti-inflammatory, antimicrobial, antioxidant	(Qadir et al., 2025)
Ajuga integrifolia	Anti-inflammatory, antiarthritic, antimicrobial, antimalarial, antioxidant, antitumor, cardiotonic	(Fereja et al., 2024; Ullah et al., 2021)

continued on following page

Table 4. Continued

WFP	Biological effect	Previous Literature
Amaranthus retroflexus	Anticancer, antioxidant, antitumor	(Babaei-Ghaghelestany et al., 2022)
Amaranthus spinosus	Anti-cholesterolemic, antidiabetic, anti-inflammatory, antioxidant, diuretic	(Rjeibi et al., 2019)
Amaranthus viridis	Anticancer, antidiabetic, anti-inflammatory, anti-phytopathogenic, antimicrobial, antinociceptive, antioxidant, antipyretic, antiviral, anthelmintic, hepatoprotective	(Fouad et al., 2024; Özdenefe et al., 2024)
Arnebia benthamii	Anticancer, antibacterial, anti-inflammatory, antifungal, wound-healing	(Ganie et al., 2014)
Bauhinia variegata	Antibacterial, antidiabetic, anti-inflammatory, antimutagenic, antitumor, antioxidant	(Ambwani et al., 2025; Sharma et al., 2025)
Berberis lyceum	Anticancer, anti-carcinogenic, anti-inflammatory, antihyperglycemic, antihyperlipidemic, anthelmintic, antioxidant, antipyretic, antitumor, antiurolithic, hypoglycemic, immunity enhancing, wound healing	(Nazir et al., 2021)
Bergenia ciliata	Antimicrobial, antioxidant	(Singh et al., 2017)
Betula utilis	Antibacterial, antidiabetic, antioxidant, hepatoprotective	(Vishwakarma et al., 2022)
Bistorta amplexicaulis	Anticancer, antimicrobial, antioxidant	(Javid et al., 2025; Tabassam et al., 2019)
Cannabis sativa	Antidiabetic, antimicrobial, antioxidant	(Shah et al., 2020a)
Capparis decidua	Antidiabetic, antimicrobial, antioxidant, arthritis, asthma, cough, cure toothache, inflammation, intermittent fevers, malaria, rheumatism, swelling	(Acar et al., 2023; Zia-Ul-Haq et al., 2011)
Carthamus oxyacantha	Anti-inflammatory, antimicrobial, antioxidant, antiworm, wound-healing	(Ikram et al., 2020)
Chenopodium album	Anti-allergic, antifungal, anti-inflammatory, anti-proliferative, anti-septic, antiviral, anticancer, giardicidal, hepatoprotective, immunomodulatory, nematicidal, spasmolytic	(Rana et al., 2020; Said et al., 2020)
Chenopodium botrys	Antioxidant, anticancer, bactericidal, fungicidal	(Maksimovic et al., 2005; Ozer et al., 2017; Rezaieseresht et al., 2020)
Chenopodium murale	Analgesic activities, anti-inflammatory, antibacterial, antifungal, antihypertensive, hepatoprotective	(El-Newary et al., 2023; Kanwal et al., 2025)
Cichorium intybus	Anticancer, antibacterial, antidiabetic, anti-inflammatory, anti-pyorrheal, antifungal, antioxidant, cardioprotective, gastroprotective, hepatoprotective, hypolipidemic, immunomodulatory, sedative	(Al-haliem et al., 2025; Azad et al., 2024; Petrovic et al., 2004)
Citrullus colocynthis	Anti-inflammatory, antimicrobial, antioxidant, immunostimulatory	(Ghahramani et al., 2024)

continued on following page

Table 4. Continued

WFP	Biological effect	Previous Literature
Cleome viscosa	Analgesic, anthelmintic, anthelmintic, anticancer, antidiarrheal, anti-inflammatory, antipyretic, antiscorbutic, antiseptic, antioxidant, carminative, cardiac stimulant, febrifuge, hepatoprotective, immune modulatory, psychopharmacological, sudorific	(Alamdari et al., 2022; Sharma & Shrivastav, 2022)
Cordia dichotoma	Anticancer, antidiabetic, anti-inflammatory, antimicrobial, antioxidant, antitumor, antiulcer, anthelmintic, hepatoprotective, hypoglycaemic, hypolipidemic, wound-healing	(El-Newary et al., 2022; Raina et al., 2025)
Cordia myxa	Analgesic, anthelmintic, anti-inflammatory, antioxidant, antiulcer, antitumor, aphrodisiac, astringent, demulcent, diuretic, emollient, expectorant, hepatoprotective, hypoglycaemic, hypolipidemic, immune-modulator, laxative, wound healing	(El-Nashar et al., 2025)
Cucurbita maxima	Analgesic, anti-inflammatory, antibiofilm, antimicrobial, antioxidant, hepatoprotective	(Anastácio et al., 2022; Laaraj et al., 2024)
Digera muricata	Anti-atherosclerotic, antidiabetic, anti-inflammatory, antimicrobial, antifungal, antioxidant	(Ravi et al., 2023)
Diospyros lotus	Anticancer, antimicrobial, antiseptic, antioxidant, astringent, antitussive, antitumor, constipation, diarrhoea, dry coughs, febrifuge, fauvism, hypertension, laxative, nephroprotective, nutritive, sedative	(Akar et al., 2020; Yunusa & Urek, 2023)
Ehretia acuminata	Analgesic effect, antidiabetic, anti-inflammatory, antimicrobial, antioxidant, antispasmodic effects, antiviral, hepatoprotective, muscle relaxant, photoprotective, wound healing	(Nguyen et al., 2024)
Elaeagnus umbellata	Anticancer, antioxidant, cardioprotective, hepatoprotective, skin protective	(Iannuzzi et al., 2020; Park et al., 2024)
Ficus palmata	Anticancer, antimicrobial, antifungal, antidiabetic, anti-tumour, antiulcer, haemorrhoids, hypoglycemic, hypolipidaemic, ringworm infestations, skin diseases, thrombolytic, wound infections	(Al-Qahtani et al., 2023; Tewari et al., 2021a; Tewari et al., 2021b)
Fragaria nubicola	Anticancer, anti-hyperlipidemic, anti-inflammatory, anti-microbial, anti-neurodegenerative, antioxidant	(Bhutia et al., 2021; Hu et al., 2024)
Gaultheria trichophylla	Anti-inflammatory, antioxidant	(Alam et al., 2017)
Geranium wallichianum	Antibacterial, antifungal, antimicrobial, cytotoxic, enzyme inhibition, insecticidal, phytotoxic	(Ismail et al., 2012; Mir et al., 2022)
Juglans regia	Antidiabetic, anti-inflammatory, antidiarrheic, antifungal, antimicrobial, antioxidant	(Altemimi et al., 2023; Bennacer et al., 2022)
Malus pumila	Anti-aging, anti-diabetes mellitus, anti-inflammatory, antiobesity, antibacterial, antioxidant, antiradiation, antitumor, hepatoprotective	(Bai et al., 2016; Lu et al., 2024)

continued on following page

Table 4. Continued

WFP	Biological effect	Previous Literature
Malva parviflora	Analgesic, antibacterial, antibacterial, antifungal, anti-inflammatory, anti-irritant, antioxidant, antidiabetic, antiulcerogenic, hepatoprotective, neuroprotective, wound healing	(Hussein et al., 2024; Munir et al., 2021)
Malva sylvestris	Anti-inflammatory, antimicrobial, antioxidant, antiproliferative, hepatoprotective, immunomodulatory, laxative, wound-healing	(Batiha et al., 2023)
Mentha arvensis	Antioxidant, asthma, bronchitis, cancer, cough, dementia, diarrhea, genoprotective, hyperglycemia, liver, nasal catarrh, piles, spleen disease	(Rashid et al., 2023)
Mentha longifolia	Anti-candida, anti-inflammatory, anti-mutagenic, antioxidant, anti-rheumatic, antispasmodic, antiviral	(Shazdehahmadi et al., 2023)
Mentha pulegium	Analgesic, antiemetic, anti-inflammatory, antimicrobial, antioxidant, antispasmodic, antitussive, carminative, diaphoretic, emmenagogue, stimulant	(Aimad et al., 2021; Alharbi et al., 2021; Ghazghazi et al., 2013)
Mentha spicata	Antibacterial, antifungal, antioxidant, antispasmodic, carminative, diuretic	(Snoussi et al., 2015)
Moringa oleifera	Antidiabetic, antidiuretic, anti-inflammatory, antimicrobial, antioxidant, cardioprotective, hepatoprotective	(Khalid et al., 2023)
Morus alba	Antiatherosclerosis, antidiabetic, antihypertension, antimicrobial, antiobesity, antiviral, liver protection	(Kim et al., 2014a)
Morus nigra	Anticancer, antidiabetic, anti-hyperlipidemic, anti-inflammatory, anti-melanogenic, antimicrobial, antifungal, antinociceptive, anti-obesity	(Al Dhanhani et al., 2025; Lim & Choi, 2019)
Mukia maderaspatana	Anti-arthritic, anti-inflammatory, antimicrobial, antioxidant, asthma, burning sensation, colic, constipation, cough, dyspepsia, flatulence, hepatoprotective, hypotensive, immunomodulatory, neuralgia, nostalgia, odontalgia, ulcers, vasodilatory, vertigo	(Devi & Gouri, 2024)
Nasturtium officinale	Anti-inflammatory, antibacterial, antioxidant, chemopreventive, hyperglycemia, hypertension, respiratory infections	(Ercan & Doğru, 2022; Mahdavi et al., 2019; Tabesh et al., 2022)
Opuntia stricta	Analgesic, antibacterial, anticancer, antidiabetic, anti-inflammatory, antioxidant, antiviral, cytotoxic, hypotensive	(Affi et al., 2024; El Mannoubi, 2023; Ganesan, 2024)
Oxalis corniculata	Analgesic, antibacterial, anticancer, anti-inflammatory, antimicrobial, antifungal, antioxidant, diuretic, insecticidal	(Pandey & Pokhrel, 2022; Rehman et al., 2015)
Physalis minima	Analgesic, anti-inflammatory, antimicrobial, antipyretic, antitumor, diuretic, laxative	(Patel et al., 2011; Yazhiniprabha et al., 2022)
Phytolacca latbenia	Anticancer, anticoagulant, anti-inflammatory, antimicrobial, diuretic, hypoglycemic, hypolipidemic	(Ismail et al., 2017; Patra et al., 2014)
Plantago major	Antibacterial, anticancer, anti-inflammatory, antimicrobial, antifungal, antiviral	(Sousa et al., 2025; Zhakipbekov et al., 2023)

continued on following page

Table 4. Continued

WFP	Biological effect	Previous Literature
Plectranthus rugosus	Anti-inflammatory, antioxidant	(Mir et al., 2024)
Polygonum plebeium	Antibacterial, antioxidant, eczema, pneumonia	(Abdelkhalek et al., 2018; Saleem et al., 2021)
Prosopis cineraria	Analgesic, antibacterial, anti-collagenase, anti-elastase, anti-inflammatory, antioxidant, antitumor, wound healing	(Yadav et al., 2018)
Prosopis juliflora	Anticancer, anti-inflammatory, antimicrobial, antifungal, antimalarial, asthma, bronchitis, conjunctivitis, insecticidal	(Saleh et al., 2023; Thakur et al., 2014)
Prunus armeniaca	Anticancer, anti-inflammatory, antimicrobial, anthelmintic, antioxidant, antipyretic, hepatoprotective, vulnerary	(El-Hajjaji et al., 2024; Kitic et al., 2022)
Pteridium aquilinum	Antibacterial, anticancer, anti-inflammatory, antiviral, hypotensive, immunomodulatory	(Zhao et al., 2025)
Punica granatum	Antibacterial, anticancer, antiatherosclerosis, anti-inflammatory, anthelmintic, antioxidant, antiviral, diuretic, immune modulation	(Feng et al., 2017; Khan et al., 2017)
Pyrus pashia	Antimicrobial, antioxidant, hypoglycemic	(Janbaz et al., 2015)
Pyrus communis	Antibacterial, antiageing, anticancer, antidiabetes, antioxidant	(Ribeiro et al., 2025; Wang et al., 2021)
Rubus fruticosus	Anxiolytic, anti-diarrheal, anti-hemorrhoidal, diuretic	(Cadi et al., 2021)
Rumex dentatus	Antibacterial, antibiofilm, anticancer, anti-dermatitis, anti-helminthic, anti-inflammatory, antimicrobial, antioxidant, antitumor, antiviral, astringent, hepato-protective, insecticidal	(Khaliq et al., 2023; Khan et al., 2024)
Rumex nepalensis	Analgesic, antialgal, antibacterial, antibradykinin, anticholinergic, antifungal, anti-inflammatory, antihistaminic, antioxidant insecticide, antiprostaglandin, antipyretic, antiviral, central nervous system depressant, genotoxic, hepatoprotective, purgative, skeletal muscle relaxant, wound healing	(Alsayadi et al., 2025; Gautam et al., 2011; Ghosh et al., 2003; Nguengang et al., 2023)
Salvadora oleoides	Analgesic, antibacterial, anti-hyperglycemic, anti-inflammatory, anti-lipidemi	(Varshney et al., 2025)
Salvadora persica	Analgesic, antidepressant, antidiabetic, anti-cariogenic, anticonvulsant, anti-gingivitis, anti-inflammatory, antimicrobial, antioxidant, anti-osteoporosis, anti-plaque, antitumor, antiulcer, anthelmintic, hypo-lipidemic, sedative, wound-healing	(Ayoub et al., 2021; Farag et al., 2021)
Sisymbrium irio	Analgesic, antimicrobial, antioxidant, antipyretic	(Chileh-Chelh et al., 2024)
Solanum surattense	Antimicrobial, antioxidant, anti-inflammatory, antitumoral, bronchitis, cardiac disorder, catarrh, chronic cough, colic constipation, dental caries, fever, hepatoprotective, hypertention, larvicidal, leprosy, rheumatis and chest pains, sore throat	(Yasir et al., 2018)

continued on following page

Table 4. Continued

WFP	Biological effect	Previous Literature
Taxus wallichiana	Analgesic, antibacterial, antifungal, anti-inflammatory, antiallergic, anticociceptive, anticonvulsant, antiosteoporotic, antiplatelet, antipyretic, antispasmodic, antimicrobial, antioxidants, immunomodulatory, vasorelaxing	(Adhikari et al., 2022; Juyal et al., 2014)
Viburnum grandiflorum	Anticancer, anti-allergic, anti-depression, anti-inflammatory, antihyperglycemic, antimicrobial, antioxidant	(Han et al., 2020; Iqbal et al., 2022a; Mughees et al., 2022)
Zanthoxylum armatum	Analgesic, antibacterial, antimicrobial, antioxidant, anti-inflammatory, antitumor, antiviral, cytotoxic, hepato-protective, insecticidal/larvicidal	(Phuyal et al., 2019; Zhuo et al., 2021)
Ziziphus mauritiana	Anticancer, antidepressant, antidiabetic, anti-diarrhoeal, anti-inflammatory, antimicrobial, antioxidant, antipyretic, cytotoxic, hepatoprotective, immunomodulator	(Khanam et al., 2025; Pauloi & Salim, 2025)
Ziziphus nummularia	Analgesic, anticancer, antidiabetic, anti-inflammatory, antimicrobial, antioxidant, anthelmintic, cardioprotective	(Mesmar et al., 2022)

5. CONCLUSION

Wild food plants (WFPs) are rich in carbohydrates, protein, fat, and fiber and are highly nutritious, making them very important for people living in high mountain regions.

WFPs are only known to the local population because they grow in high mountainous regions. Consequently, these plants are not well known, and scientific studies on them are limited. Furthermore, the harsh geographical conditions make access to these plants difficult. The parts of the plants that are used vary depending on the region.

Herbal teas are infusions made by steeping plant seeds, leaves, or roots in boiled water. They are commonly used for their relaxing, energizing, and medicinal effects. On the other hand, some ethnic groups living in high mountainous regions interact with WFPs more frequently and use them more often. Those who make a living by herding in mountain pastures have extensive experience with high-altitude wild plants in mountainous regions and frequently include these plants in their daily diet. There are various ways to consume WFPs. The most commonly used part of WFPs is the leaves, followed by the fruit, above-ground parts, rhizomes, bark, seeds, roots, stems, and tubers. The more widespread use of the leaves and fruit may be due to their rich biomolecular and antioxidant content. The active chemical components of plant parts such as stems, roots, leaves, and flowers exhibit potent pharmacological activity.

The phytochemical compound content and ratios also change when the part of the plant used, the method of use, and the solvent medium are altered. In this case, they exhibit different biological effects. Different parts can be used in the production of different natural products of biological importance. WFPs offer promising potential for practical applications in the nutraceutical and pharmaceutical industries. In conclusion, plants are very rich in phytochemicals and minerals. However, when observing the pharmacological effects of these phytochemicals, the dose-response concept should be taken into account.

Alpine plants are frequently subjected to stressful conditions such as UV rays, intense sunlight, strong winds and drought. They must adapt in order to cope with these conditions.

Although it has been noted in the literature that they possess rich phytochemical content, the biological effects of some Alpine plants, such as *Allantoparmelia alpicola, Androsace alpina, Betula nana, Bistorta emodi, H. Hara, Bryocaulon divergens, Carex curvula, Cerastium uniflorum, Cladonia macrophylla, Coelocaulon aculeatum, Erigeron uniflorus, Festuca ovina, Gentiana acaulis, Gentiana bavarica, Geum montanum, Geum reptans, Helictotrichon versicolor, Juncus trifidus, Leucanthemopsis alpina, Luzula arcuata, Nardus stricta, Poa laxa, Primula hirsuta, Salix herbacea, Saxifraga bryoides, Saxifraga moschata, Schizocodon soldanelloides, F. alpinus, Senecio incanus, Sibbaldia procumbens, Silene acaulis ssp. exscapa, Soldanella pusilla, Sphaerophorus globulus, Trifolium alpinum* and *Veronica alpina* have not been tested or have been limited to phytochemical analyses. Although some of these plants are endemic, others are widely distributed. There has been insufficient research on some species. As a result, WFPs have the potential to supply the pharmaceutical industry with lead molecules.

REFERENCES

Abdel-Mogib, M., Albar, H., & Batterjee, S. (2002). Chemistry of the genus *Plectranthus. Molecules (Basel, Switzerland)*, *7*(2), 271–301. DOI: 10.3390/70200271

Abdelkhalek, E. S., El-Hela, A. A., El-Kasaby, A. H., Sidkey, N. M., Desouky, E. M., & Abdelhaleem, H. H. (2018). Antibacterial Activity of *Polygonum plebejum* and *Euphorbia hirta* Against *Staphylococcus aureus* (MRSA). *Journal of Pure & Applied Microbiology*, *12*(4). Advance online publication. DOI: 10.22207/JPAM.12.4.60

Abidi, J., Ammar, S., Brahim, S. B., Skalicka-Woźniak, K., Ghrabi-Gammar, Z., & Bouaziz, M. (2019). Use of ultra-high-performance liquid chromatography coupled with quadrupole-time-of-flight mass spectrometry system as valuable tool for an untargeted metabolomic profiling of *Rumex tunetanus* flowers and stems and contribution to the antioxidant activity. *Journal of Pharmaceutical and Biomedical Analysis*, *162*, 66–81. DOI: 10.1016/j.jpba.2018.09.001 PMID: 30223144

Acar, N., Çelikezen, F. Ç., Şahin, İ. H., Fırat, M., Koçyiğit, R., & Kireçci, O. A. (2023). Determination of in Vitro Antioxidant, Antimicrobial Properties and COX-1/COX-2 Enzyme Inhibition Activity of *Capparis sicula. Kahramanmaras Sütçü Imam Üniversitesi Tarim ve Doga Dergisi*, *26*(6), 1221–1229. DOI: 10.18016/ksutarimdoga.vi.1139197

Adhikari, P., Joshi, K., Singh, M., & Pandey, A. (2022). Influence of altitude on secondary metabolites, antioxidants, and antimicrobial activities of Himalayan yew (*Taxus wallichiana*). *Plant Biosystems*, *156*(1), 187–195. DOI: 10.1080/11263504.2020.1845845

Affi, W., Mohamed, A. A., Gharsallah, N., Smetanska, I., & Zourgui, L. (2024). Phytochemicals, antioxidant, and antimicrobial activities of *Opuntia stricta* fruits peel. *Open Veterinary Journal*, *14*(10), 2642. DOI: 10.5455/OVJ.2024.v14.i10.14 PMID: 39545202

Agnihotri, S., Dobhal, P., Ashfaqullah, S., Oli, S., & Tamta, S. (2024). Phytochemical screening of *Zanthoxylum armatum* roots and exploring its polyphenol content and antioxidant activity. *Natural Product Research*, *38*(19), 3404–3410. DOI: 10.1080/14786419.2023.2244130 PMID: 37565659

Aimad, A., Sanae, R., Anas, F., Abdelfattah, E. M., Bourhia, M., Mohammad Salamatullah, A., Alzahrani, A., Alyahya, H. K., Albadr, N. A., Abdelkrim, A., El Barnossi, A., & Noureddine, E. (2021). Chemical characterization and antioxidant, antimicrobial, and insecticidal properties of essential oil from *Mentha pulegium* L. *Evidence-Based Complementary and Alternative Medicine : eCAM*, *2021*(1), 1108133. DOI: 10.1155/2021/1108133 PMID: 34691201

Akar, Z., Karakurt, A., Okumuş, F., Cinemre, S., Düzgün, A. Ö., Akar, B., & Can, Z. (2020). RP-HPLC-UV Analysis of the Phenolic Compounds, Antimicrobial Activity Against Multi-Drug Resistant Bacteria and Antioxidant Activity of Fruit and Seed of *Diospyros lotus* L. ctivity of Fruit and Seed of *Diospyros lotus* L. *International Journal of Secondary Metabolite*, *7*(4), 237–246. DOI: 10.21448/ijsm.714108

Akhter, K., Ghous, T., Sadaf, E., Hassan, A., Irshad, V., & Andleeb, S. (2021). Pharmacological Approach to Glycemic Treatment Using Black, Green and Herbal Tea Extracts. *Bangladesh Journal of Botany*, *50*(3), 491–498. DOI: 10.3329/bjb.v50i3.55827

Akula, R., & Ravishankar, G. A. (2011). Influence of abiotic stress signals on secondary metabolites in plants. *Plant Signaling & Behavior*, *6*(11), 1720–1731. DOI: 10.4161/psb.6.11.17613 PMID: 22041989

Al Dhanhani, A., Tsombou, F., Mirza, S., Shahin, S., & Ridouane, F. (2025). Phytochemical and antimicrobial properties of *Morus nigra* and *Morus alba* grown in Fujairah, United Arab Emirates. *Scientific Reports*, *15*(1), 34986. DOI: 10.1038/s41598-025-18872-4 PMID: 41057426

Al-haliem, S. M., Mohammed, M. J., Abedelmaksoud, T. G., Hesarinejad, M. A., & Baioumy, A. A. (2025). Chicory (*Cichorium intybus*) Leaves Extract: Phenolic Composition, Antibacterial Activity, and Antioxidant Capacity Assessment. *Food Science & Nutrition*, *13*(7), e70550. DOI: 10.1002/fsn3.70550 PMID: 40599358

Al-Jaber, N. A. (2011). Phytochemical and biological studies of *Sisymbrium irio* L. Growing in Saudi Arabia. *Journal of Saudi Chemical Society*, *15*(4), 345–350. DOI: 10.1016/j.jscs.2011.04.010

Al-Qahtani, J., Abbasi, A., Aati, H. Y., Al-Taweel, A., Al-Abdali, A., Aati, S., Yanbawi, A. N., Khan, M. A., Ghalloo, B. A., & Anwar, M. (2023). Phytochemical, antimicrobial, antidiabetic, thrombolytic, anticancer activities, and in silico studies of *Ficus palmata* Forssk. *Arabian Journal of Chemistry*, *16*(2), 104455. DOI: 10.1016/j.arabjc.2022.104455

Alam, F., Hanif, M., Rahman, A., Ali, S., & Jan, S. (2023). In vitro, in vivo and in silico evaluation of analgesic, anti-inflammatory, and anti-pyretic activity of salicylate rich fraction from *Gaultheria trichophylla* Royle (Ericaceae). *Journal of Ethnopharmacology, 301*, 115828. DOI: 10.1016/j.jep.2022.115828 PMID: 36240979

Alam, F., Saqib, Q. N., & Ashraf, M. (2017). *Gaultheria trichophylla* (Royle): A source of minerals and biologically active molecules, its antioxidant and anti-lipoxygenase activities. *BMC Complementary and Alternative Medicine, 17*(1), 3. DOI: 10.1186/s12906-016-1511-4 PMID: 28049535

Alamdari, E. G., Avarseji, Z., & Taleghani, A. (2022). Extraction, Isolation and Characterization by HPLC-ESI-MS and Evaluating Chemical Constituents of Hydroalcoholic Extracts of the Invasive Weed *Cleome viscosa* L. *Natural Product Communications, 17*(1). Advance online publication. DOI: 10.1177/1934578X2110486

Albach, D. C., Schoenswetter, P., & Tribsch, A. (2006). Comparative phylogeography of the *Veronica alpina* complex in Europe and North America. *Molecular Ecology, 15*(11), 3269–3286. DOI: 10.1111/j.1365-294X.2006.02980.x PMID: 16968270

Alhage, J., Rahman, A. A., Al-Bayssari, C., & Raad, M. T. (2025). Bioactive Compounds Isolated from the Species of the *Campanula* Genus: A Review. *Molecules (Basel, Switzerland), 30*(23), 4495. DOI: 10.3390/molecules30234495 PMID: 41375093

Alharbi, N. K., Naghmouchi, S., & Al-Zaban, M. (2021). Evaluation of antimicrobial potential and comparison of HPLC composition, secondary metabolites count, and antioxidant activity of *Mentha rotundifolia* and *Mentha pulegium* extracts. *Evidence-Based Complementary and Alternative Medicine : eCAM, 2021*(1), 9081536. DOI: 10.1155/2021/9081536 PMID: 34497659

Alhazmi, N. M. (2025). Biological and Therapeutic Applications of Gallic Acid Produced by Aspergillus terreus PQ344456 Using Submerged Fermentation. *ACS Omega, 10*(41), 48488–48499. DOI: 10.1021/acsomega.5c05913 PMID: 41141794

Alrumaihi, F., Almatroodi, S. A., Alharbi, H. O. A., Alwanian, W. M., Alharbi, F. A., Almatroudi, A., & Rahmani, A. H. (2024). Pharmacological potential of kaempferol, a flavonoid in the management of pathogenesis via modulation of inflammation and other biological activities. *Molecules (Basel, Switzerland), 29*(9), 2007. DOI: 10.3390/molecules29092007 PMID: 38731498

Alsayadi, M. M. S., Gazaem, M. A. H., Mohamad, S. A. S., Ismail, N. H., Alshoaiby, M. A., Agha, H. M., Alkamil, A. A., Salihu, M., & Salim, F. (2025). Nutritional Composition, Phytochemical contents, HPLC analysis, and Antioxidant Activity of leaf and flower extracts of *Rumex nervosus* grown in Yemen. *Biocatalysis and Agricultural Biotechnology, 103749*, 103749. Advance online publication. DOI: 10.1016/j.bcab.2025.103749

Alsharif, B., & Boylan, F. (2025). *Capparis* L.(Capparaceae): A Scoping Review of Phytochemistry, Ethnopharmacology and Pharmacological Activities. *Molecules (Basel, Switzerland), 30*(18), 3705. DOI: 10.3390/molecules30183705 PMID: 41011598

Altemimi, A. B., Al-Haliem, S. M., Alkanan, Z. T., Mohammed, M. J., Hesarinejad, M. A., Najm, M. A., Bouymajane, A., Cacciola, F., & Abedelmaksoud, T. G. (2023). Exploring the phenolic profile, antibacterial, and antioxidant properties of walnut leaves (*Juglans regia* L.). *Food Science & Nutrition, 11*(11), 6845–6853. DOI: 10.1002/fsn3.3554 PMID: 37970415

Ambwani, S., Chand, P., Kumawat, S. K., Singh, H., & Koranga, D. (2025). In vitro Analysis of Antioxidant Capacity of *Bauhinia variegata* (L.) Benth Leaf Extract. *Pharmacological Research-Natural Products, 100350*, 100350. Advance online publication. DOI: 10.1016/j.prenap.2025.100350

Anastácio, T. O., Madalozzo, E. S., Oliveira, V. S., Sandim, G. C., Moroto, B. S., Formagio, A. S. N., Cardoso, C. A. L., Sardi, J. C. O., Oliveira, C. F. R., & Macedo, M. L. R. (2022). Nutraceutical potential of *Cucurbita maxima* Duchesne pulp extract. *Food Science & Technology International, 10820132251349940*, 10820132251349940. Advance online publication. DOI: 10.1177/10820132251349940 PMID: 40538189

Aryal, B., Raut, B. K., Bhattarai, S., Bhandari, S., Tandan, P., Gyawali, K., Sharma, K., Ranabhat, D., Thapa, R., Aryal, D., Ojha, A., Devkota, H. P., & Parajuli, N. (2022). Potential therapeutic applications of plant-derived alkaloids against inflammatory and neurodegenerative diseases. *Evidence-Based Complementary and Alternative Medicine : eCAM, 2022*(1), 7299778. DOI: 10.1155/2022/7299778 PMID: 35310033

Asati, V., Deepa, P., & Sharma, P. K. (2022). Desert legume *Prosopis cineraria* as a novel source of antioxidant flavonoids/isoflavonoids: Biochemical characterization of edible pods for potential functional food development. *Biochemistry and Biophysics Reports, 29*, 101210. DOI: 10.1016/j.bbrep.2022.101210 PMID: 35198738

Augustynowicz, D., Lemieszek, M. K., Strawa, J. W., Wiater, A., & Tomczyk, M. (2023). Phytochemical profiling of extracts from rare *Potentilla* species and evaluation of their anticancer potential. *International Journal of Molecular Sciences, 24*(5), 4836. DOI: 10.3390/ijms24054836 PMID: 36902263

Ayoub, N., Badr, N., Al-Ghamdi, S. S., Alsanosi, S., Alzahrani, A. R., Abdel-Naim, A. B., Nematallah, K. A., & Swilam, N. (2021). HPLC/MSn profiling and healing activity of a muco-adhesive formula of *Salvadora persica* against acetic acid-induced oral ulcer in rats. *Nutrients, 14*(1), 28. DOI: 10.3390/nu14010028 PMID: 35010903

Azad, A., Hussain, S., Akram, H., Fida, H., Iqbal, M. A., & Butt, T. E. (2024). Antimicrobial activity of *C. intybus* leaves extract in different solvent systems: A green approach for sustainable biomedicine. *Essential Chem, 1*(1), 1–13. DOI: 10.1080/28378083.2024.2427044

Babaei-Ghaghelestany, A., Alebrahim, M. T., Farzaneh, S., & Mehrabi, M. (2022). The anticancer and antibacterial properties of aqueous and methanol extracts of weeds. *Journal of Agriculture and Food Research, 10*, 100433. DOI: 10.1016/j.jafr.2022.100433

Baban, M. M., Ahmad, S. A., Abu-Odeh, A. M., Baban, M., & Talib, W. H. (2023). Anticancer, immunomodulatory, and phytochemical screening of *Carthamus oxyacantha* M. Bieb growing in the North of Iraq. *Plants, 13*(1), 42. DOI: 10.3390/plants13010042 PMID: 38202350

Bai, L., Guo, S., Liu, Q., Cui, X., Zhang, X., Zhang, L., Yang, X., Hou, M., Ho, C.-T., & Bai, N. (2016). Characterization of nine polyphenols in fruits of *Malus pumila* Mill by high-performance liquid chromatography. *journal of food and drug analysis, 24*(2), 293-298. DOI: 10.1016/j.jfda.2015.10.002

Batiha, G. E.-S., Tene, S. T., Teibo, J. O., Shaheen, H. M., Oluwatoba, O. S., Teibo, T. K. A., Al-Kuraishy, H. M., Al-Garbee, A. I., Alexiou, A., & Papadakis, M. (2023). The phytochemical profiling, pharmacological activities, and safety of *Malva sylvestris*: A review. *Naunyn-Schmiedeberg's Archives of Pharmacology, 396*(3), 421–440. DOI: 10.1007/s00210-022-02329-w PMID: 36418467

Begum, R. F., Singh, S., & Margesan, T. (2025). Tannins as therapeutic in managing obesity and beyond. *Current Pharmacology Reports, 11*(1), 1–15. DOI: 10.1007/s40495-025-00398-y

Bennacer, A., Sahir-Halouane, F., Aitslimane-Aitkaki, S., Oukali, Z., Oliveira, I. V., Rahmouni, N., & Aissaoui, M. (2022). Structural characterization of phytochemical content, antibacterial, and antifungal activities of *Juglans regia* L. leaves cultivated in Algeria. *Biocatalysis and Agricultural Biotechnology, 40*, 102304. DOI: 10.1016/j.bcab.2022.102304

Bhutia, P. O., Kewlani, P., Pandey, A., Rawat, S., & Bhatt, I. D. (2021). Physicochemical properties and nutritional composition of fruits of the wild Himalayan strawberry (*Fragaria nubicola* Lindle.) in different ripening stages. *Journal of Berry Research, 11*(3), 481–496. DOI: 10.3233/JBR-210742

Bhutia, T. D., Valant-Vetschera, K. M., & Brecker, L. (2013). Orphan flavonoids and dihydrochalcones from Primula exudates. *Natural product communications, 8*(8), 1934578X1300800812. DOI: 10.1177/1934578X1300800812

Bojilov, D., Manolov, S., Nacheva, A., Dagnon, S., & Ivanov, I. (2023). Characterization of polyphenols from *Chenopodium botrys* after fractionation with different solvents and study of their in vitro biological activity. *Molecules (Basel, Switzerland), 28*(12), 4816. DOI: 10.3390/molecules28124816 PMID: 37375371

Bokhari, A. A., & Syed, V. (2015). Inhibition of transforming growth factor-β (TGF-β) signaling by *Scutellaria baicalensis* and *Fritillaria cirrhosa* extracts in endometrial cancer. *Journal of Cellular Biochemistry, 116*(8), 1797–1805. DOI: 10.1002/jcb.25138 PMID: 25683036

Bourhia, M., Bouothmany, K., Bakrim, H., Hadrach, S., Salamatullah, A. M., Alzahrani, A., Khalil Alyahya, H., Albadr, N. A., Gmouh, S., Laglaoui, A., El Mzibri, M., & Benbacer, L. (2021). Chemical profiling, antioxidant, antiproliferative, and antibacterial potentials of chemically characterized extract of *Citrullus colocynthis* L. seeds. *Separations, 8*(8), 114. DOI: 10.3390/separations8080114

Brahmkshatriya, P. P., & Brahmkshatriya, P. S. (2013). Terpenes: Chemistry, biological role, and therapeutic applications. In *Natural products* (pp. 2665-2691). Springer. DOI: 10.1007/978-3-642-22144-6_120

Britton, A. J., Beale, C. M., Towers, W., & Hewison, R. L. (2009). Biodiversity gains and losses: Evidence for homogenisation of Scottish alpine vegetation. *Biological Conservation, 142*(8), 1728–1739. DOI: 10.1016/j.biocon.2009.03.010

Bukvicki, D., Gottardi, D., Vannini, L., Dzamic, A., Ciric, A., Marin, P. D., & Veljic, M. (2015). Chemical composition and antimicrobial assessment of liverwort *Lophozia ventricosa* extracts. *Revista Brasileira de Botânica. Brazilian Journal of Botany, 38*(1), 25–30. DOI: 10.1007/s40415-014-0104-z

Cadi, H. E., Bouzidi, H. E., Selama, G., Ramdan, B., Majdoub, Y. O. E., Alibrando, F., Brigui, J., Altemimi, A. B., Dugo, P., Mondello, L., & Cacciola, F. (2021). Characterization of *Rubus fruticosus* L. berries growing wild in Morocco: Phytochemical screening, antioxidant activity and chromatography analysis. *European Food Research and Technology, 247*(7), 1689–1699. DOI: 10.1007/s00217-021-03739-5

Carrasco, F., Hernández, W., Castro, N., Angulo, G., Quintero, Á., Zamudio, D., Tamariz-Angeles, C., Olivera-Gonzales, P., & Echevarría-Rodríguez, D. (2023). Chemical composition and in vitro antibacterial activity of three lichen species (Usnea sp., *Thamnolia vermicularis* subsp. solida and *Ramalina asperula*) collected from the Jauja and Huaral provinces-Peru. *Revista Colombiana de Ciencias Químico-Farmacéuticas, 52*(3). Advance online publication. DOI: 10.15446/rcciquifa.v52n3.110402

Casas, M., Vallès, J., & Gras, A. (2024). Nutritional Properties of Wild Edible Plants with Traditional Use in the Catalan Linguistic Area: A First Step for Their Relevance in Food Security. *Foods, 13*(17), 2785. DOI: 10.3390/foods13172785 PMID: 39272550

Cavazos, P., Gonzalez, D., Lanorio, J., & Ynalvez, R. (2021). Secondary metabolites, antibacterial and antioxidant properties of the leaf extracts of *Acacia rigidula* benth. and *Acacia berlandieri* benth. *SN Applied Sciences, 3*(5), 522. DOI: 10.1007/s42452-021-04513-8

Chang, M.-L., Chang, J.-S., Yu, W.-Y., Cheah, K.-P., Li, J.-S., Cheng, H.-W., & Hu, C.-M. (2014). *Polygonum viviparum* L. induces vasorelaxation in the rat thoracic aorta via activation of nitric oxide synthase in endothelial cells. *BMC Complementary and Alternative Medicine, 14*(1), 150. DOI: 10.1186/1472-6882-14-150 PMID: 24885446

Chileh-Chelh, T., da Cunha-Chiamolera, T. P. L., Urrestarazu, M., Ezzaitouni, M., López-Ruiz, R., Nájera, C., Rincón-Cervera, M. Á., & Guil-Guerrero, J. L. (2024). London Rocket (*Sisymbrium irio* L.) as Healthy Green: Bioactive Compounds and Bioactivity of Plants Grown in Wild and Controlled Environments. *Molecules (Basel, Switzerland), 30*(1), 31. DOI: 10.3390/molecules30010031 PMID: 39795089

Chinnapun, D. (2025). The Structure and Biological Activities With Health Benefits of Luteolin Glycosides From Plants. *TheScientificWorldJournal, 2025*(1), 3066682. DOI: 10.1155/tswj/3066682

Colaric, M., Stampar, F., Solar, A., & Hudina, M. (2006). Influence of branch bending on sugar, organic acid and phenolic content in fruits of 'Williams' pears (*Pyrus communis* L.). *Journal of the Science of Food and Agriculture, 86*(14), 2463–2467. DOI: 10.1002/jsfa.2644

Cox-Georgian, D., Ramadoss, N., Dona, C., & Basu, C. (2019). Therapeutic and medicinal uses of terpenes. In *Medicinal plants: from farm to pharmacy* (pp. 333–359). Springer., DOI: 10.1007/978-3-030-31269-5_15

Cuendet, M., Potterat, O., & Hostettmann, K. (2001). Flavonoids and phenylpropanoid derivatives from *Campanula barbata*. *Phytochemistry, 56*(6), 631–636. DOI: 10.1016/S0031-9422(00)00423-4 PMID: 11281141

D'Angeli, F., Malfa, G. A., Garozzo, A., Li Volti, G., Genovese, C., Stivala, A., Nicolosi, D., Attanasio, F., Bellia, F., Ronsisvalle, S., & Acquaviva, R. (2021). Antimicrobial, antioxidant, and cytotoxic activities of *Juglans regia* L. pellicle extract. *Antibiotics (Basel, Switzerland), 10*(2), 159. DOI: 10.3390/antibiotics10020159 PMID: 33557378

Devi, C., Gouri, P., Subhashini, , Bavatharani, , & Bhuvaneswari, . (2024). Study on Anti-arthritic and Antimicrobial Activity of Leaves of *Mukia maderaspatana* on Membrane Stabilization Assay and Egg Albumin Assay. *SBV Journal of Basic. Clinical and Applied Health Science, 7*(3), 105–111. DOI: 10.4103/SBVJ.SBVJ_9_24

Ding, Y., Li, H., Cao, S., & Yu, Y. (2024). Effects of catechin on the malignant biological behavior of gastric cancer cells through the PI3K/Akt signaling pathway. *Toxicology and Applied Pharmacology, 490*, 117036. DOI: 10.1016/j.taap.2024.117036 PMID: 39009138

Dos Santos, E. T., Pereira, M. L. A., Da Silva, C. F. P., Souza-Neta, L. C., Geris, R., Martins, D., Santana, A. E. G., Barbosa, L. C. A., Silva, H. G. O., Freitas, G. C., Figueiredo, M., De Oliveira, F., & Batista, R. (2013). Antibacterial activity of the alkaloid-enriched extract from *Prosopis juliflora* pods and its influence on *in vitro* ruminal digestion. *International Journal of Molecular Sciences, 14*(4), 8496–8516. DOI: 10.3390/ijms14048496 PMID: 23595000

Edwards, A. C., Scalenghe, R., & Freppaz, M. (2007). Changes in the seasonal snow cover of alpine regions and its effect on soil processes: A review. *Quaternary International, 162*, 172–181. DOI: 10.1016/j.quaint.2006.10.027

El-Hajjaji, M. A., Fikri-Benbrahim, K., El Ouassete, M., Mrabti, N. N., Soulo, N., El Ghouizi, A., Lyoussi, B., & Ouaritini, Z. B. (2024). Phytochemical profiling of *Prunus armeniaca* kernel extracts and exploration of their multifaceted antioxidant and antibacterial effects through *in vitro* and *in silico* studies. *European Journal of Integrative Medicine*, *72*, 102421. DOI: 10.1016/j.eujim.2024.102421

El-Hefny, M., & Hussien, M. K. (2025). Enhancing the growth and essential oil components of *Lavandula latifolia* using *Malva parviflora* extract and humic acid as biostimulants in a field experiment. *Scientific Reports*, *15*(1), 774. DOI: 10.1038/s41598-024-82127-x PMID: 39755703

El Mannoubi, I. (2023). Impact of different solvents on extraction yield, phenolic composition, in vitro antioxidant and antibacterial activities of deseeded *Opuntia stricta* fruit. *Journal of Umm Al-Qura University for Applied Sciences*, *9*(2), 176–184. DOI: 10.1007/s43994-023-00031-y

El-Massry, K. F., Farouk, A., Mahmoud, K. F., El-Ghorab, A. H., Musa, A., Mostafa, E. M., Ghoneim, M. M., Naguib, I. A., & Abdelgawad, M. A. (2021). Chemical characteristics and targeted encapsulated *Cordia myxa* fruits extracts nanoparticles for antioxidant and cytotoxicity potentials. *Saudi Journal of Biological Sciences*, *28*(9), 5349–5358. DOI: 10.1016/j.sjbs.2021.05.064 PMID: 34466114

El-Nashar, H. A., Zengin, G., Tarek, A., & Eldahshan, O. A. (2025). Comparative metabolic profiling of *Cordia myxa* leaves and fruits based on UPLC-ESI/MS-MS analysis and investigation of antioxidant activities and enzyme inhibitory properties. *Scientific Reports*, *15*(1), 19296. DOI: 10.1038/s41598-025-04015-2 PMID: 40456930

El-Newary, S. A., Abd Elkarim, A. S., Abdelwahed, N. A., Omer, E. A., Elgamal, A. M., & ELsayed, W. M. (2023). *Chenopodium murale* juice shows anti-fungal efficacy in experimental oral candidiasis in immunosuppressed rats in relation to its chemical profile. *Molecules (Basel, Switzerland)*, *28*(11), 4304. DOI: 10.3390/molecules28114304 PMID: 37298777

El-Newary, S. A., Aly, M. S., Hameed, A. R. A. E., Kotp, M. S., Youssef, A. A., & Ali, N. A. (2022). Sperm quality and testicular histopathology of Wistar albino male rats treated with hydroethanolic extract of *Cordia dichotoma* fruits. *Pharmaceutical Biology*, *60*(1), 282–293. DOI: 10.1080/13880209.2021.2008455 PMID: 35138992

Elzaawely, A. A., & Tawata, S. (2012). Antioxidant capacity and phenolic content of *Rumex dentatus* L. grown in Egypt. *Journal of Crop Science and Biotechnology*, *15*(1), 59–64. DOI: 10.1007/s12892-011-0063-x

Ercan, L., & Doğru, M. (2022). Determination of antimicrobial activity of *Nasturtium officinale* and its content of volatile organic compounds and fatty acids. *Kahramanmaraş Sütçü İmam Üniversitesi Tarım ve Doğa Dergisi, 25*(Ek Sayı 1), 11-21. DOI: 10.18016/ksutarimdoga.vi.1001837

Erdogan-Orhan, I., & Kartal, M. (2011). Insights into research on phytochemistry and biological activities of *Prunus armeniaca* L.(apricot). *Food Research International, 44*(5), 1238–1243. DOI: 10.1016/j.foodres.2010.11.014

Erenler, R., Telci, I., Elmastaş, M., Akşit, H., Gül, F., Tüfekçi, A. R., Demírtaş, İ., & Kayir, Ö. (2018). Quantification of flavonoids isolated from\textit {*Mentha spicata*} in selected clones of Turkish mint landraces. *Turkish Journal of Chemistry, 42*(6), 1695–1705. DOI: 10.3906/kim-1712-3

Farag, M., Abdel-Mageed, W. M., El Gamal, A. A., & Basudan, O. A. (2021). *Salvadora persica* L.: Toothbrush tree with health benefits and industrial applications–An updated evidence-based review. *Saudi Pharmaceutical Journal : SPJ : the Official Publication of the Saudi Pharmaceutical Society, 29*(7), 751–763. DOI: 10.1016/j.jsps.2021.05.007 PMID: 34400870

Faysal, M., Al Amin, M., Zehravi, M., Sweilam, S. H., Arjun, U. V. N. V., Gupta, J. K., Shanmugarajan, T. S., Prakash, S. S., Dayalan, G., Kasimedu, S., Madhuri, Y. B., Reddy, K. T. K., Rab, S. O., Al Fahaid, A. A. F., & Emran, T. B. (2025). Therapeutic potential of flavonoids in neuroprotection: Brain and spinal cord injury focus. *Naunyn-Schmiedeberg's Archives of Pharmacology, 398*(7), 1–26. DOI: 10.1007/s00210-025-03888-4 PMID: 40014123

Feng, L., Yin, Y., Fang, Y., & Yang, X. (2017). Quantitative determination of punicalagin and related substances in different parts of pomegranate. *Food Analytical Methods, 10*(11), 3600–3606. DOI: 10.1007/s12161-017-0916-0

Fereja, W. M., Geleta, W. D., Desalegn, E., Moaneanda, M., Abdissa, A., & Usamo, F. B. (2024). *In vitro* antioxidant and antibacterial activities of *Ajuga integrifolia* leaf extracts obtained with different solvents. *BMC Complementary Medicine and Therapies, 24*(1), 368. DOI: 10.1186/s12906-024-04668-y PMID: 39395997

Fouad, M. S., Ghareeb, M. A., Hamed, A. A., Aidy, E. A., Tabudravu, J., Sayed, A. M., Tammam, M. A., & Mwaheb, M. A. (2024). Exploring the antioxidant, anticancer and antimicrobial potential of *Amaranthus viridis* L. collected from Fayoum depression: Phytochemical, and biological aspects. *South African Journal of Botany, 166*, 297–310. DOI: 10.1016/j.sajb.2024.01.047

Fujino, M., Sakakibara, K., Tsujimura, M., & Suzuki, K. (2023). Influence of alpine vegetation on water storage and discharge functions in an alpine headwater of Northern Japan Alps. *Journal of Hydrology: X, 18*, 100146. DOI: 10.1016/j.hydroa.2022.100146

Ganesan, P. (2024). Analysis of antiviral and antimicrobial activity of *Opuntia Stricta* (red prickly pear) extract treated cotton fabric by cationization technique for healthcare and hygiene textiles. *Journal of the Textile Institute, 115*(12), 2411–2420. DOI: 10.1080/00405000.2023.2291526

Ganie, S. A., Dar, T. A., Hamid, R., Zargar, O., Abeer, S. U., Masood, A., Amin, S., & Zargar, M. A. (2014). *In vitro* antioxidant and cytotoxic activities of *Arnebia benthamii* (Wall ex. G. Don): A critically endangered medicinal plant of Kashmir valley. *Oxidative Medicine and Cellular Longevity, 2014*(1), 792574. DOI: 10.1155/2014/792574 PMID: 24734152

Ganie, S. A., Jan, A., Muzaffar, S., Zargar, B. A., Hamid, R., & Zargar, M. A. (2012). Radical scavenging and antibacterial activity of *Arnebia benthamii* methanol extract. *Asian Pacific Journal of Tropical Medicine, 5*(10), 766–772. DOI: 10.1016/S1995-7645(12)60140-0 PMID: 23043913

Gaude, A. A., & Jalmi, S. K. (2025). Environmental stress induced biosynthesis of plant secondary metabolites-transcriptional regulation as a key. *Crop Design, 100100*(2), 100100. Advance online publication. DOI: 10.1016/j.cropd.2025.100100

Gautam, R., Srivastava, A., & Jachak, S. M. (2011). Simultaneous determination of naphthalene and anthraquinone derivatives in *Rumex nepalensis* Spreng. Roots by HPLC: Comparison of different extraction methods and validation. *Phytochemical Analysis, 22*(2), 153–157. DOI: 10.1002/pca.1261 PMID: 21046683

Ghahramani, Y., Mohammadi, N., Baghaei, S., & Jahandizi, N. G. (2024). Time-dependent antibacterial effects of *Citrullus colocynthis* seed extract compared to calcium hydroxide in teeth infected with *Enterococcus faecalis*. *Journal of Dentistry, 25*(1), 77. DOI: 10.30476/dentjods.2023.97670.2026 PMID: 38544769

Ghazghazi, H., Chedia, A., Weslati, M., Trakhna, F., Houssine, S., Abderrazak, M., & Brahim, H. (2013). Chemical Composition and *in vitro* Antimicrobial Activities of *Mentha pulegium* Leaves Extracts against Foodborne Pathogens. *Journal of Food Safety, 33*(3), 239–246. DOI: 10.1111/jfs.12045

Ghosh, L., Gayen, J. R., Sinha, S., Pal, S., Pal, M., & Saha, B. (2003). Antibacterial efficacy of *Rumex nepalensis* Spreng. roots. *Phytotherapy Research, 17*(5), 558–559. DOI: 10.1002/ptr.1162 PMID: 12748999

Gillani, S. W., Ahmad, M., Manzoor, M., Waheed, M., Iqbal, Z., Ullah, R., Pieroni, A., Zhang, L., Sulaiman, N., & Alrhmoun, M. (2024). The nexus between ecology of foraging and food security: Cross-cultural perceptions of wild food plants in Kashmir Himalaya. *Journal of Ethnobiology and Ethnomedicine, 20*(1), 77. DOI: 10.1186/s13002-024-00721-9 PMID: 39155383

Gundogdu, M., Muradoglu, F., Sensoy, R. G., & Yilmaz, H. (2011). Determination of fruit chemical properties of *Morus nigra* L., *Morus alba* L. and *Morus rubra* L. by HPLC. *Scientia Horticulturae, 132*, 37–41. DOI: 10.1016/j.scienta.2011.09.035

Han, B., Wu, J., & Huang, L. (2020). Induction of apoptosis in lung cancer cells by *Viburnum grandiflorum* via mitochondrial pathway. *Medical Science Monitor, 26*, e920265–e920261. DOI: 10.12659/MSM.920265 PMID: 31900380

Hao, J.-W., Liu, X.-Q., Chen, N.-D., & Zhu, A. (2021). Solid-phase extraction followed by direct TOF-MS-MS and HPLC analysis of ptaquiloside in *Pteridium aquilinum* from different places of China. *Journal of Food Composition and Analysis, 98*, 103845. DOI: 10.1016/j.jfca.2021.103845

Hasan, K., Sabiha, S., Islam, N., Pinto, J. F., & Silva, O. (2024). Ethnomedicinal Usage, Phytochemistry and Pharmacological Potential of *Solanum surattense* Burm. f. *Pharmaceuticals (Basel, Switzerland), 17*(7), 948. DOI: 10.3390/ph17070948 PMID: 39065797

Hasnat, H., Shompa, S. A., Islam, M. M., Alam, S., Richi, F. T., Emon, N. U., Ashrafi, S., Ahmed, N. U., Chowdhury, M. N. R., Fatema, N., Hossain, M. S., Ghosh, A., & Ahmed, F. (2024). Flavonoids: A treasure house of prospective pharmacological potentials. *Heliyon, 10*(6), e27533. Advance online publication. DOI: 10.1016/j.heliyon.2024.e27533 PMID: 38496846

Hu, Y.-K., Ma, C., Li, M.-J., Bai, X.-L., Liu, Y.-M., & Liao, X. (2024). Screening of monoamine oxidase B inhibitors in *Tibetan strawberry* by ligand fishing based on enzyme functionalized cellulose filter paper. *Microchemical Journal, 203*, 110838. DOI: 10.1016/j.microc.2024.110838 PMID: 39035460

Hua, R., Zhang, P., Wang, L., Huang, M., Hua, L., & Zhou, J. (2025). Alpine steppe vegetation communities are more sensitive to plateau pika disturbance than alpine meadows. *Frontiers in Plant Science, 16*, 1546828. DOI: 10.3389/fpls.2025.1546828 PMID: 40510176

Hussein, M. A., Islam, M. S., Ali, A. A., Mansour, M. S., Bondok, M., Salem, M. A., Amein, A. S., & Elgizawy, H. A. (2024). *Malva parviflora* seed oil; Isolation, gas chromatographic profiling and its cardioprotective activity against myocardial infraction in animal model. *Journal of King Saud University. Science, 36*(2), 103060. DOI: 10.1016/j.jksus.2023.103060

Hyun, T., Ra, J., Han, S., & Kim, J. (2018). Antioxidant, Antimicrobial, and Antidiabetic Activities of *Crowberry* Fruits. *Indian Journal of Pharmaceutical Sciences, 80*(3). Advance online publication. DOI: 10.4172/pharmaceutical-sciences.1000382

Iannuzzi, A. M., Giacomelli, C., De Leo, M., Pietrobono, D., Camangi, F., De Tommasi, N., Martini, C., Trincavelli, M. L., & Braca, A. (2020). Antioxidant activity of compounds isolated from *Elaeagnus umbellata* promotes human gingival fibroblast well-being. *Journal of Natural Products, 83*(3), 626–637. DOI: 10.1021/acs.jnatprod.9b01030 PMID: 32031808

Ikram, M., Magdy Beshbishy, A., Kifayatullah, M., Olukanni, A., Zahoor, M., Naeem, M., Amin, M., Shah, M., Abdelaziz, A. S., Ullah, R., Mothana, R. A., Siddiqui, N. A., & Batiha, G. E.-S. (2020). Chemotherapeutic potential of *Carthamus oxycantha* root extract as antidiarrheal and *in vitro* antibacterial activities. *Antibiotics (Basel, Switzerland), 9*(5), 226. DOI: 10.3390/antibiotics9050226 PMID: 32370087

Iqbal, N., Yasmin, I., Ullah, H., Rehman, F. U., Tahir, K., Bushra, , Iqbal, M., Qarnain, Z., & Khan, S. (2022a). Isolation, characterization and antibacterial studies of three new chemical constituents isolated from *Viburnum grandiflorum. Pharmaceutical Chemistry Journal, 56*(4), 480–486. DOI: 10.1007/s11094-022-02663-6

Iqbal, T., Habib, T., Hussain, K., & Khan, A. M. (2022b). Wild edible plant basket of terrestrial paradise and variations among the diverse ethnic and elevation groups: A detailed insight from the Western Himalaya of Azad Jammu and Kashmir, Pakistan. *South African Journal of Botany, 147*, 294–313. DOI: 10.1016/j.sajb.2022.01.026

Ishaq, S., Rathore, H. A., Sabir, S. M., & Maroof, M. S. (2015). Antioxidant properties of *Elaeagnus umbellata* berry solvent extracts against lipid peroxidation in mice brain and liver tissues. *Food Science and Biotechnology, 24*(2), 673–679. DOI: 10.1007/s10068-015-0088-x

Ismail, H., Rasheed, A., Haq, I., Jafri, L., Ullah, N., Dilshad, E., Sajid, M., & Mirza, B. (2017). Five Indigenous Plants of Pakistan with Antinociceptive, Anti-Inflammatory, Antidepressant, and Anticoagulant Properties in Sprague Dawley Rats. *Evidence-Based Complementary and Alternative Medicine : eCAM, 2017*(1), 7849501. DOI: 10.1155/2017/7849501 PMID: 29348774

Ismail, M., Hussain, J., Khan, A., Khan, A. L., Ali, L., Khan, F., Khan, A. Z., Niaz, U., & Lee, I.-J. (2012). Antibacterial, antifungal, cytotoxic, phytotoxic, insecticidal, and enzyme inhibitory activities of *Geranium wallichianum. Evidence-Based Complementary and Alternative Medicine : eCAM*, *2012*(1), 305906. DOI: 10.1155/2012/305906 PMID: 23049606

Jabeen, Q., Haider, S. I., Asif, A., Rasheed, R., Gul, S., & Arshad, S. (2022). *Geranium wallichianum* D. Don Ex Sweet ameliorates rheumatoid arthritis by curtailing the expression of COX-II and inflammatory cytokines as well as by alleviating the oxidative stress. *Dose-Response*, *20*(3), 15593258221112649. DOI: 10.1177/15593258221112649 PMID: 35832768

Jacques, A. C., Chaves, F. C., Zambiazi, R. C., Brasil, M. C., & Caramão, E. B. (2014). Bioactive and volatile organic compounds in Southern Brazilian blackberry (*Rubus fruticosus*) fruit cv. Tupy. *Food Science and Technology (Campinas)*, *34*(3), 636–643. DOI: 10.1590/1678-457x.6243

Janbaz, K. H., Zaeem Ahsan, M., Saqib, F., Imran, I., Zia-Ul-Haq, M., Abid Rashid, M., Jaafar, H. Z., & Moga, M. (2015). Scientific basis for use of *Pyrus pashia* Buch.-Ham. ex D. Don. fruit in gastrointestinal, respiratory and cardiovascular ailments. *PLoS One*, *10*(3), e0118605. DOI: 10.1371/journal.pone.0118605 PMID: 25786248

Javid, H., Yousuf, U., Ul Qadir, R., Magray, J. A., Wani, B. A., Islam, T., Nawchoo, I. A., & Gulzar, S. (2025). Metabolic profiling and antimicrobial activity of *Bistorta amplexicaulis* D. don by *in-vitro* implicated through computational studies. *Frontiers in Pharmacology*, *16*, 1575727. DOI: 10.3389/fphar.2025.1575727 PMID: 40667511

Juyal, D., Thawani, V., Thaledi, S., & Joshi, M. (2014). Ethnomedical properties of *Taxus wallichiana* zucc.(Himalayan yew). *Journal of Traditional and Complementary Medicine*, *4*(3), 159–161. DOI: 10.4103/2225-4110.136544 PMID: 25161920

Kanwal, S., Perveen, S., Hameed, H., Tahir, S., Rashid, R., Raish, M., Bashir, S., Zaffar, K., Irfan, A., Kotwica-Mojzych, K., Faheem, S., Ahmad, A., Kashif, M., & Bin Jardan, Y. A. (2025). Evaluating the efficacy of *Chenopodium murale* plant extract in inhibiting Streptococcus mutans and reducing dental caries risk. *Scientific Reports*, *15*(1), 15406. DOI: 10.1038/s41598-025-00127-x PMID: 40316577

Karaçelik, A. A., & Şahin, H. (2021). Determination of chemical compositions, antioxidant and enzyme inhibitory activities of naturally growing *Chenopodium album* subsp. iranicum Aellen. *Journal of the Institute of Science and Technology*, *11*(3), 2091–2101. DOI: 10.21597/jist.825781

Karakaplan, M., & Özcan, M. (2017). Determination of phenolic acids in pomegranate juices by HPLC-DAD. *Avrupa Bilim ve Teknoloji Dergisi*, *6*(10), 32–37.

Karpagasundari, C., & Kulothungan, S. (2014). Analysis of bioactive compounds in *Physalis minima* leaves using GC MS, HPLC, UV-VIS and FTIR techniques. *Journal of Pharmacognosy and Phytochemistry*, *3*(4), 196–201.

Khalid, S., Arshad, M., Mahmood, S., Siddique, F., Roobab, U., Ranjha, M. M. A. N., & Lorenzo, J. M. (2023). Extraction and quantification of *Moringa oleifera* leaf powder extracts by HPLC and FTIR. *Food Analytical Methods*, *16*(4), 787–797. DOI: 10.1007/s12161-023-02470-z

Khalid, T., Rafey, H. A., Amin, A., Kamran, M., Zargaham, M. K., Ross, S. A., & Niaz, S.-I. (2025). *In silico* and *in vitro* characterization of *Phytolacca latbenia* (Moq.): QSAR, phytochemical, and toxicological insights into antiglycation and antidiabetic potential. *Biophysical Chemistry*, *107526*. Advance online publication. DOI: 10.1016/j.bpc.2025.107526 PMID: 40983042

Khaliq, T., Akhter, S., Sultan, P., & Hassan, Q. P. (2023). Critical review on *Rumex dentatus* L. a strong pharmacophore and the future medicine: Pharmacology, phytochemical analysis and traditional uses. *Heliyon*, *9*(3), e14159. Advance online publication. DOI: 10.1016/j.heliyon.2023.e14159 PMID: 36994390

Khan, I., Khan, U., Khan, W., Alqathama, A., Riaz, M., Ahmad, R., & Alam, M. M. (2024). Antibacterial and antibiofilm potentials of *Rumex dentatus* root extract characterized by HPLC-ESI-Q-TOF-MS. *Saudi Journal of Biological Sciences*, *31*(4), 103962. DOI: 10.1016/j.sjbs.2024.103962 PMID: 38419820

Khan, I., Rahman, H., Abd El-Salam, N. M., Tawab, A., Hussain, A., Khan, T. A., Khan, U. A., Qasim, M., Adnan, M., Azizullah, A., Murad, W., Jalal, A., Muhammad, N., & Ullah, R. (2017). *Punica granatum* peel extracts: HPLC fractionation and LC MS analysis to quest compounds having activity against multidrug resistant bacteria. *BMC Complementary and Alternative Medicine*, *17*(1), 247. DOI: 10.1186/s12906-017-1766-4 PMID: 28468660

Khan, M. Z., Shabbir, M. I., Saqib, Z., Gilani, S. A., Jogezai, N. U., Kiyani, M. M., & Malik, M. A. (2020). Investigation of polyphenol profile, antioxidant activity and hepatoprotective potential of *Aconogonon alpinum* (All.) Schur roots. *Open Chemistry*, *18*(1), 516–536. DOI: 10.1515/chem-2020-0062

Khanam, A., Ijaz Hussain, A., Mohammed, E. H., Nahar, L., & Rathore, H. A. (2025). Phenolic Profile of Seedless *Ziziphus mauritiana* Fruits and Leaves Extracts with *In Vivo* Antioxidant and Anti-Inflammatory Activities: Influence on Pro-Inflammatory Mediators. *Chemistry & Biodiversity*, *22*(3), e202401728. DOI: 10.1002/cbdv.202401728 PMID: 39475065

Kim, D.-S., Ji, H. D., Rhee, M. H., Sung, Y.-Y., Yang, W.-K., Kim, S. H., & Kim, H.-K. (2014a). Antiplatelet activity of *Morus alba* leaves extract, mediated via inhibiting granule secretion and blocking the phosphorylation of extracellular-signal-regulated kinase and akt. *Evidence-Based Complementary and Alternative Medicine : eCAM, 2014*(1), 639548. DOI: 10.1155/2014/639548 PMID: 24701244

Kim, J. K., & Park, S. U. (2018). Quercetin and its role in biological functions: An updated review. *EXCLI Journal, 17*, 856. DOI: 10.17179/excli2018-1538 PMID: 30233284

Kim, M.-K., Oh, T.-J., & Park, H. (2014b). Antibacterial and antioxidant capacity of polar microorganisms isolated from Arctic lichen *Ochrolechia* sp. *Polish Journal of Microbiology, 63*(3), 317–322. DOI: 10.33073/pjm-2014-042 PMID: 25546942

Kitic, D., Miladinovic, B., Randjelovic, M., Szopa, A., Sharifi-Rad, J., Calina, D., & Seidel, V. (2022). Anticancer potential and other pharmacological properties of *Prunus armeniaca* L.: An updated overview. *Plants, 11*(14), 1885. DOI: 10.3390/plants11141885 PMID: 35890519

Koo, H., Gomes, B., Rosalen, P. L., Ambrosano, G., Park, Y. K., & Cury, J. A. (2000). In vitro antimicrobial activity of propolis and *Arnica montana* against oral pathogens. *Archives of Oral Biology, 45*(2), 141–148. DOI: 10.1016/S0003-9969(99)00117-X PMID: 10716618

Kopystecka, A., Kozioł, I., Radomska, D., Bielawski, K., Bielawska, A., & Wujec, M. (2023). *Vaccinium uliginosum* and *Vaccinium myrtillus*—two species—one used as a functional food. *Nutrients, 15*(19), 4119. DOI: 10.3390/nu15194119 PMID: 37836403

Körner, C. (2021a). Carbon investments. In *Alpine Plant Life: Functional Plant Ecology of High Mountain Ecosystems* (pp. 309-333). Springer. DOI: 10.1007/978-3-030-59538-8_12

Körner, C. (2021b). Climatic stress. In *Alpine plant life: Functional plant ecology of high mountain ecosystems* (pp. 175–201). Springer., DOI: 10.1007/978-3-030-59538-8_8

Körner, C. (2021c). Life under and in snow: protection and limitation. In *Alpine plant life: functional plant ecology of high mountain ecosystems* (pp. 89–118). Springer., DOI: 10.1007/978-3-030-59538-8_5

Körner, C. (2021d). Mineral nutrition. In *Alpine Plant Life: Functional Plant Ecology of High Mountain Ecosystems* (pp. 237-268). Springer. DOI: 10.1007/978-3-030-59538-8_10

Körner, C. (2021e). Plant biomass production. In *Alpine Plant Life: Functional Plant Ecology of High Mountain Ecosystems* (pp. 375-394). Springer. DOI: 10.1007/978-3-030-59538-8_15

Körner, C. (2021f). Plant reproduction. In *Alpine plant life: Functional plant ecology of high mountain ecosystems* (pp. 395–449). Springer., DOI: 10.1007/978-3-030-59538-8_16

Körner, C. (2023). Concepts in alpine plant ecology. *Plants*, *12*(14), 2666. DOI: 10.3390/plants12142666 PMID: 37514280

Kováč, J., Slobodníková, L., Trajčíková, E., Rendeková, K., Mučaji, P., Sychrová, A., & Bittner Fialová, S. (2022). Therapeutic potential of flavonoids and tannins in management of oral infectious diseases—A review. *Molecules (Basel, Switzerland)*, *28*(1), 158. DOI: 10.3390/molecules28010158 PMID: 36615352

Kryvtsova, M., Salamon, I., Koscova, J., & Spivak, M. (2020). Antibiofilm forming, antimicrobial activity and some biochemical properties of *Vaccinium vitis* idaea leaf and berry extracts on *Staphylococcus aureus*. *Biosystems Diversity*, *28*(3), 238–242. DOI: 10.15421/012031

Kumar, A., Kamal, A., Singh, S., Padalia, R. C., Tandon, S., Chauhan, A., Saikia, D., & Verma, R. S. (2020). Chemical composition, antimicrobial activity, kinetics and mechanism of action of Himalayan-thyme (*Thymus linearis* Benth.). *The Journal of Essential Oil Research*, *32*(1), 59–68. DOI: 10.1080/10412905.2019.1662337

Laaraj, S., Ouahidi, I., Al Moudani, N., Allali, M., & Aarab, L. (2024). Immunostimulant and Pharmacological Activities, of *Cucurbita maxima* Seeds on Humoral and Cellular Immunity Cells and their Functions. *The Open Public Health Journal*, *17*(1), e18749445274775. Advance online publication. DOI: 10.2174/0118749445274775231204095205

Laftouhi, A., Eloutassi, N., Ech-Chihbi, E., Rais, Z., Abdellaoui, A., Taleb, A., Beniken, M., Nafidi, H.-A., Salamatullah, A. M., Bourhia, M., & Taleb, M. (2023). The impact of environmental stress on the secondary metabolites and the chemical compositions of the essential oils from some medicinal plants used as food supplements. *Sustainability (Basel)*, *15*(10), 7842. DOI: 10.3390/su15107842

Landry, K. B., Azam, S., Rehman, S., Tariq, S., Iqbal, B., Abbas, M., Lembè, D. M., & Ijaz, B. (2021). Phytochemical analysis of *Berberis lyceum* methanolic extract and its antiviral activity through the restoration of MAPK signaling pathway modulated by HCV NS5A. *Asian Pacific Journal of Tropical Biomedicine*, *11*(3), 132–140. DOI: 10.4103/2221-1691.306133

Letchuman, S., Madhuranga, H., Madhurangi, B., Premarathna, A. D., & Saravanan, M. (2024). *Alkaloids unveiled: A comprehensive analysis of novel therapeutic properties, mechanisms, and plant-based innovations*. Intelligent Pharmacy., DOI: 10.1016/j.ipha.2024.09.007

Li, W., Shi, M., Wang, P., Guo, X., Li, C., & Kang, W. (2018). Efficient determination of three flavonoids in *Malus pumila* flowers by ionic liquid-HPLC. *Journal of Molecular Liquids, 263*, 139–146. DOI: 10.1016/j.molliq.2018.04.135

Lim, S. H., & Choi, C.-I. (2019). Pharmacological properties of *Morus nigra* L.(black mulberry) as a promising nutraceutical resource. *Nutrients, 11*(2), 437. DOI: 10.3390/nu11020437 PMID: 30791521

Lin, Y., Shi, R., Wang, X., & Shen, H.-M. (2008). Luteolin, a flavonoid with potential for cancer prevention and therapy. *Current Cancer Drug Targets, 8*(7), 634–646. DOI: 10.2174/156800908786241050 PMID: 18991571

Liu, S., Yang, T., Ming, T. W., Gaun, T. K. W., Zhou, T., Wang, S., & Ye, B. (2020). Isosteroid alkaloids with different chemical structures from *Fritillariae cirrhosae* bulbus alleviate LPS-induced inflammatory response in RAW 264.7 cells by MAPK signaling pathway. *International Immunopharmacology, 78*, 106047. DOI: 10.1016/j.intimp.2019.106047 PMID: 31816576

Loshali, A., Tripathi, G., Bawa, S., & Aeri, V. (2025). *Betula utilis* (Bhojpatra): A potent herb with its traditional uses, phytochemistry, clinical application, pharmacology, and toxicology. *Journal of Applied Pharmaceutical Science, 15*(5), 001-013. DOI: 10.7324/JAPS.2025.230627

Lu, Y., Liu, Y., Che, F., Gao, M., Li, A., & Wei, Y. (2024). Optimisation of isolation of polyphenols from *Malus pumila* Mill. Leaves by high-speed countercurrent chromatography using response surface methodology. *Journal of Chromatography. B, Analytical Technologies in the Biomedical and Life Sciences, 1243*, 124230. DOI: 10.1016/j.jchromb.2024.124230 PMID: 38981203

Luczaj, L., Pieroni, A., Tardío, J., Pardo-de-Santayana, M., Sõukand, R., Svanberg, I., & Kalle, R. (2012). Wild food plant use in 21st century Europe: The disappearance of old traditions and the search for new cuisines involving wild edibles. *Acta Societatis Botanicorum Poloniae, 81*(4), 359–370. DOI: 10.5586/asbp.2012.031

Mahdavi, S., Kheyrollahi, M., Sheikhloei, H., & Isazadeh, A. (2019). Antibacterial and antioxidant activities of *Nasturtium officinale* essential oil on food borne bacteria. *The Open Microbiology Journal, 13*(1), 81–85. Advance online publication. DOI: 10.2174/1874285801913010081

Maksimovic, Z. A., Dordevic, S., & Mraovic, M. (2005). Antimicrobial activity of *Chenopodium botrys* essential oil. *Fitoterapia, 76*(1), 112–114. DOI: 10.1016/j.fitote.2004.10.005 PMID: 15664473

Malík, M., Mika, O. J., Navrátilová, Z., Killi, U. K., Tlustoš, P., & Patočka, J. (2023). Health and environmental hazards of the toxic *Pteridium aquilinum* (L.) Kuhn (Bracken Fern). *Plants, 13*(1), 18. DOI: 10.3390/plants13010018 PMID: 38202326

Manojlović, N. T., Vasiljević, P., Jusković, M., Najman, S., Janković, S., & Milenković-Andjelković, A. (2010). HPLC analysis and cytotoxic potential of extracts from the lichen, *Thamnolia vermicularis* var. subuliformis. *Journal of Medicinal Plants Research, 4*(9), 817–823. DOI: 10.5897/JMPR10.104

Mashaal, K., Shabbir, A., Shahzad, M., Mobashar, A., Akhtar, T., Fatima, T., Riaz, B., Alharbi, R., Fatima, A., Alanezi, A. A., & Ahmad, A. (2023). Amelioration of rheumatoid arthritis by *Fragaria nubicola* (Wild Strawberry) via attenuation of inflammatory mediators in Sprague Dawley rats. *Medicina, 59*(11), 1917. DOI: 10.3390/medicina59111917 PMID: 38003966

Masyita, A., Sari, R. M., Astuti, A. D., Yasir, B., Rumata, N. R., Emran, T. B., Nainu, F., & Simal-Gandara, J. (2022). Terpenes and terpenoids as main bioactive compounds of essential oils, their roles in human health and potential application as natural food preservatives. *Food Chemistry: X, 13*, 100217. DOI: 10.1016/j.fochx.2022.100217 PMID: 35498985

Melodelima, C., & Lobréaux, S. (2013). Complete *Arabis alpina* chloroplast genome sequence and insight into its polymorphism. *Meta Gene, 1*, 65–75. DOI: 10.1016/j.mgene.2013.10.004 PMID: 25606376

Memon, A. A., Memon, N., Luthria, D. L., Pitafi, A. A., & Bhanger, M. I. (2012). Phenolic compounds and seed oil composition of *Ziziphus mauritiana* L. fruit. *Polish Journal of Food and Nutrition Sciences, 62*(1), 15–21. DOI: 10.2478/v10222-011-0035-3

Men, R., Li, N., Ding, C., Tang, Y., Xing, Y., Ding, W., & Ma, Z. (2016). Chemopreventive agents from *Physalis minima* function as michael reaction acceptors. *Pharmacognosy Magazine, 12*(46, Suppl 2), S231. DOI: 10.4103/0973-1296.182153 PMID: 27279713

Mesmar, J., Abdallah, R., Badran, A., Maresca, M., Shaito, A., & Baydoun, E. (2022). *Ziziphus nummularia*: A comprehensive review of its phytochemical constituents and pharmacological properties. *Molecules (Basel, Switzerland), 27*(13), 4240. DOI: 10.3390/molecules27134240 PMID: 35807485

Mi, S., Wang, W., Zhu, M., Mu, L., & Lu, Q. (2025). Utilization of *Pinus pumila* seed scales as forestry residues: Green microwave-assisted hydrodistillation for essential oil extraction and its antibacterial activity. *Journal of Chromatography. A, 466574*. Advance online publication. DOI: 10.1016/j.chroma.2025.466574 PMID: 41344058

Mir, R. H., Mohi-Ud-Din, R., Al-Keridis, L. A., Ahmad, B., Alshammari, N., Patel, M., Adnan, M., & Masoodi, M. H. (2024). Phytochemical profiling, antioxidant, cytotoxic, and anti-inflammatory activities of *Plectranthus rugosus* extract and fractions: *In vitro, in vivo*, and *in silico* approaches. *Inflammopharmacology, 32*(2), 1593–1606. DOI: 10.1007/s10787-023-01419-2 PMID: 38308794

Mir, W. R., Bhat, B. A., Rather, M. A., Muzamil, S., Almilaibary, A., Alkhanani, M., & Mir, M. A. (2022). Molecular docking analysis and evaluation of the antimicrobial properties of the constituents of *Geranium wallichianum* D. Don ex Sweet from Kashmir Himalaya. *Scientific Reports, 12*(1), 12547. DOI: 10.1038/s41598-022-16102-9 PMID: 35869098

Mohammed, F. S., Uysal, I., Mustaffa, S., Eraslan, E. C., & Akgul, H. (2024). Analysis of phenolic contents and biological activities of wild mint, *Mentha longifolia* (L.) L. [IJEB]. *Indian Journal of Experimental Biology, 62*(03), 192–198. DOI: 10.56042/ijeb.v62i03.949

Molinu, M. G., Sanna, F., D'hallewin, G., Re, G. A., Sulas, L., & Piluzza, G. (2024). Mediterranean Wild Pear Fruits as a Neglected but Valuable Source of Phenolic Compounds. *Resources, 13*(6), 72. DOI: 10.3390/resources13060072

Mughees, M., Farooq, M. A., Haq, I. U., Zeb, I., Ali, M., Hussain, Z., Shahzadi, I., & Shah, M. M. (2022). Quantification of rosmarinic acid from different plant species of lower Himalayan region and expression analysis of underlying L-Phenylalanine pathway. *Physiologia Plantarum, 174*(5), e13758. DOI: 10.1111/ppl.13758 PMID: 36281843

Mukhtar, Y., Nazir, M., Naz, S., Taj, R., Akhtar, W., Abdul, L., Kamal, A., Alrefaei, A. F., Alshehri, E., & Lackner, M. (2025). Comparative assessment of physicochemical and biological properties of silver nanoparticles (AgNPs) synthesized from aerial and underground parts of *Arnebia benthamii. Kuwait Journal of Science, 100478*. Advance online publication. DOI: 10.1016/j.kjs.2025.100478

Munir, A., Youssef, F. S., Ishtiaq, S., Kamran, S. H., Sirwi, A., Ahmed, S. A., Ashour, M. L., & Elhady, S. S. (2021). *Malva parviflora* leaves mucilage: An eco-friendly and sustainable biopolymer with antioxidant properties. *Polymers, 13*(23), 4251. DOI: 10.3390/polym13234251 PMID: 34883753

Naseer, M., & Adil, M. (2025). Phytochemical profiling, HPLC analysis, and antimicrobial potential of *Curio radicans* (L. f.) PV Heath. *Scientific Reports, 15*(1), 34753. DOI: 10.1038/s41598-025-18323-0 PMID: 41053340

Nazir, N., Rahman, A., Uddin, F., Khan Khalil, A. A., Zahoor, M., Nisar, M., Ullah, S., Ullah, R., Ezzeldin, E., & Mostafa, G. A. (2021). Quantitative ethnomedicinal status and phytochemical analysis of *Berberis lyceum* royle. *Agronomy (Basel), 11*(1), 130. DOI: 10.3390/agronomy11010130

Nazir, N., Zahoor, M., Nisar, M., Khan, I., Karim, N., Abdel-Halim, H., & Ali, A. (2018). Phytochemical analysis and antidiabetic potential of *Elaeagnus umbellata* (Thunb.) in streptozotocin-induced diabetic rats: Pharmacological and computational approach. *BMC Complementary and Alternative Medicine, 18*(1), 332. DOI: 10.1186/s12906-018-2381-8 PMID: 30545352

Nguengang, R. T., Tchegnitegni, B. T., Ateba, J. E. T., Tabekoueng, G. B., Awantu, A. F., Bankeu, J. J. K., Chouna, J. R., Nkenfou, C. N., Sewald, N., & Lenta, B. N. (2023). Antibacterial constituents of *Rumex nepalensis* spreng and its emodin derivatives. *Natural Product Research, 37*(23), 3935–3946. DOI: 10.1080/14786419.2022.2162894 PMID: 36584290

Nguyen, T.-A.-M., Phan, H.-V.-T., Dao, N.-V.-T., Nguyen, T.-T.-V., Dong, P.-S.-N., Dang, L.-T.-C., Le, H.-T.-T., Nguyen, V.-K., & Hoang, L.-T.-T.-T. (2024). A systematic review on ethnobotany, phytochemistry, and pharmacology of *Ehretia acuminata* R. Br.(Boraginaceae), a medicinal plant. *Natural Product Research*, •••, 1–19. DOI: 10.1080/14786419.2024.2446704 PMID: 39731579

Nikolova, M. (2011). Screening of radical scavenging activity and polyphenol content of Bulgarian plant species. *Pharmacognosy Research, 3*(4), 256. DOI: 10.4103/0974-8490.89746 PMID: 22224049

Noroozi, J., Khalvati, S., Nafisi, H., Kaveh, A., Nazari, B., Zare, G., Minaei, M., Vitek, E., & Schneeweiss, G. M. (2021). Endemics determine bioregionalization in the alpine zone of the Irano-Anatolian biodiversity hotspot (South-West Asia). *Alpine Botany, 131*(2), 177–186. DOI: 10.1007/s00035-021-00266-7 PMID: 34721248

Omar, N., Ismail, C. A. N., & Long, I. (2022). Tannins in the treatment of diabetic neuropathic pain: Research progress and future challenges. *Frontiers in Pharmacology, 12*, 805854. DOI: 10.3389/fphar.2021.805854 PMID: 35082680

Orhan, I., Kartal, M., Abu-Asaker, M., Şenol, F. S., Yilmaz, G., & Şener, B. (2009). Free radical scavenging properties and phenolic characterization of some edible plants. *Food Chemistry, 114*(1), 276–281. DOI: 10.1016/j.foodchem.2008.09.005

Özdenefe, M. S., Kayış, F. B., Erol, Ü. H., & Takcı, A. M. (2024). Phytochemical screening and in vitro biological activity of *Amaranthus viridis* growing in Northern Cyprus. *International Journal of Secondary Metabolite*, *11*(3), 592–603. DOI: 10.21448/ijsm.1425060

Ozer, M. S., Sarikurkcu, C., Ceylan, O., Akdeniz, I., & Tepe, B. (2017). A comprehensive study on chemical composition, antioxidant and enzyme inhibition activities of the essential oils of *Chenopodium botrys* collected from three different parts of Turkey. *Industrial Crops and Products*, *107*, 326–331. DOI: 10.1016/j.indcrop.2017.06.001

Özkan, A. İ., Gültekin, D. C., Soruklu, S., Işık, U., & Akmeşe, O. (2025). Evaluation of antibacterial, antioxidant activities and secondary metabolites profile of orchids with different dried methods. *Journal of Research in Pharmacy*, *29*(6), 2349–2360. DOI: 10.12991/jrespharm.1797852

Pandey, K., & Pokhrel, S. (2022). Phytochemical screening and biological activities of *Oxalis corniculata* leaves of Sindhupalchowk District, Nepal. *BIBECHANA*, *19*(1-2), 27–33. DOI: 10.3126/bibechana.v19i1-2.46094

Park, S.-M., Jung, C.-J., Lee, D.-G., Yu, Y.-E., Ku, T.-H., Hong, M.-S., Lim, T.-K., Paeng, K.-I., Cho, H.-K., Cho, I.-J., & Ku, S.-K. (2024). *Elaeagnus umbellata* fruit extract protects skin from ultraviolet-mediated photoaging in hairless mice. *Antioxidants*, *13*(2), 195. DOI: 10.3390/antiox13020195 PMID: 38397793

Pärnänen, P., Niikko, S., Lähteenmäki, H., Räisänen, I. T., Tervahartiala, T., Sorsa, T., & Ranki, A. (2024). Lingonberry (*Vaccinium vitis-idaea* L.) Fruit Phenolic Bioactivities—A Review of *In Vitro* and *In Vivo* Human Studies. *Microorganisms*, *12*(9), 1850. DOI: 10.3390/microorganisms12091850 PMID: 39338523

Patel, T., Shah, K., Jiwan, K., & Shrivastava, N. (2011). Study on the antibacterial potential of *Physalis minima* Linn. *Indian Journal of Pharmaceutical Sciences*, *73*(1), 111. DOI: 10.4103/0250-474X.89770 PMID: 22131635

Patonay, K., Korozs, M., Muranyi, Z., & Konya, E. P. (2017). Polyphenols in northern Hungarian *Mentha longifolia* (L.) L. treatedwith ultrasonic extraction for potential oenological uses. *Turkish Journal of Agriculture and Forestry*, *41*(3), 208–217. DOI: 10.3906/tar-1701-61

Patra, J. K., Kim, E. S., Oh, K., Kim, H.-J., Kim, Y., & Baek, K.-H. (2014). Antibacterial effect of crude extract and metabolites of *Phytolacca americana* on pathogens responsible for periodontal inflammatory diseases and dental caries. *BMC Complementary and Alternative Medicine*, *14*(1), 343. DOI: 10.1186/1472-6882-14-343 PMID: 25241105

Pauloi, L., & Salim, R. M. (2025). Exploring the antimicrobial potential from various parts of *Ziziphus mauritiana* Lam. *Scientific Reports*, *15*(1), 36762. DOI: 10.1038/s41598-025-20756-6 PMID: 41120673

Petrovic, J., Stanojkovic, A., Comic, L., & Curcic, S. (2004). Antibacterial activity of *Cichorium intybus*. *Fitoterapia*, *75*(7-8), 737–739. DOI: 10.1016/j.fitote.2004.05.001 PMID: 15567253

Phuyal, N., Jha, P. K., Raturi, P. P., & Rajbhandary, S. (2019). *Zanthoxylum armatum* DC.: Current knowledge, gaps and opportunities in Nepal. *Journal of Ethnopharmacology*, *229*, 326–341. DOI: 10.1016/j.jep.2018.08.010 PMID: 30166217

Pizzi, A. (2021). Tannins medical/pharmacological and related applications: A critical review. *Sustainable Chemistry and Pharmacy*, *22*, 100481. DOI: 10.1016/j.scp.2021.100481

Politeo, O., Bektašević, M., Carev, I., Jurin, M., & Roje, M. (2018). Phytochemical composition, antioxidant potential and cholinesterase inhibition potential of extracts from *Mentha pulegium* L. *Chemistry & Biodiversity*, *15*(12), e1800374. DOI: 10.1002/cbdv.201800374 PMID: 30253061

Popović, Z., Vidaković, V., Mijalković, T., & Krstić-Milošević, D. (2023). Population-Related Variability in Qualitative and Quantitative Secondary Metabolite Profile of *Gentianella austriaca* (A. & J. Kern.) Holub. *Plants*, *12*(13), 2434. DOI: 10.3390/plants12132434 PMID: 37446995

Popovici, V., Bucur, L., Calcan, S. I., Cucolea, E. I., Costache, T., Rambu, D., Schröder, V., Gîrd, C. E., Gherghel, D., Vochita, G., Caraiane, A., & Badea, V. (2021). Elemental analysis and *in vitro* evaluation of antibacterial and antifungal activities of *Usnea barbata* (L.) Weber ex FH Wigg from Călimani Mountains, Romania. *Plants*, *11*(1), 32. DOI: 10.3390/plants11010032 PMID: 35009036

Prakash, O., Baskaran, R., & Kudachikar, V. (2019). Characterization, quantification of free, esterified and bound phenolics in Kainth (*Pyrus pashia* Buch.-Ham. Ex D. Don) fruit pulp by UPLC-ESI-HRMS/MS and evaluation of their antioxidant activity. *Food Chemistry*, *299*, 125114. DOI: 10.1016/j.foodchem.2019.125114 PMID: 31326758

Protti, M., Brighenti, V., Battaglia, M. R., Anceschi, L., Pellati, F., & Mercolini, L. (2019). Cannabinoids from *Cannabis sativa* L.: A new tool based on HPLC–DAD–MS/MS for a rational use in medicinal chemistry. *ACS Medicinal Chemistry Letters*, *10*(4), 539–544. DOI: 10.1021/acsmedchemlett.8b00571 PMID: 30996793

Qadir, S. A., Awlqadr, F. H., Qadir, M. H., Qadir, A. M., Faraj, A. M., Salih, S. A., Saeed, M. N., Arab, K., & Hosseini, S. M. N. (2025). *Adiantum capillus-veneris*: A Comprehensive Review of Its Medicinal Properties, Bioactive Compounds, and Advanced Extraction Techniques. *Food Science & Nutrition, 13*(10), e71118. DOI: 10.1002/fsn3.71118 PMID: 41132578

Rahmani, S., Naraki, K., Roohbakhsh, A., Hayes, A. W., & Karimi, G. (2023). The protective effects of rutin on the liver, kidneys, and heart by counteracting organ toxicity caused by synthetic and natural compounds. *Food Science & Nutrition, 11*(1), 39–56. DOI: 10.1002/fsn3.3041 PMID: 36655104

Raina, S., Sheikh, Z. N., Bakshi, P., Alamri, S., Siddiqui, M. H., Islam Rather, K. U., & Ashraf, K. (2025). Functional evaluation of cis-vaccenic acid in *Cordia dichotoma* with tissue-specific biochemical insights. *Scientific Reports, 15*(1), 37943. DOI: 10.1038/s41598-025-17906-1 PMID: 41168161

Raj, N. D., & Singh, D. (2022). A critical appraisal on ferulic acid: Biological profile, biopharmaceutical challenges and nano formulations. *Health Sciences Review (Oxford, England), 5*, 100063. DOI: 10.1016/j.hsr.2022.100063

Rajasekaran, S., Rajasekar, N., & Sivanantham, A. (2021). Therapeutic potential of plant-derived tannins in non-malignant respiratory diseases. *The Journal of Nutritional Biochemistry, 94*, 108632. DOI: 10.1016/j.jnutbio.2021.108632 PMID: 33794331

Ramar, M. K., Dhayanandamoorthy, Y., Ramachandran, S. S., & Kandasamy, R. (2020). HPLC-ESI-QqQ based standardization, mutagenic and genotoxic potential of methanol extract of *Ziziphus mauritiana* Lam leaves. *Journal of Ethnopharmacology, 246*, 112216. DOI: 10.1016/j.jep.2019.112216 PMID: 31493461

Rana, S., Rahman, S., Sana, S., Biswas, T. K., Hashem, A. K. M., Parvin, S., & Mazumder, K. (2020). Anticancer potential of *Chenopodium album* leaf extract against Ehrlich ascites carcinoma cells in Swiss albino mice. *Future Journal of Pharmaceutical Sciences, 6*(1), 65. DOI: 10.1186/s43094-020-00080-8

Rashid, W., Asim, S., Rashid, S., Rafique, S., Aziz, R. S., Iftikhar, S., Rashid, M., & Arfat, Y. (2023). Antioxidant and anti-mutagenic potential of mint (*Mentha arvensis*) and its chemical characterization by HPLC. *Pakistan Journal of Medical & Health Sciences, 17*(04), 514–514. DOI: 10.53350/pjmhs2023174514

Rather, L. J., & Mohammad, F. (2015). *Acacia nilotica* (L.): A review of its traditional uses, phytochemistry, and pharmacology. *Sustainable Chemistry and Pharmacy, 2*, 12–30. DOI: 10.1016/j.scp.2015.08.002

Ravi, S., Duraisamy, P., Krishnan, M., Martin, L. C., Manikandan, B., & Ramar, M. (2023). Sitosterol-rich *Digera muricata* against 7-ketocholesterol and lipopolysaccharide-mediated atherogenic responses by modulating NF-KB/iNOS signalling pathway in macrophages. *3 Biotech, 13*(10), 331. DOI: 10.1007/s13205-023-03741-6

Rawat, P., & Singh, M. (2024). Deciphering abiotic and biotic stress effects on secondary metabolites of medicinal plants. In *Forests and Climate Change: Biological Perspectives on Impact, Adaptation, and Mitigation Strategies* (pp. 19-41). Springer. DOI: 10.1007/978-981-97-3905-9_2

Rehman, A., Rehman, A., & Ahmad, I. (2015). Antibacterial, antifungal, and insecticidal potentials of *Oxalis corniculata* and its isolated compounds. *International Journal of Analytical Chemistry, 2015*(1), 842468. DOI: 10.1155/2015/842468 PMID: 25873973

Rezaieseresht, H., Shobeiri, S. S., & Kaskani, A. (2020). *Chenopodium botrys* essential oil as a source of sesquiterpenes to induce apoptosis and G1 cell cycle arrest in cervical cancer cells. *Iranian Journal of Pharmaceutical Research : IJPR, 19*(2), 341. DOI: 10.22037/ijpr.2019.1100671 PMID: 33224241

Ribeiro, J., Silva, V., Igrejas, G., Barros, L., Heleno, S. A., Reis, F. S., & Poeta, P. (2025). Phenolic Compounds from *Pyrus communis* Residues: Mechanisms of Antibacterial Action and Therapeutic Applications. *Antibiotics (Basel, Switzerland), 14*(3), 280. DOI: 10.3390/antibiotics14030280 PMID: 40149091

Rjeibi, I., Ben Saad, A., Sdayria, J., Feriani, A., Ncib, S., Allagui, M. S., Hfaiedh, N., & Souid, S. (2019). HPLC–DAD identification of polyphenols from ethyl acetate extract of *Amaranthus spinosus* leaves and determination of their antioxidant and antinociceptive effects. *Inflammopharmacology, 27*(5), 975–984. DOI: 10.1007/s10787-018-0482-0 PMID: 29679184

Roychoudhury, S., Sinha, B., Choudhury, B. P., Jha, N. K., Palit, P., Kundu, S., Mandal, S. C., Kolesarova, A., Yousef, M. I., Ruokolainen, J., Slama, P., & Kesari, K. K. (2021). Scavenging properties of plant-derived natural biomolecule para-coumaric acid in the prevention of oxidative stress-induced diseases. *Antioxidants, 10*(8), 1205. DOI: 10.3390/antiox10081205 PMID: 34439453

Ruiz-Molina, N., Ortega-Bedoya, I., & Arias-Zabala, M. (2019). Protonema Suspension Cultures of *Polytrichum juniperinum* as a Potential Production Platform for Bioactive Compounds. *Journal of Herbs, Spices & Medicinal Plants, 25*(2), 114–127. DOI: 10.1080/10496475.2019.1577321

Said, A., Naeem, N., Siraj, S., Khan, T., Javed, A., Rasheed, H. M., Sajjad, W., Shah, K., & Wahid, F. (2020). Mechanisms underlying the wound healing and tissue regeneration properties of *Chenopodium album. 3 Biotech, 10*(10), 452. DOI: 10.1007/s13205-020-02436-6

Saklani, S., Kawra, M., Chandra, S., & Coutinho, H. D. M. (2025). GC-MS Analysis, Phytochemical Profiling and Pharmacological Assessment of *Abies pindrow* from Uttarakhand. *Food and Humanity, 100863*, 100863. Advance online publication. DOI: 10.1016/j.foohum.2025.100863

Saleem, M., Shazmeen, N., Nazir, M., Riaz, N., Zengin, G., Ataullah, H. M., Nisar, F., Mukhtar, M., & Tousif, M. I. (2021). Investigation on the phytochemical composition, antioxidant and enzyme inhibition potential of *Polygonum plebeium* R. Br: A comprehensive approach to disclose new nutraceutical and functional food ingredients. *Chemistry & Biodiversity, 18*(12), e2100706. DOI: 10.1002/cbdv.202100706 PMID: 34636484

Saleh, I. A., BiBi, A., Bibi, S., Abu-Dieyeh, M., & Al-Ghouti, M. A. (2023). Waste-to-value: Guidelines for the potential applications of *Prosopis juliflora. Bioresource Technology Reports, 24*, 101678. DOI: 10.1016/j.biteb.2023.101678

Sana, S., Aftab, T., Naeem, M., Jha, P. K., & Prasad, P. (2025). Production of secondary metabolites under challenging environments: Understanding functions and mechanisms of signalling molecules. *Frontiers in Plant Science, 16*, 1569014. DOI: 10.3389/fpls.2025.1569014 PMID: 40860743

Selvaraj, P., De Britto, A. J., & Sahayaraj, K. (2005). Phytoecdysone of *Pteridium aquilinum* (L) Kuhn (Dennstaedtiaceae) and its pesticidal property on two major pests. *Archiv für Phytopathologie und Pflanzenschutz, 38*(2), 99–105. DOI: 10.1080/03235400400007517

Shadid, K. A., Shakya, A. K., Naik, R. R., Jaradat, N., Farah, H. S., Shalan, N., Khalaf, N. A., & Oriquat, G. A. (2021). Phenolic content and antioxidant and antimicrobial activities of *Malva sylvestris* L., *Malva oxyloba* Boiss., *Malva parviflora* L., and *Malva aegyptia* L. leaves extract. *Journal of Chemistry, 2021*(1), 8867400. DOI: 10.1155/2021/8867400

Shah, S. B., Sartaj, L., Hussain, S., Ullah, N., Idrees, M., Shaheen, A., Javed, M. S., & Aslam, M. K. (2020a). *In-vitro* evaluation of antimicrobial, antioxidant, alpha-amylase inhibition and cytotoxicity properties of *Cannabis sativa. Advances in Traditional Medicine, 20*(2), 181–187. DOI: 10.1007/s13596-019-00414-9

Shah, Z., Hassan, S., Shaheen, K., Khan, S. A., Gul, T., Anwar, Y., Al-Shaeri, M. A., Khan, M., Khan, R., Haleem, M. A., & Suo, H. (2020b). Synthesis of AgNPs coated with secondary metabolites of *Acacia nilotica*: An efficient antimicrobial and detoxification agent for environmental toxic organic pollutants. *Materials Science and Engineering C, 111*, 110829. DOI: 10.1016/j.msec.2020.110829 PMID: 32279826

Shahar, B., Dolma, N., Indira, A., & Chongtham, N. (2025). Nutritional and phytochemical characterization of *Allium carolinianum* Redoute, a wild garlic species from cold Desert, Ladakh. *Proceedings of the Indian National Science Academy*, 1-14. DOI: 10.1007/s43538-025-00606-9

Shameem, N., Kamili, A. N., Parray, J. A., Hamid, R., & Bandh, S. A. (2015). Antimicrobial and antioxidant activity of methanol extracts of *Arnebia benthamii* (Wall ex. G. Don) Johnston—A critically endangered medicinal plant of North western Himalaya. *Journal of Analytical Science and Technology*, *6*(1), 36. DOI: 10.1186/s40543-015-0076-z

Sharma, A., Chandel, V., & Roy, S. (2025). Exploring the bioactive properties and diverse uses of *Bauhinia variegata*: A comprehensive review. *Discover Plants*, *2*(1), 86. DOI: 10.1007/s44372-025-00167-7

Sharma, H., & Shrivastav, A. (2022). HPLC Purification of bio-active *Cleome viscosa* and It's Anticancer Activity against Breast Cancer Cell Lines. *Oriental Journal of Chemistry*, *38*(6), 1505–1513. Advance online publication. DOI: 10.13005/ojc/380624

Sharma, S., Rasal, V. P., Patil, P. A., & Joshi, R. K. (2018). *Mentha arvensis* essential oil suppressed airway changes induced by histamine and ovalbumin in experimental animals. *Natural Product Research*, *32*(4), 468–472. DOI: 10.1080/14786419.2017.1311891 PMID: 28385073

Shazdehahmadi, F., Pournajaf, A., Kazemi, S., & Ghasempour, M. (2023). Determining the antibacterial effect of *Mentha longifolia* essential oil on cariogenic bacteria and its compounds: An *in vitro* study. *Journal of Dentistry*, *24*(1, Suppl), 146. DOI: 10.30476/dentjods.2022.92992.1688 PMID: 37051493

Sickel, H., Bilger, W., & Ohlson, M. (2012). High levels of α-tocopherol in Norwegian alpine grazing plants. *Journal of Agricultural and Food Chemistry*, *60*(31), 7573–7580. DOI: 10.1021/jf301756j PMID: 22775115

Sieniawska, E., Baj, T., Dudka, J., Gieroba, R., Swiatek, L., Rajtar, B., Glowniak, K., & Polz-Dacewicz, M. (2013a). Cytotoxicity, antioxidant activity and an effect on CYP3A4 and CYP2D6 of *Mutellina purpurea* L. extracts. *Food and Chemical Toxicology*, *52*, 188–192. DOI: 10.1016/j.fct.2012.11.017 PMID: 23182739

Sieniawska, E., Baj, T., Los, R., Skalicka-Wozniak, K., Malm, A., & Glowniak, K. (2013b). Phenolic acids content, antioxidant and antimicrobial activity of *Ligusticum mutellina* L. *Natural Product Research, 27*(12), 1108–1110. DOI: 10.1080/14786419.2012.698413 PMID: 22724452

Sierra, K., Naranjo, L., Carrillo-Hormaza, L., Franco, G., & Osorio, E. (2022). Spearmint (*Mentha spicata* L.) phytochemical profile: Impact of pre/post-harvest processing and extractive recovery. *Molecules (Basel, Switzerland), 27*(7), 2243. DOI: 10.3390/molecules27072243 PMID: 35408640

Singh, M., Pandey, N., Agnihotri, V., Singh, K., & Pandey, A. (2017). Antioxidant, antimicrobial activity and bioactive compounds of *Bergenia ciliata* Sternb.: A valuable medicinal herb of Sikkim Himalaya. *Journal of Traditional and Complementary Medicine, 7*(2), 152–157. DOI: 10.1016/j.jtcme.2016.04.002 PMID: 28417084

Snoussi, M., Noumi, E., Trabelsi, N., Flamini, G., Papetti, A., & De Feo, V. (2015). *Mentha spicata* essential oil: Chemical composition, antioxidant and antibacterial activities against planktonic and biofilm cultures of Vibrio spp. strains. *Molecules (Basel, Switzerland), 20*(8), 14402–14424. DOI: 10.3390/molecules200814402 PMID: 26262604

Soloveva, M. I., & Kuzmina, S. S. (2023). Antioxidant and antimicrobial activity of extracts of *Cetraria islandica* (L.) Ach. and *Cladonia arbuscula* (Wallr.) Flot. *Acta Biologica Sibirica, 9*, 139–146-139–146. DOI: 10.5281/zenodo.7751287

Sousa, D. S. M., Ruiz, L. S., Pereira, A. F. M., Sani, A. A., Zapata, T. B., Monari, G. P. M., Rosa, C. F., Fernandes, A., Rall, V. L. M., & Prado, D. G. (2025). *Plantago major* leaf extract against bacteria and fungi of medical importance. *Brazilian Journal of Pharmaceutical Sciences, 61*, e24143. DOI: 10.1590/s2175-97902025e24143

Stanisavljević, D. M., Stojičević, S. S., Đorđević, S. M., Zlatković, B. P., Veličković, D. T., Karabegović, I. T., & Lazić, M. L. (2012). Antioxidant activity, the content of total phenols and flavonoids in the ethanol extracts of *Mentha longifolia* (L.) Hudson dried by the use of different techniques. *Chemical Industry & Chemical Engineering Quarterly, 18*(3), 411–420. DOI: 10.2298/CICEQ110919017S

Streb, P., Aubert, S., & Bligny, R. (2003). High temperature effects on light sensitivity in the two high mountain plant species *Soldanella alpina* (L.) and *Rannunculus glacialis* (L.). *Plant Biology, 5*(04), 432–440. DOI: 10.1055/s-2003-42713

Sugier, P., Sugier, D., Miazga-Karska, M., Nurzyńska, A., Król, B., Sęczyk, Ł., & Kowalski, R. (2025). Effect of Different *Arnica montana* L. Plant Parts on the Essential Oil Composition, Antimicrobial Activity, and Synergistic Interactions with Antibiotics. *Molecules (Basel, Switzerland), 30*(18), 3812. DOI: 10.3390/molecules30183812 PMID: 41011702

Sultana, K., Ahmad, B., Ahmad, Z., Rauf, A., Al-Awthan, Y. S., Bahattab, O., Bawazeer, S., Khattak, A., & Ramadan, M. F. (2022). Effect of *Bistorta affinis* (D. Don) and *Malcolmia cabulica* (Boiss) essential oils on human and foodborne pathogens, insects and termites. *Journal of Pure & Applied Microbiology*, *5*(4), 14–21. DOI: 10.22207/JPAM.16.4.45

Tabassam, S., Anwar, M. A., Gulfraz, M., Arshad, M., Sabitaliyevich, U. Y., Nurmurzayevich, S. B., & Ahmad, M. S. (2019). Bioactivity evaluation and HPLC UV-VIS based quantification of antioxidant secondary metabolites from extract and fractions of *Bistorta amplexicaulis* rhizome. *Cellular and Molecular Biology, 65*(1), 19–26. DOI: 10.14715/cmb/2019.65.1.4 PMID: 30782289

Tabesh, M., Sh, M. E., Etemadi, M., Naddaf, F., Heidari, F., & Alizargar, J. (2022). The antibacterial activity of *Nasturtium officinale* extract on common oral pathogenic bacteria. *Nigerian Journal of Clinical Practice*, *25*(9), 1466–1475. DOI: 10.4103/njcp.njcp_1887_21 PMID: 36149206

Tapalsky, D., Petrenev, D., Khramchenkova, O., & Doroshkevich, A. (2017). Antimicrobial and antifungal activity of lichens prevalent in Belarus. *Journal of microbiology, epidemiology and immunobiology, 94*(2), 60-65. DOI: 10.36233/0372-9311-2017-2-60-65

Testolin, R., Carmona, C. P., Attorre, F., Borchardt, P., Bruelheide, H., Dolezal, J., Finckh, M., Haider, S., Hemp, A., Jandt, U., Korolyuk, A. Y., Lenoir, J., Makunina, N., Malanson, G. P., Mucina, L., Noroozi, J., Nowak, A., Peet, R. K., Peyre, G., & Jiménez-Alfaro, B. (2021). Global functional variation in alpine vegetation. *Journal of Vegetation Science*, *32*(2), e13000. DOI: 10.1111/jvs.13000

Tewari, D., Gupta, P., Bawari, S., Sah, A. N., Barreca, D., Khayatkashani, M., & Khayat Kashani, H. R. (2021a). Himalayan *Ficus palmata* L. Fruit extract showed in vivo central and peripheral analgesic activity involving COX-2 and Mu opioid receptors. *Plants*, *10*(8), 1685. DOI: 10.3390/plants10081685 PMID: 34451731

Tewari, D., Zengin, G., Ak, G., Sinan, K. I., Cziáky, Z., Mishra, S. T., & Jekő, J. (2021b). Phenolic profiling, antioxidants, multivariate, and enzyme inhibitory properties of wild Himalayan Fig (*Ficus palmata* Forssk.): A potential candidate for designing innovative nutraceuticals and related products. *Analytical Letters*, *54*(9), 1439–1456. DOI: 10.1080/00032719.2020.1804395

Thakur, P., Kumar, R., Choudhary, N., Sharma, R., & Chaudhary, A. (2023). Network pharmacology on mechanistic role of *Thymus linearis* Benth. against gastrointestinal and neurological diseases. *Phytomedicine*, *121*, 155098. DOI: 10.1016/j.phymed.2023.155098 PMID: 37757710

Thakur, R., Singh, R., Saxena, P., & Mani, A. (2014). Evaluation of antibacterial activity of *Prosopis juliflora* (SW.) DC. leaves. *African Journal of Traditional, Complementary, and Alternative Medicines : AJTCAM*, *11*(3), 182–188. DOI: 10.4314/ajtcam.v11i3.26 PMID: 25371581

Tomczyk, M., & Latté, K. P. (2009). Potentilla—A review of its phytochemical and pharmacological profile. *Journal of Ethnopharmacology*, *122*(2), 184–204. DOI: 10.1016/j.jep.2008.12.022 PMID: 19162156

Turan, K., Akcan, M., & Kalfa, O. M. (2021). Preparation, Characterization and Antibacterial Activity of *Malva sylvestris* L. Seed Extract Containing Novel Electrospun PVA Nanofibers. *Gazi University Journal of Science*, *34*(4), 987–998. DOI: 10.35378/gujs.794638

Tvrda, E., Straka, P., Galbavy, D., & Ivanic, P. (2019). Epicatechin provides antioxidant protection to bovine spermatozoa subjected to induced oxidative stress. *Molecules (Basel, Switzerland)*, *24*(18), 3226. DOI: 10.3390/molecules24183226 PMID: 31491847

Uba, A. I., Zengin, G., Montesano, D., Cakilcioglu, U., Selvi, S., Ulusan, M. D., Caprioli, G., Sagratini, G., Angeloni, S., Jugreet, S., Hasan, M. M., & Mahoomodally, M. F. (2022). Antioxidant and enzyme inhibitory properties, and HPLC–MS/MS profiles of different extracts of *Arabis carduchorum* Boiss.: An endemic plant to Turkey. *Applied Sciences (Basel, Switzerland)*, *12*(13), 6561. DOI: 10.3390/app12136561

Ullah, A., Munir, S., Badshah, S. L., Khan, N., Ghani, L., Poulson, B. G., Emwas, A.-H., & Jaremko, M. (2020). Important flavonoids and their role as a therapeutic agent. *Molecules (Basel, Switzerland)*, *25*(22), 5243. DOI: 10.3390/molecules25225243 PMID: 33187049

Ullah, H., & Badshah, L. (2023). Nutritional and mineral analysis of the ultimate wild food plants of Lotkuh, Chitral, the Eastern Hindukush Pakistan. *Heliyon*, *9*(3), e14449. Advance online publication. DOI: 10.1016/j.heliyon.2023.e14449 PMID: 36950657

Ullah, M. A., Gul, F. Z., Khan, T., Bajwa, M. N., Drouet, S., Tungmunnithum, D., Giglioli-Guivarc'h, N., Liu, C., Hano, C., & Abbasi, B. H. (2021). Differential induction of antioxidant and anti-inflammatory phytochemicals in agitated micro-shoot cultures of *Ajuga integrifolia* Buch. Ham. ex D. Don with biotic elicitors. *AMB Express*, *11*(1), 137. DOI: 10.1186/s13568-021-01297-3 PMID: 34661766

Vanneste, T., Michelsen, O., Graae, B. J., Kyrkjeeide, M. O., Holien, H., Hassel, K., Lindmo, S., Kapás, R. E., & De Frenne, P. (2017). Impact of climate change on alpine vegetation of mountain summits in Norway. *Ecological Research*, *32*(4), 579–593. DOI: 10.1007/s11284-017-1472-1

Varshney, N., Kumari, D., Janmeda, P., Meena, M., & Singh, D. (2025). *Salvadora oleoides* (Decne.): A systematic review and comprehensive botanical assessment. *Discover Plants*, *2*(1), 140. DOI: 10.1007/s44372-025-00150-2

Venditti, A., Serafini, M., Nicoletti, M., & Bianco, A. (2015). Terpenoids of *Linaria alpina* (L.) Mill. from Dolomites, Italy. *Natural Product Research*, *29*(21), 2041–2044. DOI: 10.1080/14786419.2015.1018252 PMID: 25738749

Verma, P., Dangar, R., Shah, K., Gandhi, D. M., & Suhagia, B. (2011). Pharmacognostical Potential of *Capparis decidua* Edgew. *Journal of Applied Pharmaceutical Science*(Issue), 06-11.

Verrall, B., & Pickering, C. M. (2020). Alpine vegetation in the context of climate change: A global review of past research and future directions. *The Science of the Total Environment*, *748*, 141344. DOI: 10.1016/j.scitotenv.2020.141344 PMID: 32814293

Vishwakarma, R. K., Negi, A., & Negi, D. S. (2022). Anti-oxidant, COX inhibition activity of different extract of bark of *Betula utilis* and molecular docking analysis of its phytochemicals against COX-1 and COX-2 isoenzyme. *Materials Today: Proceedings*, *57*, 251–258. DOI: 10.1016/j.matpr.2022.02.497

Vitalini, S., Iriti, M., Puricelli, C., Ciuchi, D., Segale, A., & Fico, G. (2013). Traditional knowledge on medicinal and food plants used in Val San Giacomo (Sondrio, Italy)—An alpine ethnobotanical study. *Journal of Ethnopharmacology*, *145*(2), 517–529. DOI: 10.1016/j.jep.2012.11.024 PMID: 23220197

Vohra, A., & Kaur, H. (2011). Chemical investigation of medicinal plant *Ajuga bracteosa. J Nat Prod Plant Resour, 1*(1), 37–45.

Vučić, D., Čomić, L., Petković, M., & Stefanović, O. (2009). Antibacterial effects of leaves of *Vaccinium vitis-idaea* L. *Planta Medica, 75*(09), PJ58. DOI: 10.1055/s-0029-1234863

Vysochina, G., & Khramova, E. (2010). Component composition of flavonols and their content in *Aconogonon alpinum* (All.) Schur Growing in the Altay. *Chem Sustain Dev, 18*, 477–483.

Waheed, M., Haq, S. M., Arshad, F., Bussmann, R. W., Pieroni, A., Mahmoud, E. A., Casini, R., Yessoufou, K., & Elansary, H. O. (2023). Traditional wild food plants gathered by ethnic groups living in semi-arid region of Punjab, Pakistan. *Biology (Basel), 12*(2), 269. DOI: 10.3390/biology12020269 PMID: 36829546

Wang, J., Zhao, H., Guo, Q., & Ding, H. (2022). Identification and antibacterial activity of *Thamnolia vermicularis* and *Thamnolia subuliformis. Journal of Microbiological Methods, 203*, 106628. DOI: 10.1016/j.mimet.2022.106628 PMID: 36442696

Wang, L., Pan, X., Jiang, L., Chu, Y., Gao, S., Jiang, X., Zhang, Y., Chen, Y., Luo, S., & Peng, C. (2022). The biological activity mechanism of chlorogenic acid and its applications in food industry: A review. *Frontiers in Nutrition, 9*, 943911. DOI: 10.3389/fnut.2022.943911 PMID: 35845802

Wang, Z., Barrow, C. J., Dunshea, F. R., & Suleria, H. A. (2021). A comparative investigation on phenolic composition, characterization and antioxidant potentials of five different australian grown pear varieties. *Antioxidants, 10*(2), 151. DOI: 10.3390/antiox10020151 PMID: 33498549

Weston, P. A., Gurusinghe, S., Birckhead, E., Skoneczny, D., Quinn, J. C., & Weston, L. A. (2019). Chemometric analysis of *Amaranthus retroflexus* in relation to livestock toxicity in southern Australia. *Phytochemistry, 161*, 1–10. DOI: 10.1016/j.phytochem.2019.01.016 PMID: 30776591

Wróblewska-Łuczka, P., Cabaj, J., Bargieł, J., & Łuszczki, J. J. (2023). Anticancer effect of terpenes: Focus on malignant melanoma. *Pharmacological Reports, 75*(5), 1115–1125. DOI: 10.1007/s43440-023-00512-1 PMID: 37515699

Yadav, E., Singh, D., Yadav, P., & Verma, A. (2018). Antioxidant and anti-inflammatory properties of *Prosopis cineraria* based phenolic rich ointment in wound healing. *Biomedicine and Pharmacotherapy, 108*, 1572–1583. DOI: 10.1016/j.biopha.2018.09.180 PMID: 30372859

Yaglioglu, A. S., Yaglioglu, M. S., Tosyalıoglu, N., Adem, S., & Demirtas, I. (2022). Chemical profiling, in vitro biological activities and Pearson correlation between chemical profiling and anticancer activities of four Abies species from Turkey. *South African Journal of Botany*, *151*, 600–613. DOI: 10.1016/j.sajb.2022.08.005

Yang, Y., Zhang, Y., Zhang, B., Bao, Y., Xu, S., Tang, X., Zhao, Q., Li, J., & Li, R. (2023). *In vitro* antioxidative activity of *Fritillaria cirrhosa* D. Don straw ethanolic extract and its effect on lipid, protein oxidation, and quality of Chinese-style sausage. *Journal of Food Science*, *88*(11), 4745–4772. DOI: 10.1111/1750-3841.16757 PMID: 37751083

Yasir, M., Sultana, B., & Anwar, F. (2018). LC–ESI–MS/MS based characterization of phenolic components in fruits of two species of *Solanaceae*. *Journal of Food Science and Technology*, *55*(7), 2370–2376. DOI: 10.1007/s13197-017-2702-9 PMID: 30042551

Yatoo, M. I., Dimri, U., Gopalakrishnan, A., Saxena, A., Wani, S. A., & Dhama, K. (2018). *In vitro* and *in vivo* immunomodulatory potential of *Pedicularis longiflora* and *Allium carolinianum* in alloxan-induced diabetes in rats. *Biomedicine and Pharmacotherapy*, *97*, 375–384. DOI: 10.1016/j.biopha.2017.10.133 PMID: 29091887

Yazar, M., Sevindik, M., Uysal, I., & Polat, A. O. (2025). Effects of Caffeic Acid on Human Health: Pharmacological and Therapeutic Effects, Biological Activity and Toxicity. *Pharmaceutical Chemistry Journal*, *59*(1), 1–7. DOI: 10.1007/s11094-025-03363-7

Yazhiniprabha, M., Banu, S., Ishwarya, R., Vinotha, V., Govindarajan, M., Wadaan, M. A., Mahboob, S., Nicoletti, M., & Vaseeharan, B. (2022). Biomimetically synthesized *Physalis minima* fruit extract-based zinc oxide nanoparticles as eco-friendly biomaterials for biological applications. *Journal of Drug Delivery Science and Technology*, *73*, 103475. DOI: 10.1016/j.jddst.2022.103475

Yuce, K., & Ozkan, A. I. (2020). Cys2His2 zinc finger proteins boost survival ability of plants against stress conditions. In *Plant stress physiology*. IntechOpen., DOI: 10.5772/intechopen.92590

Yunusa, U. M., & Urek, R. O. (2023). Determination of phenolic compounds in *Diospyros lotus* by RP-HPLC-DAD and evaluation of antioxidant and cytotoxic properties. *Journal of Food Measurement and Characterization*, *17*(2), 1413–1427. DOI: 10.1007/s11694-022-01696-5

Zaman, S., Al-Joufi, F. A., Zafar, M., & Zahoor, M. (2022). Phytochemical, Antimicrobial and Cytotoxic Activities of *Gaultheria trichophylla* Royle. *Applied Sciences (Basel, Switzerland)*, *12*(14), 6921. DOI: 10.3390/app12146921

Zambare, V. P., & Christopher, L. P. (2012). Biopharmaceutical potential of lichens. *Pharmaceutical Biology*, *50*(6), 778–798. DOI: 10.3109/13880209.2011.633089 PMID: 22471936

Zhakipbekov, K., Turgumbayeva, A., Issayeva, R., Kipchakbayeva, A., Kadyrbayeva, G., Tleubayeva, M., Akhayeva, T., Tastambek, K., Sainova, G., Serikbayeva, E., Tolenova, K., Makhatova, B., Anarbayeva, R., Shimirova, Z., & Tileuberdi, Y. (2023). Antimicrobial and other biomedical properties of extracts from *Plantago major*, Plantaginaceae. *Pharmaceuticals (Basel, Switzerland)*, *16*(8), 1092. DOI: 10.3390/ph16081092 PMID: 37631007

Zhang, B., Sun, X., Tang, W., Jiang, L., Liu, J., Wang, Y., Li, Y., Li, Y., & Ma, X. (2025). An effective high-performance liquid chromatography method for fingerprint and multi-component quantification of *Oxalis corniculata*. *Current Pharmaceutical Analysis*, *21*(9), 434–438. Advance online publication. DOI: 10.1016/j.cpan.2025.09.005

Zhao, Z.-H., Zhang, H.-J., Cao, X.-J., Wu, B., Ge, Y.-C., & Wang, K.-W. (2025). Preparation, characterization and biological activity of *Pteridium aquilinum* polysaccharide-Cu (II)/Zn (II) complexes in vitro. *Food Bioscience*, *107355*, 107355. Advance online publication. DOI: 10.1016/j.fbio.2025.107355

Zhou, C., Dai, S., Yin, L., Zhuo, Y., Zhao, N., Li, D., & Sun, C. (2025). Antibacterial Activity and Mechanism of *Pinus pumila* (Pall.) Pinecone Essential Oil Against *Escherichia coli*. *Waste and Biomass Valorization*, *16*(9), 1–16. DOI: 10.1007/s12649-025-02961-6

Zhuo, Z., Xu, D., Li, Y., Pu, B., & Ye, M. (2021). Fingerprint analysis of *Zanthoxylum armatum* DC. by HPLC. *Journal of Food Composition and Analysis*, *96*, 103736. DOI: 10.1016/j.jfca.2020.103736

Zia-Ul-Haq, M., Ćavar, S., Qayum, M., Imran, I., & de Feo, V. (2011). Compositional studies: Antioxidant and antidiabetic activities of *Capparis decidua* (Forsk.) Edgew. *International Journal of Molecular Sciences*, *12*(12), 8846–8861. DOI: 10.3390/ijms12128846 PMID: 22272107

Žitek, T., Postružnik, V., Knez, Ž., Golle, A., Dariš, B., & Knez Marevci, M.Supercritical Extraction Optimization for Antibiotic and Anticancer Activity. (2022). *Arnica montana* L. supercritical extraction optimization for antibiotic and anticancer activity. *Frontiers in Bioengineering and Biotechnology*, *10*, 897185. DOI: 10.3389/fbioe.2022.897185 PMID: 35620474

Chapter 12

Alpine Biodiversity at the Crossroads:
Medicinal Plants, Threats, and Nanoparticle-Based Intervention Strategies

Sajid Ali Khan Bangash
http://orcid.org/0000-0002-6936-8060
The University of Agriculture, Peshawar, Pakistan

ABSTRACT

Alpine countries possess a distinctive and varied collection of medicinal plants due to harsh environmental factors and centuries-old folk traditions. In recent decades, these species have become endangered because of climate change, habitats destruction and unsustainable harvesting practices. This chapter reviews the ecological value and pharmacological potential of major alpine medicinal plants, outlining the main threats to their survival and the resulting consequences for ecosystems and local populations. It then critically evaluates the potential role of emerging nanoparticle-based intervention strategies, such as targeted delivery systems and plant resilience enhancement, in conservation and sustainable use. By integrating ecological, ethnobotanical, and nano-biotechnological perspectives, the chapter shows how an interdisciplinary approach can help protect alpine medicinal biodiversity and support the development of sustainable solutions for future therapeutic applications.

DOI: 10.4018/979-8-2600-1207-9.ch012

1. INTRODUCTION

The amazing variety of plants and animals adjusted to high-altitude ecosystems with extremely harsh climatic conditions, such as low temperatures, intense solar radiation, and bare soils may be defined as Alpine biodiversity. These regions are ecologically important for maintaining environmental balance, regulating hydrological cycles, and serving as and serving of genetic and medicinal materials (Patra et al., 2018). Flora in the Alps, in particular, have received significant attention due to their unique biochemical adaptations, which have resulted in the accumulation of secondary metabolites with high therapeutic value.

According to the World Health Organization (WHO), approximately 80% of the population in developing nations uses alternative medicine to meet its primary healthcare requirements (Robinson & Zhang, 2011). This widespread reliance underscores the importance of medicinal plants as a universal health resource. In recent decades, there has been a surge of scientific interest in the identification, isolation, and characterization of bioactive compounds, chemical compositions, and pharmacologically active components of different plant species, with the aim of developing new bioactive ingredients that have fewer side effects compared to synthetic drugs (Atanasov et al., 2015; Beutler, 2009). Plants have been extensively used as natural treasuries of medicinal products and still hold great potential for the discovery of new and more effective medicines.

Nevertheless, identifying bioactive molecules from natural sources is not always easy. Environmental stress can substantially alter the production and composition of plant metabolites, making natural drug discovery less consistent and reproducible. As a result the pharmaceutical industry has often focused on producing synthetic compounds as an alternative (Patra et al., 2018). However, the rate at which synthetic drugs are successfully brought to market has been gradually decreasing, leading researchers to return to natural product-based compounds as an alternative (Heinrich, 2010; Namdari et al., 2017). A large proportion of the most important therapeutic agents used in medicine today originate from higher plants (Atanasov et al., 2015).

For example, Artemotil (*Artemisia annua* L.) is used to treat malaria, Reminyl (*Galanthus woronowii* Losinsk.) is used to treat Alzheimer's disease; paclitaxel (*Taxus brevifolia* Nutt.), vinblastine, and vincristine (*Catharanthus roseus* (L.) G.Don), as well as silymarin (*Silybum marianum* (L.) Gaertn.), are used to treat liver diseases. These plants have which had a long history of medicinal use (Atanasov et al., 2015).

The pharmacological actions and biochemical structures of various plant-derived molecules have been widely studied. Among the most biologically active are alkaloids, flavonoids, tannins, terpenes, saponins, steroids, and phenolic compounds (Yuan et al., 2016). However, a major limitation of these compounds is their typically low absorption and bioavailability, due to their large molecular weight and inability to pass through lipid membranes. Besides, these compounds often have a high systemic clearance rate, requiring more frequent dosing and reducing the overall therapeutic effect (Namdari et al., 2017).

Nanotechnology has introduced groundbreaking solutions to these longstanding challenges when applied to pharmacological studies (Bonifácio et al., 2014; Namdari et al., 2017). Nanotechnology offers the ability to increase the solubility, stability, and targeted delivery of bioactive molecules, thereby improving their biological functions. Nanotechnology in medicine has been utilized, and research on its applications has become widespread in recent years (Duncan & Gaspar, 2011; Patra et al., 2016). Nanoscale systems enable the incorporation of various compounds into a single formulation and allow modulation of the performance and interactions of molecules within living systems. Nanocarriers can deliver drugs to their target sites, improve bioavailability, extend the duration of therapeutic action, and minimize side effects by controlling the release process.

Moreover, since natural compounds can be attached to nanocarrier systems along with other compounds, they can assist in treating diseases that are less responsive to standard therapy and provide a competitive advantage in managing less susceptible diseases (Bonifácio et al., 2014). In addition, as natural compounds are combined with nanocarrier systems, they may help alleviate drug resistance and offer a strategic edge in treating diseases less responsive to conventional therapy (Bonifácio et al., 2014; Namdari et al., 2017). Natural-based materials used in medical applications as natural products can broadly be categorized into two groups: (1) formulations directed to the specific locations to treat particular diseases (Ramana et al., 2014; Yadav et al., 2012), and (2) biomaterials or plant extracts directly used to create nanoparticles (Wei, 2013). Cancer therapy is one of the most extensively researched medical uses of nanotechnology, as cancer remains a leading cause of death worldwide (Patra et al., 2018). Systems developed through nanotechnology provide new avenues for targeted drug delivery, providing increased selectivity and minimal damage to healthy cells in cancerous tissues. Ongoing research is also being applied to other diseases using these methods, further demonstrating the wide scope of nanomedicine (Ambesh et al., 2017; Yohan & Chithrani, 2014).

The functional efficiency of nanocarriers is determined by several major parameters, including surface charge, particle size, particle shape, dispersion stability, encapsulation capacity and biological compatibility (Obeid et al., 2017). These properties influence how nanomaterials are applied in specific biomedical uses. Nanopharmaceutics, which includes nanodiagnostics, nanobiomaterials, and nanotheranostics) is a rapidly evolving area that brings new opportunities for drug delivery, imaging, diagnostics, and combined therapy (Patra et al., 2018). In this context, the biofabrication of nanoparticles using plant extracts (*Cucurbita pepo*) can be considered an environmentally friendly and sustainable approach that integrates traditional medicine with current nanotechnology. This type of integration is highly promising not only for conserving alpine medicinal biodiversity but also for developing new therapeutic applications that address emerging biomedical needs (Gul et al., 2025).

2. THE PHARMACOLOGICAL SIGNIFICANCE OF ALPINE FLORA

2.1 The Unique Value of Alpine Medicinal Flora

The mountain ecosystems are considered the world repositories of biodiversity and support a large number of species adapted to the harsh environments. The Hindu Kush-Karakoram-Himalaya (HKH) region is particularly noteworthy particularly because it is among the most important biodiversity hotspots in the world. This vast mountain range is estimated to contain 25,000 species of flora, or 10% of the world's total. Pakistan's rocky regions are especially rich, with more than 80% of all endemic plant species in the country found there.

This is remarkable biodiversity with significant implications for human well-being. For millennia, mountain folks have developed complex traditional medicine systems based on local flora. This high level of dependence continues today, with an estimated 80% in population of the developing countries still using medicinal plants as their primary source of healthcare. Asteraceae, Lamiaceae, Rosaceae, and Fabaceae are the most common plant families in the HKH region that provide remedies for various health conditions, including gastrointestinal, respiratory, skin, and bone disorders (Ismail et al., 2022).

This remarkable biodiversity has far reaching consequences for human well-being. For millennia, mountain communities have developed elaborate traditional medicine systems based on local flora. Such strong dependence still exists today, with an estimated 80% of the population in developing nations relying on medicinal plants as their primary healthcare providers (Ismail et al., 2022). The predominant plant families in the HKH area, including Asteraceae, Lamiaceae, Rosaceae, and Fabaceae, provide remedies for various health issues, such as gastrointestinal, respiratory, skin, and bone conditions (Ismail et al., 2022). The traditional knowledge, for example, demonstrates the use of plants such as *Betula utilis* D.Don for digestive problems and *Artemisia* species for diabetes, confirming the intricate connection between alpine biodiversity and community well-being.

2.2 Biochemical and Medicinal Compounds in Alpine Plants

Uniqueness of the phytochemical profiles of alpine plants results from adaptation to severe abiotic conditions and co-evolutionary pressures. This has led to the generation of a robust array of secondary metabolites with highly potent medicinal effects. An interesting example is the alpine areas of the Japanese Alps, where the primary food source for the endangered Japanese rock ptarmigan (*Lagopus muta*) is alpine vegetation. Recently, a study by Haraguchi et al. (2024) tested the hypothesis that these plants could offer a natural defense against parasitic infections in these birds. The researchers extracted 27 different parts from eight alpine species, such as *Empetrum nigrum* var. *japonicum* R.D.Good, *Loiseleuria procumbens* (L.) Loisel., and *Vaccinium ovalifolium* Sm.. *in vitro* tests revealed that six of these components had high anticoccidial activity, with those from *L. procumbens* and *V. ovalifolium* producing the greatest reduction in parasite viability, reaching 88.3%. Subsequent analysis identified pentacyclic triterpenoids as the active compounds, including ursolic acid. The paper gives a strong illustration of alpine flora as a natural pharmacy and highlights the use of secondary metabolites that protect the plants and give pharmacological value to animals that ingest them, demonstrating the profound ecological and pharmacological importance of alpine biodiversity.

3.1 Current and Emerging Threats to Alpine Biodiversity

Although ecologically and medicinally significant, delicate alpine ecosystems such as the HKH are at risk as they face unprecedented threats from climate change and direct human stresses. These pressures are rapidly accelerating biodiversity loss and threaten the long-term sustainability of these important natural resources. The major threats impacting these ecosystems, which will be detailed in this section, are summarized in the conceptual framework in Figure 1. This framework illustrates the major climatic and anthropogenic threats to alpine medicinal plant biodiversity. The diagram highlights the direct impacts of climate change (e.g., glacial retreat) and human pressures (e.g. overharvesting, land-use change) that combine to accelerate habitat degradation and species loss.

Figure 1. Conceptual diagram of major threats and challenges

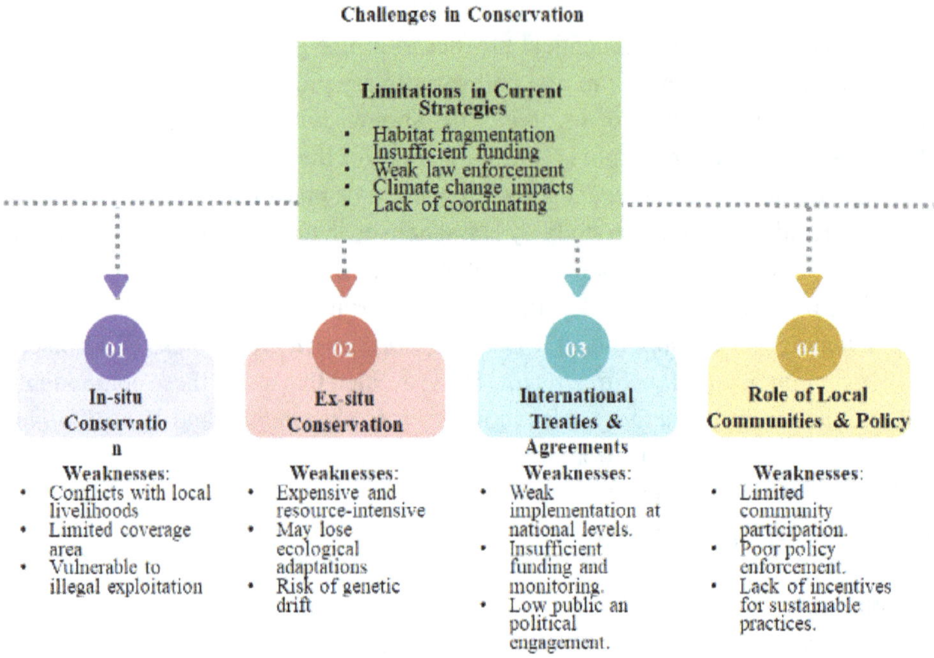

3.1.1 Glacial Retreat and Habitat Alteration

Glaciers are among the most sensitive indicators of climate change (Armstrong, 2010; Zemp, 2008). The temperature and precipitation changes affect their dynamics, with glacial length change being a slow response to climatic shifts, while mass balance provides an immediate and direct response (Haeberli & Hoelzle, 1995). In the Himalayan region glaciers have been in constant recession since the end of the Little Ice Age in about 1850 AD (Mayewski & Jeschke, 1979). This recession exposes new landscapes that enhance soil formation and primary plant succession (Haugland, 2006). The process creates a natural chronosequence, in which the distance from the glacier corresponds to the age of the terrain and the phase of ecological succession (Matthews, 1992). Even though glacier recession provides short-term habitats in where alpine vegetation can colonize new regions, long-term warming, which causes glacier recession, leads to the degradation of favorable alpine environments, threatening the flora and distribution of plant life. Factors such as temperature, moisture level, duration of exposure, and topography determine soil development in these areas (Darmody et al., 2005; Hall & Walton, 1992). Successional stages of vegetation in deglaciated terrain have been examined intensively, especially in the general case, at Glacier Bay, Alaska, where Cooper initially conducted studies (Cooper, 1923a, 1923b, 1923c; Cooper, 1931, 1939) three successional states are the pioneer community, willow- alder thicket, and spruce forest. The change between these vegetation types was strongly correlated with the accumulation of nutrients in the soil. For example, *Alnus crispa* (Aiton) Pursh (Sitka alder) made the soil much richer in nitrogen relevant and more acidic, which enabled the growth of *Picea sitchensis* (Bong.) Carrière (Sitka spruce).

Distribution of vegetation in such forelands is determined by nutrient availability, topographic differences, temperature, and length of the growing season (Kleidon & Mooney, 2000). It was observed at East Brgger Glacier in the High Arctic that early colonization occurred by scattered bryophytes and vascular plants on freshly-deglaciated moraines (Minami et al., 1996), while older sites featured well-developed vascular and bryophyte communities (Cannone et al., 2004). In the same case, plant succession in the Alexandra Fiord region of Canada occurred over about forty years in the following stages: (1) mosses, (2) graminoid-forb, (3) deciduous shrub-moss, and (4) evergreen dwarf-shrub-moss (Jones & Henry, 2003). All these studies have shown the effects of glacier retreat on soil development, vegetation succession, and eventually ecosystem structure and biodiversity in alpine environments.

3.1.2 Altered Precipitation Patterns

Alterations in precipitation patterns are another important aspect of climate change that impacts alpine ecosystems. The differences in snow and rain affect soil moisture, nutrient cycling, and the length of the growing season. Irregular precipitation patterns may lead to either drought stress or excess moisture, both of which alter plant community structure and productivity. The variable snowpacks also influence seed germination and early plant establishment, especially for alpine species accustomed to snow-covered winters.

Phenology Shifts (Flowering–Pollinator Mismatch)

Something is highly sensitive to temperature and the timing of snowmelt, as both are determinants of the short growing season and reproductive cycles in alpine ecosystems. Global studies across the Andes (Chile) (Arroyo et al., 1981), Norway (Totland, 1993), Japan (Kameyama & Kudo, 2009; Kudo, 1996, 2016), Greece (Makrodimos et al., 2008), China (Meng et al., 2016; Meng et al., 2017; Zhou et al., 2016), Switzerland (Studer et al., 2005), India (Pangtey et al., 1990; Vashistha et al., 2009), Australia (Inouye & Pyke, 1988; Venn & Morgan, 2007), the High Arctic (Semenchuk et al., 2016), and North America (Inouye, 2020) have shown that alpine flowering phenology is advancing under warming conditions. Historical herbarium records further support these trends (Gallagher et al., 2009).

Snowmelt is the main trigger for flowering, as alpine herbs typically begin to grow only after the snow has melted (Inouye, 2020). Rising temperatures have led to earlier snowmelt, which in turn causes earlier flowering and extends the growing periods (Inouye & Wielgolaski, 2025). However, in some tundra ecosystems, the duration of flowering at the community level has shortened, with late flowering species progressing more than early flowering species (Prevéy et al., 2019). These changes can be buffered or worsened by local topographic variations such as aspect, wind exposure, and elevation (Walker et al., 1993).

The dominance of alpine pollinators varies regionally: *Bombus* species are not present everywhere, flies often dominate, and flies also play an important role even when bees *(Bombus)* (Inouye et al., 2015; Lefebvre et al., 2014). Pollination is a temperature-sensitive process because most insects prefer warm flowers, a preference that can be facilitated by the plant heliotropic behavior (Cooley, 1995; Stanton & Galen, 1989).

Climate change is changing the timing of flowering and pollinator emergence, but not always in sync. This asynchrony can cause mismatches between flowers and their pollinators. For example, in the Rocky Mountains, syrphid fly activity is associated with snowmelt, while flowering can occur much faster, leading to temporal misalignment in some years. Similarly, during unusually warm seasons, bees can become active after the flowers, and different *Bombus* species show varied responses to temperature changes (Kudo, 2014). These types of mismatches may decrease the frequency of visits, pollen transfer and seed production, eventually affecting the reproductive success of plants (Inouye, 2020). Some adaptive features of alpine plants, such as prolonged flower life or morphological features that protect pollen from rain (e.g. *Gentiana algida* Pall.), help reduce pollination shortages (Bynum & Smith, 2001). Pollinators, in turn, can either evolve behaviorally or change their range and evolve gradually, for example, by simply shrinking the tongue length of Bombus species simply shrinking over decades as floral abundance decreases (Miller-Struttmann et al., 2015).

Nonetheless, varying the rate of phenological changes among plants and pollinators under ongoing climate warming poses significant risks to alpine biodiversity, particularly for species with specialized mutualistic interactions (Britten et al., 1994; Roland & Matter, 2007; Seidl, 1999).

3.2 Human Pressures

3.2.1Overharvesting and Commercial Demand

Illegal and unsustainable harvesting continues to pose a major threat to plant biodiversity. Several experts have emphasized concerns about the unregulated collection and trade of wildlife and plant materials for research, education, and commerce. When these activities exceed sustainable limits, they directly deplete species and disrupt ecological balance (Qazi et al., 2025). Overharvesting contributes to the loss of rare and endemic species, particularly those with high medicinal or ornamental value. Globally, unsustainable trade is a leading cause of biodiversity decline, and similar patterns are evident in Pakistan, where local communities depend on plant resources for their livelihoods and traditional medicine (Fierro et al., 2019).

3.2.2 Grazing Pressures

Intensive grazing, particularly in alpine and subalpine regions, alters vegetation structure, reduces regeneration potential, and facilitates soil erosion. Overgrazing limits the natural recovery of plant species and can promote the invasion of non-native plants that are more tolerant of disturbance. This has downstream consequences for endemic and slow-growing alpine flora, further compounding the effects of climate change and land-use change.

3.2.3 Land-Use Change (Deforestation, Agriculture, and Infrastructure)

Experts identified climate change as one of the most important global biodiversity threats and was the most commonly mentioned in the scientific literature (Hussain et al., 2020; Qazi et al., 2023). In Pakistan, the number of forest fires is predicted to become increasingly severe in summer and during winter floods, with the northern part of the country being most affected by these events (Tariq et al., 2022). Such fires, coupled with climate changes, negatively affect plant distribution and the survival of rare and endemic plant species.

Urbanization was ranked as the second major threat to rare flora. The impacts of urban development on plant distribution are complex, involving habitat fragmentation, soil degradation, and microclimatic changes. Land-use conversion such as transforming natural habitats into residential, industrial, or agricultural areas has drastically reduced native vegetation cover in recent decades (Hernández et al., 2016; Ong et al., 2022). Estimates suggest that Pakistan's native forests have declined by nearly one-third over the past 20 years, mainly due to the expansion of fruit tree cultivation and exotic plantations (Qazi et al., 2025).

Another key threat, considered the third most serious by experts, is invasive alien species. Approximately 73 known terrestrial and aquatic invasive species are present in Pakistan, significantly affecting the native vegetation and endangering endemic species through competition and habitat changes (Qazi et al., 2025; Wu et al., 2019).

These pressures are further compounded by agricultural, industrial and residential pollution. Both, point and non-point sources of pollution deteriorate soil and water quality, which negatively affects the biodiversity of urban and peri-urban areas. As a result, native flora is being supplanted by exotic decorative plants, and the number of birds in several Pakistani cities has decreased (Kolawole & Iyiola, 2023).

3.2.4 Tourism and Recreational Impacts

While the growth of recreational activities and ecotourism can have positive effects, there is also a risk of habitat degradation if this increase is not managed properly. The stamping, dumping and installation of tourist amenities within delicate ecosystems affect soil stratification and cause the displacement of natural vegetation. Another factor that increases the threat to already strained habitats is forest fires, which can result from human negligence during certain activities (Tariq et al., 2022). In addition to direct physical influences, indirect pressures such as weak conservation policies, ineffective management approaches, and sociopolitical turmoil also hinder biodiversity protection endeavors. Poor taxonomic information, inadequate surveillance and corruption further erode conservation systems especially in forested and mountainous areas.

3.2.5 Integration with Sustainability and Policy Frameworks

Integrating Sustainable Development Goals (SDGs) into biodiversity conservation offers a comprehensive approach to addressing threats caused by human activities. For example, agricultural intensification leads to land-use change in South East Asia, that adversely affecting SDG 15 (Life on Land) by accelerating habitat loss and fragmentation. Similarly, mining contributes to local and global greenhouse gas emissions, undermining SDG 13 (Climate Action). Knowledge of these nexuses can enable policymakers to strike a balance biodiversity planning goals with national and international sustainability goals. Experts generally agree that the two most urgent threats to endemic and rare plant species in Pakistan are climate change and urbanization. However, factors like forest fires, invasive species, pollution, and poorly controlled tourism also contribute significantly. More interdisciplinary studies are required to study the interactions among these pressures–whether synergistic, additive, or antagonistic to more effectively identify high-risk zones of biodiversity loss and support data-driven conservation prioritization in Pakistan (Qazi et al., 2025).

3.2. 6 Invasive Species and Habitat Fragmentation

The invasion of alien non-native plants introduced by human is a growing threat to the Western Himalayas, which have rich biodiversity. After establishing themselves in an area, such species interfere with native ecosystems by crowding out local flora, changing soil structure, and disrupting ecological balance (Gupta et al., 2021; Mooney & Drake, 1987). They also spread further due to habitat fragmentation caused by urbanization, tourism, and road construction (Early et al., 2016). Over 600 IAS have been reported in the Himalayan regions including Himachal Pradesh, Uttarakhand and Jammu & Kashmir These include *Lantana camara* L., *Parthenium hysterophorus* Adans., *Ageratum conyzoides* L. and *Ageratina adenophora* (Spreng.) R.M.King & H.Rob. (Jaryan et al., 2013; Khuroo et al., 2007; Sekar et al., 2015). The plants also intrude into roadsides, forest edges, and even alpine areas seriously reducing the biodiversity of native species (Gupta et al., 2021).

The species such as *L. camara* may reduce the richness of indigenous flora by up to 30%, whereas *P. hysterophorus* poses ecological and health risks (Datta et al., 2017; Kohli et al., 2006). The control of such invasions involves prompt identification, hand weeding, biological control, and restoration with native species (Gupta et al., 2021).

Overall, habitat fragmentation and the threat of invasive alien species have combined effects on the ecological integrity of the Western Himalayas. Monitoring and controlling their spread is impossible without integrated strategies that use GIS and remote sensing apparatus (Gupta et al., 2021; Woodford et al., 2017).

4.1 Limitations of Current Conservation Strategies

The preservation of alpine medicinal plants and the traditional knowledge associated with them faces numerous challenges that reflect larger global phenomena observed in other regions such as Harighal. Harsh climatic conditions, overharvesting, deforestation and unsustainable grazing in alpine ecosystems pose significant threats to the survival of many valuable plant species. The increasing commercialization of medicinal plants has led to indiscriminate harvesting by untrained collectors, often resulting in the uprooting of whole plants and the depletion of natural populations. Habitat degradation caused by forest fires, infrastructure development, and the growing human presence in sensitive mountain areas further exacerbates these threats.

Another concerning issue is the rapid decline of traditional ethnomedicinal knowledge among alpine communities. Much of this knowledge is held by older generations and local healers who possess a deep understanding of the medicinal value of alpine flora, including preparation methods and how it should be prepared as well as how sustainable harvesting practices. However, modern lifestyles, migration, and the growing preference for allopathic medicine have diminished younger generations' interest in learning or preserving these traditions. In many cases, traditional healers are also reluctant to share their expertise openly, further accelerating the loss of indigenous knowledge.

4.2 *In-situ* and *Ex-situ* Conservation Approaches

The alpine systems are unique for their biodiversity and fragile ecosystems; therefore, conservation strategies must be robust to ensure sustainability in Pakistan. *In-situ* conservation involves preserving these ecosystems in their natural locations, such as national parks and wildlife reserves, including Deosai National Park and Khunjerab National Park. These regions protect native species, sustain ecological processes, and allow natural adaptation to environmental changes. *In-situ* conservation succeeds only with community participation and effective management to minimize human disturbance and maintain habitat integrity.

Ex-situ conservation complements *in-situ* conservation by preserving genetic material and species outside their natural environments. This includes botanical gardens, seed banks and zoological facilities where threatened alpine species can be replicated and preserved. *Ex-situ* techniques serve as a precaution against extinction threats caused by habitat loss or climate change (Mahanayak, 2024). Seed bank and alpine plant nursery projects in Pakistan help preserve rare and endangered species that can be reintroduced and restored in the future (Challender et al., 2025).

4.3 International Agreements and the Role of Local Communities

To ensure effective conservation of Pakistan's ecologically rich yet vulnerable alpine areas, it is necessary to transform the inflexible, top-down governance model into a more inclusive, community based management approach. Although Pakistan has established various policy frameworks to achieve sustainability, their application, particularly in mountainous states, is still limited by institutional and social barriers (Javeed et al., 2024).

One of the most influential findings of recent assessments is that the national parks of Pakistan are structured in such a highly standardized manner that local communities are unable to engage in important conservation services. Although these communities are the major users and custodians of biodiversity resources, they are mostly excluded from the decision-making process because the current categories of biodiversity protection do not allow them to participate significantly. In the vulnerable alpine environment, true co-management that ensures the preservation of ecological integrity requires providing shared authority, responsibility, and benefits to local communities and user groups (Javeed et al., 2024).

Further gaps in enforcement exist in law. Existing laws disproportionately focus on preserving wildlife species but do not provide official safeguards for threatened and endangered plant species, which is a particularly a severe gap in botanically diverse alpine ecosystems. These deficiencies are compounded by a long term lack of resources, such as inadequate funding and insufficient trained manpower, making proper monitoring and enforcement of the law in remote high-altitude regions difficult.

Alpine ecosystems- global biodiversity hotspots are becoming increasingly sustainable by linking national conservation efforts with Multilateral Environmental Agreements (MEAs). Although global treaties like CITES promote collaboration, they are usually undermined by excessive dependence on state-based enforcement mechanisms, and little attention is given to consider the social and economic factors that motivate unlawful exploitation (Challender et al., 2025). Because alpine regions are socially complex, remote, and economically vulnerable, simple deterrence approaches are insufficient.

- The species-system Theory of Change (ToC), which focuses on the Social-Ecological Systems (SESs) where species are harvested, traded, and managed, is a more effective approach to this issue. It involves studying both formal and informal institutions, as well as conducting institutional diagnostics to redesign governance structures across the entire supply chain (Challender et al., 2025). When management rules and institutional arrangements are co-designed at the local community level, the rules gain legitimacy, compliance increases, and reliance on expensive external enforcement mechanisms is minimized. Effective alpine conservation in Pakistan also requires incorporating international conservation objectives within culturally based and community-oriented systems of governance. This alignment will require systemic, evidence-based institutional reforms that empower local actors and enhance ecological resilience in these vital high-altitude landscapes.

5. NANOPARTICLE-BASED INTERVENTION STRATEGIES

5.1 Principles of Nanoparticle Science for Conservation and Medicine

The future of alpine medicinal biodiversity depends on new, interdisciplinary approaches in response to both dual pressures of the environmental threats and the complexities of phytochemical use. Nanotechnology can serve as an effective intervention, providing a toolkit with diverse applications to address challenges on both fronts: strengthening the plants themselves and improving the delivery and effectiveness of their medicinal compounds (Tang et al., 2023). This approach transforms traditional agriculture and medicine into more precise, sustainable, and effective practices. Figure 2 conceptually illustrates the dual application of nanotechnology in both plant conservation and advanced medicine

Figure 2. Nanoparticle-based intervention strategies model for alpine biodiversity. The framework shows a two-pronged approach: (A) using nano-innovations like nano-fertilizers and nano-pesticides for in-situ and ex-situ plant conservation and resilience, and (B) using nanocarriers to enhance the therapeutic delivery and efficacy of plants phytochemicals

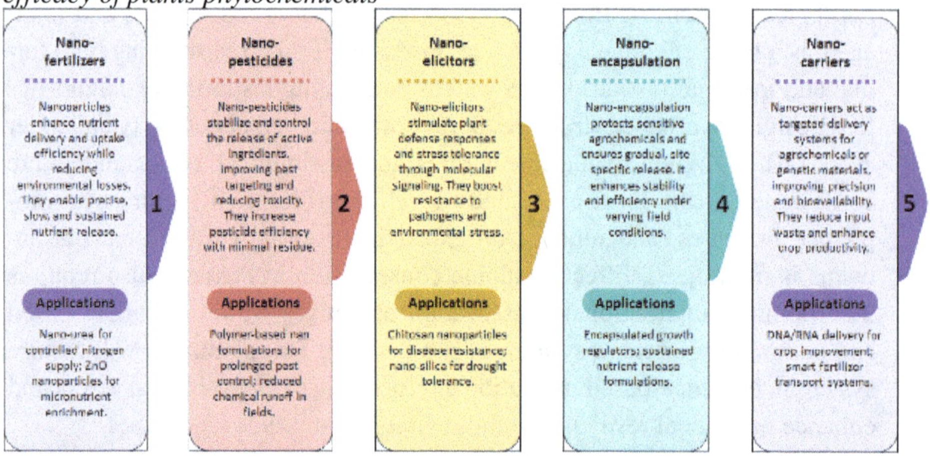

5.2. Nanotechnology for Plant Conservation and Resilience

Before phytochemicals can be used as medicine, plants must be preserved and grown sustainably, especially given climate change and habitat degradation. Nanotechnology is offers an approach to enhance plant health and productivity, which would indirectly help preserve threatened alpine species (Tyagi et al., 2023) (Ashraf et al., 2024). Of most notable innovations in this field are nano-fertilizers and nano-pesticides (Mao & Yang, 2023).

Poisonous links: Nano-fertilizers are designed to provide essential nutrients such as nitrogen, phosphorus, and potassium to plants at a very high rate. The nano-fertilizers deliver a controlled and precise discharge of nutrients, unlike traditional fertilizers, which may be washed away and harm the environment. Their small size enables greater absorption through the roots and leaves of plants, resulting in enhanced growth and increased resistance to diverse environmental stressors such as drought and cold. This precision reduces the need for chemicals and can be used to enhance the growth and biomass of slow-growing alpine plants, making their cultivation more feasible (Tang et al., 2024). Nano-pesticides involve using nanocarriers to package active compounds, enabling more sustainable protection of plants against diseases and pests. This encapsulation enhances the stability and specificity of the pesticide, so relatively lower quantities are required. As a result, fewer beneficial insects and delicate alpine ecosystems are harmed. Moreover, it is possible to implement technologies that allow real-time monitoring the health of plant health and soil conditions, enabling timely and accurate interventions with the help of nano-sensors (Tang et al., 2024).

These nanotechnological tools offer a proactive approach to protecting and conserving alpine medicinal plants by developing better growth, increasing stress resistance and enhancing plant protection to ensure these essential genetic resources are preserved.

5.3. Nanotechnology for Enhanced Phytochemical Delivery

Besides preserving plants, nanotechnology offers a groundbreaking approach to addressing the challenges of utilizing their pharmaceutical constituents. Most potent phytochemicals from alpine plants lack water solubility and bioavailability, and they break down rapidly in the body, which limits their therapeutic benefits.

Nano-based drug delivery systems are designed to address these exact problems. Encapsulating natural compounds within biocompatible nanocarriers can unlock their clinical potential.

- *Improved Bioavailability and Stability:* Encapsulation in nanocarriers like liposomes or polymeric nanoparticles protects phytochemicals from enzymatic degradation, enhances their solubility, and facilitates their absorption into the bloodstream, significantly increasing their bioavailability (Patra et al., 2018).

- *Targeted Delivery:* Nanoparticles can be engineered to accumulate in diseased tissues, such as tumors through passive targeting (the EPR effect) or active targeting (by attaching specific ligands). This results in a higher concentration of the drug at the site of action while minimizing exposure and damage to healthy cells (Patra et al., 2018).
- *Controlled Release:* These systems can be designed for sustained and controlled release of the active compound over time, maintaining a stable therapeutic concentration and reducing the need for frequent dosing (Patra et al., 2018).

Such nano- engineering of natural products not only increases their therapeutic effects but also aligns with preservation objectives. This reduces the amount of precious plant material required by making a much larger proportion of the administered dose effective. It represents a paradigm shift because, instead of simply extracting plant material future generations of phytopharmaceuticals are intelligently designed to convert the huge potential of alpine biodiversity into clinical outcomes (Tang et al., 2023). The processes employed to produce these high-tech materials can be broadly categorized as physical, chemical, or biological (green) synthesis, each with its own peculiarities and applications, as compared in Table 1 (Tang et al., 2024).

Table 1. Comparison of nanoparticle synthesis methods, key features, and applications

Classification	Typical Methods	Distinctive Features	Major Applications	Examples / Use Cases
Physical Synthesis	Vapor deposition, sputtering, laser ablation	Based solely on physical processes without involving chemical reactions; produces nanomaterials with uniform structure, high crystallinity, and purity	Fabrication of nanofilms, nanowires, and nanoparticles	Nano-fertilizers, coating materials
Chemical Synthesis	Sol–gel process, precipitation, hydrothermal or solvothermal methods	Involves controlled chemical reactions requiring solvents, catalysts, and specific reaction conditions; allows tuning of particle size and morphology	Development of nanomaterials with high purity and crystallinity	Biosensors, nanodrug delivery systems, nano-pesticides
Biological (Green) Synthesis	Use of microorganisms, plant extracts, enzymes, or biomolecules	Relies on bio-reductive or biosynthetic mechanisms driven by biological systems under eco-friendly conditions; yields biocompatible nanomaterials	Environmentally sustainable synthesis for biomedical and pharmaceutical use	Nano-drug carriers, biosensors, antimicrobial nanoparticles

5.4 Nanotechnology for Enhanced Phytochemical Delivery

While alpine medicinal plants are a rich source of bioactive compounds, their direct clinical translation is often thwarted by several well-documented limitations. Many potent phytochemicals exhibit poor aqueous solubility, low absorption, in-vivo instability, and inadequate bioavailability, which together reduce their therapeutic efficacy. These issues often require high or repeated doses, leading to systemic toxicity and making use impractical for compounds derived from rare or slow-growing high-altitude species.

Nanotechnology provides powerful tools to directly address these problems. Encapsulating or conjugating natural compounds with biocompatible nanomaterials can fundamentally alter their pharmacokinetic profiles (Patra et al., 2018).

- *Enhanced Bioavailability and Solubility:* Nanocarriers can effectively solubilize hydrophobic phytochemicals, protect them from premature degradation in the gastrointestinal tract, and facilitate their absorption into the bloodstream. For example, encapsulating thymoquinone (from *Nigella sativa* L.) in lipid nanocarriers increased its bioavailability six-fold compared to the free compound, showcasing the dramatic improvements possible.
- *Improved Stability and Controlled Release:* Encapsulation within a nanoparticle shell shields the active compound from enzymatic degradation and rapid clearance, prolonging its circulation time. These systems can also be engineered for controlled or stimuli-responsive release (e.g. in response to the acidic pH of a tumor microenvironment), ensuring the drug is released at the target site over a sustained period, thus reducing plasma fluctuations and minimizing side effects.
- *Targeted Delivery to Diseased Tissues:* Diseased tissues such as tumors, can be specifically targeted by nanoparticles due to their small size typically less than 100 nm, often 10-100 nm, by the Enhanced Permeability and Retention (EPR) effect. Leaky vasculature permits entry of nanoparticles, while poor lymphatic drainage leads to their retention. Targeting can be further improved by active targeting, in which ligands like antibodies or peptides are conjugated to the nanoparticle surface to bind receptors overexpressed on diseased cells. This two-pronged approach increases drug concentration at the area of action while sparing healthy tissues (Patra et al., 2018) (Ashraf et al., 2024).

This nano-based reformulation maximizes the therapeutic impact of the phytochemical and supports sustainability goals by enabling the use of smaller quantities of precious plant material.

5.5 Versatile Nanocarrier Systems for Phytochemical Delivery

The field of nanomedicine offers a diverse toolkit of nanocarriers, enabling the selection of an optimal system based on the physicochemical properties of the drug and the therapeutic goal (Patra et al., 2018).

Biopolymeric Nanoparticles: Chitosan and alginate are natural polymers that are highly regarded for their biocompatibility, biodegradability and mucoadhesive properties. For example, chitosan nanoparticles can penetrate the tight junctions of epithelial cells, making them ideal for delivering drugs to the mouth, nose, or eye.

Lipid-Based Nanosystems: Some clinically advanced nanocarriers include liposomes and polymeric micelles. Liposomes are spherical structures composed of lipid bilayers that can entrap both hydrophilic drugs in their aqueous core and hydrophobic in their lipid membrane. Self-assembling amphiphilic block copolymer structures, known as polymeric micelles, are particularly well suited for solubilizing poorly water-soluble compounds within their hydrophobic cores. Natural products such as curcumin, resveratrol, and quercetin formulated, in these systems have shown greatly improved stability and anticancer activity.

Inorganic and Metallic Nanoparticles: Nanoparticles made from gold, silver or iron oxide have distinct properties for both therapy (theranostics) and diagnosis. Their surfaces can be easily functionalized to transport natural compounds and their physical characteristics (such as the surface plasmon resonance of gold nanoparticles) can be used for applications such as photothermal therapy or advanced bioimaging. Green synthesis of these nanoparticles using plant extracts provides a method that increases their biocompatibility and reduces their toxicity.

5.6 Applications of Nanoparticles in Conservation and Agriculture

The application of nanotechnology to alpine medicinal flora offers a dual strategy: enhancing the resilience, growth, and conservation of the plants and improving the therapeutic efficacy of their extracted phytochemicals. This integrated approach addresses challenges at every stage, from cultivation to clinical use.

5.6.1. Enhancing Plant Growth and Stress Tolerance

Alpine plants are constantly subjected to severe abiotic stresses, including drought, extreme temperatures (cold and heat), and high salinity. Nanoparticles offer a powerful toolkit to strengthen plant defenses against these challenges.

- *Nano-fertilizers and Nutrient Uptake:* Nanomaterials can significantly improve nutrient absorption and metabolism. As an example, zinc oxide (ZnO) and iron oxide (Fe_2O_3) nanoparticles serve as highly bioavailable micronutrients that improve enzymatic processes and chlorophyll production, both of which are essential for photosynthesis and overall plant health (Kumar et al., 2025). It is especially helpful for nutrient-poor soils with slow-growing species in the Alps.
- *Drought and Salinity Resistance:* Nanoparticles can reduce drought and salinity stress by improving water use efficiency and ionic balance. Carbon-based nanomaterials, such as carbon nanotubes (CNTs) have been found to create nanochannels in root cells, enabling better water uptake. (Bharti et al., 2024). In addition, nanoparticles (e.g. silicon (Si) NPs) can induce the synthesis of osmolytes (e.g. proline), which help cells maintain turgor and protect cell structures from damage caused by osmotic pressure.
- *Pathogen Defense:* Metallic nanoparticles, especially silver nanoparticles (AgNPs) have a strong antimicrobial properties and can enhance a plant's resistance to pathogens. They work by destabilizing microbial cell membranes and producing reactive oxygen species (ROS), which create a protective barrier against bacterial and fungal infections.

5.6.2. Controlled Release and Enhanced Bioavailability

One of the biggest problems in agriculture and medicine is the effective delivery of active compounds, which is enhanced by nanotechnology.

- *Controlled Release of Agrochemicals and Extracts:* Polymeric nanomaterials based on chitosan-based may encapsulate fertilizers or pesticides release them slowly and in a controlled manner. Such targeted delivery reduces environmental runoff and decreases the overall number of chemicals required, chemicals which is vital for preserving delicate alpine ecosystems. This approach is also applicable to the regulated release of therapeutic vegetable extracts, ensuring a lasting impact

5.6.3. Improving Bioavailability of Plant-Derived Drugs:

As mentioned earlier, many potent phytochemicals are poorly soluble and have low bioavailability. These compounds cannot be degraded and their solubility is enhanced by nano-encapsulation in systems such as liposomes or polymeric nanoparticles, which enhances their absorption in the body and, consequently, greatly increases their therapeutic effect (Patra et al., 2018).

6. NANO-SENSORS FOR HEALTH AND ENVIRONMENTAL MONITORING

The integration of nanotechnology with precision agriculture enables real-time monitoring of plant and environmental health.

Nanosensors: Advanced nanosensors can detect subtle changes in soil moisture, nutrient levels, or the early presence of pathogens. Combined with intelligent technologies, such as the Internet of Things (IoT), these sensors can provide early alerts and enable targeted interventions to ensure that resources such as water and nano-fertilizers, are used only where and when necessary (Bharti et al., 2024). This accuracy is critical for the sustainable management of both wild and cultivated stocks of alpine medicinal plants.

Nanotechnology offers a gateway not only to conserve threatened alpine medicinal biodiversity but also to unlock its full therapeutic potential in a sustainable and efficient manner by utilizing its multiple uses.

6.1. Risks and Biosafety Considerations

Although nanotechnology is a promising solution, challenges remain associated with its application. The long-term ecological effects of introducing engineered nanomaterials to vulnerable alpine ecosystems are still unknown. Metal- based nanoparticles may dissolve and release ions, which can subsequently impact soil microbial communities and enter the food chain. Thus, thorough, protracted research is required to determine the environmental fate and possible toxicity of such materials before their widespread use.

6.2 Ethical, Ecological, and Socio-economic Concerns

There is a need for strong regulatory frameworks to guarantee the safety of nano-enabled products. The socio-economic aspect is also important; some nanotechnologies are expensive to manufacture and would be costly to supply to local populations in alpine regions. Moreover, public acceptance can become an obstacle because people associate agriculture with natural labels, and the term nanotechnology can be viewed skeptically. These issues must be discussed openly, and cost-effective, green, synthesis techniques for nanoparticles should be developed (Tang et al., 2023).

6.3 Future Prospects: A Nanotechnological Revolution for Alpine Medicine

6.3.1 Alpine Biodiversity at the Crossroads: Current and Emerging Threats

Although the HKH have significant ecological and medicinal significance, their delicate ecosystems are at a critical crossroads, facing unprecedented challenges from climate change and direct human pressures. These forces that are accelerating biodiversity loss and threatening the long-term sustainability of these essential natural resources (Ismail et al., 2022; Khan et al., 2022).

6.3.2 The Escalating Impact of Climate Change

Mountain areas are particularly vulnerable to climate change as warming in these regions occurs at a faster rate than the global average, a condition referred to as elevation-dependent warming. The effects in the HKH are already intense and extensive. According to scientists' projections, an increase in the global temperature by 1.5 C will make the region to lose about one-third of its glacier mass by 2100 (Wester et al., 2019). This accelerated melting of glaciers alters hydrological cycles, leading to increasingly severe climate-induced events like flash floods, landslides, and Glacial Lake Outburst Floods (GLOFs) (Ali et al., 2022).

6.3.3 Intensifying Human Pressures

These climatic changes directly threaten the existence of alpine medicinal flora. The changing patterns of precipitation and rising temperatures are causing habitat degradation and forcing several plant species to migrate to higher altitudes (Manish, 2022). Nevertheless, many species might not be able to adapt fast enough, leading to local extinctions, and a major loss of genetic diversity.

- *Deforestation and Land-Use Change:* Climate change threats are also worsened by the increasing direct pressures from human activities, which are typically driven by population growth and economic needs.
- *Overharvesting and Commercial Demand:* The increasing world demand for medicinal and aromatic plants has led to unsustainable and, in most cases, illegal harvesting of high-value species. This excessive extraction is a major cause of species decline, and the IUCN has reported that up to 15,000 species of medicinal plants worldwide could face the threat of extinction as a result (Ismail et al., 2022). This danger has become especially prominent in the

more accessible sections of the HKH, where it has been depleting natural populations and disrupting the fragile ecological balance.

- *Deforestation and Land-Use Change:* The forests of the HKH region are essential habitat for a tremendous amount of understory medicinal plants. Nevertheless, these forests are declining rapidly, with deforestation rates in the Pakistani HKH ranging from 0.2 to 0.8 percent annually. The main causes of this loss are the conversion of forest land to agriculture, the expansion of settlements, and the harvesting of timber and fuelwood. This habitat destruction not only directly eliminates plant populations but also fragments the landscape, isolating remaining populations and increasing their vulnerability.
- *Grazing Pressure and Rangeland Degradation:* High-altitude rangelands, which support a unique diversity of medicinal herbs, are under immense pressure from livestock grazing. The number of domesticated animals in regions like Gilgit-Baltistan has increased significantly, leading to overgrazing that degrades vegetation cover, accelerates soil erosion, and reduces the availability of forage for both livestock and native wildlife (Ashraf et al., 2024)

6.3.4 Future Prospects: A Nanotechnological Revolution for Alpine Medicine

Nano-intervention approaches can be revolutionary in utilizing the powerful phytochemicals found in high-altitude plants. Since the main challenges are low biomass and inefficient extraction, nanotechnology offers a crucial solution by increasing the therapeutic value of small quantities of bioactive substances. The nano-formulations also make the use of rare botanicals more sustainable and cost effective by ensuring that a larger percentage of the administered dose is put to active use.

This will be a revolution in which, rather than extracting natural products, they will be smartly engineered into highly effective next generation phytopharmaceuticals. The convergence of nanotechnology with the traditional wisdom of alpine medicine offers the prospect of producing new evidence-based treatments that can address complex diseases while preserving the less resilient biodiversity found in high-altitude environments. Only future research and prioritized policies can unlock substantial potential, which should be targeted at several areas:

Combining Nanotechnology and Climate Adaptation Strategies: Future studies should focus on developing nano-enabled solutions to help alpine plants adapt to climate change. This includes producing nano-fertilizers that increase plant resistance to drought and extreme temperatures, as well as nano-sensors that can be deployed to monitor early symptoms of climate-related stress in remote mountainous areas.

Role of Machine Learning and AI: Alpine ecosystems are highly intricate, and nanoparticle designs allow for numerous combinations that can be resolved with the help of artificial intelligence (AI) and machine learning (ML). These tools can be applied to mapping biodiversity to identify species at risk, predicting the bioactivity of new phytochemicals, and engineering nanocarriers to match certain alpine compounds and therapeutic targets.

• *Policy Recommendations for Sustainable Innovation:* To ensure these technological advancements are applied responsibly, robust policy frameworks are needed. These policies should govern the biosafety of releasing nanomaterials into sensitive ecosystems, establish clear guidelines for access and benefit-sharing under the Nagoya Protocol, and create incentives developing cost-effective, green nanotechnologies accessible to local communities.

Fostering Interdisciplinary Collaboration: The opportunities and difficulties at the intersection of alpine biodiversity and nanotechnology cannot be addressed by a single discipline. The real development will require close and sustained collaboration among ecologists, ethnobotanists, nanotechnologists, pharmacologists, and policymakers.

Through the establishment of integrated research programs and community building dialogue, the scientific community will be able to develop holistic solutions that are both technologically innovative and ecological.

Through these future paths, the potential for an unprecedented nanotechnological revolution in alpine medicine is achievable, whereby the botanical heritage of the world's high mountains will not only be preserved for the future but will also be used to improve human health.

7. CONCLUSION

Alpine biodiversity, with its unique flora adapted to harsh, high-altitude conditions such as those in the HKH, represents a significant global resource for ecological stability and healing. The heavy reliance on these medicinal plants in developing countries, supported by centuries of advanced Traditional Ecological Knowledge (TEK), offers a foundation for modern drug discovery, as seen with compounds such as artemotil and paclitaxel. However, this precious asset is at a critical crossroads. Combined effects of accelerated climate change, including glacial retreat, changes in precipitation, and phenological mismatches, along with increased human demands such as unsustainable overharvesting and land-use conversion, subject endemic alpine species to significant risks. These environmental pressures directly harm the ecological settings on which traditional ecological knowledge (TEK) is founded, threatening not only the survival of the plants themselves but also the knowledge required to use them sustainably. The interdisciplinary approach described in this chapter is essential to ensure that future generations inherit a healthy and environmentally sustainable future. A new approach involves combining ecological knowledge with developments in nanobiotechnology. Nanotechnology may increase the resilience of endangered alpine plants by delivering nutrients (nano-fertilizers) and protection (nano-pesticides), and may transform medicine by enhancing the bioavailability, stability and targeted delivery of phytochemicals with limited bio-activity. This method would maximize the effects of scarce natural resources by controlling threats at their source (plant conservation), and improving therapeutic utility (drug delivery). Finally, protecting alpine medicinal biodiversity requires integrating ethnobotanical knowledge with modern science, supported by sound policy frameworks that ensure ethical benefit sharing and sustainable use.

REFERENCES

Ali, M., Raza, M., Parveen, K., & Bashir, F. (2022). Mountains and Climate Change. *Mountain Studies: Understanding and Managing Mountains for People and Nature*, 58.

Ambesh, P., Campia, U., Obiagwu, C., Bansal, R., Shetty, V., Hollander, G., & Shani, J. (2017). Nanomedicine in coronary artery disease. *Indian Heart Journal*, *69*(2), 244–251. DOI: 10.1016/j.ihj.2017.02.007 PMID: 28460774

Armstrong, R. L. (2010). *The glaciers of the Hindu Kush-Himalayan region: a summary of the science regarding glacier melt/retreat in the Himalayan, Hindu Kush, Karakoram, Pamir, and Tien Shan mountain ranges.*

Arroyo, M. T. K., Armesto, J. J., & Villagran, C. (1981). Plant phenological patterns in the high Andean Cordillera of central Chile. *Journal of Ecology*, *69*(1), 205–223. DOI: 10.2307/2259826

Ashraf, M. V., Khan, S., Misri, S., Gaira, K. S., Rawat, S., Rawat, B., Khan, M. A. H., Shah, A. A., Asgher, M., & Ahmad, S. (2024). High-Altitude Medicinal Plants as Promising Source of Phytochemical Antioxidants to Combat Lifestyle-Associated Oxidative Stress-Induced Disorders. *Pharmaceuticals (Basel, Switzerland)*, *17*(8), 975. Advance online publication. DOI: 10.3390/ph17080975 PMID: 39204080

Atanasov, A. G., Waltenberger, B., Pferschy-Wenzig, E. M., Linder, T., Wawrosch, C., Uhrin, P., Temml, V., Wang, L., Schwaiger, S., Heiss, E. H., Rollinger, J. M., Schuster, D., Breuss, J. M., Bochkov, V., Mihovilovic, M. D., Kopp, B., Bauer, R., Dirsch, V. M., & Stuppner, H. (2015). Discovery and resupply of pharmacologically active plant-derived natural products: A review. *Biotechnology Advances*, *33*(8), 1582–1614. DOI: 10.1016/j.biotechadv.2015.08.001 PMID: 26281720

Beutler, J. A. (2009). Natural products as a foundation for drug discovery. *Current protocols in pharmacology*, *46*(1), 9.11. 11-19.11. 21.

Bharti, A., Jain, U., & Chauhan, N. (2024). From lab to field: Nano-biosensors for real-time plant nutrient tracking. *Plant Nano Biology*, *9*, 100079. DOI: 10.1016/j.plana.2024.100079

Bonifácio, B. V., da Silva, P. B., Ramos, M. A. S., Negri, K. M. S., Bauab, T. M., & Chorilli, M. (2014). Nanotechnology-based drug delivery systems and herbal medicines: A review. *International Journal of Nanomedicine*, •••, 1–15. PMID: 24363556

Britten, H. B., Brussard, P. F., & Murphy, D. D. (1994). The pending extinction of the Uncompahgre fritillary butterfly. *Conservation Biology*, *8*(1), 86–94. DOI: 10.1046/j.1523-1739.1994.08010086.x

Bynum, M. R., & Smith, W. K. (2001). Floral movements in response to thunderstorms improve reproductive effort in the alpine species Gentiana algida (Gentianaceae). *American Journal of Botany*, *88*(6), 1088–1095. DOI: 10.2307/2657092 PMID: 11410474

Cannone, N., Guglielmin, M., & Gerdol, R. (2004). Relationships between vegetation patterns and periglacial landforms in northwestern Svalbard. *Polar Biology*, *27*(9), 562–571. DOI: 10.1007/s00300-004-0622-4

Challender, D. W., 't Sas-Rolfes, M., Broad, S., Milner-Gulland, E. J. F. i. E., & Evolution. (2025). A theory of change to improve conservation outcomes through CITES. *13*, 1425267.

Cooley, J. R. (1995). Floral heat rewards and direct benefits to insect pollinators. *Annals of the Entomological Society of America*, *88*(4), 576–579. DOI: 10.1093/aesa/88.4.576

Cooper, W. S. (1923a). The recent ecological history of Glacier Bay, Alaska: permanent quadrats at Glacier Bay: an initial report upon a long-period study. *Ecology*, *4*(4), 355–365. DOI: 10.2307/1929182

Cooper, W. S. (1923b). The recent ecological history of Glacier Bay, Alaska: The interglacial forests of Glacier Bay. *Ecology*, *4*(2), 93–128. DOI: 10.2307/1929485

Cooper, W. S. (1923c). The recent ecological history of Glacier Bay, Alaska: The present vegetation cycle. *Ecology*, *4*(3), 223–246. DOI: 10.2307/1929047

Cooper, W. S. (1931). A third expedition to Glacier Bay, Alaska. *Ecology*, *12*(1), 61–95. DOI: 10.2307/1932934

Cooper, W. S. (1939). A fourth expedition to Glacier Bay, Alaska. *Ecology*, *20*(2), 130–155. DOI: 10.2307/1930735

Darmody, R., Allen, C., & Thorn, C. (2005). Soil topochronosequences at storbreen, jotunheimen, Norway. *Soil Science Society of America Journal*, *69*(4), 1275–1287. DOI: 10.2136/sssaj2004.0204

Datta, A., Kühn, I., Ahmad, M., Michalski, S., & Auge, H. (2017). Processes affecting altitudinal distribution of invasive Ageratina adenophora in western Himalaya: The role of local adaptation and the importance of different life-cycle stages. *PLoS One*, *12*(11), e0187708. DOI: 10.1371/journal.pone.0187708 PMID: 29125852

Duncan, R., & Gaspar, R. (2011). Nanomedicine(s) under the microscope. *Molecular Pharmaceutics*, *8*(6), 2101–2141. DOI: 10.1021/mp200394t PMID: 21974749

Early, R., Bradley, B. A., Dukes, J. S., Lawler, J. J., Olden, J. D., Blumenthal, D. M., Gonzalez, P., Grosholz, E. D., Ibañez, I., Miller, L. P., Sorte, C. J. B., & Tatem, A. J. (2016). Global threats from invasive alien species in the twenty-first century and national response capacities. *Nature Communications*, *7*(1), 12485. DOI: 10.1038/ncomms12485 PMID: 27549569

Fierro, P., Valdovinos, C., Arismendi, I., Díaz, G., De Gamboa, M. R., & Arriagada, L. (2019). Assessment of anthropogenic threats to Chilean Mediterranean freshwater ecosystems: Literature review and expert opinions. *Environmental Impact Assessment Review*, *77*, 114–121. DOI: 10.1016/j.eiar.2019.02.010

Gallagher, R., Hughes, L., & Leishman, M. (2009). Phenological trends among Australian alpine species: Using herbarium records to identify climate-change indicators. *Australian Journal of Botany*, *57*(1), 1–9. DOI: 10.1071/BT08051

Gul, H., Ahmed Khan, H., Nasreen, Z., Assad, N., Ali Turab, S., & Hanif, M. (2025). Exploring the Antibacterial, Anticoagulant, and Hemolytic Potential of Green-Synthesized Fe(2)O(3) Nanoparticles by Cucurbita pepo Pulp. *IEEE Transactions on Nanobioscience*, *24*(4), 411–420. DOI: 10.1109/TNB.2025.3563307 PMID: 40261780

Gupta, S., Kumar, N., Verma, D., Joshi, A., Kumar Arya, A., Joshi, K., & Bachheti, R. (2021). Impacts of invasive alien plant species on biodiversity in the regions of Western Himalayas, India: An overview. In (pp. 123-135). DOI: 10.26832/aesa-2021-bdcp-08

Haeberli, W., & Hoelzle, M. (1995). Application of inventory data for estimating characteristics of and regional climate-change effects on mountain glaciers: A pilot study with the European Alps. *Annals of Glaciology*, *21*, 206–212. DOI: 10.3189/S0260305500015834

Hall, K., & Walton, D. W. (1992). Rock weathering, soil development and colonization under a changing climate. *Philosophical Transactions of the Royal Society of London. Series B, Biological Sciences*, *338*(1285), 269–277. DOI: 10.1098/rstb.1992.0147

Haraguchi, A., Nagasawa, J., Kuramochi, K., Tsuchida, S., Kobayashi, A., Hatabu, T., Sasai, K., Ikadai, H., Ushida, K., & Matsubayashi, M. (2024). Anticoccidial activity of the secondary metabolites in alpine plants frequently ingested by wild Japanese rock ptarmigans. *International Journal for Parasitology. Parasites and Wildlife*, *25*, 100967. DOI: 10.1016/j.ijppaw.2024.100967 PMID: 39220322

Haugland, J. E. (2006). Short-term periglacial processes, vegetation succession, and soil development within sorted patterned ground: Jotunheimen, Norway. *Arctic, Antarctic, and Alpine Research, 38*(1), 82–89. DOI: 10.1657/1523-0430(2006)038[0082:SPPVSA]2.0.CO;2

Heinrich, M. (2010). Ethnopharmacology in the 21st century - grand challenges [Specialty Grand Challenge]. *Frontiers in pharmacology, Volume 1 - 2010.* DOI: 10.3389/fphar.2010.00008

Hernández, A., Miranda, M. D., Arellano, E. C., & Dobbs, C. (2016). Landscape trajectories and their effect on fragmentation for a Mediterranean semi-arid ecosystem in Central Chile. *Journal of Arid Environments, 127*, 74–81. DOI: 10.1016/j.jaridenv.2015.10.004

Hussain, M., Butt, A. R., Uzma, F., Ahmed, R., Irshad, S., Rehman, A., & Yousaf, B. (2020). A comprehensive review of climate change impacts, adaptation, and mitigation on environmental and natural calamities in Pakistan. *Environmental Monitoring and Assessment, 192*(1), 48. DOI: 10.1007/s10661-019-7956-4 PMID: 31844992

Inouye, D. W. (2020). Effects of climate change on alpine plants and their pollinators. *Annals of the New York Academy of Sciences, 1469*(1), 26–37. DOI: 10.1111/nyas.14104 PMID: 31025387

Inouye, D. W., Larson, B. M., Ssymank, A., & Kevan, P. G. (2015). Flies and flowers III: Ecology of foraging and pollination. *Journal of Pollination Ecology, 16*, 115–133. DOI: 10.26786/1920-7603(2015)15

Inouye, D. W., & Pyke, G. H. (1988). Pollination biology in the Snowy Mountains of Australia: Comparisons with montane Colorado, USA. *Australian Journal of Ecology, 13*(2), 191–205. DOI: 10.1111/j.1442-9993.1988.tb00968.x

Inouye, D. W., & Wielgolaski, F. E. (2025). Phenology at high altitudes. In *Phenology: An integrative environmental science* (pp. 281–311). Springer.

Ismail, M., Ali, S., Abbas, Q., Kamal, Z., Hussain, I., Sidiqa, S.,…Nature. (2022). Medicinal and Aromatic Plants. 141.

Jaryan, V., Uniyal, S. K., Gupta, R. C., & Singh, R. D. (2013). Alien Flora of Indian Himalayan State of Himachal Pradesh. *Environmental Monitoring and Assessment, 185*(7), 6129–6153. DOI: 10.1007/s10661-012-3013-2 PMID: 23196408

Javeed, B., Huang, D., Shangguan, D., Mukhtar, M. A., Sajjad, W., Banerjee, A., & Butt, A. Q. (2024). Assessing the effectiveness of national park's policies and laws in promoting biodiversity conservation and ecological development in Pakistan. *Frontiers in Environmental Science, 11*, 1333650. Advance online publication. DOI: 10.3389/fenvs.2023.1333650

Jones, G. A., & Henry, G. H. (2003). Primary plant succession on recently deglaciated terrain in the Canadian High Arctic. *Journal of Biogeography, 30*(2), 277–296. DOI: 10.1046/j.1365-2699.2003.00818.x

Kameyama, Y., & Kudo, G. (2009). Flowering phenology influences seed production and outcrossing rate in populations of an alpine snowbed shrub, Phyllodoce aleutica: Effects of pollinators and self-incompatibility. *Annals of Botany, 103*(9), 1385–1394. DOI: 10.1093/aob/mcp037 PMID: 19228703

Khan, M., Baig, T., Ali, S., & Shah, A. (2022). *Mountain studies: Understanding and managing mountains for people and nature.* Department of Forestry, Range & Wildlife Management Karakoram International University Gilgit 15100, Pakistan 1-214.

Khuroo, A. A., Rashid, I., Reshi, Z., Dar, G. H., & Wafai, B. A. (2007). The alien flora of Kashmir Himalaya. *Biological Invasions, 9*(3), 269–292. DOI: 10.1007/s10530-006-9032-6

Kleidon, A., & Mooney, H. A. (2000). A global distribution of biodiversity inferred from climatic constraints: Results from a process-based modelling study. *Global Change Biology, 6*(5), 507–523. DOI: 10.1046/j.1365-2486.2000.00332.x

Kohli, R. K., Batish, D. R., Singh, H. P., & Dogra, K. S. (2006). Status, invasiveness and environmental threats of three tropical American invasive weeds (*Parthenium hysterophorus* L., *Ageratum conyzoides* L., *Lantana camara* L.) in India. *Biological Invasions, 8*(7), 1501–1510. DOI: 10.1007/s10530-005-5842-1

Kolawole, A. S., & Iyiola, A. O. (2023). Environmental Pollution: Threats, Impact on Biodiversity, and Protection Strategies. In Izah, S. C., & Ogwu, M. C. (Eds.), *Sustainable Utilization and Conservation of Africa's Biological Resources and Environment* (pp. 377–409). Springer Nature Singapore., DOI: 10.1007/978-981-19-6974-4_14

Kudo, G. (1996). Effects of snowmelt timing on reproductive phenology and pollination process of alpine plants. *Memoirs of National Institute of Polar Research. Special Issue, 51*, 71–82.

Kudo, G. (2014). Vulnerability of phenological synchrony between plants and pollinators in an alpine ecosystem. *Ecological Research, 29*(4), 571–581. DOI: 10.1007/s11284-013-1108-z

Kudo, G. (2016). Landscape structure of flowering phenology in alpine ecosystems: significance of plant–pollinator interactions and evolutionary aspects. In *Structure and function of mountain ecosystems in Japan: Biodiversity and Vulnerability to Climate Change* (pp. 41-62). Springer. DOI: 10.1007/978-4-431-55954-2_3

Kumar, A., Ansari, M. Z., Banjara, R. A., Romano, R., De Luca, L., Santini, A., & Tiwari, J. K. (2025). Nanotechnology for sustainable biotic stress management in plants. *Plant Stress, 18*, 101064. DOI: 10.1016/j.stress.2025.101064

Lefebvre, V., Fontaine, C., Villemant, C., & Daugeron, C. (2014). Are empidine dance flies major flower visitors in alpine environments? A case study in the Alps, France. *Biology Letters, 10*(11), 20140742. DOI: 10.1098/rsbl.2014.0742 PMID: 25376804

Mahanayak, B. J. W. J. B. P. H. S. (2024). Ex-situ and in-situ conservation of wild life. *18*(3), 277-282.

Makrodimos, N., Blionis, G. J., Krigas, N., & Vokou, D. (2008). Flower morphology, phenology and visitor patterns in an alpine community on Mt Olympos, Greece. *Flora-Morphology, Distribution. Flora (Jena), 203*(6), 449–468. DOI: 10.1016/j.flora.2007.07.003

Manish, K. (2022). Medicinal plants in peril due to climate change in the Himalaya. *Ecological Informatics, 68*, 101546. DOI: 10.1016/j.ecoinf.2021.101546

Mao, X., & Yang, Z. (2023). Chiral nanomaterial-enabled bactericide and nanopesticide for sustainable agriculture and food safety. *Journal of Nanoparticle Research, 25*(10), 198. DOI: 10.1007/s11051-023-05845-9

Matthews, J. A. (1992). *The ecology of recently-deglaciated terrain: a geoecological approach to glacier forelands*. Cambridge University Press.

Mayewski, P. A., & Jeschke, P. A. (1979). Himalayan and Trans-Himalayan glacier fluctuations since AD 1812. *Arctic and Alpine Research, 11*(3), 267–287. DOI: 10.1080/00040851.1979.12004137

Meng, F., Cui, S., Wang, S., Duan, J., Jiang, L., Zhang, Z., & Li, X. (2016). Changes in phenological sequences of alpine communities across a natural elevation gradient. *Agricultural and Forest Meteorology, 224*, 11–16. DOI: 10.1016/j.agrformet.2016.04.013

Meng, F., Jiang, L., Zhang, Z., Cui, S., Duan, J., Wang, S., & Li, X. (2017). Changes in flowering functional group affect responses of community phenological sequences to temperature change. *Ecology*, *98*(3), 734–740. DOI: 10.1002/ecy.1685 PMID: 27984640

Miller-Struttmann, N. E., Geib, J. C., Franklin, J. D., Kevan, P. G., Holdo, R. M., Ebert-May, D., Lynn, A. M., Kettenbach, J. A., Hedrick, E., & Galen, C. (2015). Functional mismatch in a bumble bee pollination mutualism under climate change. *Science*, *349*(6255), 1541–1544. DOI: 10.1126/science.aab0868 PMID: 26404836

Minami, Y., Kanda, H., & Masuzawa, T. (1996). The relationship between distribution of bryophytes and soil conditions on deglaciated arctic terrain in Ny-Ålesund. Proceedings of the NIPR Symposium on Polar Biology.

Mooney, , H. ADrake, , J. A. (1987). The Ecology of: Biological Invasions. *Environment: Science and Policy for Sustainable Development*, *29*(5), 10–37. DOI: 10.1080/00139157.1987.9928888

Namdari, M., Eatemadi, A., Soleimaninejad, M., & Hammed, A. T. (2017). A brief review on the application of nanoparticle enclosed herbal medicine for the treatment of infective endocarditis. *Biomedicine and Pharmacotherapy*, *87*, 321–331. DOI: 10.1016/j.biopha.2016.12.099 PMID: 28064105

Obeid, M. A., Al Qaraghuli, M. M., Alsaadi, M., Alzahrani, A. R., Niwasabutra, K., & Ferro, V. A. (2017). Delivering natural products and biotherapeutics to improve drug efficacy. *Therapeutic Delivery*, *8*(11), 947–956. DOI: 10.4155/tde-2017-0060 PMID: 29061102

Ong, T. W., Lin, B. B., Lucatero, A., Cohen, H., Bichier, P., Egerer, M. H., Danieu, A., Jha, S., Philpott, S. M., & Liere, H. (2022). Rarity begets rarity: Social and environmental drivers of rare organisms in cities. *Ecological Applications*, *32*(8), e2708. DOI: 10.1002/eap.2708 PMID: 35810452

Pangtey, Y., Rawal, R., Bankoti, N., & Samant, S. (1990). Phenology of high-altitude plants of Kumaun in Central Himalaya, India. *International Journal of Biometeorology*, *34*(2), 122–127. DOI: 10.1007/BF01093457

Patra, J. K., Das, G., & Baek, K.-H. (2016). Towards a greener environment: Synthesis and applications of green nanoparticles. *Pakistan Journal of Agricultural Sciences*, *53*(2), 345–354. DOI: 10.21162/PAKJAS/16.3027

Patra, J. K., Das, G., Fraceto, L. F., Campos, E. V. R., Rodriguez-Torres, M. P., Acosta-Torres, L. S., Diaz-Torres, L. A., Grillo, R., Swamy, M. K., Sharma, S., Habtemariam, S., & Shin, H.-S. (2018). Nano based drug delivery systems: Recent developments and future prospects. *Journal of Nanobiotechnology*, *16*(1), 71. DOI: 10.1186/s12951-018-0392-8 PMID: 30231877

Prevéy, J. S., Rixen, C., Rüger, N., Høye, T. T., Bjorkman, A. D., Myers-Smith, I. H., & Chisholm, C. L. (2019). Warming shortens flowering seasons of tundra plant communities. *Nature Ecology & Evolution*, *3*(1), 45–52. DOI: 10.1038/s41559-018-0745-6 PMID: 30532048

Qazi, A. W., Haq, M. Z. U., Tariq, A., & Shitu, K. (2025). Prioritizing threats to rare and endemic species: Expertise to insights. *Sustainable Futures : An Applied Journal of Technology, Environment and Society*, *10*, 101430. DOI: 10.1016/j.sftr.2025.101430

Qazi, A. W., & Saqib, Z., Zaman-ul-Haq, M., Gardezi, S. M. H., Khan, A. M., Khan, I.,…Ahmed, I. (2023). Modelling impacts of climate change on habitat suitability of three endemic plant species in Pakistan. *Management*, *16*, 17.

Ramana, K. V., Singhal, S. S., & Reddy, A. B. (2014). Therapeutic potential of natural pharmacological agents in the treatment of human diseases. *BioMed Research International*, *573452*, 1–4. Advance online publication. DOI: 10.1155/2014/573452 PMID: 25610867

Robinson, M. M., & Zhang, X. (2011). The world medicines situation 2011, traditional medicines: Global situation, issues and challenges. *World Health Organization, Geneva, 31*, 1-2.

Roland, J., & Matter, S. F. (2007). Encroaching forests decouple alpine butterfly population dynamics. *Proceedings of the National Academy of Sciences of the United States of America*, *104*(34), 13702–13704. DOI: 10.1073/pnas.0705511104 PMID: 17699630

Seidl, A. (1999). *Population demographics and the conservation status of the uncompahgre fritillary Boloria acrocnema*. Nymphalidae.

Sekar, K. C., Pandey, A., & Giri, L. (2015). Invasive alien plants of Himachal Pradesh, India. *Indian Forester*, *141*, 520–527.

Semenchuk, P. R., Gillespie, M. A., Rumpf, S. B., Baggesen, N., Elberling, B., & Cooper, E. J. (2016). High Arctic plant phenology is determined by snowmelt patterns but duration of phenological periods is fixed: An example of periodicity. *Environmental Research Letters*, *11*(12), 125006. DOI: 10.1088/1748-9326/11/12/125006

Stanton, M. L., & Galen, C. (1989). Consequences of flower heliotropism for reproduction in an alpine buttercup (Ranunculus adoneus). *Oecologia, 78*(4), 477–485. DOI: 10.1007/BF00378737 PMID: 28312176

Studer, S., Appenzeller, C., & Defila, C. (2005). Inter-annual variability and decadal trends in alpine spring phenology: A multivariate analysis approach. *Climatic Change, 73*(3), 395–414. DOI: 10.1007/s10584-005-6886-z

Tang, S., Liang, J., Li, O., Shao, N., Jin, Y., Ni, J., Fei, X., & Li, Z. (2023). Morphology-Tailored Hydroxyapatite Nanocarrier for Rhizosphere-Targeted Phosphorus Delivery. *Small, 19*(14), 2206954. DOI: 10.1002/smll.202206954 PMID: 36599675

Tang, Y., Zhao, W., Zhu, G., Tan, Z., Huang, L., Zhang, P., Gao, L., & Rui, Y. (2024). Nano-Pesticides and Fertilizers: Solutions for Global Food Security. *Nanomaterials (Basel, Switzerland), 14*(1), 90. DOI: 10.3390/nano14010090 PMID: 38202545

Tariq, A., Shu, H., Siddiqui, S., Munir, I., Sharifi, A., Li, Q., & Lu, L. (2022). Spatio-temporal analysis of forest fire events in the Margalla Hills, Islamabad, Pakistan using socio-economic and environmental variable data with machine learning methods. *Journal of Forestry Research, 33*(1), 183–194. DOI: 10.1007/s11676-021-01354-4

Totland, Ø. (1993). Pollination in alpine Norway: Flowering phenology, insect visitors, and visitation rates in two plant communities. *Canadian Journal of Botany, 71*(8), 1072–1079. DOI: 10.1139/b93-124

Tyagi, P. K., Arya, A., Ramniwas, S., & Tyagi, S. (2023). Editorial: Recent trends in nanotechnology in precision and sustainable agriculture. *Frontiers in Plant Science, 14*, 1256319. DOI: 10.3389/fpls.2023.1256319 PMID: 37636083

Vashistha, R. K., Rawat, N., Chaturvedi, A. K., Nautiyal, B., Prasad, P., & Nautiyal, M. (2009). An exploration on the phenology of different growth forms of an alpine expanse of North-West Himalaya, India. *New York Science Journal, 2*(6), 29–41.

Venn, S. E., & Morgan, J. W. (2007). Phytomass and phenology of three alpine snowpatch species across a natural snowmelt gradient. *Australian Journal of Botany, 55*(4), 450–456. DOI: 10.1071/BT06003

Walker, D., Halfpenny, J. C., Walker, M. D., & Wessman, C. A. (1993). Long-term studies of snow-vegetation interactions. *Bioscience, 43*(5), 287–301. DOI: 10.2307/1312061

Wei, G. (2013). Green technology for nanoparticles in biomedical applications. In *Green biosynthesis of nanoparticles: mechanisms and applications* (pp. 1–12). CABI Wallingford UK.

Wester, P., Mishra, A., Mukherji, A., & Shrestha, A. B. (2019). *The Hindu Kush Himalaya assessment: mountains, climate change, sustainability and people.* Springer Nature.

Woodford, D. J., Ivey, P., Novoa, A., Shackleton, R., Richardson, D., Weyl, O., & Zengeya, T. (2017). Managing conflict-generating invasive species in South Africa: Challenges and trade-offs. *Bothalia - African Biodiversity &amp. Conservation, 47*(2), 1–11. DOI: 10.4102/abc.v47i2.2160

Wu, H., Chen, J., Xu, J., Zeng, G., Sang, L., Liu, Q., Yin, Z., Dai, J., Yin, D., Liang, J., & Ye, S. (2019). Effects of dam construction on biodiversity: A review. *Journal of Cleaner Production, 221*, 480–489. DOI: 10.1016/j.jclepro.2019.03.001

Yadav, S., Kumari, A., & Kumar, V. (2012). Nanotechnology: A tool to enhance therapeutic values of natural plant products. *Trends in Medical Research, 7*(2), 34–42. DOI: 10.3923/tmr.2012.34.42

Yohan, D., & Chithrani, B. D. (2014). Applications of nanoparticles in nanomedicine. *Journal of Biomedical Nanotechnology, 10*(9), 2371–2392. DOI: 10.1166/jbn.2014.2015 PMID: 25992462

Yuan, H., Ma, Q., Ye, L., & Piao, G. (2016). The Traditional Medicine and Modern Medicine from Natural Products. *Molecules (Basel, Switzerland), 21*(5), 559. Advance online publication. DOI: 10.3390/molecules21050559 PMID: 27136524

Zemp, M. (2008). *Global glacier changes: facts and figures.* UNEP/Earthprint.

Zhou, J., Cai, W., Qin, Y., Lai, L., Guan, T., Zhang, X., & Cong, Z. (2016). Alpine vegetation phenology dynamic over 16 years and its covariation with climate in a semi-arid region of China. *The Science of the Total Environment, 572*, 119–128. DOI: 10.1016/j.scitotenv.2016.07.206 PMID: 27494658

Chapter 13
Honey Bee Flora: Identification, Characterization, Constraints, Pollen Substitutes, and Opportunity of Beekeeping in Alpine Zones

Hansa Gul
http://orcid.org/0009-0001-5554-617X
Department of Zoology, University of Mianwali, Punjab, Pakistan

Muhammad Nauman Khan
http://orcid.org/0000-0003-3790-0902
University of Chakwal, Pakistan

Muhammad Irfan
Department of Zoology, University of Mianwali, Punjab, Pakistan

Nouman Ahmad
http://orcid.org/0009-0009-3370-2862
University of Mianwali, Punjab, Pakistan

Majid Iqbal
http://orcid.org/0000-0003-0561-8494
Institute of Geographic Sciences and Natural Resources Research, Chinese Academy of Sciences, China

ABSTRACT

The alpine regions, with their own climatic characteristic and a variety of floral resources, present rather unexploited but promising perspectives of sustainable beekeeping. It is in these areas at high altitude that this chapter addresses in quan-

DOI: 10.4018/979-8-2600-1207-9.ch013

titative detail the identification and characterization of honey bee flora, its diversity and seasonal availability of both nectar and pollen providing sources. The special emphasis is made on the classification of the types of pollen, nutritional constituents of the pollen and the part played by the pollen in maintaining bee health, productivity and disease resistance. Ecological and agricultural importance of alpine bee flora is discussed also, especially its contribution to increased pollination of crops and maintenance of biodiversity. In spite of the possibilities, a few limitations which prevail on beekeeping in the alps are climatic extremes, shortage of forage, loss of habitat, and poor infrastructure.

1. INTRODUCTION

Beekeeping in Pakistan has significant ecological and economic significance, supporting national food security, rural livelihoods, and international trade. Globally, honeybees, especially *Apis mellifera*, support an ecosystem-service economy valued at over US$200 billion annually. In Pakistan, this reliance is particularly critical, as pollination-dependent crops worth approximately US$1590 million each year rely on honeybee activity, influencing the productivity of 61 major crop species. The country is also hosts several important native species, including *Apis dorsata*, *Apis cerana*, and *Apis florea*, each of which contributes uniquely to local pollination networks (Usman et al., 2022).

Honey bees (*Apis mellifera*) play a crucial role in honey production and in pollinating agricultural crops and wild plants, making them key species for conserving biodiversity and world food security. As social insects, they significantly contribute to increasing agricultural production, maintaining vegetative biodiversity, and securing food security worldwide. In addition to pollination, honey and other hive products like beeswax and propolis are economically important and are commonly used in the food, cosmetic, and medical industries (Ahmad & Aziz, 2017; Nader et al., 2021).

The history of apiculture in Pakistan has been both strength and a weakness. Historically, beekeepers in mountainous areas kept honey-producing colonies of *Apis cerana*. However, a severe epidemic between 1980 and 1983, likely caused by tracheal mites, killed a huge percentage of these colonies (Waghchoure-Camphor & Martin, 2008). In response, programs were initiated to introduce Western honey bees (*A. mellifera*), which now make up the majority of commercial apiculture in Pakistan. In 2008, approximately 2,000 beekeepers managed 200,000 to 300,000 *A. mellifera* colonies, producing 7,500 tons of honey annually (Waghchoure-Camphor & Martin, 2008). The industry has since expanded to around 10,000 beekeepers with 0.5 to 1.1 million colonies, producing 16,000 tons of honey per year. The domestic consumption exceeds 11,000 tons, with an average of 50 g per capita. Exports total

4,000 tons, contributing about US$ 23 million to Pakistan's GDP (Usman et al., 2022). With these developments, Pakistan ranks behind only 20 countries in honey production and r 34 countries in honey export, and the average beekeeper produces 11.7 kg of honey compared to the global average of 20.6 kg. These figures reflect both the improvement and the untapped potential of the industry, especially given the ever-growing global demand for honey bee products (Usman et al., 2022).

Nevertheless, the health and biodiversity of honey bees are increasingly challenged by a complex mix of biotic and abiotic stress factors (Ahmad & Aziz, 2017; Al Naggar et al., 2021; Farooqi et al., 2021; Zaric et al., 2018). Over the past six decades, the number of managed honey bee colonies worldwide has increased by 85%, honey production by 181%, beeswax production by 116%. However, these increases are lower than the growth of the human population, which has risen by approximately 50% globally (Phiri et al., 2022) and by over 500% in Pakistan during the same period. Therefore, the ratio of colonies to population is gradually falling (Phiri et al., 2022). For example, managed colonies in Europe and the United States have dropped by approximately 30% in the last decade (Wakgari & Yigezu, 2021). Even more concerning is the situation of wild honey bee populations, which are equally threatened by pesticides, environmental pollutants, new pathogens, deforestation, and climate change (Wakgari & Yigezu, 2021). These declines reduce agricultural yields and have the potential to cause large scale food security crises.

Ecological constraints increase the magnitude of these challenges in alpine environments. Honey bees are poikilothermic insects; therefore, extreme temperatures and seasonal changes in nectar and pollen limit foraging and make the insects vulnerable to nutritional stress and even colony collapse. Alpine areas, in addition to their high levels of specialization and diversity, have long winters, brief flowering periods, and erratic weather patterns that cause prolonged periods of dearth. These periods of flower deficiency directly and negatively impact brood rearing, colony development and susceptibility to illnesses. Artificial pollen and nectar have been proven to boost survival, productivity, and hardiness of colonies in poor environments, and beekeepers everywhere use these substitutes to fill these gaps (Rueda-Uribe et al., 2025).

This is a comprehensive overview of alpine beekeeping in both the Pakistani and global context. It, first, explains the peculiar nature of alpine honey plants and the timing of their emergence. It then considers the main limitations facing alpine apiculture, including the ecological demands of climate change and the nutritional bottlenecks during intervals of dearth. The discussion then turns to management strategies, with particular emphasis is placed on pollen and nectar substitutes as essential tools for preserving colony health. Lastly, it identifies the potential for sustainable beekeeping in the alpine areas of Pakistan, reiterating their vast floral resources, economic prospects and the need to integrate traditional ecological knowledge with scientific studies to ensure sustainability.

2. IDENTIFICATION AND CHARACTERIZATION OF ALPINE HONEY BEE FLORA

Identifying the most important melliferous flora is the ultimate stage of sustainable apiculture in any area, especially in the unique ecosystems of alpine regions. Although initial insights can be gained through direct observation of bee foraging, the most conclusive and scientific method of identification is melissopalynology, the scientific discipline that ensures accurate identification of the plants pollinated by bees and provides information on the origin of honey (Daugaliyeva et al., 2024). This procedure can be performed by microscopic analysis of pollen grains collected from honey, a pollen trap, or directly from the bee to determine the precise botanical origin of the honey. This powerful scientific methodology, which dates back to the work of Pfister in 1895, goes beyond simple observation to provide empirical information about these critical resources.

2.1 Microscopic Characterization of Pollen: A Diagnostic Tool for Flora Identification and Honey Authentication

In addition to straightforward identification, the scientific field of palynology offers a microscopic set of tools for the accurate description of melliferous flora, which is necessary for clarity in both taxonomy and honey authentication. This is evidenced by research conducted by Ahmad et al. (2024), who performed a systematic morpho-palynological study of significant plants foraged by bees in Khyber Pakhtunkhwa (Ahmad et al., 2024). The research followed a meticulous procedure, recording the distinctive characteristics of each species' pollen grains using the basic acetolysis method of pollen preparation, as adopted in classical palynology. These classifications use standardized terminology for pollen grain morphology, ensuring consistency and comparability with global palynological data (Punt et al., 2007). The diagnoses showed wide differences in pollen size: *Brassica oleracea* had an equatorial diameter of 17.40 µm, while *Asphodelus tenuifolius* had a pollen size of 46.29 µm. Moreover, the exine, which forms the complex pattern on the outer wall of the pollen, was used as an identifying factor, such as the reticulate pattern of the Brassicaceae family and the smooth, non-psilate surface of *Cicer arietinum* pollen (Ahmad et al., 2024). Such detail is essential, and the study will build on previous regional investigations, such as the work at Mardan, to develop a larger palynological database for Pakistan (Nabila et al., 2022).The practical applications of this work are profound. For taxonomists, these pollen features can help resolve systematic questions within plant genera (Majeed et al., 2020). For the apiculture industry, this research forms the foundation of melissopalynology (Noor et al., 2017). By creating a reference library of local pollen types, analysts can examine honey samples to verify their botanical origin, confirm monofloral claims, and detect fraudulent products. This scientific validation enhances the market value of local honey and provides beekeepers with a powerful tool for quality assurance (Ahmad et al., 2024; Noor et al., 2017).

2.2 Seasonal Foraging Patterns and Floral Calendars in Alpine Beekeeping

The seasonal availability of floral resources is a limiting factor in honey bee foraging behavior and honey production. Mapping the flowering periods of melliferous plants in Khyber Pakhtunkhwa revealed that spring and autumn are the main honey production seasons. Spring foraging is largely supported by species of Brassicaceae and Asphodelaceae, while autumn foraging relies on Fabaceae and Poaceae, such as *Saccharum spontaneum* L. Interestingly, local beekeepers realized that autumn honey is higher in both quality and quantity (Ahmad et al., 2024).

This seasonal information forms the basis of a floral calendar, a technology that helps apiarists manage hives by coordinating colony movements with flowering transitions (Waykar & Baviskar, 2015). By using migratory practices based on such calendars, beekeepers can monitor the sequence of blooms at various elevations and districts, reduce the risk of forage shortages, and ensure a high honey yield. These and related approaches have been identified in research assessing the nectar profiles of particular plants to predict honey production potential (Na et al., 2024). Therefore, the association between floral diversity and temporal dynamics will make apiculture more resilient, increase colony survival during dearth seasons, and increase the overall economic sensitivity of alpine beekeeping.

One of the crucial results of the comprehensive floral survey conducted by Ahmad et al. (2025) is the visualization of seasonal foraging behavior, which will serve as a blueprint to guide optimal bee use in Khyber Pakhtunkhwa. The research clearly distinguishes between the two major honey seasons (spring and fall), each supported by a specific set of melliferous plants. During spring, honey bee colonies grow in size and produce honey after extensive foraging on flowers from the Brassicaceae and Asphodelaceae families. In contrast, plants from the Fabaceae and Poaceae families, such as *S. spontaneum*, maintain the fall honey flow.

An important field observation in the study is that honey produced during the fall season is often perceived by local experts as superior in both quality and quantity (Ahmad et al., 2024). This knowledge empowers beekeepers to transition from passive hive placement to a proactive, data-driven management strategy centered on a "floral calendar," a concept recognized as vital in modern apiculture (Waykar & Baviskar, 2015). With this calendar, beekeepers can plan migratory routes, moving their hives to follow the succession of blooms across different districts and elevations, thereby maximizing nectar collection and honey yield throughout the year. This approach is consistent with broader research on optimizing honey production by understanding the nectar characteristics of specific flora (Na et al., 2024). By contributing to the growing body of palynological literature on Pakistan's melliferous species, the work by Ahmad et al. (2024) provides the scientific foundation for such strategic decisions. This ultimately helps beekeepers mitigate risks associated with floral dearth, enhance colony health, and improve the economic viability of their operations, fostering a more resilient and productive regional apiculture industry (Ahmad et al., 2024).

2.3 The Dynamic Nature of Alpine Flora

The alpine regions are global biodiversity hotspots and contain exceptionally rich flora essential to pollinators. For example, the tropical Andes are estimated to host more than 28,000 known floral species, several of which are highly endemic and have evolved distinctive pollination relationships (Pérez-Escobar et al., 2022; Rueda-Uribe et al., 2025). This vast floral diversity forms the foundation for the well-being of many pollinator species and represents a tremendous, only partially utilized potential for beekeeping. Such ecosystems are often dominated by specific plants adapted to extreme environments, such as long-lived herbaceous perennials. These plants provide exclusive, high-quality sources of nectar and pollen that may lead to the production of monofloral honey with characteristic flavor, color, and healing effects.

One of the hallmarks and challenges of alpine honey bee flora is its high phenological diversity. The environmental signals that regulate flowering, which determine the timing of the nectar and pollen season, vary widely between lowland and highland systems. The annual rhythmicity of flowering in the tropical Andes, notably under rainfall-driven conditions, is the direct opposite of the temperature-based phenology found in temperate alpine areas. This climate sensitivity implies that forage availability can fluctuate dramatically each year, depending on macro-scale events such as El Ninos. This variability poses a significant problem for managed honey bees, whose survival depends on a regular and good supply of forage. Thus, the classification and description of alpine flora should consider not only the number of species, but also the variety of flowering strategies. To maintain a continuous food source, a mosaic of plants with staggered flowering periods is necessary, providing a buffer for the whole system and enabling bee colonies to survive despite the natural climatic fluctuations of high altitude habitats (Noor et al., 2017; Rueda-Uribe et al., 2025).

The description of alpine honey bee flora is not a fixed task; it should reflect the changing processes of plant flora under the influence of climate change. Long-term studies have found that earlier snowmelt is lengthening and advancing the start of growing seasons, which in turn extends flowering phenology, though this does not apply to all species. Not all plants are sensitive to environmental cues, creating a risk of phenological mismatch between plants and their pollinators. In one case, warmer temperatures compressed the community-level flowering season, with late-flowering species flowering earlier than early-flowering species. In addition, warming is causing subalpine plants and shrubs to move to higher elevations, where they may outcompete smaller native alpine herbaceous plants for resources such as light and water. This shrubification of alpine meadows is reducing the number of forageable plants available to traditional honey bees, thereby changing the composition and availability of floral resources over time (Inouye, 2020).

2.4 Constraints on Beekeeping in Alpine Zones

Frankly, the success of any beekeeping business, especially in the challenging conditions of alpine zones, depends entirely on the continuous supply of floral sources. The time of year when nectar and pollen are in short supply is called the dearth period. Extremely limited foraging behavior during this period also causes honey bee colonies to enter a partially inactive state to conserve energy. Some primary colony functions, including the queen's egg-laying and brood rearing, diminish dramatically, and the extent of this decrease depends on the amount of stored food in the hive (Paray et al., 2021).

Although migratory beekeeping has been introduced as a solution to relocate colonies to regions with more abundant flora, it is a labor-intensive, costly, and risky activity. Research has indicated that despite migration, a large portion of colonies succumbs during periods of dearth (Kumar & Agrawal, 2014). In alpine zones, this is a serious problem because the flowering season is often brief and intense, followed by a prolonged and severe period of dearth due to cold weather and snow cover. Thus, one of the most significant challenges in creating productive and sustainable beekeeping in high-altitude regions is how to support honey bee colonies during periods of dearth.

To provide nutrition to bodies under heavy nutritional stress during dearth periods, artificial feeding has become a critical management practice. To maintain colony strength, assist brood-rearing, and guarantee a healthy population for the next honey flow season, scientists and beekeepers worldwide have developed numerous diet formulations to replace natural pollen with artificial pollen sources (Paray et al., 2021). These diets are designed on the nutritional profile of natural pollen, considering acceptability, digestibility, and economic feasibility of the materials used.

3. ISSUES AND CURRENT SITUATION OF BEEKEEPING IN PAKISTAN'S ALPINE ZONES

Although the Himalayas, Karakorum, and Hindukush ranges merge in Pakistan in the alpine and sub-alpine areas, which have extremely high potential due to their rich and unique floral biodiversity, beekeeping in these high altitude regions faces a web of intertwined challenges. The paradox is that, despite enormous untapped floral resources, intensive biotic, climatic, and anthropogenic stressors restrain production and threaten the long-term sustainability of apiculture (Khan & Ghramh, 2023). These limitations are especially pronounced in alpine areas, where honey bee colonies are more exposed to extreme weather conditions. Formulating effective strategies to manage these complex challenges and realize the full potential of mountain beekeeping depends on understanding these diverse issues.

3.1 Biotic Constraints: The Dual Pressure of Pests and Species Competition

The greatest biotic stressor in Pakistan beekeeping is the constant threat of pests and diseases, a consequence of the introduction and widespread adoption of the Western honey bee (*Apis mellifera*). The native hill honey bee, *Apis cerana*, experienced a population decline between 1980 and 1983 a clear example of this susceptibility. This outbreak was believed to be caused by tracheal mites introduced by *A. mellifera*, which nearly wiped out the species best adapted to Pakistan's mountainous regions. Current *A. mellifera* beekeepers face a dual mite crisis from the ectoparasitic mites Varroa destructor (first documented in 2002) and *Tropilaelaps clareae*. The combined impact of these two mites has been estimated to reduce honey production by up to 50 percent, forcing beekeepers into a continuous and costly cycle of chemical treatments with acaricides such as flumethrin and fluvalinate (Khan & Ghramh, 2023). This biotic problem is even more severe in the Alps. Although *A. cerana* is better adapted to cold climates and does not suffer from mite infestations, its populations are small and its contribution, though highly valued (approximately 60 tons per/year), is insignificant compared to the commercial output of *A. mellifera*. This has created a high-risk dependency on the more mite-susceptible exotic species, where any lapse in management can result in catastrophic colony losses. Furthermore, the close proximity of managed *A. mellifera* and remnant native *A. cerana* populations continues to pose a risk of interspecies disease transmission. Beyond mites, beekeepers also face bacterial diseases like American Foulbrood (AFB) and fungal diseases such as chalkbrood, further straining colony health and economic viability (Khan & Ghramh, 2023).

3.2 Climatic and Forage Constraints: The Alpine Dearth Period and Phenological Shifts

The success of alpine beekeeping is basically determined by the abundance of floral resources, which are limited by a brief and intense flowering period, followed by an extended and severe winter dearth when food is scarce. Honey bees are poikilothermic insects and are therefore directly affected by extreme temperatures, which restrict their foraging activity. Long winters and the short alpine growing season provide a critical nutritional bottleneck during which colonies may starve and cease brood rearing. Although the major intervention to address this challenge is migratory beekeeping shifting colonies between lowland spring blooms and mountain summer forage this practice is labor intensive, expensive, and does not eliminate the risk of colony losses during periods of dearth (Khan & Ghramh, 2023).

These forage constraints are being aggravated by climate change. Changes in winter snowpack and reduced snowmelt expose young floral buds to late spring frosts, which can destroy much of the season's forage in a single frosty event. This unpredictability has made it challenging for beekeepers to time their migrations. Moreover, rising temperatures are causing "shrubification", the upward movement of subalpine shrubs, which outcompete and replace the native herbaceous wildflowers that bees rely on (Inouye, 2020). This environmental change is actively depleting the variety and quantity of quality forage in traditional alpine meadows, directly endangering the carrying capacity of these ecosystems for managed honey bee colonies (Table 1).

Beyond the general challenge of the dearth period, alpine beekeeping faces specific climatic constraints driven by climate change, most notably changes in winter snowpack. Snow cover is a critical environmental component in alpine ecosystems; it insulates the soil and overwintering plants from extreme cold, and its meltwater is the primary source of moisture for early-season floral growth. However, rising temperatures are causing more precipitation to fall as rain instead of snow, and the snowpack that does form melts earlier in the spring. This results in a longer, but potentially drier growing season. A particularly severe consequence of earlier snowmelt is the increased risk of frost damage to wildflowers.

Early-flowering floral buds and reproductive structures are at risk from late spring frost events. Without the protective cover of a late- melting snowpack, these frosts can cause substantial loss of potential bee forage during that season (Inouye, 2020). This increased prevalence of frost damage is an important and unforeseeable limitation on the accessibility of reliable nectar and pollen sources for honey bee colonies in alpine areas.

3.3 Habitat Loss and Fragmentation

Besides climatic changes, the physical destruction of bee habitats is a direct and increasing threat. Deforestation and the development of rangelands and forests into agricultural land are reducing the total area available to bees in the mountain valleys and foothills of Pakistan. This is a global concern in alpine ecosystems. For example, it is estimated that, by 2100, the tropical Andes will lose 15 million hectares of forest and 4 million hectares of grassland if current trends continue (Rueda-Uribe et al., 2025). This habitat fragmentation isolates pollinator populations and decreases the continuous mosaic of flowering plants needed to support bee colonies throughout the season. Even the loss of patches of native flora can eliminate important sources of nectar or pollen, leading to new or prolonged periods of starvation and further weakening colonies already struggling with pests and climatic changes.

3.4 Anthropogenic Pressures: Agrochemicals, Pollutants, and a Technological Gap

Another critical threat is pollution, which has resulted from the intrusion of agriculture into the alpine foothills and valleys. Pakistan has increased its use of agrochemicals nearly tenfold, including widespread use of bee-toxic pesticides such as neonicotinoids and organophosphates. Migratory beekeepers are particularly at risk because their colonies are exposed to a wide variety of chemicals. In addition to pesticides, recent studies also show the dangers of environmental pollutants such as heavy metals from the industrial sector, PM2.5 from air pollution, and even microplastics consumed through contaminated water and nectar, all of which can lead to gut dysbiosis and increased vulnerability to viruses. Although alpine zones may appear pristine, foraging in lowlands can result in significant exposure to chemicals and pollutants that are then carried into the mountains by the colony (Rashid et al., 2022).

Lastly, the Pakistani beekeeping industry faces a knowledge and technological gap that impedes its ability to meet growing demands. Khan & Ghramh (2023) identify the "non-implementation of new techniques" as a key challenge. This includes limited adoption of advanced practices such as artificial insemination of queen bees to genetically improve bee stock for traits like disease resistance, cold hardiness, and hygienic behavior. Although research institutions like NARC are working to bridge this gap, the widespread application of modern, molecular-based management strategies remains limited. This technological lag leaves the industry reliant on traditional methods and chemical treatments, constraining its potential for growth and its resilience against the escalating environmental and biotic challenges of the 21st century (Khan & Ghramh, 2023).

Table 1. Summary of major constraints affecting honey bee populations in alpine and associated zones

Category of Constraint	Specific Examples and Impacts	Key References
Climatic	**Altered Snowpack & Frost Risk:** Earlier snowmelt exposes floral buds to late spring frosts, destroying forage.	(Inouye, 2020)
	The Dearth Period: Short, intense flowering seasons followed by long periods of floral scarcity, causing nutritional stress and colony decline.	(Paray et al., 2021)
	Phenological Mismatches: Climate change alters the timing of flowering and pollinator emergence at different rates, disrupting synchronized relationships.	(Inouye, 2020)
Ecological	**Habitat Loss and Fragmentation:** Deforestation and land conversion for agriculture reduce the total area of available bee forage.	(Rueda-Uribe et al., 2025)
	Community Shifts ("Shrubification"): Upward migration of subalpine shrubs outcompetes and replaces native alpine herbaceous flora.	(Inouye, 2020)
	Biodiversity Loss: Over-reliance on a single managed species (*A. mellifera*) can increase competition and disease risk for native bees (*A. cerana*).	(Khan, 2020)
Chemical	**Pesticides:** Widespread use of agrochemicals (neonicotinoids, organophosphates) causes lethal and sublethal effects; residues are found in honey.	(Leska et al., 2021)
	Environmental Pollutants: Heavy metals, particulate matter (PM2.5), and ozone from industrial and urban sources impair bee health, navigation, and foraging.	(Leska et al., 2021)
	Microplastics: Ingestion of microplastics from contaminated water and nectar causes gut dysbiosis and increases susceptibility to viruses.	(Deng et al., 2021)
Biological	**Pests and Parasites:** Major threats from *Varroa destructor*, Tracheal mites (*Acarapis woodi*), and *Tropilaelaps mites*	(Fairbrother et al., 2014; Pettis et al., 2017)
	Diseases: Presence of bacterial diseases (American and European Foulbrood) and fungal diseases (Nosemosis, Chalkbrood).	(Leska et al., 2021)

Figure 1. Major constraints on honey bee population in alpine zones: A causal pathway

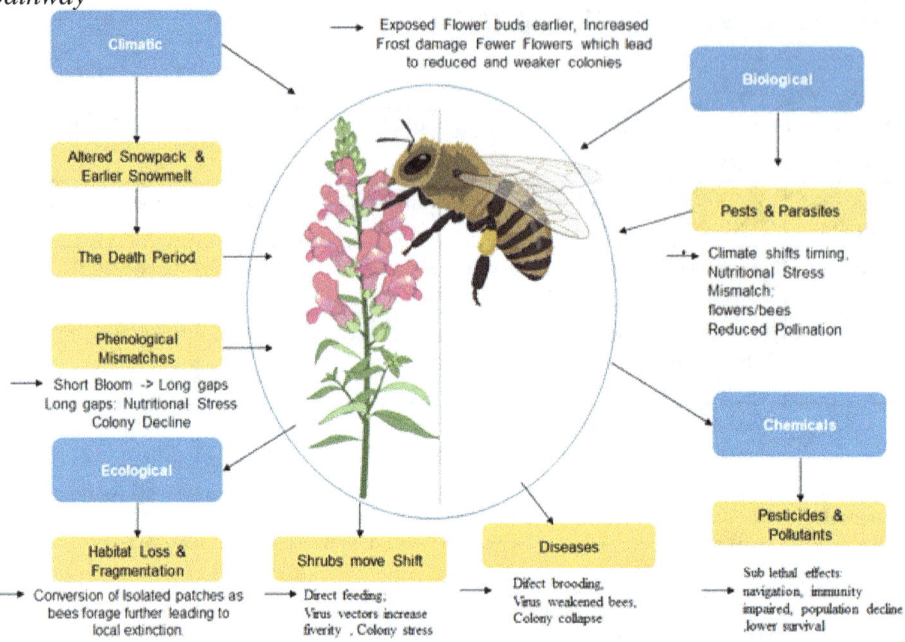

4. MANAGEMENT STRATEGIES: POLLEN SUBSTITUTES AND SUPPLEMENTAL FEEDING

To mitigate the nutritional stress caused by dearth periods, supplemental feeding is a critical management practice. Artificial diets are formulated to mimic the nutritional profile of natural nectar and pollen, ensuring colony survival and preparing colonies for the next honey flow season.

4.1 Protein-Based Diets:

Protein-based diets are the most important type of supplemental feeding since natural pollen is the major source of protein, lipids, vitamins, and minerals needed to rear brood. Protein deficiency directly affects the life expectancy of workers and the growth of a colony (Doull, 1980; Saffari et al., 2004). The tested ingredients include a wide range such as soybean flour, brewer's yeast, parched gram flour, skimmed milk powder, and casein. Recipes using these ingredients have been reported to stimulate brood rearing and result in satisfactory honey production, with some research findings indicating that fed colonies are much more likely to survive the dearth period than unfed colonies (Kim et al., 2024; Mishra et al., 1981).

4.2 Carbohydrate-Based Feeding:

During a nectar dearth, colonies can face starvation. The ideal carbohydrate supplement is a sucrose solution, typically mixed at a 2:1 ratio (sugar to water) for winter or 1:1 for spring stimulation (Herbert & Hill, 1992). This feeding not only prevents starvation but also enhances comb building, brood rearing, and overall foraging activity. Unrefined sugars like jaggery are generally not recommended as they can cause dysentery in bees (Paray et al., 2021).

4.3 Physiological Impact on Hypopharyngeal Glands:

The effectiveness of protein-based pollen substitutes can be understood by their impact on the hypopharyngeal glands of worker bees. These glands produce the protein-rich royal jelly essential for feeding the queen and larvae. The development of these glands depends directly on the quality of protein consumed. Studies have shown that bees fed high-quality protein diets have significantly larger hypopharyngeal glands than those fed lower-quality substitutes (Al-Ghamdi et al., 2011). A well-developed gland is a direct indicator of a bee's nutritional status and its capacity to rear brood effectively (Zahra & Talal, 2008).

Beyond the general challenge of the dearth period, alpine beekeeping faces specific climatic factors driven by climate change, most notably changes in winter snowpack. Snow cover is a critical environmental component in alpine ecosystems; it insulates soil and overwintering plants from extreme cold and its meltwater is the primary source of moisture for early-season floral growth. However, rising temperatures are causing more precipitation to fall as rain instead of snow, and the snowpack that does form melts earlier in the spring. This results in a longer, but potentially drier, growing season. A particularly severe consequence of earlier snowmelt is the increased risk of frost damage to wildflowers. Without the protective insulation of a late-lying snowpack, early-emerging floral buds and reproductive structures are more vulnerable to late spring frosts, which can destroy a significant portion of the available bee forage for that season. This increased frequency of frost damage is a major and unpredictable constraint on the availability of reliable nectar and pollen sources for honey bee colonies in alpine zones (Inouye, 2020).

The floral resources essential for alpine beekeeping are seriously threatened as habitat loss and climate change rapidly accelerate. A clear example of this issue is found in the tropical Andes. According to a recent thorough review, by 2100 the region might lose 15 million hectares of unmanaged forest and 4 million hectares of unmanaged grasslands due to continued fossil fuel consumption under current estimates. At the same time, climate change is causing plants and pollinators to migrate to higher altitudes, leading to a two-fold crisis: local extinctions at mountain tops and disruption of flower-pollinator interactions due to spatiotemporal disparities (Rueda-Uribe et al., 2025). This case study illustrates a universal problem, as the loss of alpine forage directly limits to the sustainability of both wild pollinators and managed honey bee populations.

5. OPPORTUNITIES FOR SUSTAINABLE ALPINE BEEKEEPING

One particularly crucial opportunity to advance alpine beekeeping is to bridge the gap between science and traditional ecological knowledge. Knowledge of high-altitude pollination is often incomplete and taxonomically biased.

5.1 Integrating Traditional and Scientific Knowledge

This lack of engagement is underscored by studies carried out in the tropical Andes, where a review of more than 470 scientific publications revealed that only 3% of the studies used local knowledge into their research design (Rueda-Uribe et al., 2025). This omission overlooks a critical source of information held by indigenous and local communities, who possess a strong, multi-generational understanding of the phenology of important bee flora, the locations of the best forage, and historical changes in plant communities. Integrating this knowledge should not be seen as an additive approach, but rather as a main approach to developing resilient and culturally relevant beekeeping programs. When local communities participate in research and management, it is possible to co- develop a more ecologically-responsive strategy that aligns with the specific ecological rhythms of alpine environments, thereby increasing both conservation outcomes and livelihood opportunities.

5.2 Leveraging Native Pollinators and Biodiversity

Although this work concentrates on the managed honey bee, *Apis mellifera*, it is necessary to know the native pollinator community to determine the opportunities and challenges of bee keeping. In alpine ecosystems, a specific set of pollinators has developed under severe conditions. Bumble bees (*Bombus* sp.) are typically the most prominent and significant pollinators in these areas because they can forage at relatively lower temperatures than other insects. Nevertheless, flies (*Diptera*) from the *Syrphidae* (hoverflies), *Muscidae*, and *Empididae* families can be the most important high-altitude pollinators in most alpine areas (especially those without bumble bees) (Inouye, 2020). These indigenous pollinators too are also adapting to climate change, with most species moving to higher elevations. The introduction of managed *Apis mellifera* colonies into these vulnerable ecosystems offers the potential for increased pollination, but also poses the risk of greater competition for floral resources with these dedicated native pollinators. Sustainable beekeeping in the Alps should therefore consider the environment's carrying capacity and seek to complement, rather than replace, the essential pollination services provided by native fauna.

To provide a real-life example of the application of alpine beekeeping principles, we refer to a detailed case study from Pakistan. The geographical peculiarity of the country's unique geography, where the Himalayas, Karakorum, and Hindukush mountain ranges converge, creates a mosaic of alpine and sub-alpine ecosystems and supports specialization in honey production. The land use data show that rangelands (27% of the total area) and forests (5.4%) are extensive and serve as prime bee forage (Khan, 2020). The scale of these floral resources is remarkable; it has been calculated that Pakistan's bee flora can sustain up to 3.5 million bee colonies, a capacity that is currently highly underutilized and represents significant potential for the growth of the beekeeping sector. Coniferous sub-alpine forests (3,350-3,800 m) and dry temperate forests (1,525-3,350 m) are the primary types and abundant sources of bee flora, including *Abies pindrow* (Royle ex D.Don) Royle (fir), *Pinus wallichiana* (Wall. ex D.Don) A.B.Jacks. (blue pine), *Salix* (willow), *Pyrus* (pear) and *Prunus* (plum/cherry) (Khan, 2020).

5.3 The Case of Pakistan: A Model for Alpine Apiculture

The beekeeping landscape in Pakistan is defined four distinct honey bee species of the genus *Apis*. Three of these species *Apis cerana* (the Eastern honey bee), *Apis florea* (the dwarf honey bee), and *Apis dorsata* (the giant honey bee) are endemic to the region.

Historically, traditional beekeepers have mostly used *Apis cerana* for honey production because of its exceptional ability to adjust to local environmental conditions and survive on native plants and trees. At present, Pakistan produces an average of 7 to 9 kg of honey per colony per year, contributing approximately 60 tons of honey to the country. The fourth species, *Apis mellifera* (the Western honey bee), is an exotic species introduced into Pakistan. Although it has become a key component of commercial beekeeping, its introduction has brought challenges, including the potential transfer of mites to the native *A. cerana* in the 1980s. Understanding the nature and foraging preferences of both native and introduced species is essential for developing sustainable management strategies that maximize honey production while preserving local biodiversity (Khan, 2020).

Beekeeping in Pakistan is a big and expanding business prospect, especially due to the superior quality and distinct flavor of honey produced from its diverse plants. Traditionally, beekeeping has been focused in the Khyber Pakhtunkhwa province and northern and central Punjab, but it is now rapidly expanding across the nation as its commercial potential is fulfilled. Honey produced in Pakistan, especially the raw and pure varieties from mountainous and woodland areas, has gained a strong reputation in global markets, especially in the Middle East. Each year, Pakistan exports around 4,000 tons of honey, generating approximately 23 million dollars in revenue. This export achievement highlights the economic sustainability of beekeeping as a non-timber forest business that can play a major role in rural livelihoods and the national economy (Khan, 2020). By fully utilizing the bee forage available in Pakistan through modern beekeeping methods and effective marketing, the country could further strengthen its position in the international honey market (Figure 2).

Figure 2. The diverse range of honey bee products and their uses. This figure illustrates the various valuable products derived from honey bee colonies, which form the basis of the economic opportunities in alpine apiculture. These include food items (honey, pollen), medicinal and cosmetic ingredients (propolis, royal jelly, beeswax), and even apitherapy applications (bee venom)

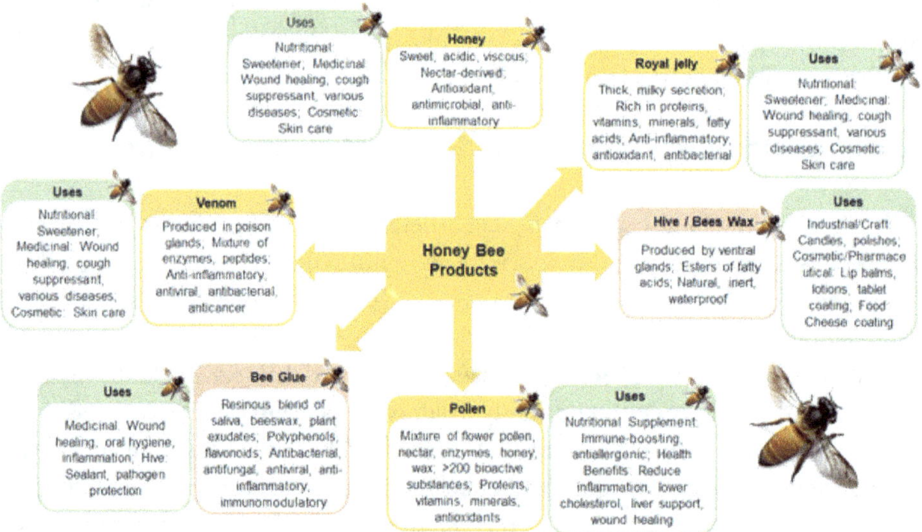

5.5 The Native Pollinator Community in Alpine Zones

Although this chapter focuses on the honey bee, *Apis mellifera*, a managed species, it is imperative to understand the indigenous pollinator community to recognize opportunities and barriers to beekeeping. The Alpine ecosystems have a specialized set of pollinators adapted to severe conditions. Where present, bumble bees (*Bombus* sp.) are often the dominant and most important pollinators, as they can forage at colder temperatures than many other insects. However, in many alpine regions, particularly where bumble bees are absent (like New Zealand and Australia) or less common, flies (*Diptera*) are the most crucial high-altitude pollinators. Common families include Syrphidae (hoverflies), Muscidae, and Empididae (Inouye, 2020). These native pollinators are also responding to climate change, with many species shifting their ranges upward in elevation. The introduction of managed *Apis mellifera* colonies into these sensitive ecosystems presents both an opportunity for enhanced pollination and a potential constraint through increased competition for floral resources with these specialized native pollinators. Sustainable beekeeping in alpine zones must therefore environment's carrying capacity and aim to supplement, rather than displace, the vital pollination services already provided by the native fauna.

6. FUTURE DIRECTIONS AND CONCLUSION

Further work on long-term monitoring and experimental research should be prioritized to make beekeeping in alpine locations viable over time. According to Inouye (2020), because of the natural spatial and temporal heterogeneity of the alpine ecosystem, single plot, or single-season studies are of limited value. Multi-year studies in varied microhabitats are urgently required to provide dependable data on trends in flowering phenology, pollinator populations, and their interactions. The use of modern technologies, such as remote automated cameras to record phenology and visitation, and ultrasonic snow sensors, can be effective in these remote and logistically challenging regions. Moreover, experimental methods, including snowmelt experiments and the use of open-topped chambers (OTCs) to model warming, can help in understanding how alpine plant communities will respond to future climate change (Inouye, 2020).

The potential for beekeeping in alpine areas is significant but comes with serious obstacles, the most important of which relates to floral resources. The unique and elevated flora available to honeybees provides a large but short-lived resource, leading to long term scarcity that limits the survival and productivity of colonies. Habitat loss and climate change only worsen these natural constraints. Nevertheless, as demonstrated in this chapter, these problems can be addressed through strategic interventions. Man-made sources of pollen and nectar are crucial for colony survival of the colonies during lean seasons. The case study of Pakistan discloses that, despite these limitations, there is substantial economic opportunity due to honey production from exotic alpine flora. The future of alpine beekeeping lies in a holistic approach that integrates modern management methods, such as supplemental feeding, with the preservation of native flora, the invaluable ecological experience.

REFERENCES

Ahmad, S., & Aziz, M. A. (2017). A review: Risk assessment of pesticides on honey bee and pollination of agriculture crops in Pakistan. *Asian Journal of Agriculture and Biology*, *5*(03).

Ahmad, S., Zafar, M., Ahmad, M., Alamer, K. H., Abid, A., Rozina, R.,…Shafeeq, U. J. P. J. E. S. h. d. o. (2024). Insect pollinators and plant interactions; identification of selected melliferous plants using pollen morphological features. *15244*.

Al-Ghamdi, A. A., Al-Khaibari, A. M., & Omar, M. O. (2011). Consumption rate of some proteinic diets affecting hypopharyngeal glands development in honeybee workers. *Saudi Journal of Biological Sciences*, *18*(1), 73–77. DOI: 10.1016/j.sjbs.2010.10.001 PMID: 23961106

Al Naggar, Y., Brinkmann, M., Sayes, C. M., Al-Kahtani, S. N., Dar, S. A., El-Seedi, H. R., Grünewald, B., & Giesy, J. P. (2021). Are honey bees at risk from microplastics? *Toxics*, *9*(5), 109. DOI: 10.3390/toxics9050109 PMID: 34063384

Daugaliyeva, S., Daugaliyeva, A., Amirova, K., Yelubayeva, A., Kapar, A., Abiti, A.,…Peletto, S. J. I. (2024). Foraging of Honeybees from Different Ecological Areas Determined through Melissopalynological Analysis and DNA Metabarcoding. *15*(9), 674.

Deng, Y., Jiang, X., Zhao, H., Yang, S., Gao, J., Wu, Y., Diao, Q., & Hou, C. (2021). Microplastic Polystyrene Ingestion Promotes the Susceptibility of Honeybee to Viral Infection. *Environmental Science & Technology*, *55*(17), 11680–11692. DOI: 10.1021/acs.est.1c01619 PMID: 34374532

Doull, K. M. (1980). Relationships between consumption of a pollen supplement, honey production and broodrearing in colonies of honeybees *Apis mellifera* L. II. *Apidologie*, *11*(4), 367–374. DOI: 10.1051/apido:19800405

Fairbrother, A., Purdy, J., Anderson, T., & Fell, R. (2014). Risks of neonicotinoid insecticides to honeybees. *Environmental Toxicology and Chemistry*, *33*(4), 719–731. DOI: 10.1002/etc.2527 PMID: 24692231

Farooqi, M. A., Aslam, M. N., Sajjad, A., Akram, W., & Maqsood, A. (2021). Impact of abiotic factors on the foraging behavior of two honeybee species on canola in Bahawalpur, Punjab-Pakistan. *Asian Journal of Agriculture and Biology*, *2021*(01).

Herbert, E., & Hill, D. (1992). Honey bee nutrition. *The hive and the honey bee*, *3*, 197-233.

Inouye, D. W. (2020). Effects of climate change on alpine plants and their pollinators. *Annals of the New York Academy of Sciences, 1469*(1), 26–37. DOI: 10.1111/nyas.14104 PMID: 31025387

Khan, K., & Ghramh, H. (2023). Beekeeping in Pakistan: History, Potential, and Current Status. *Pakistan Journal of Zoology, 56*(1). DOI: 10.17582/journal.pjz/20230419090432

Kim, H., Frunze, O., Maigoro, A. Y., Lee, M. L., Lee, J. H., & Kwon, H. W. (2024). Comparative Study of the Effect of Pollen Substitute Diets on Honey Bees during Early Spring. *Insects, 15*(2), 101. Advance online publication. DOI: 10.3390/insects15020101 PMID: 38392520

Kumar, R., & Agrawal, O. (2014). Comparative performance of honey bee colonies fed with artificial diets in Gwalior and Panchkula region. *Journal of Entomology and Zoology Studies, 2*(4), 104–107.

Leska, A., Nowak, A., Nowak, I., & Górczyńska, A. (2021). Effects of Insecticides and Microbiological Contaminants on Apis mellifera Health. *Molecules (Basel, Switzerland), 26*(16), 5080. Advance online publication. DOI: 10.3390/molecules26165080 PMID: 34443668

Majeed, S., Zafar, M., Ahmad, M., Kilic, O., Sultana, S., Raza, J. (2020). Pollen morphological investigations of family Cactaceae and its taxonomic implication by light microscopy and scanning electron microscopy. *83*(7), 767-777.

Mishra, R., Dogra, G., & Gupta, P. (1981). Apicultural activities in Himachal Pradesh. *Indian Bee Journal, 41*(3/4), 29-31

Na, S.-J., Kim, Y.-K., & Park, J.-M. J. P. (2024). Nectar characteristics and honey production potential of five rapeseed cultivars and two wildflower species in South Korea. *13*(3), 419.

Nabila, A.M., Ali, M., Zafar, M., Sultana, S., Majeed, S., ... & Ahmad, S. (2022). Palynological Diversity of Melliferous flora around Apiaries from District Mardan Khyber Pakhtunkhwa-Pakistan: Nabila et al. *The Botanical Review, 88*(3), 299-332.

Nader, R. A., Mackieh, R., Wehbe, R., El Obeid, D., Sabatier, J. M., & Fajloun, Z. (2021). Beehive products as antibacterial agents: A review. *Antibiotics (Basel, Switzerland), 10*(6), 717. DOI: 10.3390/antibiotics10060717 PMID: 34203716

Noor, M. J., Ahmad, M., Zafar, M., & Ashraf, M. A. (2017). Palynological studies of melliferous and allergenic flora of Pakistan: A key to pollen and spore identification. *Journal of Apicultural Research, 56*(3), 300–309.

Paray, B. A., Kumari, I., Hajam, Y. A., Sharma, B., Kumar, R., Albeshr, M. F., Farah, M. A., & Khan, J. M. (2021). Honeybee nutrition and pollen substitutes: A review. *Saudi Journal of Biological Sciences*, *28*(1), 1167–1176. https://doi.org/https://doi .org/10.1016/j.sjbs.2020.11.053. DOI: 10.1016/j.sjbs.2020.11.053 PMID: 33424413

Pérez-Escobar, O. A., Zizka, A., Bermúdez, M. A., Meseguer, A. S., Condamine, F. L., Hoorn, C.,…Boschman, L. M. J. T. i. P. S. (2022). The Andes through time: evolution and distribution of Andean floras. *27*(4), 364-378.

Pettis, J. S., Rose, R., & Chaimanee, V. (2017). Chemical and cultural control of Tropilaelaps mercedesae mites in honeybee (Apis mellifera) colonies in Northern Thailand. *PLoS One*, *12*(11), e0188063.

Phiri, B. J., Fèvre, D., & Hidano, A. (2022). Uptrend in global managed honey bee colonies and production based on a six-decade viewpoint, 1961–2017. *Scientific Reports*, *12*(1), 21298. DOI: 10.1038/s41598-022-25290-3 PMID: 36494404

Punt, W., Hoen, P. P., Blackmore, S., Nilsson, S., Le Thomas, A. J. R. o. p., & palynology. (2007). Glossary of pollen and spore terminology. *143*(1-2), 1-81.

Rashid, S., Wajid, R., Supe Tulcan, R., & Huang, H. (2022). Use, exposure, and environmental impacts of pesticides in Pakistan: A critical review. *Environmental Science and Pollution Research International*, *29*(29), 43675–43689. Advance online publication. DOI: 10.1007/s11356-022-20164-7 PMID: 35435556

Rueda-Uribe, C., Chautá, A., Woodman, T. L., Lasso, E., Pelayo, R. C., Manrique-Garzón, L. M.,…Bocedi, G. J. B. R. (2025). Pollination ecology in the tropical Andes: moving towards a cross-scale approach.

Rueda-Uribe, C., Chautá, A., Woodman, T. L., Lasso, E., Pelayo, R. C., Manrique-Garzón, L. M., & Travis, J. M. (2025). Pollination ecology in the tropical Andes: Moving towards a cross-scale approach. *Biological Reviews of the Cambridge Philosophical Society*, *100*(6), 2312–2345.

Saffari, A. M., Kevan, P. G., & Atkinson, J. L. (2004). A promising pollen substitute for honey bees. *American Bee Journal*, *144*(3), 230–231.

Usman, M., Hasnain, M., Banaras, S., Akram, M., Abbas, Q., Shah, J. A.,…Khan, M. N. (2022). Potential emerging constraints and management strategies of different honeybee species in Pakistan: A review. *CABI Reviews*(2022).

Waghchoure-Camphor, E., & Martin, S. (2008). Beekeeping in Pakistan: A bright future in a troubled land. *American Bee Journal*, *148*(8), 726–728.

Wakgari, M., & Yigezu, G. (2021). Honeybee keeping constraints and future prospects. *Cogent Food & Agriculture*, *7*(1), 1872192. DOI: 10.1080/23311932.2021.1872192

Waykar, B., & Baviskar, R. K. (2015). Diversity of bee foraging flora and floral calendar of Paithan taluka of Aurangabad district (Maharashtra), India. *Journal of Applied Horticulture*, *17*(2), 155–159.

Zahra, A., & Talal, M. (2008). Impact of pollen supplements and vitamins on the development of hypopharyngeal glands and on brood number in honey bees. *Journal of Apicultural Science*, *52*(2).

Zaric, N. M., Deljanin, I., Ilijević, K., Stanisavljević, L., Ristić, M., & Gržetić, I. (2018). Assessment of spatial and temporal variations in trace element concentrations using honeybees (*Apis mellifera*) as bioindicators. *PeerJ*, *6*, e5197. DOI: 10.7717/peerj.5197 PMID: 30038861

Compilation of References

Abbasi, A. M., Shah, M. H., & Khan, M. A. (2015). *Wild edible vegetables of lesser Himalayas*. Springer. DOI: 10.1007/978-3-319-09543-1

Abbas, Z., Bussmann, R. W., Khan, S. M., & Abbasi, A. M. (2022). A review of current trends and future directions in the medical ethnobotany of Gilgit-Baltistan (Northern Pakistan). *Ethnobotany Research and Applications, 24*, 1–16. DOI: 10.32859/era.24.18.1-16

Abbas, Z., Khan, S. M., Alam, J., Khan, S. W., & Abbasi, A. M. (2017). Medicinal plants used by inhabitants of the Shigar Valley, Baltistan region of Karakorum range-Pakistan. *Journal of Ethnobiology and Ethnomedicine, 13*(1), 53. DOI: 10.1186/s13002-017-0172-9 PMID: 28946889

Abbas, Z., Kousar, S., Aziz, M. A., Pieroni, A., Aldosari, A. A., Bussmann, R. W., Raza, G., & Abbasi, A. M. (2021). Comparative Assessment of Medicinal Plant Utilization among Balti and Shina Communities in the Periphery of Deosai National Park, Pakistan. *Biology (Basel), 10*(5), 434. DOI: 10.3390/biology10050434 PMID: 34068859

Abbet, C., Mayor, R., Roguet, D., Spichiger, R., Hamburger, M., & Potterat, O. (2014). Ethnobotanical survey on wild alpine food plants in Lower and Central Valais (Switzerland). *Journal of Ethnopharmacology, 151*(1), 624–634. DOI: 10.1016/j.jep.2013.11.022 PMID: 24269246

Abdelkhalek, E. S., El-Hela, A. A., El-Kasaby, A. H., Sidkey, N. M., Desouky, E. M., & Abdelhaleem, H. H. (2018). Antibacterial Activity of *Polygonum plebejum* and *Euphorbia hirta* Against *Staphylococcus aureus* (MRSA). *Journal of Pure & Applied Microbiology, 12*(4). Advance online publication. DOI: 10.22207/JPAM.12.4.60

Abdel-Mogib, M., Albar, H., & Batterjee, S. (2002). Chemistry of the genus *Plectranthus*. *Molecules (Basel, Switzerland), 7*(2), 271–301. DOI: 10.3390/70200271

Abeli, T., Barni, E., Siniscalco, C., Amosso, C., & Rossi, G. (2012). A cost-effective model for preliminary site evaluation for the reintroduction of a threatened quillwort. *Aquatic Conservation*, *22*(1), 66–73. DOI: 10.1002/aqc.1246

Abidi, J., Ammar, S., Brahim, S. B., Skalicka-Woźniak, K., Ghrabi-Gammar, Z., & Bouaziz, M. (2019). Use of ultra-high-performance liquid chromatography coupled with quadrupole-time-of-flight mass spectrometry system as valuable tool for an untargeted metabolomic profiling of *Rumex tunetanus* flowers and stems and contribution to the antioxidant activity. *Journal of Pharmaceutical and Biomedical Analysis*, *162*, 66–81. DOI: 10.1016/j.jpba.2018.09.001 PMID: 30223144

Acar, N., Çelikezen, F. Ç., Şahin, İ. H., Fırat, M., Koçyiğit, R., & Kireçci, O. A. (2023). Determination of in Vitro Antioxidant, Antimicrobial Properties and COX-1/COX-2 Enzyme Inhibition Activity of *Capparis sicula. Kahramanmaras Sütçü Imam Üniversitesi Tarim ve Doga Dergisi*, *26*(6), 1221–1229. DOI: 10.18016/ksutarimdoga.vi.1139197

Adhikari, B. S., & Kumar, R. (2020). Effect of snowmelt regime on phenology of herbaceous species at and around treeline in Western Himalaya, India. *Notulae Scientia Biologicae*, *12*(4), 901–919. DOI: 10.15835/nsb12410716

Adhikari, P., Joshi, K., Singh, M., & Pandey, A. (2022). Influence of altitude on secondary metabolites, antioxidants, and antimicrobial activities of Himalayan yew (*Taxus wallichiana*). *Plant Biosystems*, *156*(1), 187–195. DOI: 10.1080/11263504.2020.1845845

Affi, W., Mohamed, A. A., Gharsallah, N., Smetanska, I., & Zourgui, L. (2024). Phytochemicals, antioxidant, and antimicrobial activities of *Opuntia stricta* fruits peel. *Open Veterinary Journal*, *14*(10), 2642. DOI: 10.5455/OVJ.2024.v14.i10.14 PMID: 39545202

Agbeno, E. K. (2025). Ethnobotanical Knowledge and Medicinal Plant Use Among Indigenous Communities: Conservation and Cultural Preservation Strategies. *International Journal of Indigenous Medicine*, 15-58.

Agnihotri, S., Dobhal, P., Ashfaqullah, S., Oli, S., & Tamta, S. (2024). Phytochemical screening of *Zanthoxylum armatum* roots and exploring its polyphenol content and antioxidant activity. *Natural Product Research*, *38*(19), 3404–3410. DOI: 10.1080/14786419.2023.2244130 PMID: 37565659

Agrawal, A., & Ostrom, E. (2001). Collective action, property rights, and decentralization in resource use in India and Nepal. *Politics & Society*, *29*(4), 485–514. DOI: 10.1177/0032329201029004002

Agrawal, K., Gupta, V. K., & Verma, P. (2022). Microbial cell factories a new dimension in bio-nanotechnology: Exploring the robustness of nature. *Critical Reviews in Microbiology*, *48*(4), 397–427. DOI: 10.1080/1040841X.2021.1977779 PMID: 34555291

Ahmad, S., Zafar, M., Ahmad, M., Alamer, K. H., Abid, A., Rozina, R.,...Shafeeq, U. J. P. J. E. S. h. d. o. (2024). Insect pollinators and plant interactions; identification of selected melliferous plants using pollen morphological features. *15244*.

Ahmad, K., Hameed, M., Fatima, S., Ashraf, M., Ahmad, F., Naseer, M., & Akhtar, N. (2016). Morpho-anatomical and physiological adaptations to high altitude in some Aveneae grasses from Neelum Valley, Western Himalayan Kashmir. *Acta Physiologiae Plantarum*, *38*(4), 93. Advance online publication. DOI: 10.1007/s11738-016-2114-x

Ahmad, M., Luo, Y.-H., Rathee, S., Spicer, R. A., Zhang, J., Wambulwa, M. C., Zhu, G.-F., Cadotte, M. W., Wu, Z.-Y., Khan, S. M., Maity, D., Li, D.-Z., & Liu, J. (2025). Multifaceted plant diversity patterns across the Himalaya: Status and outlook. *Plant Diversity*, *47*(4), 529–543. DOI: 10.1016/j.pld.2025.04.003 PMID: 40734839

Ahmad, M., Sultana, S., Fazl-i-Hadi, S., ben Hadda, T., Rashid, S., Zafar, M., Khan, M. A., Khan, M. P. Z., & Yaseen, G. (2014a). An Ethnobotanical study of Medicinal Plants in high mountainous region of Chail valley (District Swat- Pakistan). *Journal of Ethnobiology and Ethnomedicine*, *10*(1), 36. DOI: 10.1186/1746-4269-10-36 PMID: 24739524

Ahmad, S., & Aziz, M. A. (2017). A review: Risk assessment of pesticides on honey bee and pollination of agriculture crops in Pakistan. *Asian Journal of Agriculture and Biology*, *5*(03).

Aimad, A., Sanae, R., Anas, F., Abdelfattah, E. M., Bourhia, M., Mohammad Salamatullah, A., Alzahrani, A., Alyahya, H. K., Albadr, N. A., Abdelkrim, A., El Barnossi, A., & Noureddine, E. (2021). Chemical characterization and antioxidant, antimicrobial, and insecticidal properties of essential oil from *Mentha pulegium* L. *Evidence-Based Complementary and Alternative Medicine : eCAM*, *2021*(1), 1108133. DOI: 10.1155/2021/1108133 PMID: 34691201

Aïtcin, P.-C. (2000). Cements of yesterday and today: Concrete of tomorrow. *Cement and Concrete Research*, *30*(9), 1349–1359. DOI: 10.1016/S0008-8846(00)00365-3

Akar, Z., Karakurt, A., Okumuş, F., Cinemre, S., Düzgün, A. Ö., Akar, B., & Can, Z. (2020). RP-HPLC-UV Analysis of the Phenolic Compounds, Antimicrobial Activity Against Multi-Drug Resistant Bacteria and Antioxidant Activity of Fruit and Seed of *Diospyros lotus* L. ctivity of Fruit and Seed of *Diospyros lotus* L. *International Journal of Secondary Metabolite*, *7*(4), 237–246. DOI: 10.21448/ijsm.714108

Akhter, K., Ghous, T., Sadaf, E., Hassan, A., Irshad, V., & Andleeb, S. (2021). Pharmacological Approach to Glycemic Treatment Using Black, Green and Herbal Tea Extracts. *Bangladesh Journal of Botany*, *50*(3), 491–498. DOI: 10.3329/bjb.v50i3.55827

Akhter, S. M. H., Mahmood, Z., Ahmad, S., & Mohammad, F. (2018). Plant-mediated green synthesis of zinc oxide nanoparticles using Swertia chirayita leaf extract, characterization and its antibacterial efficacy against some common pathogenic bacteria. *BioNanoScience*, *8*(3), 811–817. DOI: 10.1007/s12668-018-0549-9

Akula, R., & Ravishankar, G. A. (2011). Influence of abiotic stress signals on secondary metabolites in plants. *Plant Signaling & Behavior*, *6*(11), 1720–1731. DOI: 10.4161/psb.6.11.17613 PMID: 22041989

Al Dhanhani, A., Tsombou, F., Mirza, S., Shahin, S., & Ridouane, F. (2025). Phytochemical and antimicrobial properties of *Morus nigra* and *Morus alba* grown in Fujairah, United Arab Emirates. *Scientific Reports*, *15*(1), 34986. DOI: 10.1038/s41598-025-18872-4 PMID: 41057426

Al Naggar, Y., Brinkmann, M., Sayes, C. M., Al-Kahtani, S. N., Dar, S. A., El-Seedi, H. R., Grünewald, B., & Giesy, J. P. (2021). Are honey bees at risk from microplastics? *Toxics*, *9*(5), 109. DOI: 10.3390/toxics9050109 PMID: 34063384

Alamdari, E. G., Avarseji, Z., & Taleghani, A. (2022). Extraction, Isolation and Characterization by HPLC-ESI-MS and Evaluating Chemical Constituents of Hydroalcoholic Extracts of the Invasive Weed *Cleome viscosa* L. *Natural Product Communications*, *17*(1). Advance online publication. DOI: 10.1177/1934578X2110486

Alam, F., Hanif, M., Rahman, A., Ali, S., & Jan, S. (2023). In vitro, in vivo and in silico evaluation of analgesic, anti-inflammatory, and anti-pyretic activity of salicylate rich fraction from *Gaultheria trichophylla* Royle (Ericaceae). *Journal of Ethnopharmacology*, *301*, 115828. DOI: 10.1016/j.jep.2022.115828 PMID: 36240979

Alam, F., Saqib, Q. N., & Ashraf, M. (2017). *Gaultheria trichophylla* (Royle): A source of minerals and biologically active molecules, its antioxidant and anti-lipoxygenase activities. *BMC Complementary and Alternative Medicine*, *17*(1), 3. DOI: 10.1186/s12906-016-1511-4 PMID: 28049535

Alami, M. M., Guo, S., Mei, Z., Yang, G., & Wang, X. (2024). Environmental factors on secondary metabolism in medicinal plants: Exploring accelerating factors. *Medicinal Plant Biology*, *3*(1), 0. DOI: 10.48130/mpb-0024-0016

Alam, N. M., Shaheen, H., Manzoor, M., Tinghong, T., Arfan, M., & Idrees, M. (2023). Spatial distribution and population structure of Himalayan Fir (*Abies pindrow* (Royle ex D. Don) Royle) in moist temperate forests of the Kashmir region. *Forests*, *14*(3), 482. DOI: 10.3390/f14030482

Albach, D. C., Schoenswetter, P., & Tribsch, A. (2006). Comparative phylogeography of the *Veronica alpina* complex in Europe and North America. *Molecular Ecology*, *15*(11), 3269–3286. DOI: 10.1111/j.1365-294X.2006.02980.x PMID: 16968270

Alber, E. (2017). South Tyrol's negotiated autonomy. *The Journal of Ethnic Studies*, *78*, 41–58.

Al-Brahim, J. S. (2023). Saussurea costus extract as bio mediator in synthesis iron oxide nanoparticles (IONPs) and their antimicrobial ability. *PLoS One*, *18*(3), e0282443. DOI: 10.1371/journal.pone.0282443 PMID: 36893115

Alcorn, J. B., & Toledo, V. M. (1998). Resilient resource management in Mexico's forest ecosystems: the contribution of property rights. *Linking Social and Ecological Systems: Management Practices and Social Mechanisms for Building Resilience*, 216-249.

Alexander, J. M., Lembrechts, J. J., Cavieres, L. A., Daehler, C., Haider, S., Kueffer, C., Liu, G., McDougall, K., Milbau, A., Pauchard, A., Rew, L. J., & Seipel, T. (2016). Plant invasions into mountains and alpine ecosystems: Current status and future challenges. *Alpine Botany*, *126*(2), 89–103. DOI: 10.1007/s00035-016-0172-8

Al-Ghamdi, A. A., Al-Khaibari, A. M., & Omar, M. O. (2011). Consumption rate of some proteinic diets affecting hypopharyngeal glands development in honeybee workers. *Saudi Journal of Biological Sciences*, *18*(1), 73–77. DOI: 10.1016/j.sjbs.2010.10.001 PMID: 23961106

Alhage, J., Rahman, A. A., Al-Bayssari, C., & Raad, M. T. (2025). Bioactive Compounds Isolated from the Species of the *Campanula* Genus: A Review. *Molecules (Basel, Switzerland)*, *30*(23), 4495. DOI: 10.3390/molecules30234495 PMID: 41375093

Al-haliem, S. M., Mohammed, M. J., Abedelmaksoud, T. G., Hesarinejad, M. A., & Baioumy, A. A. (2025). Chicory (*Cichorium intybus*) Leaves Extract: Phenolic Composition, Antibacterial Activity, and Antioxidant Capacity Assessment. *Food Science & Nutrition*, *13*(7), e70550. DOI: 10.1002/fsn3.70550 PMID: 40599358

Alharbi, N. K., Naghmouchi, S., & Al-Zaban, M. (2021). Evaluation of antimicrobial potential and comparison of HPLC composition, secondary metabolites count, and antioxidant activity of *Mentha rotundifolia* and *Mentha pulegium* extracts. *Evidence-Based Complementary and Alternative Medicine : eCAM*, *2021*(1), 9081536. DOI: 10.1155/2021/9081536 PMID: 34497659

Alhazmi, N. M. (2025). Biological and Therapeutic Applications of Gallic Acid Produced by Aspergillus terreus PQ344456 Using Submerged Fermentation. *ACS Omega*, *10*(41), 48488–48499. DOI: 10.1021/acsomega.5c05913 PMID: 41141794

Ali, M., Raza, M., Parveen, K., & Bashir, F. (2022). Mountains and Climate Change. *Mountain Studies: Understanding and Managing Mountains for People and Nature*, 58.

Ali, A., Khan, M. A., Ashraf, M. I., & Goheer, M. A. (2021). Economic valuation of ecosystem services in alpine pastures of western Himalayas: A case study of Kaghan valley, Pakistan. *Pakistan Journal of Botany*, *53*(5), 1801–1806. DOI: 10.30848/PJB2021-5(18)

Ali, M., Aldosari, A., Tng, D. Y., Ullah, M., Hussain, W., Ahmad, M., & Sher, H. (2019). Traditional uses of plants by indigenous communities for veterinary practices at Kurram District, Pakistan. *Ethnobotany Research and Applications*, *18*, 1–19. DOI: 10.32859/era.18.24.1-19

Al-Jaber, N. A. (2011). Phytochemical and biological studies of *Sisymbrium irio* L. Growing in Saudi Arabia. *Journal of Saudi Chemical Society*, *15*(4), 345–350. DOI: 10.1016/j.jscs.2011.04.010

Al-Qahtani, J., Abbasi, A., Aati, H. Y., Al-Taweel, A., Al-Abdali, A., Aati, S., Yanbawi, A. N., Khan, M. A., Ghalloo, B. A., & Anwar, M. (2023). Phytochemical, antimicrobial, antidiabetic, thrombolytic, anticancer activities, and in silico studies of *Ficus palmata* Forssk. *Arabian Journal of Chemistry*, *16*(2), 104455. DOI: 10.1016/j.arabjc.2022.104455

Alrhmoun, M., Romano, A., Sulaiman, N., & Pieroni, A. (2025). Old Plants for New Food Products? The Diachronic Human Ecology of Wild Herbs in the Western Alps. *Plants*, *14*(1), 122. DOI: 10.3390/plants14010122 PMID: 39795381

Alrumaihi, F., Almatroodi, S. A., Alharbi, H. O. A., Alwanian, W. M., Alharbi, F. A., Almatroudi, A., & Rahmani, A. H. (2024). Pharmacological potential of kaempferol, a flavonoid in the management of pathogenesis via modulation of inflammation and other biological activities. *Molecules (Basel, Switzerland)*, *29*(9), 2007. DOI: 10.3390/molecules29092007 PMID: 38731498

Alsayadi, M. M. S., Gazaem, M. A. H., Mohamad, S. A. S., Ismail, N. H., Alshoaiby, M. A., Agha, H. M., Alkamil, A. A., Salihu, M., & Salim, F. (2025). Nutritional Composition, Phytochemical contents, HPLC analysis, and Antioxidant Activity of leaf and flower extracts of *Rumex nervosus* grown in Yemen. *Biocatalysis and Agricultural Biotechnology, 103749*, 103749. Advance online publication. DOI: 10.1016/j.bcab.2025.103749

Alsharif, B., & Boylan, F. (2025). *Capparis* L.(Capparaceae): A Scoping Review of Phytochemistry, Ethnopharmacology and Pharmacological Activities. *Molecules (Basel, Switzerland), 30*(18), 3705. DOI: 10.3390/molecules30183705 PMID: 41011598

Altemimi, A. B., Al-Haliem, S. M., Alkanan, Z. T., Mohammed, M. J., Hesarinejad, M. A., Najm, M. A., Bouymajane, A., Cacciola, F., & Abedelmaksoud, T. G. (2023). Exploring the phenolic profile, antibacterial, and antioxidant properties of walnut leaves (*Juglans regia* L.). *Food Science & Nutrition, 11*(11), 6845–6853. DOI: 10.1002/fsn3.3554 PMID: 37970415

Alum, E. U. (2024). The role of indigenous knowledge in advancing the therapeutic use of medicinal plants: Challenges and opportunities. *Plant Signaling & Behavior, 19*(1), 2439255. DOI: 10.1080/15592324.2024.2439255 PMID: 39652401

Ambesh, P., Campia, U., Obiagwu, C., Bansal, R., Shetty, V., Hollander, G., & Shani, J. (2017). Nanomedicine in coronary artery disease. *Indian Heart Journal, 69*(2), 244–251. DOI: 10.1016/j.ihj.2017.02.007 PMID: 28460774

Ambwani, S., Chand, P., Kumawat, S. K., Singh, H., & Koranga, D. (2025). In vitro Analysis of Antioxidant Capacity of *Bauhinia variegata* (L.) Benth Leaf Extract. *Pharmacological Research-Natural Products, 100350*, 100350. Advance online publication. DOI: 10.1016/j.prenap.2025.100350

Amin, M., Aziz, M. A., Manduzai, A. K., Pieroni, A., Alkahtani, J., AbdelGawwad, M. R., Gafforov, Y., Nazeer, A., & Abbasi, A. M. (2024). Standard medical ethnobotany of Kohistan, North Pakistan. *Journal of Ethnobiology and Ethnomedicine, 20*(1), 64. DOI: 10.1186/s13002-024-00704-w PMID: 38978072

Anastácio, T. O., Madalozzo, E. S., Oliveira, V. S., Sandim, G. C., Moroto, B. S., Formagio, A. S. N., Cardoso, C. A. L., Sardi, J. C. O., Oliveira, C. F. R., & Macedo, M. L. R. (2022). Nutraceutical potential of *Cucurbita maxima* Duchesne pulp extract. *Food Science & Technology International, 10820132251349940*, 10820132251349940. Advance online publication. DOI: 10.1177/10820132251349940 PMID: 40538189

Antonelli, A., Kissling, W. D., Flantua, S. G., Bermúdez, M. A., Mulch, A., Muellner-Riehl, A. N., & Fjeldså, J. (2018). Geological and climatic influences on mountain biodiversity. *Nature Geoscience, 11*(10), 718–725. DOI: 10.1038/s41561-018-0236-z

Appadurai, A. (1996). *Modernity at large: Cultural dimensions of globalization* (Vol. 1). U of Minnesota Press.

Armstrong, R. L. (2010). *The glaciers of the Hindu Kush-Himalayan region: a summary of the science regarding glacier melt/retreat in the Himalayan, Hindu Kush, Karakoram, Pamir, and Tien Shan mountain ranges.*

Arnold, P. A., Wang, S., Catling, A. A., Kruuk, L. E., & Nicotra, A. B. (2022). Patterns of phenotypic plasticity along a thermal gradient differ by trait type in an alpine plant. *Functional Ecology, 36*(9), 2412–2428. DOI: 10.1111/1365-2435.14128

Arroyo, M. T. K., Armesto, J. J., & Villagran, C. (1981). Plant phenological patterns in the high Andean Cordillera of central Chile. *Journal of Ecology, 69*(1), 205–223. DOI: 10.2307/2259826

Aryal, B., Raut, B. K., Bhattarai, S., Bhandari, S., Tandan, P., Gyawali, K., Sharma, K., Ranabhat, D., Thapa, R., Aryal, D., Ojha, A., Devkota, H. P., & Parajuli, N. (2022). Potential therapeutic applications of plant-derived alkaloids against inflammatory and neurodegenerative diseases. *Evidence-Based Complementary and Alternative Medicine : eCAM, 2022*(1), 7299778. DOI: 10.1155/2022/7299778 PMID: 35310033

Asati, V., Deepa, P., & Sharma, P. K. (2022). Desert legume *Prosopis cineraria* as a novel source of antioxidant flavonoids/isoflavonoids: Biochemical characterization of edible pods for potential functional food development. *Biochemistry and Biophysics Reports, 29*, 101210. DOI: 10.1016/j.bbrep.2022.101210 PMID: 35198738

Ashcroft, M. B., & Gollan, J. R. (2013). The sensitivity of topoclimatic models to fine-scale microclimatic variability and the relevance for ecological studies. *Theoretical and Applied Climatology, 114*(1), 281–289. DOI: 10.1007/s00704-013-0841-0

Ashraf, A., Naz, R., & Roohi, R. (2012). Monitoring and estimation of glacial resource of Azad Jammu and Kashmir using remote sensing and GIS techniques. *Pakistan Journal of Meteorology, 8*(16), 31–41.

Ashraf, M. V., Khan, S., Misri, S., Gaira, K. S., Rawat, S., Rawat, B., Khan, M. A. H., Shah, A. A., Asgher, M., & Ahmad, S. (2024). High-Altitude Medicinal Plants as Promising Source of Phytochemical Antioxidants to Combat Lifestyle-Associated Oxidative Stress-Induced Disorders. *Pharmaceuticals (Basel, Switzerland), 17*(8), 975. Advance online publication. DOI: 10.3390/ph17080975 PMID: 39204080

Aslam, M., Fakher, B., Ashraf, M. A., Cheng, Y., Wang, B., & Qin, Y. (2022). Plant low-temperature stress: Signaling and response. *Agronomy (Basel)*, *12*(3), 702. DOI: 10.3390/agronomy12030702

Assad, N. Ansa, Laila, M. B., Ullah, S., Adnan, M., Wahab, S.,...Ullah, B. (2025). Nanotechnology in Oilseed Crops: A New Frontier for Abiotic Stress Adaptation. In *Oilseed Crops Under Abiotic Stress: Mitigation Strategies and Future Perspectives* (pp. 507-536). Springer.

Assad, N. (2023). Aconitum lycoctonum L.(Ranunculaceae) mediated biogenic synthesis of silver nanoparticles as potential antioxidant, anti-inflammatory, antimicrobial and antidiabetic agents. *BMC Chemistry*, *17*(1), 128. DOI: 10.1186/s13065-023-01047-5 PMID: 37770921

Assad, N., Abbas, A., Fayyaz ur Rehman, M., & Naeem-ul-Hassan, M. (2024). Photocatalytic and biological applications of phyto-functionalized zinc oxide nanoparticles synthesized using a polar extract of Equisetum diffusum D. *RSC Advances*, *14*(31), 22344–22358. DOI: 10.1039/D4RA03573A PMID: 39010906

Assad, N., Laila, M. B., Hassan, M. N. U., Rehman, M. F., Ali, L., Mustaqeem, M., Ullah, B., Khan, M. N., Iqbal, M., Ercişli, S., Alarfaj, A. A., Ansari, M. J., & Malik, T. (2025). Eco-friendly synthesis of gold nanoparticles using Equisetum diffusum D. Don. with broad-spectrum antibacterial, anticancer, antidiabetic, and antioxidant potentials. *Scientific Reports*, *15*(1), 19246. DOI: 10.1038/s41598-025-02450-9 PMID: 40456810

Assad, N., Naeem-ul-Hassan, M., Ajaz Hussain, M., Abbas, A., Sher, M., Muhammad, G., Assad, Y., & Farid-ul-Haq, M. (2025). Diffused sunlight assisted green synthesis of silver nanoparticles using Cotoneaster nummularia polar extract for antimicrobial and wound healing applications. *Natural Product Research*, *39*(8), 2203–2217. DOI: 10.1080/14786419.2023.2295936 PMID: 38146228

Atanasov, A. G., Waltenberger, B., Pferschy-Wenzig, E. M., Linder, T., Wawrosch, C., Uhrin, P., Temml, V., Wang, L., Schwaiger, S., Heiss, E. H., Rollinger, J. M., Schuster, D., Breuss, J. M., Bochkov, V., Mihovilovic, M. D., Kopp, B., Bauer, R., Dirsch, V. M., & Stuppner, H. (2015). Discovery and resupply of pharmacologically active plant-derived natural products: A review. *Biotechnology Advances*, *33*(8), 1582–1614. DOI: 10.1016/j.biotechadv.2015.08.001 PMID: 26281720

Augustynowicz, D., Lemieszek, M. K., Strawa, J. W., Wiater, A., & Tomczyk, M. (2023). Phytochemical profiling of extracts from rare *Potentilla* species and evaluation of their anticancer potential. *International Journal of Molecular Sciences*, *24*(5), 4836. DOI: 10.3390/ijms24054836 PMID: 36902263

Ayoub, N., Badr, N., Al-Ghamdi, S. S., Alsanosi, S., Alzahrani, A. R., Abdel-Naim, A. B., Nematallah, K. A., & Swilam, N. (2021). HPLC/MSn profiling and healing activity of a muco-adhesive formula of *Salvadora persica* against acetic acid-induced oral ulcer in rats. *Nutrients, 14*(1), 28. DOI: 10.3390/nu14010028 PMID: 35010903

Azad, A., Hussain, S., Akram, H., Fida, H., Iqbal, M. A., & Butt, T. E. (2024). Antimicrobial activity of *C. intybus* leaves extract in different solvent systems: A green approach for sustainable biomedicine. *Essential Chem, 1*(1), 1–13. DOI: 10.1080/28378083.2024.2427044

Aziz, M., Ullah, Z., & Pieroni, A. (2020). Wild food plant gathering among Kalasha, Yidgha, Nuristani and Khowar speakers in Chitral NW Pakistan. [In.]. *Sustainability (Basel), 12*(21), 9176. DOI: 10.3390/su12219176

Babaei-Ghaghelestany, A., Alebrahim, M. T., Farzaneh, S., & Mehrabi, M. (2022). The anticancer and antibacterial properties of aqueous and methanol extracts of weeds. *Journal of Agriculture and Food Research, 10*, 100433. DOI: 10.1016/j.jafr.2022.100433

Baban, M. M., Ahmad, S. A., Abu-Odeh, A. M., Baban, M., & Talib, W. H. (2023). Anticancer, immunomodulatory, and phytochemical screening of *Carthamus oxyacantha* M. Bieb growing in the North of Iraq. *Plants, 13*(1), 42. DOI: 10.3390/plants13010042 PMID: 38202350

Babatunde, D. E., Denwigwe, I. H., Babatunde, O. M., Gbadamosi, S. L., Babalola, I. P., & Agboola, O. (2019). Environmental and societal impact of nanotechnology. *IEEE Access : Practical Innovations, Open Solutions, 8*, 4640–4667. DOI: 10.1109/ACCESS.2019.2961513

Bai, L., Guo, S., Liu, Q., Cui, X., Zhang, X., Zhang, L., Yang, X., Hou, M., Ho, C.-T., & Bai, N. (2016). Characterization of nine polyphenols in fruits of *Malus pumila* Mill by high-performance liquid chromatography. *journal of food and drug analysis, 24*(2), 293-298. DOI: 10.1016/j.jfda.2015.10.002

Bai, Y., Peng, Y., Zhang, D., Yang, G., Chen, L., Kang, L., Zhou, W., Wei, B., Xie, Y., & Yang, Y. (2025). Heating up the roof of the world: Tracing the impacts of in-situ warming on carbon cycle in alpine grasslands on the Tibetan Plateau. *National Science Review, 12*(2), nwae371. DOI: 10.1093/nsr/nwae371 PMID: 39872222

Balian, E. V., Segers, H., Lévêque, C., & Martens, K. (2008). The Freshwater Animal Diversity Assessment: An overview of the results. *Hydrobiologia, 595*(1), 627–637. DOI: 10.1007/s10750-007-9246-3

Balunas, M. J., & Kinghorn, A. D. (2005). Drug discovery from medicinal plants. *Life Sciences, 78*(5), 431–441. DOI: 10.1016/j.lfs.2005.09.012 PMID: 16198377

Bandh, S. A., Shafi, S., Peerzada, M., Rehman, T., Bashir, S., Wani, S. A., & Dar, R. (2021). Multidimensional analysis of global climate change: A review. *Environmental Science and Pollution Research International, 28*(20), 24872–24888. DOI: 10.1007/s11356-021-13139-7 PMID: 33763833

Banerjee, S., Rathore, N., Semerád, J., Cajthaml, T., Münzbergová, Z., & Thakur, D. (2025). Soil moisture and its interaction with temperature determine root metabolomes of a Himalayan alpine shrub. *Physiologia Plantarum, 177*(4), e70444. DOI: 10.1111/ppl.70444 PMID: 40798898

Banks-Leite, C., Ewers, R. M., Folkard-Tapp, H., & Fraser, A. (2020). Countering the effects of habitat loss, fragmentation, and degradation through habitat restoration. *One Earth, 3*(6), 672–676. DOI: 10.1016/j.oneear.2020.11.016

Barinotto, L. (2012). Piante medicinali e di uso tradizionale in Valle Orco.

Barredo, J. I., Mauri, A., & Caudullo, G. (2020). Alpine tundra contraction under future warming scenarios in Europe. *Atmosphere (Basel), 11*(7), 698. DOI: 10.3390/atmos11070698

Barry, R. G. (2013). *Mountain weather and climate.* Routledge. DOI: 10.4324/9780203416020

Barthel, S., Crumley, C., & Svedin, U. (2013). Bio-cultural refugia-Safeguarding diversity of practices for food security and biodiversity. *Global Environmental Change, 23*(5), 1142–1152. DOI: 10.1016/j.gloenvcha.2013.05.001

Barthlott, W., Lauer, W., & Placke, A. (1996). Global distribution of species diversity in vascular plants: Towards a world map of phytodiversity (globale verteilung der artenvielfalt höherer pflanzen: Vorarbeiten zu einer weltkarte der phytodiversität). *Erdkunde*, •••, 317–327.

Batiha, G. E.-S., Tene, S. T., Teibo, J. O., Shaheen, H. M., Oluwatoba, O. S., Teibo, T. K. A., Al-Kuraishy, H. M., Al-Garbee, A. I., Alexiou, A., & Papadakis, M. (2023). The phytochemical profiling, pharmacological activities, and safety of *Malva sylvestris*: A review. *Naunyn-Schmiedeberg's Archives of Pharmacology, 396*(3), 421–440. DOI: 10.1007/s00210-022-02329-w PMID: 36418467

Batisti, R. (2018). Espressioni dell'eros infelice in due poeti italiani del nuovo millennio. *TICONTRE, 10*, 181–202.

Begum, F. (2002). *Ex situ conservation of medicinal plants for primary healthcare in Laxmipur District, Bangladesh.* Paper presented at the Paper at a Workshop on Wise Practices and Experiential Learning in the Conservation and Management of Himalayan Medicinal Plants, Kathmandu, Nepal.

Begum, R. F., Singh, S., & Margesan, T. (2025). Tannins as therapeutic in managing obesity and beyond. *Current Pharmacology Reports*, *11*(1), 1–15. DOI: 10.1007/s40495-025-00398-y

Behera, M. D., Behera, S. K., & Sharma, S. (2019). Recent advances in biodiversity and climate change studies in India. *Biodiversity and Conservation*, *28*(8), 1943–1951. DOI: 10.1007/s10531-019-01781-0

Bellard, C., Thuiller, W., Leroy, B., Genovesi, P., Bakkenes, M., & Courchamp, F. (2013). Will climate change promote future invasions? *Global Change Biology*, *19*(12), 3740–3748. DOI: 10.1111/gcb.12344 PMID: 23913552

Benito Alonso, J. L., & Villar Pérez, L. (2012). Continent-wide response of mountain vegetation to climate change. *Nature climate change*, 1-31.

Bennacer, A., Sahir-Halouane, F., Aitslimane-Aitkaki, S., Oukali, Z., Oliveira, I. V., Rahmouni, N., & Aissaoui, M. (2022). Structural characterization of phytochemical content, antibacterial, and antifungal activities of *Juglans regia* L. leaves cultivated in Algeria. *Biocatalysis and Agricultural Biotechnology*, *40*, 102304. DOI: 10.1016/j.bcab.2022.102304

Bentley, R., & Trimen, H. (1880). Medicinal plants: Being descriptions with original figures of the principal plants employed in medicine and an account of the characters, properties, and uses of their parts and products of medicinal value. Vol. 4.

Berkes, F., Colding, J., & Folke, C. (2000). Rediscovery of traditional ecological knowledge as adaptive management. *Ecological Applications*, *10*(5), 1251–1262. DOI: 10.1890/1051-0761(2000)010[1251:ROTEKA]2.0.CO;2

Bertani, I., Primicerio, R., & Rossetti, G. (2016). Extreme climatic event triggers a lake regime shift that propagates across multiple trophic levels. *Ecosystems (New York, N.Y.)*, *19*(1), 16–31. DOI: 10.1007/s10021-015-9914-5

Bertness, M. D., & Callaway, R. (1994). Positive interactions in communities. *Trends in Ecology & Evolution*, *9*(5), 191–193. DOI: 10.1016/0169-5347(94)90088-4 PMID: 21236818

Bertsch-Hoermann, B., Egger, C., Gaube, V., & Gingrich, S. (2021). Agroforestry trade-offs between biomass provision and aboveground carbon sequestration in the alpine Eisenwurzen region, Austria. *Regional Environmental Change, 21*(3), 77. DOI: 10.1007/s10113-021-01794-y PMID: 34720739

Beutler, J. A. (2009). Natural products as a foundation for drug discovery. *Current protocols in pharmacology, 46*(1), 9.11. 11-19.11. 21.

Bharti, A., Jain, U., & Chauhan, N. (2024). From lab to field: Nano-biosensors for real-time plant nutrient tracking. *Plant Nano Biology, 9*, 100079. DOI: 10.1016/j.plana.2024.100079

Bhattacharya, A. (2022). Effect of low temperature stress on photosynthesis and allied traits: A review. In *Physiological processes in plants under low temperature stress* (pp. 199–297). Springer. DOI: 10.1007/978-981-16-9037-2_3

Bhattacharyya, M., Thattantavide, A., & Kumar, A. (2023). Ethnic mountain foods of western and eastern Himalayas, India. In *Wild food plants for zero hunger and resilient agriculture* (pp. 181–205). Springer. DOI: 10.1007/978-981-19-6502-9_8

Bhutia, T. D., Valant-Vetschera, K. M., & Brecker, L. (2013). Orphan flavonoids and dihydrochalcones from Primula exudates. *Natural product communications, 8*(8), 1934578X1300800812. DOI: 10.1177/1934578X1300800812

Bhutia, P. O., Kewlani, P., Pandey, A., Rawat, S., & Bhatt, I. D. (2021). Physico-chemical properties and nutritional composition of fruits of the wild Himalayan strawberry (*Fragaria nubicola* Lindle.) in different ripening stages. *Journal of Berry Research, 11*(3), 481–496. DOI: 10.3233/JBR-210742

Billings, W. (1973). Arctic and alpine vegetations: Similarities, differences, and susceptibility to disturbance. *Bioscience, 23*(12), 697–704. DOI: 10.2307/1296827

Billings, W. D., & Mooney, H. A. (1968). The ecology of arctic and alpine plants. *Biological Reviews of the Cambridge Philosophical Society, 43*(4), 481–529. DOI: 10.1111/j.1469-185X.1968.tb00968.x

Binter, J., Macek, M., & Dolezal, J. (2025). Integrating morphological, anatomical, and physiological traits to explain elevational distributions in Himalayan steppe and alpine plants. *Journal of Integrative Plant Biology, n/a*(n/a). https://doi.org/https://doi.org/10.1111/jipb.13971

Binter, J., Macek, M., & Dolezal, J. (2025). Integrating morphological, anatomical, and physiological traits to explain elevational distributions in Himalayan steppe and alpine plants. *Journal of Integrative Plant Biology, 67*(10), 2643–2657. DOI: 10.1111/jipb.13971 PMID: 40662507

Biológica-CBD, C. P. D. (1992). *Convention on Biological Diversity*. Nações Unidas.

Bison, M., Asse, D., Vitasse, Y., Chuine, I., Delpierre, N., Yoccoz, N. G., & Randin, C. F. (2019). Warmer winters reduce the advance of tree spring phenology induced by warmer springs in the Alps. *Agricultural and Forest Meteorology*, *252*, 220–230.

Blachnik, T., & Saller, R. (2015). In situ-Vermehrung von Arnica montana–Ergebnisse und Handlungsempfehlungen für die Artenschutz-Praxis. *ANLiegen Natur*, *37*(1), 31–41. DOI: 10.63653/ztzr3600

BöHmer. H. J. (2011). Vulnerability of tropical montane rain forest ecosystems due to climate change. In *Coping with Global Environmental Change, Disasters and Security: Threats, Challenges, Vulnerabilities and Risks* (pp. 789-802). Springer.

Böhm, R., Auer, I., Brunetti, M., Maugeri, M., Nanni, T., & Schöner, W. (2001). Regional temperature variability in the European Alps: 1760–1998 from homogenized instrumental time series. *International Journal of Climatology*, *21*(14), 1779–1801. DOI: 10.1002/joc.689

Bojilov, D., Manolov, S., Nacheva, A., Dagnon, S., & Ivanov, I. (2023). Characterization of polyphenols from *Chenopodium botrys* after fractionation with different solvents and study of their in vitro biological activity. *Molecules (Basel, Switzerland)*, *28*(12), 4816. DOI: 10.3390/molecules28124816 PMID: 37375371

Bokhari, A. A., & Syed, V. (2015). Inhibition of transforming growth factor-β (TGF-β) signaling by *Scutellaria baicalensis* and *Fritillaria cirrhosa* extracts in endometrial cancer. *Journal of Cellular Biochemistry*, *116*(8), 1797–1805. DOI: 10.1002/jcb.25138 PMID: 25683036

Bona, F., Bo, T., Doretto, A., Falasco, E., Zoppi, M., & Fenoglio, S. (2022). Are protected areas effective in preserving alpine stream morphology and biodiversity? An experience in the first Italian National Park. *Authorea Preprints*. DOI: 10.22541/au.166987995.50028967/v1

Bona, F., Bo, T., Doretto, A., Falasco, E., Zoppi, M., & Fenoglio, S. (2023). Are protected areas effective in preserving Alpine stream morphology and biodiversity? A field study in the oldest Italian National Park. *River Research and Applications*, *39*(5), 942–953. DOI: 10.1002/rra.4124

Bona, F., Falasco, E., Fenoglio, S., Iorio, L., & Badino, G. (2008). Response of macroinvertebrate and diatom communities to human-induced physical alteration in mountain streams. *River Research and Applications*, *24*(8), 1068–1081. DOI: 10.1002/rra.1110

Bonifácio, B. V., da Silva, P. B., Ramos, M. A. S., Negri, K. M. S., Bauab, T. M., & Chorilli, M. (2014). Nanotechnology-based drug delivery systems and herbal medicines: A review. *International Journal of Nanomedicine*, •••, 1–15. PMID: 24363556

Boucher, F. C., Dentant, C., Ibanez, S., Capblancq, T., Boleda, M., Boulangeat, L., Smyčka, J., Roquet, C., & Lavergne, S. (2021). Discovery of cryptic plant diversity on the rooftops of the Alps. *Scientific Reports*, *11*(1), 11128. DOI: 10.1038/s41598-021-90612-w PMID: 34045566

Bourhia, M., Bouothmany, K., Bakrim, H., Hadrach, S., Salamatullah, A. M., Alzahrani, A., Khalil Alyahya, H., Albadr, N. A., Gmouh, S., Laglaoui, A., El Mzibri, M., & Benbacer, L. (2021). Chemical profiling, antioxidant, antiproliferative, and antibacterial potentials of chemically characterized extract of *Citrullus colocynthis* L. seeds. *Separations*, *8*(8), 114. DOI: 10.3390/separations8080114

Bowler, D. E., Mant, R., Orr, H., Hannah, D. M., & Pullin, A. S. (2012). What are the effects of wooded riparian zones on stream temperature? *Environmental Evidence*, *1*(1), 3. DOI: 10.1186/2047-2382-1-3

Brahmkshatriya, P. P., & Brahmkshatriya, P. S. (2013). Terpenes: Chemistry, biological role, and therapeutic applications. In *Natural products* (pp. 2665-2691). Springer. DOI: 10.1007/978-3-642-22144-6_120

Brambilla, M., Rubolini, D., Appukuttan, O., Calvi, G., Karger, D. N., Kmecl, P., Mihelič, T., Sattler, T., Seaman, B., Teufelbauer, N., Wahl, J., & Celada, C. (2022). Identifying climate refugia for high-elevation Alpine birds under current climate warming predictions. *Global Change Biology*, *28*(14), 4276–4291. DOI: 10.1111/gcb.16187 PMID: 35441422

Breman, E., Ballesteros, D., Castillo-Lorenzo, E., Cockel, C., Dickie, J., Faruk, A., & Sharrock, S. (2021). Plant diversity conservation challenges and prospects—The perspective of botanic gardens and the Millennium Seed Bank. *Plants*, *10*(11), 2371. DOI: 10.3390/plants10112371 PMID: 34834734

Bridgewater, P., & Rotherham, I. D. (2019). A critical perspective on the concept of biocultural diversity and its emerging role in nature and heritage conservation. *People and Nature*, *1*(3), 291–304. DOI: 10.1002/pan3.10040

Brighenti, S., Tolotti, M., Bruno, M. C., Wharton, G., Pusch, M. T., & Bertoldi, W. (2019). Ecosystem shifts in Alpine streams under glacier retreat and rock glacier thaw: A review. *The Science of the Total Environment*, *675*, 542–559. DOI: 10.1016/j.scitotenv.2019.04.221 PMID: 31030160

Britten, H. B., Brussard, P. F., & Murphy, D. D. (1994). The pending extinction of the Uncompahgre fritillary butterfly. *Conservation Biology*, *8*(1), 86–94. DOI: 10.1046/j.1523-1739.1994.08010086.x

Britton, A. J., Beale, C. M., Towers, W., & Hewison, R. L. (2009). Biodiversity gains and losses: Evidence for homogenisation of Scottish alpine vegetation. *Biological Conservation*, *142*(8), 1728–1739. DOI: 10.1016/j.biocon.2009.03.010

Broadbent, A. A., Snell, H. S., Michas, A., Pritchard, W. J., Newbold, L., Cordero, I., Goodall, T., Schallhart, N., Kaufmann, R., Griffiths, R. I., Schloter, M., Bahn, M., & Bardgett, R. D. (2021). Climate change alters temporal dynamics of alpine soil microbial functioning and biogeochemical cycling via earlier snowmelt. *The ISME Journal*, *15*(8), 2264–2275. DOI: 10.1038/s41396-021-00922-0 PMID: 33619353

Brondízio, E., Settele, J., Diaz, S., Ngo, H. T., Experts, G., & Mohamed, A. (2021). *Global assessment report on biodiversity and ecosystem services of the Intergovernmental Science-Policy Platform on Biodiversity and Ecosystem Services*. https://doi.org/DOI: 10.5281/zenodo.383188

Bueno de Mesquita, C. P., Elmendorf, S. C., Smith, J. G., & Suding, K. N. (2024). Shifting alpine plant distributions with global change: Testing the environmental matching hypothesis. *Arctic, Antarctic, and Alpine Research*, *56*(1), 2393443. DOI: 10.1080/15230430.2024.2393443

Bukvicki, D., Gottardi, D., Vannini, L., Dzamic, A., Ciric, A., Marin, P. D., & Veljic, M. (2015). Chemical composition and antimicrobial assessment of liverwort *Lophozia ventricosa* extracts. *Revista Brasileira de Botânica. Brazilian Journal of Botany*, *38*(1), 25–30. DOI: 10.1007/s40415-014-0104-z

Buytaert, W., Cuesta-Camacho, F., & Tobón, C. (2011). Potential impacts of climate change on the environmental services of humid tropical alpine regions. *Global Ecology and Biogeography*, *20*(1), 19–33. DOI: 10.1111/j.1466-8238.2010.00585.x

Bynum, M. R., & Smith, W. K. (2001). Floral movements in response to thunderstorms improve reproductive effort in the alpine species Gentiana algida (Gentianaceae). *American Journal of Botany*, *88*(6), 1088–1095. DOI: 10.2307/2657092 PMID: 11410474

Bystrzejewska-Piotrowska, G., Golimowski, J., & Urban, P. L. (2009). Nanoparticles: Their potential toxicity, waste and environmental management. *Waste Management (New York, N.Y.)*, *29*(9), 2587–2595. DOI: 10.1016/j.wasman.2009.04.001 PMID: 19427190

Cadi, H. E., Bouzidi, H. E., Selama, G., Ramdan, B., Majdoub, Y. O. E., Alibrando, F., Brigui, J., Altemimi, A. B., Dugo, P., Mondello, L., & Cacciola, F. (2021). Characterization of *Rubus fruticosus* L. berries growing wild in Morocco: Phytochemical screening, antioxidant activity and chromatography analysis. *European Food Research and Technology, 247*(7), 1689–1699. DOI: 10.1007/s00217-021-03739-5

Caf, N., Sabatier, P., Cucinotta, J., Rapuc, W., Dolenec, M., & Šmuc, A. (2025). Human influence at the Holocene treeline in the Julian Alps, Slovenia. *The Holocene, 35*(4), 446–459. DOI: 10.1177/09596836241307300

Cahill, A. E., Aiello-Lammens, M. E., Fisher-Reid, M. C., Hua, X., Karanewsky, C. J., Yeong Ryu, H., Sbeglia, G. C., Spagnolo, F., Waldron, J. B., Warsi, O., & Wiens, J. J. (2013). How does climate change cause extinction? *Proceedings of the Royal Society B: Biological Sciences*, 280(1750), Article 20121890. DOI: 10.1098/rspb.2012.1890

Cai, L., Zhao, J., & Chen, J. (2025). Coevolutionary dynamics in the grass-livestock social-ecological system of China's alpine pastoral areas: A case study of the Qilian Mountains region in China. *PLoS One, 20*(1), e0317769. DOI: 10.1371/journal.pone.0317769 PMID: 39883742

Callaway, R. M., Brooker, R. W., Choler, P., Kikvidze, Z., Lortie, C. J., Michalet, R., Paolini, L., Pugnaire, F. I., Newingham, B., Aschehoug, E. T., Armas, C., Kikodze, D., & Cook, B. J. (2002). Positive interactions among alpine plants increase with stress. *Nature, 417*(6891), 844–848. DOI: 10.1038/nature00812 PMID: 12075350

Canella, M., Beltran-Sanz, N., Gröger, A., Pungaršek, Š., Rome, M., Natale, S., Bonomi, C., Wiesinger, H. C., Mainetti, A., Scapin, A., Valecic, M., Sensato, S., Sommacal, M., Piutti, E., Bottelli, F., Pizzato, M. T., Senn, J., Monod, A., La Rocca, N., & Dal Grande, F. (2025). The role of Alpine botanical gardens in integrating germplasm bank collections and mission. *PLANTS, PEOPLE, PLANET*, ppp3.70120. DOI: 10.1002/ppp3.70120

Canella, M., Beltran-Sanz, N., Gröger, A., Pungaršek, Š., Rome, M., Natale, S., & Scapin, A. (2025). The role of Alpine botanical gardens in integrating germplasm bank collections and mission. *Plants, People, Planet.*

Caneva, G., Pieroni, A., & Guarrera, P. (2013). *Etnobotanica: conservazione di un patrimonio culturale come risorsa per uno sviluppo sostenibile*. Edipuglia.

Cannone, N., Guglielmin, M., & Gerdol, R. (2004). Relationships between vegetation patterns and periglacial landforms in northwestern Svalbard. *Polar Biology, 27*(9), 562–571. DOI: 10.1007/s00300-004-0622-4

Carlà, A. (2015). Societal security in South Tyrol: A model to deal with ethnic conflicts. *European Yearbook of Minority Issues Online, 12*(1), 56–93. DOI: 10.1163/9789004306134_004

Carolli, M., Geneletti, D., Majone, B., Bellin, A., & Zolezzi, G. (2025). Effects of climate change and anthropic water uses on ecosystem services provided by an Alpine river. *Environmental Research Letters, 20*(8), 084036. DOI: 10.1088/1748-9326/ade902

Carrasco, F., Hernández, W., Castro, N., Angulo, G., Quintero, Á., Zamudio, D., Tamariz-Angeles, C., Olivera-Gonzales, P., & Echevarría-Rodríguez, D. (2023). Chemical composition and in vitro antibacterial activity of three lichen species (Usnea sp., *Thamnolia vermicularis* subsp. solida and *Ramalina asperula*) collected from the Jauja and Huaral provinces-Peru. *Revista Colombiana de Ciencias Químico-Farmacéuticas, 52*(3). Advance online publication. DOI: 10.15446/rcciquifa.v52n3.110402

Carrer, F., Walsh, K., & Mocci, F. (2020). Ecology, Economy, and Upland Landscapes: Socio-Ecological Dynamics in the Alps during the Transition to Modernity. *Human Ecology: an Interdisciplinary Journal, 48*(1), 69–84. DOI: 10.1007/s10745-020-00130-y

Carruthers, T., Moerland, M. S., Ebersbach, J., Favre, A., Folk, R. A., Hawkins, J. A., Muellner-Riehl, A. N., Röser, M., Soltis, D. E., Tkach, N., Baker, W. J., De Vos, J. M., & Eiserhardt, W. L. (2024). Repeated upslope biome shifts in Saxifraga during late-Cenozoic climate cooling. *Nature Communications, 15*(1), 1100. DOI: 10.1038/s41467-024-45289-w PMID: 38321017

Casas, M., Vallès, J., & Gras, A. (2024). Nutritional Properties of Wild Edible Plants with Traditional Use in the Catalan Linguistic Area: A First Step for Their Relevance in Food Security. *Foods, 13*(17), 2785. DOI: 10.3390/foods13172785 PMID: 39272550

Case, R. J., Pauli, G. F., & Soejarto, D. D. (2005). Factors in maintaining indigenous knowledge among ethnic communities of Manus Island. *Economic Botany, 59*(4), 356–365. DOI: 10.1663/0013-0001(2005)059[0356:FIMIKA]2.0.CO;2

Catalan, J., Ninot, J., & Aniz, M. (2017). *The High Mountain Conservation in a Changing World.* In (pp. 3-36). DOI: 10.1007/978-3-319-55982-7

Cavazos, P., Gonzalez, D., Lanorio, J., & Ynalvez, R. (2021). Secondary metabolites, antibacterial and antioxidant properties of the leaf extracts of *Acacia rigidula* benth. and *Acacia berlandieri* benth. *SN Applied Sciences, 3*(5), 522. DOI: 10.1007/s42452-021-04513-8

Challender, D. W., 't Sas-Rolfes, M., Broad, S., Milner-Gulland, E. J. F. i. E., & Evolution. (2025). A theory of change to improve conservation outcomes through CITES. *13*, 1425267.

Chamberlain, D. E., Negro, M., Caprio, E., & Rolando, A. (2013). Assessing the sensitivity of alpine birds to potential future changes in habitat and climate to inform management strategies. *Biological Conservation*, *167*, 127–135. DOI: 10.1016/j.biocon.2013.07.036

Chandra, S., & Lata, H. (2025) High Altitude Medicinal Plants. Eds. 01. 1-561. Springer Cham. https://doi.org/10.1007/978-3-031-99406-7

Chandra, S., Trivedi, V., Purohit, V., Nautiyal, M., & Chandra, S. (2025). Climatic Change and Physiological Adaptations in Alpine Medicinal Plants. In *High Altitude Medicinal Plants: Botany, Conservation and Cultivation* (pp. 11-35). Springer.

Chandra, N., Singh, G., Lingwal, S., Bisht, M., Tewari, L. M., & Joshi, V. C. (2022). Ecological status of alpine medicinal and aromatic plants of Western Himalaya. *Journal of Herbs, Spices & Medicinal Plants*, *28*(1), 73–88. DOI: 10.1080/10496475.2021.1963903

Chandra, N., Singh, G., Lingwal, S., Jalal, J., Bisht, M., Pal, V., Bisht, M., Rawat, B., & Tiwari, L. (2022). Ecological niche modeling and status of threatened alpine medicinal plant Dactylorhiza hatagirea D. Don in Western Himalaya. *Journal of Sustainable Forestry*, *41*(10), 1029–1045. DOI: 10.1080/10549811.2021.1923530

Chandra, S., Trivedi, V., Purohit, V., Nautiyal, M., & Chandra, S. (2025). Climatic change and physiological adaptations in alpine medicinal plants. In *High altitude medicinal plants: Botany, conservation and cultivation* (pp. 11–35). Springer.

Chang, M.-L., Chang, J.-S., Yu, W.-Y., Cheah, K.-P., Li, J.-S., Cheng, H.-W., & Hu, C.-M. (2014). *Polygonum viviparum* L. induces vasorelaxation in the rat thoracic aorta via activation of nitric oxide synthase in endothelial cells. *BMC Complementary and Alternative Medicine*, *14*(1), 150. DOI: 10.1186/1472-6882-14-150 PMID: 24885446

Chapman, L. (2007). Transport and climate change: A review. *Journal of Transport Geography*, *15*(5), 354–367. DOI: 10.1016/j.jtrangeo.2006.11.008

Chardon, N. I., Stone, P., Hilbert, C., Maclachlan, T., Ragsdale, B., Zhao, A., Goodwin, K., Collins, C. G., Hewitt, N., & Elphinstone, C. (2023). Species-Specific Responses to Human Trampling Indicate Alpine Plant Size Is More Sensitive than Reproduction to Disturbance. *Plants*, *12*(17), 3040. DOI: 10.3390/plants12173040 PMID: 37687287

Chauvier-Mendes, Y., Pollock, L. J., Verburg, P. H., Karger, D. N., Pellissier, L., Lavergne, S., Zimmermann, N. E., & Thuiller, W. (2024). Transnational conservation to anticipate future plant shifts in Europe. *Nature Ecology & Evolution, 8*(3), 454–466. DOI: 10.1038/s41559-023-02287-3 PMID: 38253754

Chen, B., Ma, X., Zhou, X., Zhang, X., Wang, X., Li, Z., Yang, X., Lu, S., & Du, W. (2025). Medium-Term Effect of Livestock Grazing Intensities on the Vegetation Dynamics in Alpine Meadow Ecosystems. *Land (Basel), 14*(3), 591. DOI: 10.3390/land14030591

Chen, W., Ding, H., Li, J., Fu, F., Li, Y., Xiao, S., Xu, D., Lu, J., & Fang, J. (2023). How do montane plants manage to survive? Inferring from non-structural carbohydrates. *Trees (Berlin), 37*(2), 331–348. DOI: 10.1007/s00468-022-02351-8

Che, T., Li, X., Liu, S., Li, H., Xu, Z., Tan, J., Zhang, Y., Ren, Z., Xiao, L., Deng, J., Jin, R., Ma, M., Wang, J., & Yang, X. (2019). Integrated hydrometeorological, snow and frozen-ground observations in the alpine region of the Heihe River Basin, China. *Earth System Science Data, 11*(3), 1483–1499. DOI: 10.5194/essd-11-1483-2019

Chileh-Chelh, T., da Cunha-Chiamolera, T. P. L., Urrestarazu, M., Ezzaitouni, M., López-Ruiz, R., Nájera, C., Rincón-Cervera, M. Á., & Guil-Guerrero, J. L. (2024). London Rocket (*Sisymbrium irio* L.) as Healthy Green: Bioactive Compounds and Bioactivity of Plants Grown in Wild and Controlled Environments. *Molecules (Basel, Switzerland), 30*(1), 31. DOI: 10.3390/molecules30010031 PMID: 39795089

Chinnapun, D. (2025). The Structure and Biological Activities With Health Benefits of Luteolin Glycosides From Plants. *TheScientificWorldJournal, 2025*(1), 3066682. DOI: 10.1155/tswj/3066682

Chiriac, E. R., Chiţescu, C. L., Geană, E.-I., Gird, C. E., Socoteanu, R. P., & Boscencu, R. (2021). Advanced Analytical Approaches for the Analysis of Polyphenols in Plants Matrices—A Review. *Separations, 8*(5), 65. DOI: 10.3390/separations8050065

Choler, P., Bayle, A., Fort, N., & Gascoin, S. (2025). Waning snowfields have transformed into hotspots of greening within the alpine zone. *Nature Climate Change, 15*(1), 80–85. DOI: 10.1038/s41558-024-02177-x

Colaric, M., Stampar, F., Solar, A., & Hudina, M. (2006). Influence of branch bending on sugar, organic acid and phenolic content in fruits of 'Williams' pears (*Pyrus communis* L.). *Journal of the Science of Food and Agriculture, 86*(14), 2463–2467. DOI: 10.1002/jsfa.2644

Colwell, R. K., & Lees, D. C. (2000). The mid-domain effect: Geometric constraints on the geography of species richness. *Trends in Ecology & Evolution, 15*(2), 70–76. DOI: 10.1016/S0169-5347(99)01767-X PMID: 10652559

Conservancy, S. L., Treasury, C. A., Jackson, R., & Jain, N. (2006). Mountain cultures, keystone species: exploring the role of cultural keystone species in Central Asia. 1-47.

Cook, J., & Lewandowsky, S. (2016). Rational irrationality: Modeling climate change belief polarization using Bayesian networks. *Topics in Cognitive Science, 8*(1), 160–179. DOI: 10.1111/tops.12186 PMID: 26749179

Cooley, J. R. (1995). Floral heat rewards and direct benefits to insect pollinators. *Annals of the Entomological Society of America, 88*(4), 576–579. DOI: 10.1093/aesa/88.4.576

Cooper, W. S. (1923a). The recent ecological history of Glacier Bay, Alaska: permanent quadrats at Glacier Bay: an initial report upon a long-period study. *Ecology, 4*(4), 355–365. DOI: 10.2307/1929182

Cooper, W. S. (1923b). The recent ecological history of Glacier Bay, Alaska: The interglacial forests of Glacier Bay. *Ecology, 4*(2), 93–128. DOI: 10.2307/1929485

Cooper, W. S. (1923c). The recent ecological history of Glacier Bay, Alaska: The present vegetation cycle. *Ecology, 4*(3), 223–246. DOI: 10.2307/1929047

Cooper, W. S. (1931). A third expedition to Glacier Bay, Alaska. *Ecology, 12*(1), 61–95. DOI: 10.2307/1932934

Cooper, W. S. (1939). A fourth expedition to Glacier Bay, Alaska. *Ecology, 20*(2), 130–155. DOI: 10.2307/1930735

Corlett, R. T. (2016). Plant diversity in a changing world: Status, trends, and conservation needs. *Plant Diversity, 38*(1), 10–16. DOI: 10.1016/j.pld.2016.01.001 PMID: 30159445

Costanza, R. (2016). Ecosystem services in theory and practice. In *Routledge handbook of ecosystem services* (pp. 15–24). Routledge. DOI: 10.4324/9781315775302-3

Cout, F. (1999). *Secret: formule di guarigione in uso in Valle d'Aosta*. Priuli & Verlucca.

Cowie, R. H., Bouchet, P., & Fontaine, B. (2022). The Sixth Mass Extinction: Fact, fiction or speculation? *Biological Reviews of the Cambridge Philosophical Society, 97*(2), 640–663. DOI: 10.1111/brv.12816 PMID: 35014169

Cox, G. W. (1996). Laboratory manual of general ecology (7th ed.). Wm. C. *Brown Publishers*.

Cox-Georgian, D., Ramadoss, N., Dona, C., & Basu, C. (2019). Therapeutic and medicinal uses of terpenes. In *Medicinal plants: from farm to pharmacy* (pp. 333–359). Springer., DOI: 10.1007/978-3-030-31269-5_15

Crivellaro, A., & Büntgen, U. (2020). New evidence of thermally constrained plant cell wall lignification. *Trends in Plant Science*, *25*(4), 322–324. DOI: 10.1016/j.tplants.2020.01.011 PMID: 32191868

Cruz García, G. S. (2006). The mother–child nexus. Knowledge and valuation of wild food plants in Wayanad, Western Ghats, India. *Journal of Ethnobiology and Ethnomedicine*, *2*(1), 39. DOI: 10.1186/1746-4269-2-39 PMID: 16968535

Cuendet, M., Potterat, O., & Hostettmann, K. (2001). Flavonoids and phenylpropanoid derivatives from *Campanula barbata*. *Phytochemistry*, *56*(6), 631–636. DOI: 10.1016/S0031-9422(00)00423-4 PMID: 11281141

Cunningham, M., Cunningham, A., & Schippmann, U. (1997). *Trade in Prunus africana and the implementation of CITES*. German Federal Agency for Nature Conservation.

D'Angeli, F., Malfa, G. A., Garozzo, A., Li Volti, G., Genovese, C., Stivala, A., Nicolosi, D., Attanasio, F., Bellia, F., Ronsisvalle, S., & Acquaviva, R. (2021). Antimicrobial, antioxidant, and cytotoxic activities of *Juglans regia* L. pellicle extract. *Antibiotics (Basel, Switzerland)*, *10*(2), 159. DOI: 10.3390/antibiotics10020159 PMID: 33557378

Danna, C. (2024). *By-products from medicinal and aromatic plants: from tradition to innovation*. PhD Thesis submitted to Catalogo prodotti della ricerca Università di Genova.

Danna, C., Poggio, L., Smeriglio, A., Mariotti, M., & Cornara, L. (2022). Ethnomedicinal and Ethnobotanical Survey in the Aosta Valley Side of the Gran Paradiso National Park (Western Alps, Italy). *Plants*, *11*(2), 170. DOI: 10.3390/plants11020170 PMID: 35050058

Darmody, R., Allen, C., & Thorn, C. (2005). Soil topochronosequences at storbreen, jotunheimen, Norway. *Soil Science Society of America Journal*, *69*(4), 1275–1287. DOI: 10.2136/sssaj2004.0204

Das, K., Suple, S. D., Hossain, E., Kumar, S., & Kumar, R. (2025). Ethnobotany: a bridge between traditional knowledge and modern pharmacy. In book: Ethnobotany of India, Volume 1Publisher: Ambika Prasad Research Foundation.pp.110-126.

Datta, A., Kühn, I., Ahmad, M., Michalski, S., & Auge, H. (2017). Processes affecting altitudinal distribution of invasive Ageratina adenophora in western Himalaya: The role of local adaptation and the importance of different life-cycle stages. *PLoS One, 12*(11), e0187708. DOI: 10.1371/journal.pone.0187708 PMID: 29125852

Daugaliyeva, S., Daugaliyeva, A., Amirova, K., Yelubayeva, A., Kapar, A., Abiti, A.,...Peletto, S. J. I. (2024). Foraging of Honeybees from Different Ecological Areas Determined through Melissopalynological Analysis and DNA Metabarcoding. *15*(9), 674.

de Lima, E. P., Laurindo, L. F., Catharin, V. C. S., Direito, R., Tanaka, M., Jasmin Santos German, I., & Fiorini, A. M. R. (2025). Polyphenols, alkaloids, and terpenoids against neurodegeneration: Evaluating the neuroprotective effects of phytocompounds through a comprehensive review of the current evidence. *Metabolites, 15*(2), 124. DOI: 10.3390/metabo15020124 PMID: 39997749

De Villemereuil, P., Mouterde, M., Gaggiotti, O. E., & Till-Bottraud, I. (2018). Patterns of phenotypic plasticity and local adaptation in the wide elevation range of the alpine plant *Arabis alpina. Journal of Ecology, 106*(5), 1952–1971. DOI: 10.1111/1365-2745.12955

Dean, M. (2024). Exploring ethnobotanical knowledge: Qualitative insights into the therapeutic potential of medicinal plants. *Golden Ratio of Data in Summary, 4*(2), 154–166. DOI: 10.52970/grdis.v4i2.491

Del Curto, D., Garzulino, A., Menini, G., & Schiesaro, C. (2022). Sustainable Conservation and Management of a 20th-Century Landscape in the Alps: The Former Sanatorium Village of Sondalo. *Sustainability (Basel), 14*(12), 7424. DOI: 10.3390/su14127424

Deng, Y., Jiang, X., Zhao, H., Yang, S., Gao, J., Wu, Y., Diao, Q., & Hou, C. (2021). Microplastic Polystyrene Ingestion Promotes the Susceptibility of Honeybee to Viral Infection. *Environmental Science & Technology, 55*(17), 11680–11692. DOI: 10.1021/acs.est.1c01619 PMID: 34374532

Devi, C., Gouri, P., Subhashini, , Bavatharani, , & Bhuvaneswari, . (2024). Study on Anti-arthritic and Antimicrobial Activity of Leaves of *Mukia maderaspatana* on Membrane Stabilization Assay and Egg Albumin Assay. *SBV Journal of Basic. Clinical and Applied Health Science, 7*(3), 105–111. DOI: 10.4103/SBVJ.SBVJ_9_24

Dhar, U. (2002). Conservation implications of plant endemism in high-altitude Himalaya. *Current Science*, •••, 141–148.

Dietrich, L., & Körner, C. (2014). Thermal imaging reveals massive heat accumulation in flowers across a broad spectrum of alpine taxa. *Alpine Botany*, *124*(1), 27–35. DOI: 10.1007/s00035-014-0123-1

Ding, J., Ge, W., Liu, Q., Wang, Q., Kong, D., & Yin, H. (2023). Temperature drives the coordination between above-ground nutrient conservation and below-ground nutrient acquisition in alpine coniferous forests. *Functional Ecology*, *37*(6), 1674–1687. DOI: 10.1111/1365-2435.14330

Ding, Y., Li, H., Cao, S., & Yu, Y. (2024). Effects of catechin on the malignant biological behavior of gastric cancer cells through the PI3K/Akt signaling pathway. *Toxicology and Applied Pharmacology*, *490*, 117036. DOI: 10.1016/j.taap.2024.117036 PMID: 39009138

Dirnböck, T., Essl, F., & Rabitsch, W. (2011). Disproportional risk for habitat loss of high-altitude endemic species under climate change. *Global Change Biology*, *17*(2), 990–996. DOI: 10.1111/j.1365-2486.2010.02266.x

Dobrucka, R., & Długaszewska, J. (2015). Antimicrobial activities of silver nanoparticles synthesized by using water extract of Arnicae anthodium. *Indian Journal of Microbiology*, *55*(2), 168–174. DOI: 10.1007/s12088-015-0516-x PMID: 25805903

Dokulil, M. T. (2022). Impacts of climate warming on Alpine lakes. In *Oxford research encyclopedia of climate science*. DOI: 10.1093/acrefore/9780190228620.013.874

Domingo-Fernández, D., Gadiya, Y., Mubeen, S., Bollerman, T. J., Healy, M. D., Chanana, S., Sadovsky, R. G., Healey, D., & Colluru, V. (2023). Modern drug discovery using ethnobotany: A large-scale cross-cultural analysis of traditional medicine reveals common therapeutic uses. *iScience*, *26*(9), 107729. DOI: 10.1016/j.isci.2023.107729 PMID: 37701812

Dong, S., Shang, Z., Gao, J., & Boone, R. B. (2020). Enhancing sustainability of grassland ecosystems through ecological restoration and grazing management in an era of climate change on Qinghai-Tibetan Plateau. *Agriculture, Ecosystems & Environment*, *287*, 106684. DOI: 10.1016/j.agee.2019.106684

Donhauser, J., & Frey, B. (2018). Alpine soil microbial ecology in a changing world. *FEMS Microbiology Ecology*, *94*(9), fiy099. Advance online publication. DOI: 10.1093/femsec/fiy099 PMID: 30032189

Doretto, A., Bona, F., Falasco, E., Morandini, D., Piano, E., & Fenoglio, S. (2020). Stay with the flow: How macroinvertebrate communities recover during the rewetting phase in Alpine streams affected by an exceptional drought. *River Research and Applications*, *36*(1), 91–101. DOI: 10.1002/rra.3563

Dos Santos, E. T., Pereira, M. L. A., Da Silva, C. F. P., Souza-Neta, L. C., Geris, R., Martins, D., Santana, A. E. G., Barbosa, L. C. A., Silva, H. G. O., Freitas, G. C., Figueiredo, M., De Oliveira, F., & Batista, R. (2013). Antibacterial activity of the alkaloid-enriched extract from *Prosopis juliflora* pods and its influence on *in vitro* ruminal digestion. *International Journal of Molecular Sciences*, *14*(4), 8496–8516. DOI: 10.3390/ijms14048496 PMID: 23595000

dos Santos, R. F., Antunes, P., Ring, I., & Clemente, P. (2015). Engaging local private and public actors in biodiversity conservation: The role of agri-environmental schemes and ecological fiscal transfers. *Environmental Policy and Governance*, *25*(2), 83–96. DOI: 10.1002/eet.1661

Doull, K. M. (1980). Relationships between consumption of a pollen supplement, honey production and broodrearing in colonies of honeybees *Apis mellifera* L. II. *Apidologie*, *11*(4), 367–374. DOI: 10.1051/apido:19800405

Duchicela, S. A., Llambí, L. D., Bonnesoeur, V., & Román-Dañobeytia, F. (2024). Pastoralism in the high tropical Andes: A review of the effect of grazing intensity on plant diversity and ecosystem services. *Applied Vegetation Science*, *27*(3), e12791. DOI: 10.1111/avsc.12791

Du, E., de Vries, W., Collalti, A., & De Marco, A. (2025). Climate warming alters nutrient cycling and its constraint on CO2 fertilization in global forests. *Current Climate Change Reports*, *11*(1), 1–11. DOI: 10.1007/s40641-025-00201-6

Du, J., Wei, P., Bahadur, A., & Chen, S. (2025). The Impacts of Revegetation on Ecosystem Services in the Extremely Degraded Alpine Grassland of Permafrost Regions. *Sustainability (Basel)*, *17*(4), 1512. DOI: 10.3390/su17041512

Dullinger, S., Gattringer, A., Thuiller, W., Moser, D., Zimmermann, N. E., Guisan, A., Willner, W., Plutzar, C., Leitner, M., Mang, T., Caccianiga, M., Dirnböck, T., Ertl, S., Fischer, A., Lenoir, J., Svenning, J.-C., Psomas, A., Schmatz, D. R., Silc, U., & Hülber, K. (2012). Extinction debt of high-mountain plants under twenty-first-century climate change. *Nature Climate Change*, *2*(8), 619–622. DOI: 10.1038/nclimate1514

Dullinger, S., Kleinbauer, I., Pauli, H., Gottfried, M., Brooker, R., Nagy, L., & Benito, J. L. (2007). Weak and variable relationships between environmental severity and small-scale co-occurrence in alpine plant communities. *Journal of Ecology*, *95*(6), 1284–1295. DOI: 10.1111/j.1365-2745.2007.01288.x

Duncan, R., & Gaspar, R. (2011). Nanomedicine(s) under the microscope. *Molecular Pharmaceutics*, *8*(6), 2101–2141. DOI: 10.1021/mp200394t PMID: 21974749

Early, R., Bradley, B. A., Dukes, J. S., Lawler, J. J., Olden, J. D., Blumenthal, D. M., Gonzalez, P., Grosholz, E. D., Ibañez, I., Miller, L. P., Sorte, C. J. B., & Tatem, A. J. (2016). Global threats from invasive alien species in the twenty-first century and national response capacities. *Nature Communications*, *7*(1), 12485. DOI: 10.1038/ncomms12485 PMID: 27549569

Ebner, M., Fontana, V., Schirpke, U., & Tappeiner, U. (2022). Stakeholder perspectives on ecosystem services of mountain lakes in the European Alps. *Ecosystem Services*, *53*, 101386. DOI: 10.1016/j.ecoser.2021.101386

Edwards, A. C., Scalenghe, R., & Freppaz, M. (2007). Changes in the seasonal snow cover of alpine regions and its effect on soil processes: A review. *Quaternary International*, *162*, 172–181. DOI: 10.1016/j.quaint.2006.10.027

Egarter Vigl, L., Depellegrin, D., Pereira, P., De Groot, R., & Tappeiner, U. (2017). Mapping the ecosystem service delivery chain: Capacity, flow, and demand pertaining to aesthetic experiences in mountain landscapes. *The Science of the Total Environment*, *574*, 422–436. DOI: 10.1016/j.scitotenv.2016.08.209 PMID: 27639477

Egarter Vigl, L., Marsoner, T., Schirpke, U., Tscholl, S., Candiago, S., & Depellegrin, D. (2021). A multi-pressure analysis of ecosystem services for conservation planning in the Alps. *Ecosystem Services*, *47*, 101230. DOI: 10.1016/j.ecoser.2020.101230

Egarter Vigl, L., Schirpke, U., Tasser, E., & Tappeiner, U. (2016). Linking long-term landscape dynamics to the multiple interactions among ecosystem services in the European Alps. *Landscape Ecology*, *31*(9), 1903–1918. DOI: 10.1007/s10980-016-0389-3

El Mannoubi, I. (2023). Impact of different solvents on extraction yield, phenolic composition, in vitro antioxidant and antibacterial activities of deseeded *Opuntia stricta* fruit. *Journal of Umm Al-Qura University for Applied Sciences*, *9*(2), 176–184. DOI: 10.1007/s43994-023-00031-y

El-Hajjaji, M. A., Fikri-Benbrahim, K., El Ouassete, M., Mrabti, N. N., Soulo, N., El Ghouizi, A., Lyoussi, B., & Ouaritini, Z. B. (2024). Phytochemical profiling of *Prunus armeniaca* kernel extracts and exploration of their multifaceted antioxidant and antibacterial effects through *in vitro* and *in silico* studies. *European Journal of Integrative Medicine*, *72*, 102421. DOI: 10.1016/j.eujim.2024.102421

El-Hefny, M., & Hussien, M. K. (2025). Enhancing the growth and essential oil components of *Lavandula latifolia* using *Malva parviflora* extract and humic acid as biostimulants in a field experiment. *Scientific Reports*, *15*(1), 774. DOI: 10.1038/s41598-024-82127-x PMID: 39755703

El-Massry, K. F., Farouk, A., Mahmoud, K. F., El-Ghorab, A. H., Musa, A., Mostafa, E. M., Ghoneim, M. M., Naguib, I. A., & Abdelgawad, M. A. (2021). Chemical characteristics and targeted encapsulated *Cordia myxa* fruits extracts nanoparticles for antioxidant and cytotoxicity potentials. *Saudi Journal of Biological Sciences*, *28*(9), 5349–5358. DOI: 10.1016/j.sjbs.2021.05.064 PMID: 34466114

El-Nashar, H. A., Zengin, G., Tarek, A., & Eldahshan, O. A. (2025). Comparative metabolic profiling of *Cordia myxa* leaves and fruits based on UPLC-ESI/MS-MS analysis and investigation of antioxidant activities and enzyme inhibitory properties. *Scientific Reports*, *15*(1), 19296. DOI: 10.1038/s41598-025-04015-2 PMID: 40456930

El-Newary, S. A., Abd Elkarim, A. S., Abdelwahed, N. A., Omer, E. A., Elgamal, A. M., & ELsayed, W. M. (2023). *Chenopodium murale* juice shows anti-fungal efficacy in experimental oral candidiasis in immunosuppressed rats in relation to its chemical profile. *Molecules (Basel, Switzerland)*, *28*(11), 4304. DOI: 10.3390/molecules28114304 PMID: 37298777

El-Newary, S. A., Aly, M. S., Hameed, A. R. A. E., Kotp, M. S., Youssef, A. A., & Ali, N. A. (2022). Sperm quality and testicular histopathology of Wistar albino male rats treated with hydroethanolic extract of *Cordia dichotoma* fruits. *Pharmaceutical Biology*, *60*(1), 282–293. DOI: 10.1080/13880209.2021.2008455 PMID: 35138992

Elsen, P. R., & Tingley, M. W. (2015). Global mountain topography and the fate of montane species under climate change. *Nature Climate Change*, *5*(8), 772–776. DOI: 10.1038/nclimate2656

Elzaawely, A. A., & Tawata, S. (2012). Antioxidant capacity and phenolic content of *Rumex dentatus* L. grown in Egypt. *Journal of Crop Science and Biotechnology*, *15*(1), 59–64. DOI: 10.1007/s12892-011-0063-x

Engelmann, F. (2011). Use of biotechnologies for the conservation of plant biodiversity. *In Vitro Cellular & Developmental Biology. Plant*, *47*(1), 5–16. DOI: 10.1007/s11627-010-9327-2

Engler, R., Randin, C. F., Thuiller, W., Dullinger, S., Zimmermann, N. E., Araújo, M. B., Pearman, P. B., Le Lay, G., Piedallu, C., Albert, C. H., Choler, P., Coldea, G., De Lamo, X., Dirnböck, T., Gégout, J.-C., Gómez-García, D., Grytnes, J.-A., Heegaard, E., Høistad, F., & Guisan, A. (2011). 21st century climate change threatens mountain flora unequally across Europe. *Global Change Biology*, *17*(7), 2330–2341. DOI: 10.1111/j.1365-2486.2010.02393.x

Engler, R., Randin, C. F., Vittoz, P., Czáka, T., Beniston, M., Zimmermann, N. E., & Guisan, A. (2009). Predicting future distributions of mountain plants under climate change: Does dispersal capacity matter? *Ecography, 32*(1), 34–45. DOI: 10.1111/j.1600-0587.2009.05789.x

Ercan, L., & Doğru, M. (2022). Determination of antimicrobial activity of *Nasturtium officinale* and its content of volatile organic compounds and fatty acids. *Kahramanmaraş Sütçü İmam Üniversitesi Tarım ve Doğa Dergisi, 25*(Ek Sayı 1), 11-21. DOI: 10.18016/ksutarimdoga.vi.1001837

Erdogan-Orhan, I., & Kartal, M. (2011). Insights into research on phytochemistry and biological activities of *Prunus armeniaca* L.(apricot). *Food Research International, 44*(5), 1238–1243. DOI: 10.1016/j.foodres.2010.11.014

Erenler, R., Telci, I., Elmastaş, M., Akşit, H., Gül, F., Tüfekçi, A. R., Demirtaş, İ., & Kayir, Ö. (2018). Quantification of flavonoids isolated from\textit {*Mentha spicata*} in selected clones of Turkish mint landraces. *Turkish Journal of Chemistry, 42*(6), 1695–1705. DOI: 10.3906/kim-1712-3

Essl, F., Dullinger, S., Plutzar, C., Willner, W., & Rabitsch, W. (2011). Imprints of glacial history and current environment on correlations between endemic plant and invertebrate species richness. *Journal of Biogeography, 38*(4), 604–614. DOI: 10.1111/j.1365-2699.2010.02425.x

Essl, F., Staudinger, M., Stöhr, O., Schratt-Ehrendorfer, L., Rabitsch, W., & Niklfeld, H. (2009). Distribution patterns, range size and niche breadth of Austrian endemic plants. *Biological Conservation, 142*(11), 2547–2558. DOI: 10.1016/j.biocon.2009.05.027

European Commission. (n.d.). *Consequences of climate change*. Climate Action. Retrieved January 18, 2026, from https://climate.ec.europa.eu/climate-change/consequences-climate-change_en

Fahy, J. C., Demierre, E., & Oertli, B. (2024). Long-term monitoring of water temperature and macroinvertebrates highlights climate change threat to alpine ponds in protected areas. *Biological Conservation, 290*, 110461. DOI: 10.1016/j.biocon.2024.110461

Fairbrother, A., Purdy, J., Anderson, T., & Fell, R. (2014). Risks of neonicotinoid insecticides to honeybees. *Environmental Toxicology and Chemistry, 33*(4), 719–731. DOI: 10.1002/etc.2527 PMID: 24692231

Falasco, E., Piano, E., & Bona, F. (2016). Diatom flora in Mediterranean streams: Flow intermittency threatens endangered species. *Biodiversity and Conservation, 25*(14), 2965–2986. DOI: 10.1007/s10531-016-1213-8

Fantasma, F., Samukha, V., Saviano, G., Chini, M. G., Iorizzi, M., & Caprari, C. (2024). Nutraceutical Aspects of Selected Wild Edible Plants of the Italian Central Apennines. *Nutraceuticals, 4*(2), 190–231. DOI: 10.3390/nutraceuticals4020013

Farag, M., Abdel-Mageed, W. M., El Gamal, A. A., & Basudan, O. A. (2021). *Salvadora persica* L.: Toothbrush tree with health benefits and industrial applications–An updated evidence-based review. *Saudi Pharmaceutical Journal : SPJ : the Official Publication of the Saudi Pharmaceutical Society, 29*(7), 751–763. DOI: 10.1016/j.jsps.2021.05.007 PMID: 34400870

Farooqi, M. A., Aslam, M. N., Sajjad, A., Akram, W., & Maqsood, A. (2021). Impact of abiotic factors on the foraging behavior of two honeybee species on canola in Bahawalpur, Punjab-Pakistan. *Asian Journal of Agriculture and Biology, 2021*(01).

Faysal, M., Al Amin, M., Zehravi, M., Sweilam, S. H., Arjun, U. V. N. V., Gupta, J. K., Shanmugarajan, T. S., Prakash, S. S., Dayalan, G., Kasimedu, S., Madhuri, Y. B., Reddy, K. T. K., Rab, S. O., Al Fahaid, A. A. F., & Emran, T. B. (2025). Therapeutic potential of flavonoids in neuroprotection: Brain and spinal cord injury focus. *Naunyn-Schmiedeberg's Archives of Pharmacology, 398*(7), 1–26. DOI: 10.1007/s00210-025-03888-4 PMID: 40014123

Fedrigotti, C., Aschonitis, V., & Fano, E. A. (2016). Effects of forest expansion and land abandonment on ecosystem services of alpine environments: Case study in ledro valley (Italy) for the period 1859–2011. *Global NEST Journal, 18*(4), 875–884. DOI: 10.30955/gnj.001770

Feng, M. (2025). *Functional diversity and metabolic capabilities of cold-adapted microbial communities in the Swiss Alps* [Doctoral dissertation, University of Zurich].

Feng, B., Zhou, J., Hu, L., Liu, Z., Yang, Y., Yang, S., Ni, J., Bai, W., & Zhao, S. (2025). Land Use and Ecosystem Service Value Spatiotemporal Dynamics, Topographic Gradient Effect and Their Driving Factors in Typical Alpine Ecosystems of the East Qinghai-Tibet Plateau: Implications for Conservation and Development. *Ecology and Evolution, 15*(3), e71125. DOI: 10.1002/ece3.71125 PMID: 40099216

Feng, L., Yin, Y., Fang, Y., & Yang, X. (2017). Quantitative determination of punicalagin and related substances in different parts of pomegranate. *Food Analytical Methods, 10*(11), 3600–3606. DOI: 10.1007/s12161-017-0916-0

Fereja, W. M., Geleta, W. D., Desalegn, E., Moaneanda, M., Abdissa, A., & Usamo, F. B. (2024). *In vitro* antioxidant and antibacterial activities of *Ajuga integrifolia* leaf extracts obtained with different solvents. *BMC Complementary Medicine and Therapies*, *24*(1), 368. DOI: 10.1186/s12906-024-04668-y PMID: 39395997

Field, C. B., & Barros, V. R. (Eds.). (2014). *Climate change 2014–impacts, adaptation and vulnerability: Global and sectoral aspects*. Cambridge University Press. DOI: 10.1017/CBO9781107415379

Fierro, P., Valdovinos, C., Arismendi, I., Díaz, G., De Gamboa, M. R., & Arriagada, L. (2019). Assessment of anthropogenic threats to Chilean Mediterranean freshwater ecosystems: Literature review and expert opinions. *Environmental Impact Assessment Review*, *77*, 114–121. DOI: 10.1016/j.eiar.2019.02.010

Fina, A., Koljančić, N., Squara, S., Ferrara, D., Bicchi, C., Špánik, I., & Cordero, C. (2023). Food metabolomics by GCXGC-TOF MS and tandem ionization: understanding the impact of climate events on edible crops quality. XIII Congresso Nazionale di Chimica degli Alimenti Libro degli abstracts, Grabherr, G. (2009). Biodiversity in the high ranges of the Alps: ethnobotanical and climate change perspectives. *Global Environmental Change*, *19*(2), 167–172.

Fontefrancesco, M. F., & Pieroni, A. (2020). Renegotiating situativity: Transformations of local herbal knowledge in a Western Alpine valley during the past 40 years. *Journal of Ethnobiology and Ethnomedicine*, *16*(1), 58. DOI: 10.1186/s13002-020-00402-3 PMID: 32998765

Fouad, M. S., Ghareeb, M. A., Hamed, A. A., Aidy, E. A., Tabudravu, J., Sayed, A. M., Tammam, M. A., & Mwaheb, M. A. (2024). Exploring the antioxidant, anticancer and antimicrobial potential of *Amaranthus viridis* L. collected from Fayoum depression: Phytochemical, and biological aspects. *South African Journal of Botany*, *166*, 297–310. DOI: 10.1016/j.sajb.2024.01.047

Fugh-Berman, A., & Ernst, E. J. B. J. O. C. P. (2002). 'Herb-drug interactions: Review and assessment of report reliability'(vol 52, pg 587, 2001). *53*(4), 449P-449P.

Fujino, M., Sakakibara, K., Tsujimura, M., & Suzuki, K. (2023). Influence of alpine vegetation on water storage and discharge functions in an alpine headwater of Northern Japan Alps. *Journal of Hydrology: X*, *18*, 100146. DOI: 10.1016/j.hydroa.2022.100146

Gallagher, R., Hughes, L., & Leishman, M. (2009). Phenological trends among Australian alpine species: Using herbarium records to identify climate-change indicators. *Australian Journal of Botany*, *57*(1), 1–9. DOI: 10.1071/BT08051

Ganesan, P. (2024). Analysis of antiviral and antimicrobial activity of *Opuntia Stricta* (red prickly pear) extract treated cotton fabric by cationization technique for healthcare and hygiene textiles. *Journal of the Textile Institute*, *115*(12), 2411–2420. DOI: 10.1080/00405000.2023.2291526

Ganie, S. A., Dar, T. A., Hamid, R., Zargar, O., Abeer, S. U., Masood, A., Amin, S., & Zargar, M. A. (2014). *In vitro* antioxidant and cytotoxic activities of *Arnebia benthamii* (Wall ex. G. Don): A critically endangered medicinal plant of Kashmir valley. *Oxidative Medicine and Cellular Longevity*, *2014*(1), 792574. DOI: 10.1155/2014/792574 PMID: 24734152

Ganie, S. A., Jan, A., Muzaffar, S., Zargar, B. A., Hamid, R., & Zargar, M. A. (2012). Radical scavenging and antibacterial activity of *Arnebia benthamii* methanol extract. *Asian Pacific Journal of Tropical Medicine*, *5*(10), 766–772. DOI: 10.1016/S1995-7645(12)60140-0 PMID: 23043913

Gao, Q., Guo, Y., Xu, H., Ganjurjav, H., Li, Y., Wan, Y., Qin, X., Ma, X., & Liu, S. (2016). Climate change and its impacts on vegetation distribution and net primary productivity of the alpine ecosystem in the Qinghai-Tibetan Plateau. *The Science of the Total Environment*, *554–555*, 34–41. DOI: 10.1016/j.scitotenv.2016.02.131 PMID: 26950617

Garcés-Pastor, S., Coissac, E., Lavergne, S., Schwoerer, C., Theurillat, J.-P., Heintzman, P., Wangensteen, O., Tinner, W., Rey, F., & Heer, M. (2021). High resolution ancient sedimentary DNA shows that alpine plant biodiversity is a result of human land use.

Garcés-Pastor, S., Coissac, E., Lavergne, S., Schwörer, C., Theurillat, J.-P., Heintzman, P. D., Wangensteen, O. S., Tinner, W., Rey, F., Heer, M., Rutzer, A., Walsh, K., Lammers, Y., Brown, A. G., Goslar, T., Rijal, D. P., Karger, D. N., Pellissier, L., Pouchon, C., & Alsos, I. G. (2022). High resolution ancient sedimentary DNA shows that alpine plant diversity is associated with human land use and climate change. *Nature Communications*, *13*(1), 6559. DOI: 10.1038/s41467-022-34010-4 PMID: 36333301

García-Navas, V., Sattler, T., Schmid, H., & Ozgul, A. (2020). Temporal homogenization of functional and beta diversity in bird communities of the Swiss Alps. *Diversity & Distributions*, *26*(8), 900–911. DOI: 10.1111/ddi.13076

Gardner, J. S., & Dekens, J. (2007). Mountain hazards and the resilience of social–ecological systems: Lessons learned in India and Canada. *Natural Hazards*, *41*(2), 317–336. DOI: 10.1007/s11069-006-9038-5

Gatt, A. R., Vella Bonanno, P., & Zammit, R. (2024). Ethical considerations in the regulation and use of herbal medicines in the European Union. *Frontiers in Medical Technology*, *6*, 1358956. DOI: 10.3389/fmedt.2024.1358956 PMID: 38948354

Gaude, A. A., & Jalmi, S. K. (2025). Environmental stress induced biosynthesis of plant secondary metabolites-transcriptional regulation as a key. *Crop Design*, *100100*(2), 100100. Advance online publication. DOI: 10.1016/j.cropd.2025.100100

Gauslaa, Y. (1984). Heat resistance and energy budget in different Scandinavian plants. *Ecography*, *7*(1), 5–6. DOI: 10.1111/j.1600-0587.1984.tb01098.x

Gautam, R., Srivastava, A., & Jachak, S. M. (2011). Simultaneous determination of naphthalene and anthraquinone derivatives in *Rumex nepalensis* Spreng. Roots by HPLC: Comparison of different extraction methods and validation. *Phytochemical Analysis*, *22*(2), 153–157. DOI: 10.1002/pca.1261 PMID: 21046683

Geneletti, D., Scolozzi, R., & Adem Esmail, B. (2018). Assessing ecosystem services and biodiversity tradeoffs across agricultural landscapes in a mountain region. *The International Journal of Biodiversity Science, Ecosystem Services & Management*, *14*(1), 188–208. DOI: 10.1080/21513732.2018.1526214

Ghaffar, S., Abbas, A., Naeem-ul-Hassan, M., Assad, N., Sher, M., Ullah, S., Al-hazmi, H. A., Najmi, A., Zoghebi, K., Al Bratty, M., Hanbashi, A., Makeen, H. A., & Amin, H. M. A. (2023). Improved photocatalytic and antioxidant activity of olive fruit extract-mediated ZnO nanoparticles. *Antioxidants*, *12*(6), 1201. DOI: 10.3390/antiox12061201 PMID: 37371931

Ghahramani, Y., Mohammadi, N., Baghaei, S., & Jahandizi, N. G. (2024). Time-dependent antibacterial effects of *Citrullus colocynthis* seed extract compared to calcium hydroxide in teeth infected with *Enterococcus faecalis*. *Journal of Dentistry*, *25*(1), 77. DOI: 10.30476/dentjods.2023.97670.2026 PMID: 38544769

Ghazghazi, H., Chedia, A., Weslati, M., Trakhna, F., Houssine, S., Abderrazak, M., & Brahim, H. (2013). Chemical Composition and *in vitro* Antimicrobial Activities of *Mentha pulegium* Leaves Extracts against Foodborne Pathogens. *Journal of Food Safety*, *33*(3), 239–246. DOI: 10.1111/jfs.12045

Ghosh, L., Gayen, J. R., Sinha, S., Pal, S., Pal, M., & Saha, B. (2003). Antibacterial efficacy of *Rumex nepalensis* Spreng. roots. *Phytotherapy Research*, *17*(5), 558–559. DOI: 10.1002/ptr.1162 PMID: 12748999

Gillani, S. W., Ahmad, M., Manzoor, M., Al-Andal, A., Khan, R. W. A., Ayoola, J. O., & Waheed, M. (2025a). Integrating environmental variables and biotic interactions to predict the future distribution of the endangered *Aconitum heterophyllum* Wall. ex Royle. *BMC Plant Biology, 26*(1), 45. DOI: 10.1186/s12870-025-07859-y PMID: 41339780

Gillani, S. W., Ahmad, M., Manzoor, M., Waheed, M., Iqbal, Z., Ullah, R., Pieroni, A., Zhang, L., Sulaiman, N., & Alrhmoun, M. (2024). The nexus between ecology of foraging and food security: Cross-cultural perceptions of wild food plants in Kashmir Himalaya. *Journal of Ethnobiology and Ethnomedicine, 20*(1), 77. DOI: 10.1186/s13002-024-00721-9 PMID: 39155383

Gillani, S. W., Ahmad, M., Manzoor, M., Waheed, M., Khan, R. W. A., & Sohail, A. (2026). Geographic distribution patterns, population dynamics, and habitat vulnerability of endangered and endemic *Trillium govanianum. Environmental Monitoring and Assessment, 198*(2), 162. DOI: 10.1007/s10661-026-15013-w PMID: 41569443

Gillani, S. W., Ahmad, M., Manzoor, M., Zafar, M., Shaheen, H., Hussain, J., Al Farraj, D. A., Elshikh, M. S., & Abbas, N. (2025b). Local perceptions of ecosystem services and anthropogenic disturbances on endangered *Taxus wallichiana*: Implications for conservation and sustainable management. *Polish Journal of Environmental Studies, 34*(4), 1–14. DOI: 10.15244/pjoes/188236

Gillani, S. W., Ahmad, M., Zafar, M., Manzoor, M., Shah, G. M., Shaheen, H., Zaman, W., Sultana, S., Sadia, B., & Khishlatovna, K. K. (2024). Ethnobotanical Exploration of Traditional Medicinal Plants Among the Rural Inhabitants of District Muzaffarabad, Kashmir Himalayan Region. *Plant Science Today, 11*(sp1). Advance online publication. DOI: 10.14719/pst.3265

GoAJK. (2017). *District disaster management plan.* District Disaster Management Authority.

Grabherr, G. (2009). Biodiversity in the high ranges of the Alps: Ethnobotanical and climate change perspectives. *Global Environmental Change, 19*(2), 167–172. DOI: 10.1016/j.gloenvcha.2009.01.007

Grabherr, G., Gottfried, M., & Pauli, H. (2010). Climate change impacts in alpine environments. *Geography Compass, 4*(8), 1133–1153. DOI: 10.1111/j.1749-8198.2010.00356.x

Greiber, T. (2012). *An explanatory guide to the Nagoya Protocol on access and benefit-sharing*. Iucn.

Grêt-Regamey, A., Brunner, S. H., & Kienast, F. (2012). Mountain ecosystem services: Who cares? *Mountain Research and Development*, *32*(S1), S23–S34. DOI: 10.1659/MRD-JOURNAL-D-10-00115.S1

Grime, J. P. (1974). Vegetation classification by reference to strategies. *Nature*, *250*(5461), 26–31. DOI: 10.1038/250026a0

Guarrera, P. M. (2006). *Usi e tradizioni della flora italiana: medicina popolare ed etnobotanica*.

Gul, H., Ahmed Khan, H., Nasreen, Z., Assad, N., Ali Turab, S., & Hanif, M. (2025). Exploring the Antibacterial, Anticoagulant, and Hemolytic Potential of Green-Synthesized Fe(2)O(3) Nanoparticles by Cucurbita pepo Pulp. *IEEE Transactions on Nanobioscience*, *24*(4), 411–420. DOI: 10.1109/TNB.2025.3563307 PMID: 40261780

Gundogdu, M., Muradoglu, F., Sensoy, R. G., & Yilmaz, H. (2011). Determination of fruit chemical properties of *Morus nigra* L., *Morus alba* L. and *Morus rubra* L. by HPLC. *Scientia Horticulturae*, *132*, 37–41. DOI: 10.1016/j.scienta.2011.09.035

Gupta, S., Kumar, N., Verma, D., Joshi, A., Kumar Arya, A., Joshi, K., & Bachheti, R. (2021). Impacts of invasive alien plant species on biodiversity in the regions of Western Himalayas, India: An overview. In (pp. 123-135). DOI: 10.26832/aesa-2021-bdcp-08

Gupta, H., Nishi, M., & Gasparatos, A. (2022). Community-based responses for tackling environmental and socio-economic change and impacts in mountain social–ecological systems. *Ambio*, *51*(5), 1123–1142. DOI: 10.1007/s13280-021-01651-6 PMID: 34784008

Gupta, S., Jasrotia, S., & Purohit, M. (2023). Conservation, cultivation, and sustainable use of high altitude medicinal and aromatic plants for socio-economic development: A review. *Journal of Mountain Research*, *18*(1), 201–209. DOI: 10.51220/jmr.v18i1.22

Güsewell, S. (2014). Long-term trends in spring phenology of plants in Switzerland. *International Journal of Biometeorology*, *58*(1), 1–14. PMID: 23800836

Haddad, N. M., Brudvig, L. A., Clobert, J., Davies, K. F., Gonzalez, A., Holt, R. D., Lovejoy, T. E., Sexton, J. O., Austin, M. P., Collins, C. D., Cook, W. M., Damschen, E. I., Ewers, R. M., Foster, B. L., Jenkins, C. N., King, A. J., Laurance, W. F., Levey, D. J., Margules, C. R., & Townshend, J. R. (2015). Habitat fragmentation and its lasting impact on Earth's ecosystems. *Science Advances*, *1*(2), e1500052. DOI: 10.1126/sciadv.1500052 PMID: 26601154

Haeberli, W., & Hoelzle, M. (1995). Application of inventory data for estimating characteristics of and regional climate-change effects on mountain glaciers: A pilot study with the European Alps. *Annals of Glaciology*, *21*, 206–212. DOI: 10.3189/S0260305500015834

Hagedorn, F., Shiyatov, S. G., Mazepa, V. S., Devi, N. M., Grigor'ev, A. A., Bartysh, A. A., Fomin, V. V., Kapralov, D. S., Terent'ev, M., Bugman, H., Rigling, A., & Moiseev, P. A. (2014). Treeline advances along the Urals mountain range – driven by improved winter conditions? *Global Change Biology*, *20*(11), 3530–3543. DOI: 10.1111/gcb.12613 PMID: 24756980

Hall, K., & Walton, D. W. (1992). Rock weathering, soil development and colonization under a changing climate. *Philosophical Transactions of the Royal Society of London. Series B, Biological Sciences*, *338*(1285), 269–277. DOI: 10.1098/rstb.1992.0147

Hamid, M., Gulzar, A., Dar, F. A., Singh, C. P., Malik, A. H., Kamili, A. N., & Khuroo, A. A. (2023). Microclimate heterogeneity modulates fine-scale edaphic and vegetation patterns on the Himalayan treelines: Implications under climate change. *Agricultural and Forest Meteorology*, *341*, 109688. DOI: 10.1016/j.agrformet.2023.109688

Hamid, M., Khuroo, A. A., Malik, A. H., Ahmad, R., & Singh, C. P. (2020). Assessment of alpine summit flora in Kashmir Himalaya and its implications for long-term monitoring of climate change impacts. *Journal of Mountain Science*, *17*(8), 1974–1988. DOI: 10.1007/s11629-019-5924-7

Hamilton, A. J. I. P. C. U., WWF-UK. (2003). Medicinal plants and conservation: issues and approaches. *51*.

Hanazaki, N., Herbst, D. F., Marques, M. S., & Vandebroek, I. (2013). Evidence of the shifting baseline syndrome in ethnobotanical research. *Journal of Ethnobiology and Ethnomedicine*, *9*(1), 75. DOI: 10.1186/1746-4269-9-75 PMID: 24229063

Han, B., Wu, J., & Huang, L. (2020). Induction of apoptosis in lung cancer cells by *Viburnum grandiflorum* via mitochondrial pathway. *Medical Science Monitor*, *26*, e920265–e920261. DOI: 10.12659/MSM.920265 PMID: 31900380

Hao, J.-W., Liu, X.-Q., Chen, N.-D., & Zhu, A. (2021). Solid-phase extraction followed by direct TOF-MS-MS and HPLC analysis of ptaquiloside in *Pteridium aquilinum* from different places of China. *Journal of Food Composition and Analysis*, *98*, 103845. DOI: 10.1016/j.jfca.2021.103845

Haq, S. M., Khoja, A. A., Waheed, M., Pieroni, A., Siddiqui, M. H., & Bussmann, R. W. (2024). Plant cultural indicators of forest resources from the Himalayan high mountains: Implications for improving agricultural resilience, subsistence, and forest restoration. *Journal of Ethnobiology and Ethnomedicine*, *20*(1), 44. DOI: 10.1186/s13002-024-00685-w PMID: 38659048

Haraguchi, A., Nagasawa, J., Kuramochi, K., Tsuchida, S., Kobayashi, A., Hatabu, T., Sasai, K., Ikadai, H., Ushida, K., & Matsubayashi, M. (2024). Anticoccidial activity of the secondary metabolites in alpine plants frequently ingested by wild Japanese rock ptarmigans. *International Journal for Parasitology. Parasites and Wildlife*, *25*, 100967. DOI: 10.1016/j.ijppaw.2024.100967 PMID: 39220322

Harsch, M. A., Hulme, P. E., McGlone, M. S., & Duncan, R. P. (2009). Are treelines advancing? A global meta-analysis of treeline response to climate warming. *Ecology Letters*, *12*(10), 1040–1049. DOI: 10.1111/j.1461-0248.2009.01355.x PMID: 19682007

Hasan, K., Sabiha, S., Islam, N., Pinto, J. F., & Silva, O. (2024). Ethnomedicinal Usage, Phytochemistry and Pharmacological Potential of *Solanum surattense* Burm. f. *Pharmaceuticals (Basel, Switzerland)*, *17*(7), 948. DOI: 10.3390/ph17070948 PMID: 39065797

Hashmi, S. S., Ibrahim, M., Adnan, M., Ullah, A., Khan, M. N., Kamal, A., & Elshikh, M. S. (2024). Green synthesis of silver nanoparticles from Olea europaea L. extracted polysaccharides, characterization, and its assessment as an antimicrobial agent against multiple pathogenic microbes. *Open Chemistry*, *22*(1), 20240016. DOI: 10.1515/chem-2024-0016

Hasnat, H., Shompa, S. A., Islam, M. M., Alam, S., Richi, F. T., Emon, N. U., Ashrafi, S., Ahmed, N. U., Chowdhury, M. N. R., Fatema, N., Hossain, M. S., Ghosh, A., & Ahmed, F. (2024). Flavonoids: A treasure house of prospective pharmacological potentials. *Heliyon*, *10*(6), e27533. Advance online publication. DOI: 10.1016/j.heliyon.2024.e27533 PMID: 38496846

Haugland, J. E. (2006). Short-term periglacial processes, vegetation succession, and soil development within sorted patterned ground: Jotunheimen, Norway. *Arctic, Antarctic, and Alpine Research*, *38*(1), 82–89. DOI: 10.1657/1523-0430(2006)038[0082:SPPVSA]2.0.CO;2

Haynes, K. R., Friedman, J., Stella, J. C., & Leopold, D. J. (2021). Assessing climate change tolerance and the niche breadth-range size hypothesis in rare and widespread alpine plants. *Oecologia*, *196*(4), 1233–1245. DOI: 10.1007/s00442-021-05003-9 PMID: 34331105

Hazarika, A., Nath, A. J., Reang, D., Pandey, R., Sileshi, G. W., & Das, A. K. (2024). Climate change vulnerability and adaptation among farmers practicing shifting agriculture in the Indian Himalayas. *Environmental and Sustainability Indicators, 23*, 100430. DOI: 10.1016/j.indic.2024.100430

Heinrich, M. (2010). Ethnopharmacology in the 21st century - grand challenges [Specialty Grand Challenge]. *Frontiers in pharmacology, Volume 1 - 2010*. DOI: 10.3389/fphar.2010.00008

Herbert, E., & Hill, D. (1992). Honey bee nutrition. *The hive and the honey bee, 3*, 197-233.

Hernández, A., Miranda, M. D., Arellano, E. C., & Dobbs, C. (2016). Landscape trajectories and their effect on fragmentation for a Mediterranean semi-arid ecosystem in Central Chile. *Journal of Arid Environments, 127*, 74–81. DOI: 10.1016/j.jaridenv.2015.10.004

Heywood, V. H. (2014). An overview of in situ conservation of plant species in the Mediterranean. *Flora Mediterranea, 24*, 5–24. DOI: 10.7320/FlMedit24.005

Hiltbrunner, E., Arnaiz, J., & Körner, C. (2021). Biomass allocation and seasonal non-structural carbohydrate dynamics do not explain the success of tall forbs in short alpine grassland. *Oecologia, 197*(4), 1063–1077. DOI: 10.1007/s00442-021-04950-7 PMID: 34047842

Hoad, J. (2003). *Guarire con le erbe*. MIR.

Hobohm, C. (2014). *Endemism in vascular plants* (Vol. 9). Springer. DOI: 10.1007/978-94-007-6913-7

Hock, R., Rasul, G., Adler, C., Cáceres, B., Gruber, S., Hirabayashi, Y., Jackson, M., Kääb, A., Kang, S., Kutuzov, S., Milner, A., Molau, U., Morin, S., Orlove, B., & Steltzer, H. (2019). High mountain areas. In Pörtner, H.-O., Roberts, D. C., Masson-Delmotte, V., Zhai, P., Tignor, M., Poloczanska, E., Mintenbeck, K., Alegría, A., Nicolai, M., Okem, A., Petzold, J., Rama, B., & Weyer, N. M. (Eds.), *IPCC special report on the ocean and cryosphere in a changing climate* (pp. 131–202). IPCC.

Hofmann, G., Lange-Bertalot, H., Werum, M., Klee, R., Kusber, W., Metzeltin, D., & Hajek, G. (2018). Rote Liste und Gesamtartenliste der limnischen Kieselalgen (Bacillariophyta) Deutschlands. *Rote Liste gefährdeter Tiere. Pflanzen und Pilze Deutschlands, 7*, 601–708.

Holopainen, J. K., Himanen, S. J., Yuan, J. S., Chen, F., & Stewart, C. N., Jr. (2025). Ecological functions of terpenoids in changing climates. In *Natural Products: Phytochemistry, Botany, Metabolism of Alkaloids, Phenolics and Terpenes* (pp. 1-44). Springer. DOI: 10.1007/978-3-642-36202-6_129-1

Hong, T., & Chongjian, Y. (2024). Temporal and Spatial Evolution of Ecosystem Service Supply and Demand in the Tibetan Plateau: Implications for Land Use Patterns and Relationships. *Polish Journal of Environmental Studies*, *33*(3), 2679–2691. DOI: 10.15244/pjoes/175059

Hoorn, C., Perrigo, A., & Antonelli, A. (Eds.). (2018). *Mountains, climate and biodiversity*. Wiley Blackwell.

Hopping, K. A., Knapp, A. K., Dorji, T., & Klein, J. A. (2018). Warming and land use change concurrently erode ecosystem services in Tibet. *Global Change Biology*, *24*(11), 5534–5548. DOI: 10.1111/gcb.14417 PMID: 30086187

Hotaling, S., Finn, D. S., Joseph Giersch, J., Weisrock, D. W., & Jacobsen, D. (2017). Climate change and alpine stream biology: Progress, challenges, and opportunities for the future. *Biological Reviews of the Cambridge Philosophical Society*, *92*(4), 2024–2045. DOI: 10.1111/brv.12319 PMID: 28105701

Hou, M., Leng, C., Zhu, J., Yang, M., Yin, Y., Xing, Y., & Chen, J. (2024). Alpine and subalpine plant microbiome mediated plants adapt to the cold environment: A systematic review. *Environmental Microbiome*, *19*(1), 82. DOI: 10.1186/s40793-024-00614-0 PMID: 39487507

Howard, T. G., White, K., & Goren, J. (2021). Thirteen years of rare plant population changes in the Adirondack alpine. *Northeastern Naturalist*, *28*(sp11), 29–46. DOI: 10.1656/045.028.s1103

Huang, L., Cao, W., Xu, X., Fan, J., & Wang, J. (2018). Linking the benefits of ecosystem services to sustainable spatial planning of ecological conservation strategies. *Journal of Environmental Management*, *222*, 385–395. DOI: 10.1016/j.jenvman.2018.05.066 PMID: 29870967

Hua, R., Zhang, P., Wang, L., Huang, M., Hua, L., & Zhou, J. (2025). Alpine steppe vegetation communities are more sensitive to plateau pika disturbance than alpine meadows. *Frontiers in Plant Science*, *16*, 1546828. DOI: 10.3389/fpls.2025.1546828 PMID: 40510176

Huber, L., Schirpke, U., Marsoner, T., Tasser, E., & Leitinger, G. (2020). Does socioeconomic diversification enhance multifunctionality of mountain landscapes? *Ecosystem Services*, *44*, 101122. DOI: 10.1016/j.ecoser.2020.101122

Hu, D., Yang, X., Chen, W., Feng, Z., Hu, C., Yan, F., Chen, X., Qu, D., & Chen, Z. (2021). Rhodiola rosea rhizome extract-mediated green synthesis of silver nanoparticles and evaluation of their potential antioxidant and catalytic reduction activities. *ACS Omega*, *6*(38), 24450–24461. DOI: 10.1021/acsomega.1c02843 PMID: 34604627

Humphries, H. C. (2020). Alpine ecosystems in temperate mountains of North America.

Hurtado, P., de Los Ríos, A., & Arias-Real, R. (2024). (Dis) connecting the Globe Through Water-Driven Ecological and Biogeochemical Corridors in the Polar-Alpine Biome. *Global Change Biology*, *30*(12), e17606. DOI: 10.1111/gcb.17606 PMID: 39614421

Hussain, M., Butt, A. R., Uzma, F., Ahmed, R., Irshad, S., Rehman, A., & Yousaf, B. (2020). A comprehensive review of climate change impacts, adaptation, and mitigation on environmental and natural calamities in Pakistan. *Environmental Monitoring and Assessment*, *192*(1), 48. DOI: 10.1007/s10661-019-7956-4 PMID: 31844992

Hussein, M. A., Islam, M. S., Ali, A. A., Mansour, M. S., Bondok, M., Salem, M. A., Amein, A. S., & Elgizawy, H. A. (2024). *Malva parviflora* seed oil; Isolation, gas chromatographic profiling and its cardioprotective activity against myocardial infraction in animal model. *Journal of King Saud University. Science*, *36*(2), 103060. DOI: 10.1016/j.jksus.2023.103060

Huss, M., Bookhagen, B., Huggel, C., Jacobsen, D., Bradley, R. S., Clague, J. J., Vuille, M., Buytaert, W., Cayan, D. R., Greenwood, G., Mark, B. G., Milner, A. M., Weingartner, R., & Winder, M. (2017). Toward mountains without permanent snow and ice. *Earth's Future*, *5*(5), 418–435. DOI: 10.1002/2016EF000514

Hu, Y.-K., Ma, C., Li, M.-J., Bai, X.-L., Liu, Y.-M., & Liao, X. (2024). Screening of monoamine oxidase B inhibitors in *Tibetan strawberry* by ligand fishing based on enzyme functionalized cellulose filter paper. *Microchemical Journal*, *203*, 110838. DOI: 10.1016/j.microc.2024.110838 PMID: 39035460

Hyun, T., Ra, J., Han, S., & Kim, J. (2018). Antioxidant, Antimicrobial, and Antidiabetic Activities of *Crowberry* Fruits. *Indian Journal of Pharmaceutical Sciences*, *80*(3). Advance online publication. DOI: 10.4172/pharmaceutical-sciences.1000382

Ianni, E., Geneletti, D., & Ciolli, M. (2015). Revitalizing Traditional Ecological Knowledge: A Study in an Alpine Rural Community. *Environmental Management*, *56*(1), 144–156. DOI: 10.1007/s00267-015-0479-z PMID: 25847107

Iannuzzi, A. M., Giacomelli, C., De Leo, M., Pietrobono, D., Camangi, F., De Tommasi, N., Martini, C., Trincavelli, M. L., & Braca, A. (2020). Antioxidant activity of compounds isolated from *Elaeagnus umbellata* promotes human gingival fibroblast well-being. *Journal of Natural Products, 83*(3), 626–637. DOI: 10.1021/acs.jnatprod.9b01030 PMID: 32031808

Ibarra, J. T., Caviedes, J., Marchant, C., Mathez-Stiefel, S.-L., Navarro-Manquilef, S., & Sarmiento, F. O. (2023). Mountain social-ecological resilience requires transdisciplinarity with Indigenous and local worldviews. *Trends in Ecology & Evolution, 38*(11), 1005–1009. DOI: 10.1016/j.tree.2023.07.004 PMID: 37574394

Ikram, M., Magdy Beshbishy, A., Kifayatullah, M., Olukanni, A., Zahoor, M., Naeem, M., Amin, M., Shah, M., Abdelaziz, A. S., Ullah, R., Mothana, R. A., Siddiqui, N. A., & Batiha, G. E.-S. (2020). Chemotherapeutic potential of *Carthamus oxycantha* root extract as antidiarrheal and *in vitro* antibacterial activities. *Antibiotics (Basel, Switzerland), 9*(5), 226. DOI: 10.3390/antibiotics9050226 PMID: 32370087

Ilea, R. C. (2009). Intensive livestock farming: Global trends, increased environmental concerns, and ethical solutions. *Journal of Agricultural & Environmental Ethics, 22*(2), 153–167. DOI: 10.1007/s10806-008-9136-3

Ingty, T. (2021). Pastoralism in the highest peaks: Role of the traditional grazing systems in maintaining biodiversity and ecosystem function in the alpine Himalaya. *PLoS One, 16*(1), e0245221. DOI: 10.1371/journal.pone.0245221 PMID: 33411837

Inouye, D. W. (2020). Effects of climate change on alpine plants and their pollinators. *Annals of the New York Academy of Sciences, 1469*(1), 26–37. DOI: 10.1111/nyas.14104 PMID: 31025387

Inouye, D. W. (2022). Climate change and phenology. *Wiley Interdisciplinary Reviews: Climate Change, 13*(3), e764. DOI: 10.1002/wcc.764

Inouye, D. W., Larson, B. M., Ssymank, A., & Kevan, P. G. (2015). Flies and flowers III: Ecology of foraging and pollination. *Journal of Pollination Ecology, 16*, 115–133. DOI: 10.26786/1920-7603(2015)15

Inouye, D. W., & Pyke, G. H. (1988). Pollination biology in the Snowy Mountains of Australia: Comparisons with montane Colorado, USA. *Australian Journal of Ecology, 13*(2), 191–205. DOI: 10.1111/j.1442-9993.1988.tb00968.x

Inouye, D. W., & Wielgolaski, F. E. (2025). Phenology at high altitudes. In *Phenology: An integrative environmental science* (pp. 281–311). Springer.

Intergovernmental Panel on Climate Change. (2001). *Climate change 2001: Synthesis report*. Cambridge University Press. https://www.ipcc.ch/2001/

Intergovernmental Panel on Climate Change. (2013). Summary for policymakers. In Stocker, T. F., Qin, D., Plattner, G.-K., Tignor, M., Allen, S. K., Boschung, J., Nauels, A., Xia, Y., Bex, V., & Midgley, P. M. (Eds.), *Climate change 2013: The physical science basis. Contribution of Working Group I to the fifth assessment report of the Intergovernmental Panel on Climate Change*. Cambridge University Press.

Intergovernmental Panel on Climate Change. (2023). *Climate change 2023: Synthesis report. Contribution of Working Groups I, II and III to the sixth assessment report of the Intergovernmental Panel on Climate Change* [Core Writing Team, H. Lee & J. Romero (Eds.)]. IPCC. https://www.ipcc.ch/report/ar6/syr/

Iqbal, M. Z., Gill, D., & Shafiq, M. (1998). Plant communities along the sewage effluents channels of Lyari river in Pakistan. *Taiwania, 43*(1), 1–11.

Iqbal, M. Z., Shafiq, M., & Kabir, M., & Farooqi, Z.-u.-R. (2018). Impact of anthropogenic activities on biodiversity in Pakistan: A review. *Bioscience Research, 15*(1), 481–490.

Iqbal, M., Sherwani, A., & Shafiq, M. (1999). Vegetation characteristics and trace metals (Cu, Zn and Pb) in soils along the super highways near Karachi, Pakistan. *Studia Botanica Hungarica, 29*, 79–86.

Iqbal, N., Yasmin, I., Ullah, H., Rehman, F. U., Tahir, K., Bushra, , Iqbal, M., Qarnain, Z., & Khan, S. (2022a). Isolation, characterization and antibacterial studies of three new chemical constituents isolated from *Viburnum grandiflorum. Pharmaceutical Chemistry Journal, 56*(4), 480–486. DOI: 10.1007/s11094-022-02663-6

Iqbal, T., Habib, T., Hussain, K., & Khan, A. M. (2022b). Wild edible plant basket of terrestrial paradise and variations among the diverse ethnic and elevation groups: A detailed insight from the Western Himalaya of Azad Jammu and Kashmir, Pakistan. *South African Journal of Botany, 147*, 294–313. DOI: 10.1016/j.sajb.2022.01.026

Ishaq, S., Khan, M., Begum, F., Hussain, K., Amir, R., Hussain, A., & Ali, S. (2016). Climate change impact on mountain biodiversity: A special reference to Gilgit-Baltistan of Pakistan. *Journal of Mountain Area Research, 1*, 53–63. DOI: 10.53874/jmar.v1i0.12

Ishaq, S., Rathore, H. A., Sabir, S. M., & Maroof, M. S. (2015). Antioxidant properties of *Elaeagnus umbellata* berry solvent extracts against lipid peroxidation in mice brain and liver tissues. *Food Science and Biotechnology, 24*(2), 673–679. DOI: 10.1007/s10068-015-0088-x

Ishtiaq, M., Sardar, T., Hussain, I., Maqbool, M., Mazhar, M. W., Parveen, A., Ajaib, M., Bhatti, K. H., Hussain, T., Gul, A., Azeem, M., Khanum, H., Moussa, I. M., Ullah, F., Elansary, H. O., Thind, S., & Sridhara, S. (2024). Traditional ethnobotanical knowledge of important local plants in Sudhnoti, Azad Kashmir, Pakistan. *Scientific Reports*, *14*(1), 22165. DOI: 10.1038/s41598-024-73431-7 PMID: 39333361

Islam, R., Sun, L., & Zhang, L. (2021). Biomedical applications of Chinese herb-synthesized silver nanoparticles by phytonanotechnology. *Nanomaterials (Basel, Switzerland)*, *11*(10), 2757. DOI: 10.3390/nano11102757 PMID: 34685197

Ismail, M., Ali, S., Abbas, Q., Kamal, Z., Hussain, I., Sidiqa, S.,…Nature. (2022). Medicinal and Aromatic Plants. 141.

Ismail, H., Rasheed, A., Haq, I., Jafri, L., Ullah, N., Dilshad, E., Sajid, M., & Mirza, B. (2017). Five Indigenous Plants of Pakistan with Antinociceptive, Anti-Inflammatory, Antidepressant, and Anticoagulant Properties in Sprague Dawley Rats. *Evidence-Based Complementary and Alternative Medicine : eCAM*, *2017*(1), 7849501. DOI: 10.1155/2017/7849501 PMID: 29348774

Ismail, M., Hussain, J., Khan, A., Khan, A. L., Ali, L., Khan, F., Khan, A. Z., Niaz, U., & Lee, I.-J. (2012). Antibacterial, antifungal, cytotoxic, phytotoxic, insecticidal, and enzyme inhibitory activities of *Geranium wallichianum*. *Evidence-Based Complementary and Alternative Medicine : eCAM*, *2012*(1), 305906. DOI: 10.1155/2012/305906 PMID: 23049606

Jabbar, A., Abbas, A., Assad, N., Naeem-ul-Hassan, M., Alhazmi, H. A., Najmi, A., Zoghebi, K., Al Bratty, M., Hanbashi, A., & Amin, H. M. A. (2023). A highly selective Hg 2+ colorimetric sensor and antimicrobial agent based on green synthesized silver nanoparticles using Equisetum diffusum extract. *RSC Advances*, *13*(41), 28666–28675. DOI: 10.1039/D3RA05070J PMID: 37790097

Jabeen, Q., Haider, S. I., Asif, A., Rasheed, R., Gul, S., & Arshad, S. (2022). *Geranium wallichianum* D. Don Ex Sweet ameliorates rheumatoid arthritis by curtailing the expression of COX-II and inflammatory cytokines as well as by alleviating the oxidative stress. *Dose-Response*, *20*(3), 15593258221112649. DOI: 10.1177/15593258221112649 PMID: 35832768

Jabis, M. D., Winkler, D. E., & Kueppers, L. M. (2020). Warming acts through earlier snowmelt to advance but not extend alpine community flowering. *Ecology*, *101*(9), e03108. DOI: 10.1002/ecy.3108 PMID: 32455489

Jacquemart, M., Weber, S., Chiarle, M., Chmiel, M., Cicoira, A., Corona, C., Deline, P., & Stoffel, M. (2024). Detecting the impact of climate change on alpine mass movements in observational records from the European Alps. *Earth-Science Reviews*, *258*, 104886. DOI: 10.1016/j.earscirev.2024.104886

Jacques, A. C., Chaves, F. C., Zambiazi, R. C., Brasil, M. C., & Caramão, E. B. (2014). Bioactive and volatile organic compounds in Southern Brazilian blackberry (*Rubus fruticosus*) fruit cv. Tupy. *Food Science and Technology (Campinas)*, *34*(3), 636–643. DOI: 10.1590/1678-457x.6243

Jaganathan, G. K., Dalrymple, S. E., & Pritchard, H. W. (2020). Seed survival at low temperatures: A potential selecting factor influencing community level changes in high altitudes under climate change. *Critical Reviews in Plant Sciences*, *39*(6), 479–492. DOI: 10.1080/07352689.2020.1848277

Jan, A. (2024). *Predicting species distribution in freshwaters facing climate change and biological invasions using ecological niche models.*

Janbaz, K. H., Zaeem Ahsan, M., Saqib, F., Imran, I., Zia-Ul-Haq, M., Abid Rashid, M., Jaafar, H. Z., & Moga, M. (2015). Scientific basis for use of *Pyrus pashia* Buch.-Ham. ex D. Don. fruit in gastrointestinal, respiratory and cardiovascular ailments. *PLoS One*, *10*(3), e0118605. DOI: 10.1371/journal.pone.0118605 PMID: 25786248

Jaroszynska, F., Althuizen, I., Halbritter, A. H., Klanderud, K., Lee, H., Telford, R. J., & Vandvik, V. (2023). Bryophytes dominate plant regulation of soil microclimate in alpine grasslands. *Oikos*, *2023*(12), e10091. DOI: 10.1111/oik.10091

Jaryan, V., Uniyal, S. K., Gupta, R. C., & Singh, R. D. (2013). Alien Flora of Indian Himalayan State of Himachal Pradesh. *Environmental Monitoring and Assessment*, *185*(7), 6129–6153. DOI: 10.1007/s10661-012-3013-2 PMID: 23196408

Javed, M., & Hayat, S. (1999). Heavy metal toxicity of river Ravi aquatic ecosystem. *Pakistan Journal of Agricultural Sciences*, *36*, 1–9.

Javeed, B., Huang, D., Shangguan, D., Mukhtar, M. A., Sajjad, W., Banerjee, A., & Butt, A. Q. (2024). Assessing the effectiveness of national park's policies and laws in promoting biodiversity conservation and ecological development in Pakistan. *Frontiers in Environmental Science*, *11*, 1333650. Advance online publication. DOI: 10.3389/fenvs.2023.1333650

Javid, H., Yousuf, U., Ul Qadir, R., Magray, J. A., Wani, B. A., Islam, T., Nawchoo, I. A., & Gulzar, S. (2025). Metabolic profiling and antimicrobial activity of *Bistorta amplexicaulis* D. don by *in-vitro* implicated through computational studies. *Frontiers in Pharmacology*, *16*, 1575727. DOI: 10.3389/fphar.2025.1575727 PMID: 40667511

Jenkins, C. N., & Joppa, L. (2009). Expansion of the global terrestrial protected area system. *Biological Conservation, 142*(10), 2166–2174. DOI: 10.1016/j.biocon.2009.04.016

Jerome, D. K., Petry, W. K., Mooney, K. A., & Iler, A. M. (2021). Snow melt timing acts independently and in conjunction with temperature accumulation to drive subalpine plant phenology. *Global Change Biology, 27*(20), 5054–5069. DOI: 10.1111/gcb.15803 PMID: 34265142

Jha, S. K., Negi, A. K., Alatalo, J. M., & Negi, R. S. (2021). Socio-ecological vulnerability and resilience of mountain communities residing in capital-constrained environments. *Mitigation and Adaptation Strategies for Global Change, 26*(8), 38. DOI: 10.1007/s11027-021-09974-1 PMID: 34744484

Ji, Z., Zou, S., Zhang, W., Song, F., Yuan, T., & Xu, B. (2024). Optimizing zoning for ecological management in alpine region by combining ecosystem service supply and demand with ecosystem resilience. *Journal of Environmental Management, 365*, 121508. DOI: 10.1016/j.jenvman.2024.121508 PMID: 38924889

Jones, G. A., & Henry, G. H. (2003). Primary plant succession on recently deglaciated terrain in the Canadian High Arctic. *Journal of Biogeography, 30*(2), 277–296. DOI: 10.1046/j.1365-2699.2003.00818.x

Juyal, D., Thawani, V., Thaledi, S., & Joshi, M. (2014). Ethnomedical properties of *Taxus wallichiana* zucc.(Himalayan yew). *Journal of Traditional and Complementary Medicine, 4*(3), 159–161. DOI: 10.4103/2225-4110.136544 PMID: 25161920

Kaltenborn, B. P., Kaltenborn, E. F., & Linnell, J. D. (2019). It's All About the Scenery. *Arctic, 72*(1), 1–12. DOI: 10.14430/arctic67944

Kameyama, Y., & Kudo, G. (2009). Flowering phenology influences seed production and outcrossing rate in populations of an alpine snowbed shrub, Phyllodoce aleutica: Effects of pollinators and self-incompatibility. *Annals of Botany, 103*(9), 1385–1394. DOI: 10.1093/aob/mcp037 PMID: 19228703

Kanwal, H., & Sherazi, B. A. (2017). Herbal medicine: Trend of practice, perspective, and limitations in Pakistan. *Asian Pacific Journal of Health Sciences, 4*(4), 6–8. DOI: 10.21276/apjhs.2017.4.4.2

Kanwal, M., Sher, M., Abbas, A., Akhtar, S., Siddique, A. B., Hasan, M. N., Assad, N., Alhazmi, H. A., & Amin, H. M. A. (2025). Dual colorimetric sensing of Hg (II) and Fe (III) using sulfanilamide-stabilized silver nanoparticles and evaluating their photodegradation and antibacterial properties. *Journal of Water Process Engineering, 75*, 107981. DOI: 10.1016/j.jwpe.2025.107981

Kanwal, S., Perveen, S., Hameed, H., Tahir, S., Rashid, R., Raish, M., Bashir, S., Zaffar, K., Irfan, A., Kotwica-Mojzych, K., Faheem, S., Ahmad, A., Kashif, M., & Bin Jardan, Y. A. (2025). Evaluating the efficacy of *Chenopodium murale* plant extract in inhibiting Streptococcus mutans and reducing dental caries risk. *Scientific Reports*, *15*(1), 15406. DOI: 10.1038/s41598-025-00127-x PMID: 40316577

Karaçelik, A. A., & Şahin, H. (2021). Determination of chemical compositions, antioxidant and enzyme inhibitory activities of naturally growing *Chenopodium album* subsp. iranicum Aellen. *Journal of the Institute of Science and Technology*, *11*(3), 2091–2101. DOI: 10.21597/jist.825781

Karakaplan, M., & Özcan, M. (2017). Determination of phenolic acids in pomegranate juices by HPLC-DAD. *Avrupa Bilim ve Teknoloji Dergisi*, *6*(10), 32–37.

Karanth, K. K., Nichols, J. D., Karanth, K. U., Hines, J. E., & Christensen, N. L., Jr. (2010). The shrinking ark: patterns of large mammal extinctions in India. *Proceedings of the Royal Society B: Biological Sciences, 277*(1690), 1971-1979. DOI: 10.1098/rspb.2010.0171

Karimi, S., Nawaz, M. A., Naseem, S., & Ali, Z. (2025). Influence of open-top chambers induced climate warming on secondary metabolic profile of culturally and medicinally important plants of Himalaya, Karakoram and Hindukush. *PLoS One*, *20*(5), e0322480. DOI: 10.1371/journal.pone.0322480 PMID: 40367051

Karpagasundari, C., & Kulothungan, S. (2014). Analysis of bioactive compounds in *Physalis minima* leaves using GC MS, HPLC, UV-VIS and FTIR techniques. *Journal of Pharmacognosy and Phytochemistry*, *3*(4), 196–201.

Kaur, D., Tiwana, A. S., Kaur, S., & Gupta, S. (2022). Climate change: Concerns and influences on biodiversity of the Indian Himalayas. In Kaur, D., Tiwana, A. S., Kaur, S., & Gupta, S. (Eds.), *Climate change: Impacts, responses and sustainability in the Indian Himalaya* (pp. 265–281). Springer. DOI: 10.1007/978-3-030-92782-0_13

Kayani, S., Ahmad, , Gillani, S. W., Manzoor, M., Rehman, F., Jabeen, S., & Shah, S. A. H. (2024a). Ethnomedicinal appraisal of the medicinal flora among the sub-alpine and alpine Iindigenous communities of Palas Valley Kohistan, Northern Pakistan. *Ethnobotany Research and Applications*, *28*. Advance online publication. DOI: 10.32859/era.28.9.1-29

Kayani, S., Ahmad, M., Sultana, S., Shinwari, Z. K., Zafar, M., Yaseen, G., & Bibi, T. (2015). Ethnobotany of medicinal plants among the communities of Alpine and Sub-alpine regions of Pakistan. *Journal of Ethnopharmacology*, *164*, 186–202. DOI: 10.1016/j.jep.2015.02.004 PMID: 25680839

Keiler, M., Knight, J., & Harrison, S. (2010). Climate change and geomorphological hazards in the eastern European Alps. *Philosophical Transactions of the Royal Society A: Mathematical, Physical and Engineering Sciences, 368*(1919), 2461-2479.

Khalid, S., Arshad, M., Mahmood, S., Siddique, F., Roobab, U., Ranjha, M. M. A. N., & Lorenzo, J. M. (2023). Extraction and quantification of *Moringa oleifera* leaf powder extracts by HPLC and FTIR. *Food Analytical Methods, 16*(4), 787–797. DOI: 10.1007/s12161-023-02470-z

Khalid, T., Rafey, H. A., Amin, A., Kamran, M., Zargaham, M. K., Ross, S. A., & Niaz, S.-I. (2025). *In silico* and *in vitro* characterization of *Phytolacca latbenia* (Moq.): QSAR, phytochemical, and toxicological insights into antiglycation and antidiabetic potential. *Biophysical Chemistry, 107526*. Advance online publication. DOI: 10.1016/j.bpc.2025.107526 PMID: 40983042

Khaliq, T., Akhter, S., Sultan, P., & Hassan, Q. P. (2023). Critical review on *Rumex dentatus* L. a strong pharmacophore and the future medicine: Pharmacology, phytochemical analysis and traditional uses. *Heliyon, 9*(3), e14159. Advance online publication. DOI: 10.1016/j.heliyon.2023.e14159 PMID: 36994390

Khan, M. N., & Ali, S. RAZAK, S. A., Zaman, A., Iqbal, M., & Shah, S. N. J. B. J. o. B. D. (2022). Assessment of floristic diversity in the mountain ecosystem of Marghazar Valley, Hindukush Range, Swat, Pakistan. *23*(2).

Khan, M. N., Ali, S., Yaseen, T., Ullah, S., Zaman, A., Iqbal, M.,…Sciences, A. (2019). Eco-taxonomic study of family Poaceae (Gramineae). *10*(2), 63-75.

Khan, M., Baig, T., Ali, S., & Shah, A. (2022). *Mountain studies: Understanding and managing mountains for people and nature.* Department of Forestry, Range & Wildlife Management Karakoram International University Gilgit 15100, Pakistan 1-214.

Khan, , R. S. MAkhter, , THussain, , M. (2021). Ethno-veterinary practice for the treatment of animal diseases in Neelum Valley, Kashmir Himalaya, Pakistan. *PLoS One, 16*(4), e0250114. DOI: 10.1371/journal.pone.0250114 PMID: 33930052

Khanam, A., Ijaz Hussain, A., Mohammed, E. H., Nahar, L., & Rathore, H. A. (2025). Phenolic Profile of Seedless *Ziziphus mauritiana* Fruits and Leaves Extracts with *In Vivo* Antioxidant and Anti-Inflammatory Activities: Influence on Pro-Inflammatory Mediators. *Chemistry & Biodiversity, 22*(3), e202401728. DOI: 10.1002/cbdv.202401728 PMID: 39475065

Khan, H., & Baig, S. (2017). High altitude wetlands of the HKH region of northern Pakistan–status of current knowledge, challenges and research opportunities. *Wetlands*, *37*(2), 371–380. DOI: 10.1007/s13157-016-0868-y

Khan, I., Khan, U., Khan, W., Alqathama, A., Riaz, M., Ahmad, R., & Alam, M. M. (2024). Antibacterial and antibiofilm potentials of *Rumex dentatus* root extract characterized by HPLC-ESI-Q-TOF-MS. *Saudi Journal of Biological Sciences*, *31*(4), 103962. DOI: 10.1016/j.sjbs.2024.103962 PMID: 38419820

Khan, I., Rahman, H., Abd El-Salam, N. M., Tawab, A., Hussain, A., Khan, T. A., Khan, U. A., Qasim, M., Adnan, M., Azizullah, A., Murad, W., Jalal, A., Muhammad, N., & Ullah, R. (2017). *Punica granatum* peel extracts: HPLC fractionation and LC MS analysis to quest compounds having activity against multidrug resistant bacteria. *BMC Complementary and Alternative Medicine*, *17*(1), 247. DOI: 10.1186/s12906-017-1766-4 PMID: 28468660

Khan, K., & Ghramh, H. (2023). Beekeeping in Pakistan: History, Potential, and Current Status. *Pakistan Journal of Zoology*, *56*(1). DOI: 10.17582/journal.pjz/20230419090432

Khan, M. N., & Badshah, L. (2019). Floristic diversity and utility of flora of district Charsadda, Khyber Pakhtunkhwa. *Acta Ecologica Sinica*, *39*(4), 306–320. https://doi.org/https://doi.org/10.1016/j.chnaes.2018.10.003. DOI: 10.1016/j.chnaes.2018.10.003

Khan, M. S., Khan, S. M., Abdullah, , Liu, J., Wu, Z.-Y., Hussain, J., Zeb, S. A., Mohammad, N., Batool, Z., Saqib, Z., Afza, R., Manan, F., & Ali, S. (2025). Ecological assessment of Iris hookeriana across subalpine and alpine regions of the Hindu-Himalayas. *Frontiers in Forests and Global Change*, *8*, 1539025. DOI: 10.3389/ffgc.2025.1539025

Khan, M. Z., Shabbir, M. I., Saqib, Z., Gilani, S. A., Jogezai, N. U., Kiyani, M. M., & Malik, M. A. (2020). Investigation of polyphenol profile, antioxidant activity and hepatoprotective potential of *Aconogonon alpinum* (All.) Schur roots. *Open Chemistry*, *18*(1), 516–536. DOI: 10.1515/chem-2020-0062

Khan, S. M., Page, S. E., Ahmad, H., & Harper, D. M. (2013). Sustainable utilization and conservation of plant biodiversity in montane ecosystems: The western Himalayas as a case study. *Annals of Botany*, *112*(3), 479–501. DOI: 10.1093/aob/mct125 PMID: 23825353

Khan, S., Nabi, G., Ullah, M. W., Yousaf, M., Manan, S., Siddique, R., & Hou, H. (2016). Overview on the role of advance genomics in conservation biology of endangered species. *International Journal of Genomics*, *2016*(1), 3460416. DOI: 10.1155/2016/3460416 PMID: 28025636

Khazieva, E., Verburg, P. H., & Pazúr, R. (2022). Grassland degradation by shrub encroachment: Mapping patterns and drivers of encroachment in Kyrgyzstan. *Journal of Arid Environments*, *207*, 104849. DOI: 10.1016/j.jaridenv.2022.104849

Khuroo, A. A., Rashid, I., Reshi, Z., Dar, G. H., & Wafai, B. A. (2007). The alien flora of Kashmir Himalaya. *Biological Invasions*, *9*(3), 269–292. DOI: 10.1007/s10530-006-9032-6

Kiese, R., Fersch, B., Baessler, C., Brosy, C., Butterbach-Bahl, K., Chwala, C., Dannenmann, M., Fu, J., Gasche, R., Grote, R., Jahn, C., Klatt, J., Kunstmann, H., Mauder, M., Rödiger, T., Smiatek, G., Soltani, M., Steinbrecher, R., Völksch, I., & Schmid, H. P. (2018). The TERENO Pre-Alpine Observatory: Integrating Meteorological, Hydrological, and Biogeochemical Measurements and Modeling. *Vadose Zone Journal*, *17*(1), 1–17. DOI: 10.2136/vzj2018.03.0060

Kim, D.-S., Ji, H. D., Rhee, M. H., Sung, Y.-Y., Yang, W.-K., Kim, S. H., & Kim, H.-K. (2014a). Antiplatelet activity of *Morus alba* leaves extract, mediated via inhibiting granule secretion and blocking the phosphorylation of extracellular-signal-regulated kinase and akt. *Evidence-Based Complementary and Alternative Medicine : eCAM*, *2014*(1), 639548. DOI: 10.1155/2014/639548 PMID: 24701244

Kim, H., Frunze, O., Maigoro, A. Y., Lee, M. L., Lee, J. H., & Kwon, H. W. (2024). Comparative Study of the Effect of Pollen Substitute Diets on Honey Bees during Early Spring. *Insects*, *15*(2), 101. Advance online publication. DOI: 10.3390/insects15020101 PMID: 38392520

Kim, J. K., & Park, S. U. (2018). Quercetin and its role in biological functions: An updated review. *EXCLI Journal*, *17*, 856. DOI: 10.17179/excli2018-1538 PMID: 30233284

Kim, M.-K., Oh, T.-J., & Park, H. (2014b). Antibacterial and antioxidant capacity of polar microorganisms isolated from Arctic lichen *Ochrolechia* sp. *Polish Journal of Microbiology*, *63*(3), 317–322. DOI: 10.33073/pjm-2014-042 PMID: 25546942

Kirchner, M., Schönhart, M., & Schmid, E. (2016). Spatial impacts of the CAP post-2013 and climate change scenarios on agricultural intensification and environment in Austria. *Ecological Economics*, *123*, 35–56. DOI: 10.1016/j.ecolecon.2015.12.009

Kisku, S., & Barik, D. K. L. (2015). Ecofloristic Studies on Sacred Groves of Bamanghaty Subdivision in Mayurbhanj District, Odisha and its Conservation Strategies. *Indian Journal of Research*, *4*(6), 482–483.

Kitic, D., Miladinovic, B., Randjelovic, M., Szopa, A., Sharifi-Rad, J., Calina, D., & Seidel, V. (2022). Anticancer potential and other pharmacological properties of *Prunus armeniaca* L.: An updated overview. *Plants*, *11*(14), 1885. DOI: 10.3390/plants11141885 PMID: 35890519

Klanderud, K. (2005). Climate change effects on species interactions in an alpine plant community. *Journal of Ecology*, *93*(1), 127–137. DOI: 10.1111/j.1365-2745.2004.00944.x

Kleidon, A., & Mooney, H. A. (2000). A global distribution of biodiversity inferred from climatic constraints: Results from a process-based modelling study. *Global Change Biology*, *6*(5), 507–523. DOI: 10.1046/j.1365-2486.2000.00332.x

Klein, J. A., Tucker, C. M., Nolin, A. W., Hopping, K. A., Reid, R. S., Steger, C., Grêt-Regamey, A., Lavorel, S., Müller, B., Yeh, E. T., Boone, R. B., Bourgeron, P., Butsic, V., Castellanos, E., Chen, X., Dong, S. K., Greenwood, G., Keiler, M., Marchant, R., & Yager, K. (2019). Catalyzing transformations to sustainability in the world's mountains. *Earth's Future*, *7*(5), 547–557. DOI: 10.1029/2018EF001024

Kohli, R. K., Batish, D. R., Singh, H. P., & Dogra, K. S. (2006). Status, invasiveness and environmental threats of three tropical American invasive weeds (*Parthenium hysterophorus* L., *Ageratum conyzoides* L., *Lantana camara* L.) in India. *Biological Invasions*, *8*(7), 1501–1510. DOI: 10.1007/s10530-005-5842-1

Kolawole, A. S., & Iyiola, A. O. (2023). Environmental Pollution: Threats, Impact on Biodiversity, and Protection Strategies. In Izah, S. C., & Ogwu, M. C. (Eds.), *Sustainable Utilization and Conservation of Africa's Biological Resources and Environment* (pp. 377–409). Springer Nature Singapore., DOI: 10.1007/978-981-19-6974-4_14

Koo, H., Gomes, B., Rosalen, P. L., Ambrosano, G., Park, Y. K., & Cury, J. A. (2000). In vitro antimicrobial activity of propolis and *Arnica montana* against oral pathogens. *Archives of Oral Biology*, *45*(2), 141–148. DOI: 10.1016/S0003-9969(99)00117-X PMID: 10716618

Kopystecka, A., Kozioł, I., Radomska, D., Bielawski, K., Bielawska, A., & Wujec, M. (2023). *Vaccinium uliginosum* and *Vaccinium myrtillus*—two species—one used as a functional food. *Nutrients*, *15*(19), 4119. DOI: 10.3390/nu15194119 PMID: 37836403

Körner, C. (2021a). Carbon investments. In *Alpine Plant Life: Functional Plant Ecology of High Mountain Ecosystems* (pp. 309-333). Springer. DOI: 10.1007/978-3-030-59538-8_12

Körner, C. (2021d). Mineral nutrition. In *Alpine Plant Life: Functional Plant Ecology of High Mountain Ecosystems* (pp. 237-268). Springer. DOI: 10.1007/978-3-030-59538-8_10

Körner, C. (2021e). Plant biomass production. In *Alpine Plant Life: Functional Plant Ecology of High Mountain Ecosystems* (pp. 375-394). Springer. DOI: 10.1007/978-3-030-59538-8_15

Körner, C. (2023). Concepts in Alpine Plant Ecology. Plants 2023, 12, 2666. *Arctic and Alpine Plants*, 5.

Körner, C., & Hiltbrunner, E. J. D. (2021). Why is the alpine flora comparatively robust against climatic warming?, *13*(8), 383.

Körner, C. (1999). *Alpine plant life: Functional plant ecology of high mountain ecosystems* (Vol. 4). Springer. DOI: 10.1007/978-3-642-98018-3

Körner, C. (1999). Alpine plants: stressed or adapted? Physiological Plant Ecology: 39th Symposium of the British Ecological Society, Körner, C. (2003). Limitation and stress–always or never? *Journal of Vegetation Science*, *14*(2), 141–143.

Körner, C. (2004). Mountain biodiversity, its causes and function. *Ambio*, *33*(sp13), 11–17. DOI: 10.1007/0044-7447-33.sp13.11 PMID: 15575177

Körner, C. (2012). *Alpine treelines: Functional ecology of the global high elevation tree limits*. Springer. DOI: 10.1007/978-3-0348-0396-0

Körner, C. (2021). *Alpine Plant Life: Functional Plant Ecology of High Mountain Ecosystems*. Springer International Publishing.

Körner, C. (2021c). Life under and in snow: protection and limitation. In *Alpine plant life: functional plant ecology of high mountain ecosystems* (pp. 89–118). Springer., DOI: 10.1007/978-3-030-59538-8_5

Körner, C. (2023). Concepts in alpine plant ecology. *Plants*, *12*(14), 2666. DOI: 10.3390/plants12142666 PMID: 37514280

Körner, C., Berninger, U.-G., Daim, A., Eberl, T., Mendoza, F. F., Füreder, L., & Meyer, E. (2022). Long-term monitoring of high-elevation terrestrial and aquatic ecosystems in the Alps–a five-year synthesis. *Eco.Mont*, *14*(2), 48–69. DOI: 10.1553/eco.mont-14-2s48

Körner, C., & Hiltbrunner, E. (2018). The 90 ways to describe plant temperature. *Perspectives in Plant Ecology, Evolution and Systematics*, *30*, 16–21. DOI: 10.1016/j.ppees.2017.04.004

Körner, C., & Hiltbrunner, E. (2021). Why is the alpine flora comparatively robust against climatic warming? *Diversity*, *13*(8), 383. DOI: 10.3390/d13080383

Körner, C., & Hoch, G. (2023). Not every high-latitude or high-elevation forest edge is a treeline. *Journal of Biogeography*, *50*(5), 838–845. DOI: 10.1111/jbi.14593

Körner, C., Jetz, W., Paulsen, J., Payne, D., Rudmann-Maurer, K., & Spehn, M., E. (. (2017). A global inventory of mountains for bio-geographical applications. *Alpine Botany*, *127*(1), 1–15. DOI: 10.1007/s00035-016-0182-6

Körner, C., Paulsen, J., & Spehn, E. M. (2011). A definition of mountains and their bioclimatic belts for global comparisons of biodiversity data. *Alpine Botany*, *121*(2), 73–78. DOI: 10.1007/s00035-011-0094-4

Körner, C., & Renhardt, U. (1987). Dry matter partitioning and root length/leaf area ratios in herbaceous perennial plants with diverse altitudinal distribution. *Oecologia*, *74*(3), 411–418. DOI: 10.1007/BF00378938 PMID: 28312481

Körner, C., Riedl, S., Keplinger, T., Richter, A., Wiesenbauer, J., Schweingruber, F., & Hiltbrunner, E. (2019). Life at 0 C: The biology of the alpine snowbed plant Soldanella pusilla. *Alpine Botany*, *129*(2), 63–80. DOI: 10.1007/s00035-019-00220-8

Kou, X., Zhao, J., & Sang, W. (2024). Impact of Typical Land Use Expansion Induced by Ecological Restoration and Protection Projects on Landscape Patterns. *Land (2012)*, *13*(9).

Kougioumoutzis, K., Tsakiri, M., Kokkoris, I. P., Trigas, P., Iatrou, G., Lamari, F. N., Tzanoudakis, D., Koumoutsou, E., Dimopoulos, P., Strid, A., & Panitsa, M. (2024). Assessing the vulnerability of medicinal and aromatic plants to climate and land-use changes in a Mediterranean biodiversity hotspot. *Land (Basel)*, *13*(2), 133. DOI: 10.3390/land13020133

Kouznetsov, V. V. (2025). *Chemical Defense in Nature: Structure, Function, and Synthesis of Toxic Secondary Metabolites from Plants and Animals*. CRC Press. DOI: 10.1201/9781003335771

Kováč, J., Slobodníková, L., Trajčíková, E., Rendeková, K., Mučaji, P., Sychrová, A., & Bittner Fialová, S. (2022). Therapeutic potential of flavonoids and tannins in management of oral infectious diseases—A review. *Molecules (Basel, Switzerland)*, *28*(1), 158. DOI: 10.3390/molecules28010158 PMID: 36615352

Kowarik, I. (2011). Novel urban ecosystems, biodiversity, and conservation. *Environmental Pollution, 159*(8-9), 1974–1983. DOI: 10.1016/j.envpol.2011.02.022 PMID: 21435761

Kozlowski, G. (2024). Understanding Arctic-Alpine Plants from Ecological and Evolutionary Perspectives. *Plants, 13*(11), 1560. Advance online publication. DOI: 10.3390/plants13111560 PMID: 38891368

Kryvtsova, M., Salamon, I., Koscova, J., & Spivak, M. (2020). Antibiofilm forming, antimicrobial activity and some biochemical properties of *Vaccinium vitis* idaea leaf and berry extracts on *Staphylococcus aureus*. *Biosystems Diversity, 28*(3), 238–242. DOI: 10.15421/012031

Kudo, G. (2016). Landscape structure of flowering phenology in alpine ecosystems: significance of plant–pollinator interactions and evolutionary aspects. In *Structure and function of mountain ecosystems in Japan: Biodiversity and Vulnerability to Climate Change* (pp. 41-62). Springer. DOI: 10.1007/978-4-431-55954-2_3

Kudo, G., & Cooper, E. J. (2019). When spring ephemerals fail to meet pollinators: Mechanism of phenological mismatch and its impact on plant reproduction. *Proceedings of the Royal Society B, 286*(1904), 20190573. DOI: 10.1098/rspb.2019.0573

Kudo, G. (1996). Effects of snowmelt timing on reproductive phenology and pollination process of alpine plants. *Memoirs of National Institute of Polar Research. Special Issue, 51*, 71–82.

Kudo, G. (2014). Vulnerability of phenological synchrony between plants and pollinators in an alpine ecosystem. *Ecological Research, 29*(4), 571–581. DOI: 10.1007/s11284-013-1108-z

Kudo, G. (2020). Dynamics of flowering phenology of alpine plant communities in response to temperature and snowmelt time: Analysis of a nine-year phenological record collected by citizen volunteers. *Environmental and Experimental Botany, 170*, 103843. DOI: 10.1016/j.envexpbot.2019.103843

Kuipers, K. J. J., May, R., & Verones, F. (2021). Considering habitat conversion and fragmentation in characterisation factors for land-use impacts on vertebrate species richness. *The Science of the Total Environment, 801*, 149737. DOI: 10.1016/j.scitotenv.2021.149737 PMID: 34525717

Kullman, L. (2002). Rapid recent range-margin rise of tree and shrub species in the Swedish Scandes. *Journal of Ecology, 90*(1), 68–77. DOI: 10.1046/j.0022-0477.2001.00630.x

Kulonen, A., Imboden, R. A., Rixen, C., Maier, S. B., & Wipf, S. (2018). Enough space in a warmer world? Microhabitat diversity and small-scale distribution of alpine plants on mountain summits. *Diversity & Distributions*, *24*(2), 252–261. DOI: 10.1111/ddi.12673

Kumar, A., Ansari, M. Z., Banjara, R. A., Romano, R., De Luca, L., Santini, A., & Tiwari, J. K. (2025). Nanotechnology for sustainable biotic stress management in plants. *Plant Stress*, *18*, 101064. DOI: 10.1016/j.stress.2025.101064

Kumar, A., Kamal, A., Singh, S., Padalia, R. C., Tandon, S., Chauhan, A., Saikia, D., & Verma, R. S. (2020). Chemical composition, antimicrobial activity, kinetics and mechanism of action of Himalayan-thyme (*Thymus linearis* Benth.). *The Journal of Essential Oil Research*, *32*(1), 59–68. DOI: 10.1080/10412905.2019.1662337

Kumari, A., Joshi, R., & Yadav, S. K. (2025). Prospects and potential for sustainable food and nutritional security in North Western Himalayan region. *Indian Journal of Plant Genetic Resources*, *38*(02), 145–154.

Kumar, R., & Agrawal, O. (2014). Comparative performance of honey bee colonies fed with artificial diets in Gwalior and Panchkula region. *Journal of Entomology and Zoology Studies*, *2*(4), 104–107.

Kumar, R., Joshi, R., Kumar, R., Srivatsan, V., Satyakam, , Chawla, A., Patial, V., & Kumar, S. (2022). Nutritional quality evaluation and proteome profile of forage species of Western Himalaya. *Grassland Science*, *68*(3), 214–225. DOI: 10.1111/grs.12357

Kumar, S. (2025). Data collection from literature for biological sciences, medicinal plants research, ethnobotany, and pharmacology: A methodological overview. *Journal of Biodiversity and Conservation*, *9*(2), 167–169.

Kumar, V., Kumar, P., Bhargava, B., Sharma, R., Irfan, M., & Chandora, R. (2023). Transcriptomic and metabolomic reprogramming to explore the high-altitude adaptation of medicinal plants: A review. *Journal of Plant Growth Regulation*, *42*(12), 7315–7329. DOI: 10.1007/s00344-023-11018-8

Kumeh, E. M. (2023). Contestations, counteractions and equitable conservation–a case study of Ghana's Krokosua Hills Forest Reserve. *Forest Policy and Economics*, *157*, 103090. DOI: 10.1016/j.forpol.2023.103090

Kunwar, R. M., Baral, B., Luintel, S., Uprety, Y., Poudel, R. C., Adhikari, B., & Poudel, P. (2022). Ethnomedicinal landscape: Distribution of used medicinal plant species in Nepal. *Journal of Ethnobiology and Ethnomedicine*, *18*(1), 34. DOI: 10.1186/s13002-022-00531-x PMID: 35436921

La Rosa, A., Cornara, L., Saitta, A., Salam, A. M., Grammatico, S., Caputo, M., La Mantia, T., & Quave, C. L. (2021). Ethnobotany of the Aegadian Islands: Safeguarding biocultural refugia in the Mediterranean. *Journal of Ethnobiology and Ethnomedicine, 17*(1), 47. DOI: 10.1186/s13002-021-00470-z PMID: 34321030

Laaraj, S., Ouahidi, I., Al Moudani, N., Allali, M., & Aarab, L. (2024). Immunostimulant and Pharmacological Activities, of *Cucurbita maxima* Seeds on Humoral and Cellular Immunity Cells and their Functions. *The Open Public Health Journal, 17*(1), e18749445274775. Advance online publication. DOI: 10.2174/0118749445274775231204095205

Ladol, S., Kumar, A., & Sharma, D. (2023). Nutritional and Health Benefits of High Altitude Wild Food Plant, Hippophae rhamnoides for the Himalayan Communities. In Kumar, A., Singh, P., Singh, S., & Singh, B. (Eds.), *Wild Food Plants for Zero Hunger and Resilient Agriculture* (pp. 149–161). Springer Nature Singapore., DOI: 10.1007/978-981-19-6502-9_6

Laftouhi, A., Eloutassi, N., Ech-Chihbi, E., Rais, Z., Abdellaoui, A., Taleb, A., Beniken, M., Nafidi, H.-A., Salamatullah, A. M., Bourhia, M., & Taleb, M. (2023). The impact of environmental stress on the secondary metabolites and the chemical compositions of the essential oils from some medicinal plants used as food supplements. *Sustainability (Basel), 15*(10), 7842. DOI: 10.3390/su15107842

Lamouille-Hébert, M., Arthaud, F., & Datry, T. (2024). Climate change and the biodiversity of alpine ponds: Challenges and perspectives. *Ecology and Evolution, 14*(2), e10883. DOI: 10.1002/ece3.10883 PMID: 38327685

Lamprecht, A., Semenchuk, P. R., Steinbauer, K., Winkler, M., & Pauli, H. (2018). Climate change leads to accelerated transformation of high-elevation vegetation in the central Alps. *The New Phytologist, 220*(2), 447–459. DOI: 10.1111/nph.15290 PMID: 29938796

Landry, K. B., Azam, S., Rehman, S., Tariq, S., Iqbal, B., Abbas, M., Lembè, D. M., & Ijaz, B. (2021). Phytochemical analysis of *Berberis lyceum* methanolic extract and its antiviral activity through the restoration of MAPK signaling pathway modulated by HCV NS5A. *Asian Pacific Journal of Tropical Biomedicine, 11*(3), 132–140. DOI: 10.4103/2221-1691.306133

Lave, J., & Wenger, E. (1991). *Situated learning: Legitimate peripheral participation.* Cambridge university press. DOI: 10.1017/CBO9780511815355

Lavorel, S., Colloff, M. J., Locatelli, B., Gorddard, R., Prober, S. M., Gabillet, M., Devaux, C., Laforgue, D., & Peyrache-Gadeau, V. (2019). Mustering the power of ecosystems for adaptation to climate change. *Environmental Science & Policy*, *92*, 87–97. DOI: 10.1016/j.envsci.2018.11.010

Lavorel, S., Quetier, F., Thebault, A., Liancourt, P., Daigney, S., Davies, I., & de Chazal, J. (2006). Vulnerability to land use change of services provided by alpine landscapes. *Open Science Conference on Global Change in Mountain Regions (GLOCHAMORE 2005)*

Lee, S., Jung, H., & Choi, J. (2020). Projecting the impact of climate change on the spatial distribution of six subalpine tree species in South Korea using a multi-model ensemble approach. *Forests*, *12*(1), 37. DOI: 10.3390/f12010037

Lefebvre, V., Fontaine, C., Villemant, C., & Daugeron, C. (2014). Are empidine dance flies major flower visitors in alpine environments? A case study in the Alps, France. *Biology Letters*, *10*(11), 20140742. DOI: 10.1098/rsbl.2014.0742 PMID: 25376804

Legendre, P., & De Cáceres, M. (2013). Beta diversity as the variance of community data: Dissimilarity coefficients and partitioning. *Ecology Letters*, *16*(8), 951–963. DOI: 10.1111/ele.12141 PMID: 23809147

Leonti, M. (2022). The relevance of quantitative ethnobotanical indices for ethnopharmacology and ethnobotany. *Journal of Ethnopharmacology*, *288*, 115008. DOI: 10.1016/j.jep.2022.115008 PMID: 35066067

Lesica, P. (2023). What good are herbaria? *Sage (Atlanta, Ga.)*, *45*(3), 2.

Leska, A., Nowak, A., Nowak, I., & Górczyńska, A. (2021). Effects of Insecticides and Microbiological Contaminants on Apis mellifera Health. *Molecules (Basel, Switzerland)*, *26*(16), 5080. Advance online publication. DOI: 10.3390/molecules26165080 PMID: 34443668

Letchuman, S., Madhuranga, H., Madhurangi, B., Premarathna, A. D., & Saravanan, M. (2024). *Alkaloids unveiled: A comprehensive analysis of novel therapeutic properties, mechanisms, and plant-based innovations*. Intelligent Pharmacy., DOI: 10.1016/j.ipha.2024.09.007

Liao, Z., Zhang, L., Nobis, M. P., Wu, X., Pan, K., Wang, K., Dakhil, M. A., Du, M., Xiong, Q., Pandey, B., & Tian, X. (2020). Climate change jointly with migration ability affect future range shifts of dominant fir species in Southwest China. *Diversity & Distributions*, *26*(3), 352–367. DOI: 10.1111/ddi.13018

Li, F., Shah, D. N., Pauls, S. U., Qu, X., Cai, Q., & Tachamo Shah, R. D. (2016). Elevational shifts of freshwater communities cannot catch up climate warming in the Himalaya. *Water (Basel)*, *8*(8), 327. DOI: 10.3390/w8080327

Li, K.-J., Liu, X.-F., Yang, L., & Shen, S.-K. (2024). Alpine Rhododendron population contractions lead to spatial distribution mismatch with their pollinators under climate change. *The Science of the Total Environment*, *926*, 171832. DOI: 10.1016/j.scitotenv.2024.171832 PMID: 38521263

Lim, S. H., & Choi, C.-I. (2019). Pharmacological properties of *Morus nigra* L.(black mulberry) as a promising nutraceutical resource. *Nutrients*, *11*(2), 437. DOI: 10.3390/nu11020437 PMID: 30791521

Lindholm, M., Stordal, F., Moe, S. J., Hessen, D. O., & Aass, P. (2012). Climate-driven range retraction of an Arctic freshwater crustacean. *Freshwater Biology*, *57*(12), 2591–2601. DOI: 10.1111/fwb.12030

Lin, Y., Shi, R., Wang, X., & Shen, H.-M. (2008). Luteolin, a flavonoid with potential for cancer prevention and therapy. *Current Cancer Drug Targets*, *8*(7), 634–646. DOI: 10.2174/156800908786241050 PMID: 18991571

Li, S., Dong, S., Fu, Y., Zhou, B., Liu, S., Shen, H., Xu, Y., Gao, X., Xiao, J., Wu, S., & Li, F. (2022). Air or soil temperature matters the responses of alpine plants in biomass accumulation to climate warming. *The Science of the Total Environment*, *844*, 157141. DOI: 10.1016/j.scitotenv.2022.157141 PMID: 35798113

Liu, D., He, X.-Q., Wu, D.-T., Li, H.-B., Feng, Y.-B., Zou, L., . . . chemistry, f. (2022). Elderberry (Sambucus nigra L.): Bioactive compounds, health functions, and applications. *70*(14), 4202-4220.

Liu, H., Mi, Z., Lin, L., Wang, Y., Zhang, Z., Zhang, F., & Cao, G. (2018). Shifting plant species composition in response to climate change stabilizes grassland primary production. *Proceedings of the National Academy of Sciences of the United States of America*, *115*(16), 4051–4056. DOI: 10.1073/pnas.1700299114 PMID: 29666319

Liu, J., Wang, X., Miao, H., & Yang, Y. (2025). Community Beta Diversity as a Catalyst: Mediating Livestock Grazing Effects on Ecosystem Multifunctionality in the Qinghai-Tibet Plateau. *Land Degradation & Development*.

Liu, S., Yang, T., Ming, T. W., Gaun, T. K. W., Zhou, T., Wang, S., & Ye, B. (2020). Isosteroid alkaloids with different chemical structures from *Fritillariae cirrhosae* bulbus alleviate LPS-induced inflammatory response in RAW 264.7 cells by MAPK signaling pathway. *International Immunopharmacology*, *78*, 106047. DOI: 10.1016/j.intimp.2019.106047 PMID: 31816576

Liu, W., Zheng, L., & Qi, D. (2020). Variation in leaf traits at different altitudes reflects the adaptive strategy of plants to environmental changes. *Ecology and Evolution, 10*(15), 8166–8175. DOI: 10.1002/ece3.6519 PMID: 32788969

Liu, X., Ding, J., & Zhao, W. (2023). Divergent responses of ecosystem services to afforestation and grassland restoration in the Tibetan Plateau. *Journal of Environmental Management, 344*, 118471. DOI: 10.1016/j.jenvman.2023.118471 PMID: 37364488

Liu, Y., Lü, Y., Zhao, M., & Fu, B. (2023). Multiple pressures and vegetation conditions shape the spatiotemporal variations of ecosystem services in the Qinghai-Tibet Plateau. *Frontiers in Plant Science, 14*, 1127808. DOI: 10.3389/fpls.2023.1127808 PMID: 36743496

Li, W., Shi, M., Wang, P., Guo, X., Li, C., & Kang, W. (2018). Efficient determination of three flavonoids in *Malus pumila* flowers by ionic liquid-HPLC. *Journal of Molecular Liquids, 263*, 139–146. DOI: 10.1016/j.molliq.2018.04.135

Li, Y., Zhang, X., Shao, Q., Fan, J., Chen, Z., Dong, J., Hu, Z., & Zhan, Y. (2022). Community composition and structure affect ecosystem and canopy water use efficiency across three typical alpine ecosystems. *Frontiers in Plant Science, 12*, 771424.

Losapio, G., Schöb, C., Staniczenko, P. P. A., Carrara, F., Palamara, G. M., De Moraes, C. M., Mescher, M. C., Brooker, R. W., Butterfield, B. J., Callaway, R. M., Cavieres, L. A., Kikvidze, Z., Lortie, C. J., Michalet, R., Pugnaire, F. I., & Bascompte, J. (2021). Network motifs involving both competition and facilitation predict biodiversity in alpine plant communities. *Proceedings of the National Academy of Sciences of the United States of America, 118*(6), e2005759118. DOI: 10.1073/pnas.2005759118 PMID: 33526655

Loshali, A., Tripathi, G., Bawa, S., & Aeri, V. (2025). *Betula utilis* (Bhojpatra): A potent herb with its traditional uses, phytochemistry, clinical application, pharmacology, and toxicology. *Journal of Applied Pharmaceutical Science, 15*(5), 001-013. DOI: 10.7324/JAPS.2025.230627

Luczaj, L., Pieroni, A., Tardío, J., Pardo-de-Santayana, M., Sõukand, R., Svanberg, I., & Kalle, R. (2012). Wild food plant use in 21st century Europe: The disappearance of old traditions and the search for new cuisines involving wild edibles. *Acta Societatis Botanicorum Poloniae, 81*(4), 359–370. DOI: 10.5586/asbp.2012.031

Ludovicy, S., Noroozi, J., Semenchuk, P., Moser, D., Wessely, J., Talebi, A., & Dullinger, S. (2022). Protected area network insufficiently represents climatic niches of endemic plants in a Global Biodiversity Hotspot. *Biological Conservation, 275*, 109768. DOI: 10.1016/j.biocon.2022.109768

Luterbacher, J., Dietrich, D., Xoplaki, E., Grosjean, M., & Wanner, H. (2004). European seasonal and annual temperature variability, trends, and extremes since 1500. *Science, 303*(5663), 1499–1503. DOI: 10.1126/science.1093877 PMID: 15001774

Lu, X., Zheng, X., Liang, E., Piao, S., Babst, F., Elliott, G. P., Sigdel, S. R., Wang, T., Wang, Y., Li, X., Gao, S., Zhang, L., Sun, J., Li, J., Zhu, H., Rossi, S., Peñuelas, J., & Camarero, J. J. (2025). Patterns, dynamics and drivers of alpine treelines and shrublines. *Nature Reviews. Earth & Environment*, *6*(8), 489–502. DOI: 10.1038/s43017-025-00703-9

Lu, Y., Liu, Y., Che, F., Gao, M., Li, A., & Wei, Y. (2024). Optimisation of isolation of polyphenols from *Malus pumila* Mill. Leaves by high-speed countercurrent chromatography using response surface methodology. *Journal of Chromatography. B, Analytical Technologies in the Biomedical and Life Sciences*, *1243*, 124230. DOI: 10.1016/j.jchromb.2024.124230 PMID: 38981203

Ma, B., Singh, R. B., & Hietala, R. (2012). Human driving forces for ecosystem services in the Himalayan region. *Environment and Ecology*, (3, Iss. 1), 53–57.

Macicasan, V., Ocrain, A., Balc, R., Dicu, T., Bodmer, M. D., & Roba, C. (2024). How can land use management in traditional cultural landscapes become a policy instrument for soil organic carbon sequestration and climate change mitigation? A Transylvanian case study. *Applied Sciences (Basel, Switzerland)*, *14*(21), 9851. DOI: 10.3390/app14219851

Mack, A., Choffnes, E. R., Hamburg, M. A., & Relman, D. A. (Eds.). (2008). *Global climate change and extreme weather events: Understanding the contributions to infectious disease emergence: Workshop summary*. National Academies Press.

Magaña Ugarte, R., Escudero, A., & Gavilán, R. G. (2019). Metabolic and physiological responses of Mediterranean high-mountain and alpine plants to combined abiotic stresses. *Physiologia Plantarum*, *165*(2), 403–412. DOI: 10.1111/ppl.12898 PMID: 30536685

Mahanayak, B. J. W. J. B. P. H. S. (2024). Ex-situ and in-situ conservation of wild life. *18*(3), 277-282.

Mahdavi, S., Kheyrollahi, M., Sheikhloei, H., & Isazadeh, A. (2019). Antibacterial and antioxidant activities of *Nasturtium officinale* essential oil on food borne bacteria. *The Open Microbiology Journal*, *13*(1), 81–85. Advance online publication. DOI: 10.2174/1874285801913010081

Maiti, P., Kuniyal, J. C., Sekar, K. C., Satish, K. V., Singh, D., Bisht, N., Kumar, A., Arya, S. C., Nand, M., & Sundriyal, R. C. (2022). Landscape level ecological assessment and eco-restoration strategies for alpine and sub-alpine regions of the Central Himalaya. *Ecological Engineering*, *180*, 106674. DOI: 10.1016/j.ecoleng.2022.106674

Majeed, S., Zafar, M., Ahmad, M., Kilic, O., Sultana, S., Raza, J. (2020). Pollen morphological investigations of family Cactaceae and its taxonomic implication by light microscopy and scanning electron microscopy. *83*(7), 767-777.

Makrodimos, N., Blionis, G. J., Krigas, N., & Vokou, D. (2008). Flower morphology, phenology and visitor patterns in an alpine community on Mt Olympos, Greece. *Flora-Morphology, Distribution. Flora (Jena)*, *203*(6), 449–468. DOI: 10.1016/j.flora.2007.07.003

Maksimovic, Z. A., Dordevic, S., & Mraovic, M. (2005). Antimicrobial activity of *Chenopodium botrys* essential oil. *Fitoterapia*, *76*(1), 112–114. DOI: 10.1016/j.fitote.2004.10.005 PMID: 15664473

Malanson, G. P., Fagre, D. B., Butler, D. R., & Shen, Z. (2024). Alpine plant communities and current topographic microrefugia vary with regional climates. *Geomorphology*, *458*, 109241. DOI: 10.1016/j.geomorph.2024.109241

Malík, M., Mika, O. J., Navrátilová, Z., Killi, U. K., Tlustoš, P., & Patočka, J. (2023). Health and environmental hazards of the toxic *Pteridium aquilinum* (L.) Kuhn (Bracken Fern). *Plants*, *13*(1), 18. DOI: 10.3390/plants13010018 PMID: 38202326

Mammola, S., Pétillon, J., Hacala, A., Monsimet, J., Marti, S., Cardoso, P., & Lafage, D. (2021). Challenges and opportunities of species distribution modelling of terrestrial arthropod predators. *Diversity & Distributions*, *27*(12), 2596–2614. DOI: 10.1111/ddi.13434

Ma, N., Zhang, Z., Liao, F., Jiang, T., & Tu, Y. (2020). The birth of artemisinin. *Pharmacology & Therapeutics*, *216*, 107658. DOI: 10.1016/j.pharmthera.2020.107658 PMID: 32777330

Manish, K. (2022). Medicinal plants in peril due to climate change in the Himalaya. *Ecological Informatics*, *68*, 101546. DOI: 10.1016/j.ecoinf.2021.101546

Mann, S., & Lanz, S. (2013). Happy Tinbergen: Switzerland's new direct payment system. *EuroChoices (Uckfield)*, *12*(3), 24–28. DOI: 10.1111/1746-692X.12036

Manojlović, N. T., Vasiljević, P., Jusković, M., Najman, S., Janković, S., & Milenković-Andjelković, A. (2010). HPLC analysis and cytotoxic potential of extracts from the lichen, *Thamnolia vermicularis* var. subuliformis. *Journal of Medicinal Plants Research, 4*(9), 817–823. DOI: 10.5897/JMPR10.104

Manzoor, M., Ahmad, M., Gillani, S. W., Waheed, M., Shaheen, H., Haq, S. M., & Mustafa, A. (2026). Alpine endemism under threat: Ecological patterns and conservation strategies in the changing Himalayas. *Global Ecology and Conservation, 60*, e04051. DOI: 10.1016/j.gecco.2025.e04051

Manzoor, M., Ahmad, M., Gillani, S. W., Waheed, M., Shaheen, H., Mehmood, A. B., Fonge, B. A., & Al-Andal, A. (2025). Population dynamics, threat assessment, and conservation strategies for critically endangered Meconopsis aculeata in alpine zone. *BMC Plant Biology, 25*(1), 358. DOI: 10.1186/s12870-025-06361-9 PMID: 40102752

Manzoor, M., Shaheen, H., Attique, A., Khan, R. W. A., Gillani, S. W., & Khan, M. S. (2024). Distribution pattern, population structure and habitat geography of *Betula utilis* subalpine forests of Kashmir region in the western Himalayas: Implications for conservation and sustainable management. *Environment, Development and Sustainability, 26*(1), 1–26. DOI: 10.1007/s10668-024-05049-8

Mao, X., & Yang, Z. (2023). Chiral nanomaterial-enabled bactericide and nanopesticide for sustainable agriculture and food safety. *Journal of Nanoparticle Research, 25*(10), 198. DOI: 10.1007/s11051-023-05845-9

Mashaal, K., Shabbir, A., Shahzad, M., Mobashar, A., Akhtar, T., Fatima, T., Riaz, B., Alharbi, R., Fatima, A., Alanezi, A. A., & Ahmad, A. (2023). Amelioration of rheumatoid arthritis by *Fragaria nubicola* (Wild Strawberry) via attenuation of inflammatory mediators in Sprague Dawley rats. *Medicina, 59*(11), 1917. DOI: 10.3390/medicina59111917 PMID: 38003966

Mason, N. W., Mouillot, D., Lee, W. G., & Wilson, J. B. (2005). Functional richness, functional evenness and functional divergence: The primary components of functional diversity. *Oikos, 111*(1), 112–118. DOI: 10.1111/j.0030-1299.2005.13886.x

Mastrotheodoros, T., Pappas, C., Molnar, P., Burlando, P., Hadjidoukas, P., & Fatichi, S. (2019). Ecohydrological dynamics in the Alps: Insights from a modelling analysis of the spatial variability. *Ecohydrology, 12*(1), e2054. DOI: 10.1002/eco.2054

Masyita, A., Sari, R. M., Astuti, A. D., Yasir, B., Rumata, N. R., Emran, T. B., Nainu, F., & Simal-Gandara, J. (2022). Terpenes and terpenoids as main bioactive compounds of essential oils, their roles in human health and potential application as natural food preservatives. *Food Chemistry: X, 13*, 100217. DOI: 10.1016/j.fochx.2022.100217 PMID: 35498985

Matthews, J. A. (1992). *The ecology of recently-deglaciated terrain: a geoecological approach to glacier forelands*. Cambridge University Press.

Mayewski, P. A., & Jeschke, P. A. (1979). Himalayan and Trans-Himalayan glacier fluctuations since AD 1812. *Arctic and Alpine Research, 11*(3), 267–287. DOI: 10.1080/00040851.1979.12004137

Mehmood, A., & Shah, A. (2015). CULTURAL USES OF PLANTS AMONG BASIKHEL TRIBE OF DISTRICT TOR GHAR, KHYBER PAKHTUNKHWA, PAKISTAN. *Pakistan Journal of Botany*.

Mehta, S., Sharma, A. K., & Singh, R. K. (2021). Advances in Ethnobotany, Synthetic Phytochemistry and Pharmacology of Endangered Herb Picrorhiza kurroa (Kutki): A Comprehensive Review (2010-2020). *Mini-Reviews in Medicinal Chemistry, 21*(19), 2976–2995. DOI: 10.2174/1389557521666210401090028 PMID: 33797375

Melodelima, C., & Lobréaux, S. (2013). Complete *Arabis alpina* chloroplast genome sequence and insight into its polymorphism. *Meta Gene, 1*, 65–75. DOI: 10.1016/j.mgene.2013.10.004 PMID: 25606376

Memon, A. A., Memon, N., Luthria, D. L., Pitafi, A. A., & Bhanger, M. I. (2012). Phenolic compounds and seed oil composition of *Ziziphus mauritiana* L. fruit. *Polish Journal of Food and Nutrition Sciences, 62*(1), 15–21. DOI: 10.2478/v10222-011-0035-3

Meng, F., Cui, S., Wang, S., Duan, J., Jiang, L., Zhang, Z., & Li, X. (2016). Changes in phenological sequences of alpine communities across a natural elevation gradient. *Agricultural and Forest Meteorology, 224*, 11–16. DOI: 10.1016/j.agrformet.2016.04.013

Meng, F., Jiang, L., Zhang, Z., Cui, S., Duan, J., Wang, S., & Li, X. (2017). Changes in flowering functional group affect responses of community phenological sequences to temperature change. *Ecology, 98*(3), 734–740. DOI: 10.1002/ecy.1685 PMID: 27984640

Mengist, W., Soromessa, T., & Legese, G. (2020). Ecosystem services research in mountainous regions: A systematic literature review on current knowledge and research gaps. *The Science of the Total Environment, 702*, 134581. DOI: 10.1016/j.scitotenv.2019.134581 PMID: 31734614

Men, R., Li, N., Ding, C., Tang, Y., Xing, Y., Ding, W., & Ma, Z. (2016). Chemopreventive agents from *Physalis minima* function as michael reaction acceptors. *Pharmacognosy Magazine, 12*(46, Suppl 2), S231. DOI: 10.4103/0973-1296.182153 PMID: 27279713

Mesmar, J., Abdallah, R., Badran, A., Maresca, M., Shaito, A., & Baydoun, E. (2022). *Ziziphus nummularia*: A comprehensive review of its phytochemical constituents and pharmacological properties. *Molecules (Basel, Switzerland), 27*(13), 4240. DOI: 10.3390/molecules27134240 PMID: 35807485

Mességué, M. (2017). *Il Mio Erbario: Con 100 piante illustrate e il prontuario per la cura di oltre 150 patologie*. Macro Edizioni.

Messerli, P. (2015). Die Alpen–Geschichte und Zukunft einer europäischen Kulturlandschaft [The Alps—History and Future of a European Cultural Landscape]. *Mountain Research and Development, 35*(4), 428–429. DOI: 10.1659/mrd.mm171

Meyfroidt, P., De Bremond, A., Ryan, C. M., Archer, E., Aspinall, R., Chhabra, A., Camara, G., Corbera, E., DeFries, R., Díaz, S., Dong, J., Ellis, E. C., Erb, K.-H., Fisher, J. A., Garrett, R. D., Golubiewski, N. E., Grau, H. R., Grove, J. M., Haberl, H., & zu Ermgassen, E. K. H. J. (2022). Ten facts about land systems for sustainability. *Proceedings of the National Academy of Sciences of the United States of America, 119*(7), e2109217118. DOI: 10.1073/pnas.2109217118 PMID: 35131937

Meza-Joya, F. L., Morgan-Richards, M., Koot, E. M., & Trewick, S. A. (2023). Global warming leads to habitat loss and genetic erosion of alpine biodiversity. *Journal of Biogeography, 50*(5), 961–975. DOI: 10.1111/jbi.14590

Meza-Joya, F. L., Morgan-Richards, M., & Trewick, S. A. (2025). Forecasting Range Shifts in Terrestrial Alpine Insects Under Global Warming. *Ecology and Evolution, 15*(1), e70810. DOI: 10.1002/ece3.70810 PMID: 39803196

Milani, F., Bottoni, M., Giuliani, C., Colombo, L., Casiraghi, M. C., Colombo, P. S., Bruschi, P., Erba, D., & Fico, G. (2023). Alpine Diet in Valmalenco (Lombardy, Italy): Nutritional Features of Spontaneous Plants and Traditional Dishes. *Nutrients, 15*(8), 1988. DOI: 10.3390/nu15081988 PMID: 37111208

Miller, R. M., Rodríguez, J. P., Aniskowicz-Fowler, T., Bambaradeniya, C., Boles, R., Eaton, M. A., & Walker, S. (2006). Extinction risk and conservation priorities. *Science, 313*(5786), 441–441. DOI: 10.1126/science.313.5786.441a PMID: 16873627

Miller-Struttmann, N. E., Geib, J. C., Franklin, J. D., Kevan, P. G., Holdo, R. M., Ebert-May, D., Lynn, A. M., Kettenbach, J. A., Hedrick, E., & Galen, C. (2015). Functional mismatch in a bumble bee pollination mutualism under climate change. *Science, 349*(6255), 1541–1544. DOI: 10.1126/science.aab0868 PMID: 26404836

Minami, Y., Kanda, H., & Masuzawa, T. (1996). The relationship between distribution of bryophytes and soil conditions on deglaciated arctic terrain in Ny-Ålesund. Proceedings of the NIPR Symposium on Polar Biology.

Mir, R. H., Mohi-Ud-Din, R., Al-Keridis, L. A., Ahmad, B., Alshammari, N., Patel, M., Adnan, M., & Masoodi, M. H. (2024). Phytochemical profiling, antioxidant, cytotoxic, and anti-inflammatory activities of *Plectranthus rugosus* extract and fractions: *In vitro, in vivo*, and *in silico* approaches. *Inflammopharmacology, 32*(2), 1593–1606. DOI: 10.1007/s10787-023-01419-2 PMID: 38308794

Mir, W. R., Bhat, B. A., Rather, M. A., Muzamil, S., Almilaibary, A., Alkhanani, M., & Mir, M. A. (2022). Molecular docking analysis and evaluation of the antimicrobial properties of the constituents of *Geranium wallichianum* D. Don ex Sweet from Kashmir Himalaya. *Scientific Reports, 12*(1), 12547. DOI: 10.1038/s41598-022-16102-9 PMID: 35869098

Mi, S., Wang, W., Zhu, M., Mu, L., & Lu, Q. (2025). Utilization of *Pinus pumila* seed scales as forestry residues: Green microwave-assisted hydrodistillation for essential oil extraction and its antibacterial activity. *Journal of Chromatography. A, 466574.* Advance online publication. DOI: 10.1016/j.chroma.2025.466574 PMID: 41344058

Mishra, R., Dogra, G., & Gupta, P. (1981). Apicultural activities in Himachal Pradesh. *Indian Bee Journal, 41*(3/4), 29-31

Mittermeier, R. A., Turner, W. R., Larsen, F. W., Brooks, T. M., & Gascon, C. (2011). Global biodiversity conservation: The critical role of hotspots. In Zachos, F. E., & Habel, J. C. (Eds.), *Biodiversity hotspots: Distribution and protection of conservation priority areas* (pp. 3–22). Springer. DOI: 10.1007/978-3-642-20992-5_1

Mohammed, F. S., Uysal, I., Mustaffa, S., Eraslan, E. C., & Akgul, H. (2024). Analysis of phenolic contents and biological activities of wild mint, *Mentha longifolia* (L.) L. [IJEB]. *Indian Journal of Experimental Biology, 62*(03), 192–198. DOI: 10.56042/ijeb.v62i03.949

Möhl, P., Von Büren, R. S., & Hiltbrunner, E. (2022). Growth of alpine grassland will start and stop earlier under climate warming. *Nature Communications*, *13*(1), 7398. DOI: 10.1038/s41467-022-35194-5 PMID: 36456572

Molinu, M. G., Sanna, F., D'hallewin, G., Re, G. A., Sulas, L., & Piluzza, G. (2024). Mediterranean Wild Pear Fruits as a Neglected but Valuable Source of Phenolic Compounds. *Resources*, *13*(6), 72. DOI: 10.3390/resources13060072

Molnár, Z., Kelemen, A., Kun, R., Máté, J., Sáfián, L., Provenza, F., & Máté, A. (2020). Knowledge co-production with traditional herders on cattle grazing behaviour for better management of species-rich grasslands. *Journal of Applied Ecology*, *57*(9), 1677–1687. DOI: 10.1111/1365-2664.13664

Monteiro, J. A. F., & Körner, C. (2013). Leaf turnover and herbivory in the tall tussock grass Festuca orthophylla in the Andean Altiplano. *Alpine Botany*, *123*(1), 13–20. DOI: 10.1007/s00035-013-0113-8

Mooney, , H. ADrake, , J. A. (1987). The Ecology of: Biological Invasions. *Environment: Science and Policy for Sustainable Development*, *29*(5), 10–37. DOI: 10.1080/00139157.1987.9928888

Morales, L. V., Sierra-Almeida, A., Sandoval-Urzúa, C., & Arroyo, M. T. (2025). Thermal decoupling may promote cooling and avoid heat stress in alpine plants. *Plants*, *14*(13), 2023. DOI: 10.3390/plants14132023 PMID: 40648032

Morschel, F., Arduino, S., & Morschel, F. (2004). *The Alps: a unique natural heritage: a common vision for the conservation of their biodiversity*. WWF Germany for the WWF European Alpine Programme.

Moser, B., Jaeger, J. A., Tappeiner, U., Tasser, E., & Eiselt, B. (2007). Modification of the effective mesh size for measuring landscape fragmentation to solve the boundary problem. *Landscape Ecology*, *22*(3), 447–459. DOI: 10.1007/s10980-006-9023-0

Mounce, R., Smith, P., & Brockington, S. (2017). Ex situ conservation of plant diversity in the world's botanic gardens. *Nature Plants*, *3*(10), 795–802. DOI: 10.1038/s41477-017-0019-3 PMID: 28947807

Muddassir, M., Raza, A., Munir, S., Basirat, A., Ahmed, M., Butt, M. S., Dar, O. A., Ahmed, S. S., Shamim, S., & Naqvi, S. Z. H. (2022). Antibacterial efficacy of silver nanoparticles (AgNPs) against metallo-β-lactamase and extended spectrum β-lactamase producing clinically procured isolates of Pseudomonas aeruginosa. *Scientific Reports*, *12*(1), 20685. DOI: 10.1038/s41598-022-24531-9 PMID: 36450765

Mughal, B., Zaidi, S. Z. J., Zhang, X., & Hassan, S. U. (2021). Biogenic nanoparticles: Synthesis, characterisation and applications. *Applied Sciences (Basel, Switzerland), 11*(6), 2598. DOI: 10.3390/app11062598

Mughees, M., Farooq, M. A., Haq, I. U., Zeb, I., Ali, M., Hussain, Z., Shahzadi, I., & Shah, M. M. (2022). Quantification of rosmarinic acid from different plant species of lower Himalayan region and expression analysis of underlying L-Phenylalanine pathway. *Physiologia Plantarum, 174*(5), e13758. DOI: 10.1111/ppl.13758 PMID: 36281843

Mukhtar, Y., Nazir, M., Naz, S., Taj, R., Akhtar, W., Abdul, L., Kamal, A., Alrefaei, A. F., Alshehri, E., & Lackner, M. (2025). Comparative assessment of physicochemical and biological properties of silver nanoparticles (AgNPs) synthesized from aerial and underground parts of *Arnebia benthamii. Kuwait Journal of Science, 100478.* Advance online publication. DOI: 10.1016/j.kjs.2025.100478

Mulliken, T. (2000). Implementing CITES for Himalayan medicinal plants Nardostachys grandiflora and Picrorhiza kurrooa. *TRAFFIC BULLETIN-CAMBRIDGE-TRAFFIC INTERNATIONAL, 18*(2), 63–72.

Munir, A., Youssef, F. S., Ishtiaq, S., Kamran, S. H., Sirwi, A., Ahmed, S. A., Ashour, M. L., & Elhady, S. S. (2021). *Malva parviflora* leaves mucilage: An eco-friendly and sustainable biopolymer with antioxidant properties. *Polymers, 13*(23), 4251. DOI: 10.3390/polym13234251 PMID: 34883753

Myers, N. (1988). Threatened biotas:" hot spots" in tropical forests. *The Environmentalist, 8*(3), 187–208. DOI: 10.1007/BF02240252 PMID: 12322582

Na, S.-J., Kim, Y.-K., & Park, J.-M. J. P. (2024). Nectar characteristics and honey production potential of five rapeseed cultivars and two wildflower species in South Korea. *13*(3), 419.

Nabila, A.M., Ali, M., Zafar, M., Sultana, S., Majeed, S., ... & Ahmad, S. (2022). Palynological Diversity of Melliferous flora around Apiaries from District Mardan Khyber Pakhtunkhwa-Pakistan: Nabila et al. *The Botanical Review, 88*(3), 299-332.

Nader, R. A., Mackieh, R., Wehbe, R., El Obeid, D., Sabatier, J. M., & Fajloun, Z. (2021). Beehive products as antibacterial agents: A review. *Antibiotics (Basel, Switzerland), 10*(6), 717. DOI: 10.3390/antibiotics10060717 PMID: 34203716

Nagajyothi, P. C., Prabhakar Vattikuti, S. V., Devarayapalli, K. C., Yoo, K., Shim, J., & Sreekanth, T. V. M. (2020). Green synthesis: Photocatalytic degradation of textile dyes using metal and metal oxide nanoparticles-latest trends and advancements. *Critical Reviews in Environmental Science and Technology*, *50*(24), 2617–2723. DOI: 10.1080/10643389.2019.1705103

Nagy, L., & Grabherr, G. (2009). *The biology of alpine habitats*. Oxford University Press. DOI: 10.1093/oso/9780198567035.001.0001

Nagy, L., Grabherr, G., Körner, C., & Thompson, D. B. (2012). *Alpine biodiversity in Europe* (Vol. 167). Springer Science & Business Media.

Naiel, M. A.-E., Shehata, A. M., Paswan, V. K., Said, M. B., & El-Tarabily, K. A. (2024). Utilizing the potential of Saussurea lappa in aquaculture industry: A review on immune enhancement and pollution remediation. *Aquaculture International*, *32*(5), 5513–5550. DOI: 10.1007/s10499-024-01435-1

Namdari, M., Eatemadi, A., Soleimaninejad, M., & Hammed, A. T. (2017). A brief review on the application of nanoparticle enclosed herbal medicine for the treatment of infective endocarditis. *Biomedicine and Pharmacotherapy*, *87*, 321–331. DOI: 10.1016/j.biopha.2016.12.099 PMID: 28064105

Naseer, M., & Adil, M. (2025). Phytochemical profiling, HPLC analysis, and antimicrobial potential of *Curio radicans* (L. f.) PV Heath. *Scientific Reports*, *15*(1), 34753. DOI: 10.1038/s41598-025-18323-0 PMID: 41053340

National Aeronautics and Space Administration. (2024, October 23). *The effects of climate change*. NASA Science. https://science.nasa.gov/climate-change/effects/

Nazir, N., Rahman, A., Uddin, F., Khan Khalil, A. A., Zahoor, M., Nisar, M., Ullah, S., Ullah, R., Ezzeldin, E., & Mostafa, G. A. (2021). Quantitative ethnomedicinal status and phytochemical analysis of *Berberis lyceum* royle. *Agronomy (Basel)*, *11*(1), 130. DOI: 10.3390/agronomy11010130

Nazir, N., Zahoor, M., Nisar, M., Khan, I., Karim, N., Abdel-Halim, H., & Ali, A. (2018). Phytochemical analysis and antidiabetic potential of *Elaeagnus umbellata* (Thunb.) in streptozotocin-induced diabetic rats: Pharmacological and computational approach. *BMC Complementary and Alternative Medicine*, *18*(1), 332. DOI: 10.1186/s12906-018-2381-8 PMID: 30545352

Negi, V. S., Joshi, B. C., Pathak, R., Rawal, R. S., & Sekar, K. C. (2018). Assessment of fuelwood diversity and consumption patterns in cold desert part of Indian Himalaya: Implication for conservation and quality of life. *Journal of Cleaner Production*, *196*, 23–31. DOI: 10.1016/j.jclepro.2018.05.237

Negi, V. S., & Maikhuri, R. (2017). Forest resources consumption pattern in Govind wildlife sanctuary, western Himalaya, India. *Journal of Environmental Planning and Management*, *60*(7), 1235–1252. DOI: 10.1080/09640568.2016.1213707

Negi, V. S., Tiwari, D. C., Singh, L., Thakur, S., & Bhatt, I. D. (2022). Review and synthesis of climate change studies in the Himalayan region. *Environment, Development and Sustainability*, *24*(9), 10471–10502. DOI: 10.1007/s10668-021-01880-5

Newbold, T., Hudson, L. N., Hill, S. L. L., Contu, S., Lysenko, I., Senior, R. A., Börger, L., Bennett, D. J., Choimes, A., Collen, B., Day, J., De Palma, A., Díaz, S., Echeverria-Londoño, S., Edgar, M. J., Feldman, A., Garon, M., Harrison, M. L. K., Alhusseini, T., & Purvis, A. (2015). Global effects of land use on local terrestrial biodiversity. *Nature*, *520*(7545), 45–50. DOI: 10.1038/nature14324 PMID: 25832402

Nguengang, R. T., Tchegnitegni, B. T., Ateba, J. E. T., Tabekoueng, G. B., Awantu, A. F., Bankeu, J. J. K., Chouna, J. R., Nkenfou, C. N., Sewald, N., & Lenta, B. N. (2023). Antibacterial constituents of *Rumex nepalensis* spreng and its emodin derivatives. *Natural Product Research*, *37*(23), 3935–3946. DOI: 10.1080/14786419.2022.2162894 PMID: 36584290

Nguyen, T.-A.-M., Phan, H.-V.-T., Dao, N.-V.-T., Nguyen, T.-T.-V., Dong, P.-S.-N., Dang, L.-T.-C., Le, H.-T.-T., Nguyen, V.-K., & Hoang, L.-T.-T.-T. (2024). A systematic review on ethnobotany, phytochemistry, and pharmacology of *Ehretia acuminata* R. Br.(Boraginaceae), a medicinal plant. *Natural Product Research*, •••, 1–19. DOI: 10.1080/14786419.2024.2446704 PMID: 39731579

Niesenbaum, R. A. (2019). *The integration of conservation, biodiversity, and sustainability* (Vol. 11). MDPI.

Nieuwschepen, M. (2024, July 3). *Climate change and shifting ecosystems: Understanding the impact of altitudinal and latitudinal shifts*. European Wilderness Society. https://wilderness-society.org/climate-change-and-shifting-ecosystems-understanding-the-impact-of-altitudinal-and-latitudinal-shifts/

Nigam, M., Atanassova, M., Mishra, A. P., Pezzani, R., Devkota, H. P., Plygun, S.,...Sharifi-Rad, J. (2019). Bioactive Compounds and Health Benefits of Artemisia Species. *Natural Product Communications, 14*(7), 1934578X19850354. DOI: 10.1177/1934578X19850354

Nikolova, M. (2011). Screening of radical scavenging activity and polyphenol content of Bulgarian plant species. *Pharmacognosy Research*, *3*(4), 256. DOI: 10.4103/0974-8490.89746 PMID: 22224049

Nogués-Bravo, D., Araújo, M. B., Errea, M. P., & Martínez-Rica, J. P. (2007). Exposure of global mountain systems to climate warming during the 21st century. *Global Environmental Change, 17*(3–4), 420–428. DOI: 10.1016/j.gloenvcha.2006.11.007

Noor, M. J., Ahmad, M., Zafar, M., & Ashraf, M. A. (2017). Palynological studies of melliferous and allergenic flora of Pakistan: A key to pollen and spore identification. *Journal of Apicultural Research, 56*(3), 300–309.

Noroozi, J., Akhani, H., & Breckle, S.-W. (2008). Biodiversity and phytogeography of the alpine flora of Iran. *Biodiversity and Conservation, 17*(3), 493–521. DOI: 10.1007/s10531-007-9246-7

Noroozi, J., Khalvati, S., Nafisi, H., Kaveh, A., Nazari, B., Zare, G., Minaei, M., Vitek, E., & Schneeweiss, G. M. (2021). Endemics determine bioregionalization in the alpine zone of the Irano-Anatolian biodiversity hotspot (South-West Asia). *Alpine Botany, 131*(2), 177–186. DOI: 10.1007/s00035-021-00266-7 PMID: 34721248

Noroozi, J., & Körner, C. (2018). A bioclimatic characterization of high elevation habitats in the Alborz mountains of Iran. *Alpine Botany, 128*(1), 1–11. DOI: 10.1007/s00035-018-0202-9 PMID: 29576762

Noroozi, J., Minaei, M., Khalvati, S., Kaveh, A., Nafisi, H., Nazari, B., Zare, G., Vitek, E., Hülber, K., & Schneeweiss, G. M. (2023). Hotspots of (sub) alpine plants in the Irano-Anatolian global biodiversity hotspot are insufficiently protected. *Diversity & Distributions, 29*(2), 244–253. DOI: 10.1111/ddi.13656

Nygren, A. (1999). Local Knowledge in the Environment–Development Discourse:- From dichotomies to situated knowledges. *Critique of Anthropology, 19*(3), 267–288. DOI: 10.1177/0308275X9901900304

Obeid, M. A., Al Qaraghuli, M. M., Alsaadi, M., Alzahrani, A. R., Niwasabutra, K., & Ferro, V. A. (2017). Delivering natural products and biotherapeutics to improve drug efficacy. *Therapeutic Delivery, 8*(11), 947–956. DOI: 10.4155/tde-2017-0060 PMID: 29061102

Oberhuber, W., Dobler, A. L., Heinzle, T., Scandurra, F., Gruber, A., & Wieser, G. (2023). Climate Overrides the Influence of Microsite Conditions on Radial Growth of the Tall Multi-Stemmed Shrub Alnus alnobetula at Treeline. *Plants, 12*(8), 1708. Advance online publication. DOI: 10.3390/plants12081708 PMID: 37111935

Oberprieler, S. K., Andersen, A. N., Gillespie, G. R., & Einoder, L. D. (2019a). Vertebrates are poor umbrellas for invertebrates: Cross-taxon congruence in an Australian tropical savanna. *Ecosphere, 10*(6), e02755. DOI: 10.1002/ecs2.2755

Oehl, F., & Körner, C. (2014). Multiple mycorrhization at the coldest place known for Angiosperm plant life. *Alpine Botany*, *124*(2), 193–198. DOI: 10.1007/s00035-014-0138-7

Oertli, B. (2010). The local species richness of Dragonflies in mountain waterbodies: An indicator of climate warming? *BioRisk*, *5*, 243–251. DOI: 10.3897/biorisk.5.853

Ohler, L. M., Lechleitner, M., & Junker, R. R. (2020). Microclimatic effects on alpine plant communities and flower-visitor interactions. *Scientific Reports*, *10*(1), 1366. DOI: 10.1038/s41598-020-58388-7 PMID: 31992825

Okamoto, R., Ide, R., & Oguma, H. (2024). Automatically drawing vegetation classification maps using digital time-lapse cameras in alpine ecosystems. *Remote Sensing in Ecology and Conservation*, *10*(2), 188–202. DOI: 10.1002/rse2.364

Oldekop, J. A., Holmes, G., Harris, W. E., & Evans, K. L. (2016). A global assessment of the social and conservation outcomes of protected areas. *Conservation Biology*, *30*(1), 133–141. DOI: 10.1111/cobi.12568 PMID: 26096222

Omar, N., Ismail, C. A. N., & Long, I. (2022). Tannins in the treatment of diabetic neuropathic pain: Research progress and future challenges. *Frontiers in Pharmacology*, *12*, 805854. DOI: 10.3389/fphar.2021.805854 PMID: 35082680

Ong, T. W., Lin, B. B., Lucatero, A., Cohen, H., Bichier, P., Egerer, M. H., Danieu, A., Jha, S., Philpott, S. M., & Liere, H. (2022). Rarity begets rarity: Social and environmental drivers of rare organisms in cities. *Ecological Applications*, *32*(8), e2708. DOI: 10.1002/eap.2708 PMID: 35810452

Onipchenko, V. G., Semenova, G. V., & van der Maarel, E. (1998). Population strategies in severe environments: Alpine plants in the northwestern Caucasus. *Journal of Vegetation Science*, *9*(1), 27–40. DOI: 10.2307/3237220

Orhan, I., Kartal, M., Abu-Asaker, M., Şenol, F. S., Yilmaz, G., & Şener, B. (2009). Free radical scavenging properties and phenolic characterization of some edible plants. *Food Chemistry*, *114*(1), 276–281. DOI: 10.1016/j.foodchem.2008.09.005

Osano, P. M., Said, M. Y., de Leeuw, J., Ndiwa, N., Kaelo, D., Schomers, S., Birner, R., & Ogutu, J. O. (2013). Why keep lions instead of livestock? Assessing wildlife tourism-based payment for ecosystem services involving herders in the M aasai M ara, K enya. Natural Resources Forum, Pauli, H., & Halloy, S. R. (2019). High mountain ecosystems under climate change. In *Oxford research encyclopedia of climate science*.

Özdenefe, M. S., Kayış, F. B., Erol, Ü. H., & Takcı, A. M. (2024). Phytochemical screening and in vitro biological activity of *Amaranthus viridis* growing in Northern Cyprus. *International Journal of Secondary Metabolite, 11*(3), 592–603. DOI: 10.21448/ijsm.1425060

Ozer, M. S., Sarikurkcu, C., Ceylan, O., Akdeniz, I., & Tepe, B. (2017). A comprehensive study on chemical composition, antioxidant and enzyme inhibition activities of the essential oils of *Chenopodium botrys* collected from three different parts of Turkey. *Industrial Crops and Products, 107*, 326–331. DOI: 10.1016/j.indcrop.2017.06.001

Ozinga, W. A., Römermann, C., Bekker, R. M., Prinzing, A., Tamis, W. L. M., Schaminée, J. H. J., Hennekens, S. M., Thompson, K., Poschlod, P., Kleyer, M., Bakker, J. P., & van Groenendael, J. M. (2009). Dispersal failure contributes to plant losses in NW Europe. *Ecology Letters, 12*(1), 66–74. DOI: 10.1111/j.1461-0248.2008.01261.x PMID: 19016826

Özkan, A. İ., Gültekin, D. C., Soruklu, S., Işık, U., & Akmeşe, O. (2025). Evaluation of antibacterial, antioxidant activities and secondary metabolites profile of orchids with different dried methods. *Journal of Research in Pharmacy, 29*(6), 2349–2360. DOI: 10.12991/jrespharm.1797852

Painuli, S., Semwal, P., Cruz-Martins, N., & Bachheti, R. K. (2021). Medicinal Plants of Himalayan Forests. In Husen, A., Bachheti, R. K., & Bachheti, A. (Eds.), *Non-Timber Forest Products: Food, Healthcare and Industrial Applications* (pp. 175–212). Springer International Publishing., DOI: 10.1007/978-3-030-73077-2_8

Pal, D. (2011). Sunflower (Helianthus annuus L.) Seeds in Health and Nutrition. In Preedy, V. R., Watson, R. R., & Patel, V. B. (Eds.), *Nuts and Seeds in Health and Disease Prevention* (pp. 1097–1105). Academic Press., DOI: 10.1016/B978-0-12-375688-6.10130-6

Pallarés, S., Millán, A., Mirón, J. M., Velasco, J., Sánchez-Fernández, D., Botella-Cruz, M., & Abellán, P. (2020). Assessing the capacity of endemic alpine water beetles to face climate change. *Insect Conservation and Diversity, 13*(3), 271–282. DOI: 10.1111/icad.12394

Panda, D., Bisoi, S. S., & Palita, S. (2014). Floral diversity conservation through sacred groves in Koraput district, Odisha, India: a case study.

Panda, T., Mishra, N., Rahimuddin, S., Pradhan, B. K., Apollo, M., Kar, M. K., & Mohanty, R. B. (2023). Diversity and Conservation Significance of Angiosperm Climbers in Bhadrak District of Odisha, India. [JTBC]. *Journal of Tropical Biology & Conservation, 20*, 135–160. DOI: 10.51200/jtbc.v20i.4652

Pandey, K., & Pokhrel, S. (2022). Phytochemical screening and biological activities of *Oxalis corniculata* leaves of Sindhupalchowk District, Nepal. *BIBECHANA, 19*(1-2), 27–33. DOI: 10.3126/bibechana.v19i1-2.46094

Pandit, M. K. (2013). The Himalayas must be protected. *Nature, 501*(7467), 283. DOI: 10.1038/501283a PMID: 24048033

Pangtey, Y., Rawal, R., Bankoti, N., & Samant, S. (1990). Phenology of high-altitude plants of Kumaun in Central Himalaya, India. *International Journal of Biometeorology, 34*(2), 122–127. DOI: 10.1007/BF01093457

Paniagua-Zambrana, N. Y., Camara-Lerét, R., Bussmann, R. W., & Macía, M. J. (2014). The influence of socioeconomic factors on traditional knowledge: A cross scale comparison of palm use in northwestern South America. *Ecology and Society, 19*(4), art9. DOI: 10.5751/ES-06934-190409

Pant, P., Pandey, S., & Dall'Acqua, S. (2021). The influence of environmental conditions on secondary metabolites in medicinal plants: A literature review. *Chemistry & Biodiversity, 18*(11), e2100345. DOI: 10.1002/cbdv.202100345 PMID: 34533273

Paray, B. A., Kumari, I., Hajam, Y. A., Sharma, B., Kumar, R., Albeshr, M. F., Farah, M. A., & Khan, J. M. (2021). Honeybee nutrition and pollen substitutes: A review. *Saudi Journal of Biological Sciences, 28*(1), 1167–1176. https://doi.org/https://doi .org/10.1016/j.sjbs.2020.11.053. DOI: 10.1016/j.sjbs.2020.11.053 PMID: 33424413

Park, S.-M., Jung, C.-J., Lee, D.-G., Yu, Y.-E., Ku, T.-H., Hong, M.-S., Lim, T.-K., Paeng, K.-I., Cho, H.-K., Cho, I.-J., & Ku, S.-K. (2024). *Elaeagnus umbellata* fruit extract protects skin from ultraviolet-mediated photoaging in hairless mice. *Antioxidants, 13*(2), 195. DOI: 10.3390/antiox13020195 PMID: 38397793

Parmesan, C., Duarte, C., Poloczanska, E., Richardson, A. J., & Singer, M. C. (2011). Overstretching attribution. *Nature Climate Change, 1*(1), 2–4. DOI: 10.1038/nclimate1056

Pärnänen, P., Niikko, S., Lähteenmäki, H., Räisänen, I. T., Tervahartiala, T., Sorsa, T., & Ranki, A. (2024). Lingonberry (*Vaccinium vitis-idaea* L.) Fruit Phenolic Bioactivities—A Review of *In Vitro* and *In Vivo* Human Studies. *Microorganisms, 12*(9), 1850. DOI: 10.3390/microorganisms12091850 PMID: 39338523

Parvin, S., Reza, A., Das, S., Miah, M. M. U., & Karim, S. (2023). Potential role and international trade of medicinal and aromatic plants in the world. *European Journal of Agriculture and Food Sciences, 5*(5), 89–99. DOI: 10.24018/ejfood.2023.5.5.701

Patel, T., Shah, K., Jiwan, K., & Shrivastava, N. (2011). Study on the antibacterial potential of *Physalis minima* Linn. *Indian Journal of Pharmaceutical Sciences*, *73*(1), 111. DOI: 10.4103/0250-474X.89770 PMID: 22131635

Pathak, A., Gupta, A. P., & Pandey, P. (2024). *Herbal Medicine and Sustainable Development Challenges and Opportunities.* In (pp. 1-26).

Patonay, K., Korozs, M., Muranyi, Z., & Konya, E. P. (2017). Polyphenols in northern Hungarian *Mentha longifolia* (L.) L. treatedwith ultrasonic extraction for potential oenological uses. *Turkish Journal of Agriculture and Forestry*, *41*(3), 208–217. DOI: 10.3906/tar-1701-61

Patra, C., Ahmad, I., Ayaz, M., Khalil, A. T., Mukherjee, S., & Ovais, M. (2021). *Biogenic nanoparticles for cancer theranostics*. Elsevier.

Patra, J. K., Das, G., & Baek, K.-H. (2016). Towards a greener environment: Synthesis and applications of green nanoparticles. *Pakistan Journal of Agricultural Sciences*, *53*(2), 345–354. DOI: 10.21162/PAKJAS/16.3027

Patra, J. K., Das, G., Fraceto, L. F., Campos, E. V. R., Rodriguez-Torres, M. P., Acosta-Torres, L. S., Diaz-Torres, L. A., Grillo, R., Swamy, M. K., Sharma, S., Habtemariam, S., & Shin, H.-S. (2018). Nano based drug delivery systems: Recent developments and future prospects. *Journal of Nanobiotechnology*, *16*(1), 71. DOI: 10.1186/s12951-018-0392-8 PMID: 30231877

Patra, J. K., Kim, E. S., Oh, K., Kim, H.-J., Kim, Y., & Baek, K.-H. (2014). Antibacterial effect of crude extract and metabolites of *Phytolacca americana* on pathogens responsible for periodontal inflammatory diseases and dental caries. *BMC Complementary and Alternative Medicine*, *14*(1), 343. DOI: 10.1186/1472-6882-14-343 PMID: 25241105

Patty, L., Halloy, S. R., Hiltbrunner, E., & Körner, C. (2010). Biomass allocation in herbaceous plants under grazing impact in the high semi-arid Andes. *Flora-Morphology, Distribution. Flora (Jena)*, *205*(10), 695–703. DOI: 10.1016/j.flora.2009.12.039

Pauli, H., Gottfried, M., & Grabherr, G. (2003). Effects of climate change on the alpine and nival vegetation of the Alps. *Journal of Mountain Ecology*, *7*(Suppl), 9–12.

Pauli, H., & Halloy, S. R. (2019). High mountain ecosystems under climate change. In *Oxford research encyclopedia of climate science*. Oxford University Press. DOI: 10.1093/acrefore/9780190228620.013.764

Pauloi, L., & Salim, R. M. (2025). Exploring the antimicrobial potential from various parts of *Ziziphus mauritiana* Lam. *Scientific Reports*, *15*(1), 36762. DOI: 10.1038/s41598-025-20756-6 PMID: 41120673

Paulsen, J., & Körner, C. (2014). A climate-based model to predict potential treeline position around the globe. *Alpine Botany*, *124*(1), 1–12. DOI: 10.1007/s00035-014-0124-0

Pe'er, G., Dicks, L. V., Visconti, P., Arlettaz, R., Báldi, A., Benton, T. G., & Hartig, F. (2014). EU agricultural reform fails on biodiversity. *Science*, *344*(6188), 1090–1092. DOI: 10.1126/science.1253425 PMID: 24904142

Pe'Er, G., Finn, J. A., Díaz, M., Birkenstock, M., Lakner, S., Röder, N., & Concepción, E. D. (2022). How can the European Common Agricultural Policy help halt biodiversity loss? Recommendations by over 300 experts. *Conservation Letters*, *15*(6), e12901. DOI: 10.1111/conl.12901

Pecchi, M., Marchi, M., Moriondo, M., Forzieri, G., Ammoniaci, M., Bernetti, I., Bindi, M., & Chirici, G. (2020). Potential impact of climate change on the forest coverage and the spatial distribution of 19 key forest tree species in Italy under RCP4.5 IPCC trajectory for 2050s. *Forests*, *11*(9), 934. DOI: 10.3390/f11090934

Pence, V. C., Ballesteros, D., Walters, C., Reed, B. M., Philpott, M., Dixon, K. W., Pritchard, H. W., Culley, T. M., & Vanhove, A.-C. (2020). Cryobiotechnologies: Tools for expanding long-term ex situ conservation to all plant species. *Biological Conservation*, *250*, 108736. DOI: 10.1016/j.biocon.2020.108736

Pereira, H. M., Leadley, P. W., Proença, V., Alkemade, R., Scharlemann, J. P. W., Fernandez-Manjarrés, J. F., Araújo, M. B., Balvanera, P., Biggs, R., Cheung, W. W. L., Chini, L., Cooper, H. D., Gilman, E. L., Guénette, S., Hurtt, G. C., Huntington, H. P., Mace, G. M., Oberdorff, T., Revenga, C., & Walpole, M. (2010). Scenarios for global biodiversity in the 21st century. *Science*, *330*(6010), 1496–1501. DOI: 10.1126/science.1196624 PMID: 20978282

Pereira, H. M., Navarro, L. M., & Martins, I. S. (2012). Global biodiversity change: The bad, the good, and the unknown. *Annual Review of Environment and Resources*, *37*(1), 25–50. DOI: 10.1146/annurev-environ-042911-093511

Pérez-Escobar, O. A., Zizka, A., Bermúdez, M. A., Meseguer, A. S., Condamine, F. L., Hoorn, C.,…Boschman, L. M. J. T. i. P. S. (2022). The Andes through time: evolution and distribution of Andean floras. *27*(4), 364-378.

Perrigo, A., Hoorn, C., & Antonelli, A. (2020). Why mountains matter for biodiversity. *Journal of Biogeography*, *47*(2), 315–325. DOI: 10.1111/jbi.13731

Peszek, Ł., Kawecka, B., & Robinson, C. T. (2022). Long-term response of diatoms in high-elevation streams influenced by rock glaciers. *Ecological Indicators*, *144*, 109515. DOI: 10.1016/j.ecolind.2022.109515

Petelka, J., Bonari, G., Säumel, I., Plagg, B., & Zerbe, S. (2022). Conservation with local people: Medicinal plants as cultural keystone species in the Southern Alps. *Ecology and Society*, *27*(4), art14. DOI: 10.5751/ES-13510-270414

Petelka, J., Plagg, B., Säumel, I., & Zerbe, S. (2020). Traditional medicinal plants in South Tyrol (northern Italy, southern Alps): Biodiversity and use. *Journal of Ethnobiology and Ethnomedicine*, *16*(1), 74. DOI: 10.1186/s13002-020-00419-8 PMID: 33243238

Petrovic, J., Stanojkovic, A., Comic, L., & Curcic, S. (2004). Antibacterial activity of *Cichorium intybus*. *Fitoterapia*, *75*(7-8), 737–739. DOI: 10.1016/j.fitote.2004.05.001 PMID: 15567253

Pettis, J. S., Rose, R., & Chaimanee, V. (2017). Chemical and cultural control of Tropilaelaps mercedesae mites in honeybee (Apis mellifera) colonies in Northern Thailand. *PLoS One*, *12*(11), e0188063.

PhenoForest. (2026). *PhenoForest database: Phenological trends of forest tree species in Switzerland* [Data set]. Swiss Federal Institute for Forest, Snow and Landscape Research (WSL). Retrieved January 15, 2026.

Phiri, B. J., Fèvre, D., & Hidano, A. (2022). Uptrend in global managed honey bee colonies and production based on a six-decade viewpoint, 1961–2017. *Scientific Reports*, *12*(1), 21298. DOI: 10.1038/s41598-022-25290-3 PMID: 36494404

Phuyal, N., Jha, P. K., Raturi, P. P., & Rajbhandary, S. (2019). *Zanthoxylum armatum* DC.: Current knowledge, gaps and opportunities in Nepal. *Journal of Ethnopharmacology*, *229*, 326–341. DOI: 10.1016/j.jep.2018.08.010 PMID: 30166217

Piccinelli, S., & Cannone, N. (2025). Divergent Responses of Alpine Rock Glaciers to Climate Change: A Review of Ecological and Abiotic Dynamics. *Permafrost and Periglacial Processes*, *36*(3), 438–450. DOI: 10.1002/ppp.2278

Pieroni, A., & Giusti, M. E. (2009). Alpine ethnobotany in Italy: Traditional knowledge of gastronomic and medicinal plants among the Occitans of the upper Varaita valley, Piedmont. *Journal of Ethnobiology and Ethnomedicine*, *5*(1), 32. DOI: 10.1186/1746-4269-5-32 PMID: 19895681

Pieroni, A., Giusti, M. E., & Quave, C. L. (2011). Cross-cultural ethnobiology in the Western Balkans: Medical ethnobotany and ethnozoology among Albanians and Serbs in the Pešter Plateau, Sandžak, South-Western Serbia. *Human Ecology: an Interdisciplinary Journal, 39*(3), 333–349. DOI: 10.1007/s10745-011-9401-3

Pieroni, A., Nedelcheva, A., Hajdari, A., Mustafa, B., Scaltriti, B., Cianfaglione, K., & Quave, C. L. (2014). Local knowledge on plants and domestic remedies in the mountain villages of Peshkopia (Eastern Albania). *Journal of Mountain Science, 11*(1), 180–193. DOI: 10.1007/s11629-013-2651-3

Pieroni, A., Pawera, L., & Shah, G. M. (2016). Gastronomic ethnobiology. In *Introduction to ethnobiology* (pp. 53–62). Springer. DOI: 10.1007/978-3-319-28155-1_9

Pimm, S. L., Russell, G. J., Gittleman, J. L., & Brooks, T. M. (1995). The future of biodiversity. *Science, 269*(5222), 347–350. DOI: 10.1126/science.269.5222.347 PMID: 17841251

Pizzi, A. (2021). Tannins medical/pharmacological and related applications: A critical review. *Sustainable Chemistry and Pharmacy, 22*, 100481. DOI: 10.1016/j.scp.2021.100481

Politeo, O., Bektašević, M., Carev, I., Jurin, M., & Roje, M. (2018). Phytochemical composition, antioxidant potential and cholinesterase inhibition potential of extracts from *Mentha pulegium* L. *Chemistry & Biodiversity, 15*(12), e1800374. DOI: 10.1002/cbdv.201800374 PMID: 30253061

Popovici, V., Bucur, L., Calcan, S. I., Cucolea, E. I., Costache, T., Rambu, D., Schröder, V., Gîrd, C. E., Gherghel, D., Vochita, G., Caraiane, A., & Badea, V. (2021). Elemental analysis and *in vitro* evaluation of antibacterial and antifungal activities of *Usnea barbata* (L.) Weber ex FH Wigg from Călimani Mountains, Romania. *Plants, 11*(1), 32. DOI: 10.3390/plants11010032 PMID: 35009036

Popović, Z., Vidaković, V., Mijalković, T., & Krstić-Milošević, D. (2023). Population-Related Variability in Qualitative and Quantitative Secondary Metabolite Profile of *Gentianella austriaca* (A. & J. Kern.) Holub. *Plants, 12*(13), 2434. DOI: 10.3390/plants12132434 PMID: 37446995

Posch, B. C., Amoanimaa-Dede, H., Aparecido, L. M., Atkin, O. K., Bison, N. N., Blonder, B. W., Coast, O., Doughty, C. E., Guo, J. S., & Van Haren, J. (2025). High-temperature acclimation of photosystem II in land plants. *The New Phytologist*. Advance online publication. PMID: 41178112

Prabhakar, S. (2021). A succinct review and analysis of drivers and impacts of agricultural land transformations in Asia. *Land Use Policy, 102*, 105238. DOI: 10.1016/j.landusepol.2020.105238

Pradhan, S., Mohanty, J., & Kumar, S. (2025). Ethnomedicinal Plants Used by Santhal Community of India. *Annals of Agri Bio Research, 30*(1), 86–97. DOI: 10.53941/agrbio.2025.1000011

Prakash, O., Baskaran, R., & Kudachikar, V. (2019). Characterization, quantification of free, esterified and bound phenolics in Kainth (*Pyrus pashia* Buch.-Ham. Ex D. Don) fruit pulp by UPLC-ESI-HRMS/MS and evaluation of their antioxidant activity. *Food Chemistry, 299*, 125114. DOI: 10.1016/j.foodchem.2019.125114 PMID: 31326758

Prevéy, J. S., Rixen, C., Rüger, N., Høye, T. T., Bjorkman, A. D., Myers-Smith, I. H., & Chisholm, C. L. (2019). Warming shortens flowering seasons of tundra plant communities. *Nature Ecology & Evolution, 3*(1), 45–52. DOI: 10.1038/s41559-018-0745-6 PMID: 30532048

Previšić, B. (2025). Cultures of the Alps: The intersection of sectors as the main tool for future. *Archalp, 13*(2024). DOI: 10.30682/aa2413c

Protti, M., Brighenti, V., Battaglia, M. R., Anceschi, L., Pellati, F., & Mercolini, L. (2019). Cannabinoids from *Cannabis sativa* L.: A new tool based on HPLC–DAD–MS/MS for a rational use in medicinal chemistry. *ACS Medicinal Chemistry Letters, 10*(4), 539–544. DOI: 10.1021/acsmedchemlett.8b00571 PMID: 30996793

Punt, W., Hoen, P. P., Blackmore, S., Nilsson, S., Le Thomas, A. J. R. o. p., & palynology. (2007). Glossary of pollen and spore terminology. *143*(1-2), 1-81.

Qaderi, M. M., Martel, A. B., & Strugnell, C. A. (2023). Environmental factors regulate plant secondary metabolites. *Plants, 12*(3), 447. DOI: 10.3390/plants12030447 PMID: 36771531

Qadir, A. (2025) VULNERABILITY CHALLENGES TO HIGH ALTITUDE WETLANDS AND THEIR FEEDING GLACIERS IN PAKISTAN REGARDING CLIMATE CHANGE.

Qadir, S. A., Awlqadr, F. H., Qadir, M. H., Qadir, A. M., Faraj, A. M., Salih, S. A., Saeed, M. N., Arab, K., & Hosseini, S. M. N. (2025). *Adiantum capillus-veneris*: A Comprehensive Review of Its Medicinal Properties, Bioactive Compounds, and Advanced Extraction Techniques. *Food Science & Nutrition, 13*(10), e71118. DOI: 10.1002/fsn3.71118 PMID: 41132578

Qazi, A. W., Saqib, Z., Zaman-ul-Haq, M. J. P. o. t. P. A. o. S. B. L., & Sciences, E. (2023). Diversity and distribution of endemic flora in Pakistan. *60*(2), 165-179.

Qazi, A. W., Haq, M. Z. U., Tariq, A., & Shitu, K. (2025). Prioritizing threats to rare and endemic species: Expertise to insights. *Sustainable Futures : An Applied Journal of Technology, Environment and Society*, *10*, 101430. DOI: 10.1016/j.sftr.2025.101430

Qazi, A. W., & Saqib, Z., Zaman-ul-Haq, M., Gardezi, S. M. H., Khan, A. M., Khan, I.,...Ahmed, I. (2023). Modelling impacts of climate change on habitat suitability of three endemic plant species in Pakistan. *Management*, *16*, 17.

Qian, D., Fan, B., Lan, Y., Si, M., Li, Q., & Guo, X. (2023). Ecosystem service relationships, drivers, and regulation strategies in a degraded alpine shrub meadow on the northeastern Qinghai-Tibetan plateau. *Diversity*, *15*(5), 596. DOI: 10.3390/d15050596

Qian, D., Li, Q., Guo, X., Fan, B., Lan, Y., Si, M., & Cao, G. (2023). Ecosystem services relationship characteristics of the degraded alpine shrub meadow on the Qinghai-Tibetan Plateau. *Ecology and Evolution*, *13*(7), e10351. DOI: 10.1002/ece3.10351 PMID: 37492459

Qiu, J., Queiroz, C., Bennett, E. M., Cord, A. F., Crouzat, E., Lavorel, S., Maes, J., Meacham, M., Norström, A. V., Peterson, G. D., Seppelt, R., & Turner, M. G. (2021). Land-use intensity mediates ecosystem service tradeoffs across regional social-ecological systems. *Ecosystems and People (Abingdon, England)*, *17*(1), 264–278. DOI: 10.1080/26395916.2021.1925743

Qu, R., He, L., He, Z., Wang, B., Lyu, P., Wang, J., Kang, G., & Bai, W. (2022). A study of carbon stock changes in the Alpine Grassland Ecosystem of Zoigê, China, 2000–2020. *Land (Basel)*, *11*(8), 1232. DOI: 10.3390/land11081232

Rahbek, C., Borregaard, M. K., Antonelli, A., Colwell, R. K., Holt, B. G., Nogues-Bravo, D., Rasmussen, C. M. Ø., Richardson, K., Rosing, M. T., Whittaker, R. J., & Fjeldså, J. (2019). Building mountain biodiversity: Geological and evolutionary processes. *Science*, *365*(6458), 1114–1119. DOI: 10.1126/science.aax0151 PMID: 31515384

Rahbek, C., Borregaard, M. K., Colwell, R. K., Dalsgaard, B., Holt, B. G., Morueta-Holme, N., Nogues-Bravo, D., Whittaker, R. J., & Fjeldså, J. (2019). Humboldt's enigma: What causes global patterns of mountain biodiversity? *Science*, *365*(6458), 1108–1113. DOI: 10.1126/science.aax0149 PMID: 31515383

Rahmani, S., Naraki, K., Roohbakhsh, A., Hayes, A. W., & Karimi, G. (2023). The protective effects of rutin on the liver, kidneys, and heart by counteracting organ toxicity caused by synthetic and natural compounds. *Food Science & Nutrition, 11*(1), 39–56. DOI: 10.1002/fsn3.3041 PMID: 36655104

Raina, S., Sheikh, Z. N., Bakshi, P., Alamri, S., Siddiqui, M. H., Islam Rather, K. U., & Ashraf, K. (2025). Functional evaluation of cis-vaccenic acid in *Cordia dichotoma* with tissue-specific biochemical insights. *Scientific Reports, 15*(1), 37943. DOI: 10.1038/s41598-025-17906-1 PMID: 41168161

Rajasekaran, S., Rajasekar, N., & Sivanantham, A. (2021). Therapeutic potential of plant-derived tannins in non-malignant respiratory diseases. *The Journal of Nutritional Biochemistry, 94*, 108632. DOI: 10.1016/j.jnutbio.2021.108632 PMID: 33794331

Raj, N. D., & Singh, D. (2022). A critical appraisal on ferulic acid: Biological profile, biopharmaceutical challenges and nano formulations. *Health Sciences Review (Oxford, England), 5*, 100063. DOI: 10.1016/j.hsr.2022.100063

Ramana, K. V., Singhal, S. S., & Reddy, A. B. (2014). Therapeutic potential of natural pharmacological agents in the treatment of human diseases. *BioMed Research International, 573452*, 1–4. Advance online publication. DOI: 10.1155/2014/573452 PMID: 25610867

Ramanathan, V., & Carmichael, G. (2008). Global and regional climate changes due to black carbon. *Nature Geoscience, 1*(4), 221–227. DOI: 10.1038/ngeo156

Ramankutty, N., Gibbs, H. K., Achard, F., Defries, R., Foley, J. A., & Houghton, R. (2007). Challenges to estimating carbon emissions from tropical deforestation. *Global Change Biology, 13*(1), 51–66. DOI: 10.1111/j.1365-2486.2006.01272.x

Ramar, M. K., Dhayanandamoorthy, Y., Ramachandran, S. S., & Kandasamy, R. (2020). HPLC-ESI-QqQ based standardization, mutagenic and genotoxic potential of methanol extract of *Ziziphus mauritiana* Lam leaves. *Journal of Ethnopharmacology, 246*, 112216. DOI: 10.1016/j.jep.2019.112216 PMID: 31493461

Ramel, C., Rey, P.-L., Fernandes, R., Vincent, C., Cardoso, A. R., Broennimann, O., Pellissier, L., Pradervand, J.-N., Ursenbacher, S., Schmidt, B. R., & Guisan, A. (2020). Integrating ecosystem services within spatial biodiversity conservation prioritization in the Alps. *Ecosystem Services, 45*, 101186. DOI: 10.1016/j.ecoser.2020.101186

Rammig, S. P. D. A. (2023). *Local and regional responses of alpine flora to climate change induced warming*.

Ramskogler, K., Lepesant, L., & Tasser, E. (2025). Understanding the long-term dynamics of vegetation since 1953 in high-mountain regions. *Journal of Ecology, 113*(2), 485–503. DOI: 10.1111/1365-2745.14472

Rana, S. K., Rawal, R. S., Dangwal, B., Bhatt, I. D., & Price, T. D. (2021). 200 years of research on Himalayan biodiversity: Trends, gaps, and policy implications. *Frontiers in Ecology and Evolution, 8*, 603422. DOI: 10.3389/fevo.2020.603422

Rana, S., Rahman, S., Sana, S., Biswas, T. K., Hashem, A. K. M., Parvin, S., & Mazumder, K. (2020). Anticancer potential of *Chenopodium album* leaf extract against Ehrlich ascites carcinoma cells in Swiss albino mice. *Future Journal of Pharmaceutical Sciences, 6*(1), 65. DOI: 10.1186/s43094-020-00080-8

Rao, P. V., Nallappan, D., Madhavi, K., Rahman, S., & Wei, J. (2016). L., & Gan, S. H. (2016). Phytochemicals and biogenic metallic nanoparticles as anticancer agents. *Oxidative Medicine and Cellular Longevity, 2016*(1), 3685671. DOI: 10.1155/2016/3685671 PMID: 27057273

Rashid, N., Gbedomon, R. C., Ahmad, M., Salako, V. K., Zafar, M., & Malik, K. (2018). Traditional knowledge on herbal drinks among indigenous communities in Azad Jammu and Kashmir, Pakistan. *Journal of Ethnobiology and Ethnomedicine, 14*(1), 16. DOI: 10.1186/s13002-018-0217-8 PMID: 29467005

Rashid, S., Wajid, R., Supe Tulcan, R., & Huang, H. (2022). Use, exposure, and environmental impacts of pesticides in Pakistan: A critical review. *Environmental Science and Pollution Research International, 29*(29), 43675–43689. Advance online publication. DOI: 10.1007/s11356-022-20164-7 PMID: 35435556

Rashid, W., Asim, S., Rashid, S., Rafique, S., Aziz, R. S., Iftikhar, S., Rashid, M., & Arfat, Y. (2023). Antioxidant and anti-mutagenic potential of mint (*Mentha arvensis*) and its chemical characterization by HPLC. *Pakistan Journal of Medical & Health Sciences, 17*(04), 514–514. DOI: 10.53350/pjmhs2023174514

Rather, L. J., & Mohammad, F. (2015). *Acacia nilotica* (L.): A review of its traditional uses, phytochemistry, and pharmacology. *Sustainable Chemistry and Pharmacy, 2*, 12–30. DOI: 10.1016/j.scp.2015.08.002

Rather, Z. A., Ahmad, R., & Khuroo, A. A. (2022). Ensemble modelling enables identification of suitable sites for habitat restoration of threatened biodiversity under climate change: A case study of *Himalayan Trillium. Ecological Engineering, 176*, 106534. DOI: 10.1016/j.ecoleng.2021.106534

Rathore, P., Roy, A., & Karnatak, H. (2022). Predicting the future of species assemblages under climate and land use land cover changes in Himalaya: A geo-spatial modelling approach. *Climate Change Ecology, 3*, 100048. DOI: 10.1016/j.ecochg.2022.100048

Ravetto Enri, S., Falla, N. M., Demasi, S., Bianchi, D. M., Squadrone, S., Lombardi, G., & Scariot, V. (2024). Exploring the potential of wild leafy vegetables widespread in European Alps as functional food. *Journal of Agriculture and Food Research, 18*, 101368. DOI: 10.1016/j.jafr.2024.101368

Ravi, S., Duraisamy, P., Krishnan, M., Martin, L. C., Manikandan, B., & Ramar, M. (2023). Sitosterol-rich *Digera muricata* against 7-ketocholesterol and lipopoly-saccharide-mediated atherogenic responses by modulating NF-KB/iNOS signalling pathway in macrophages. *3 Biotech, 13*(10), 331. DOI: 10.1007/s13205-023-03741-6

Rawat, P., & Singh, M. (2024). Deciphering abiotic and biotic stress effects on secondary metabolites of medicinal plants. In *Forests and Climate Change: Biological Perspectives on Impact, Adaptation, and Mitigation Strategies* (pp. 19-41). Springer. DOI: 10.1007/978-981-97-3905-9_2

Reddy, P., & Lakshmikumaran, M. (2015). Protecting traditional knowledge related to biological resources: Is scientific research going to become more bureaucratized? *Cold Spring Harbor Perspectives in Medicine, 5*(10), a020974. DOI: 10.1101/cshperspect.a020974 PMID: 26101205

Redmond, L. E. (2018). Alpine limnology of the Rocky Mountains of Canada and the USA in the context of environmental change. *Environmental Reviews, 26*(3), 231–238. DOI: 10.1139/er-2017-0046

Reed, M. S. (2008). Stakeholder participation for environmental management: A literature review. *Biological Conservation, 141*(10), 2417–2431. DOI: 10.1016/j.biocon.2008.07.014

Rehman, A., Rehman, A., & Ahmad, I. (2015). Antibacterial, antifungal, and insecticidal potentials of *Oxalis corniculata* and its isolated compounds. *International Journal of Analytical Chemistry, 2015*(1), 842468. DOI: 10.1155/2015/842468 PMID: 25873973

Reyes-García, V. (2010). The relevance of traditional knowledge systems for ethnopharmacological research: Theoretical and methodological contributions. *Journal of Ethnobiology and Ethnomedicine, 6*(1), 32. DOI: 10.1186/1746-4269-6-32 PMID: 21083913

Reyes-García, V., Marti, N., McDade, T., Tanner, S., & Vadez, V. (2007). Concepts and methods in studies measuring individual ethnobotanical knowledge. *Journal of Ethnobiology, 27*(2), 182–203. DOI: 10.2993/0278-0771_2007_27_182_camism_2.0.co_2

Rezaieseresht, H., Shobeiri, S. S., & Kaskani, A. (2020). *Chenopodium botrys* essential oil as a source of sesquiterpenes to induce apoptosis and G1 cell cycle arrest in cervical cancer cells. *Iranian Journal of Pharmaceutical Research : IJPR, 19*(2), 341. DOI: 10.22037/ijpr.2019.1100671 PMID: 33224241

Ribeiro, J., Silva, V., Igrejas, G., Barros, L., Heleno, S. A., Reis, F. S., & Poeta, P. (2025). Phenolic Compounds from *Pyrus communis* Residues: Mechanisms of Antibacterial Action and Therapeutic Applications. *Antibiotics (Basel, Switzerland), 14*(3), 280. DOI: 10.3390/antibiotics14030280 PMID: 40149091

Rindt, O., Rosinger, C., Bonkowski, M., Rixen, C., Brüggemann, N., Urich, T., & Fiore-Donno, A. M. (2023). Biogeochemical dynamics during snowmelt and in summer in the Alps. *Biogeochemistry, 162*(2), 257–266. DOI: 10.1007/s10533-022-01005-8

Rivera, A., Movalia, S., & Rutkowski, E. (2024). *Global greenhouse gas emissions: 1990–2022 and preliminary 2023 estimates*. Rhodium Group.

Rixen, C., Høye, T. T., Macek, P., Aerts, R., Alatalo, J. M., Anderson, J. T., Arnold, P. A., Barrio, I. C., Bjerke, J. W., Björkman, M. P., Blok, D., Blume-Werry, G., Boike, J., Bokhorst, S., Carbognani, M., Christiansen, C. T., Convey, P., Cooper, E. J., Cornelissen, J. H. C., & Zong, S. (2022). Winters are changing: Snow effects on Arctic and alpine tundra ecosystems. *Arctic Science, 8*(3), 572–608. DOI: 10.1139/as-2020-0058

Rjeibi, I., Ben Saad, A., Sdayria, J., Feriani, A., Ncib, S., Allagui, M. S., Hfaiedh, N., & Souid, S. (2019). HPLC–DAD identification of polyphenols from ethyl acetate extract of *Amaranthus spinosus* leaves and determination of their antioxidant and antinociceptive effects. *Inflammopharmacology, 27*(5), 975–984. DOI: 10.1007/s10787-018-0482-0 PMID: 29679184

Robinson, M. M., & Zhang, X. (2011). The world medicines situation 2011, traditional medicines: Global situation, issues and challenges. *World Health Organization, Geneva, 31*, 1-2.

Robinson, C. T., & Oertli, B. (2009). Long-term biomonitoring of alpine waters in the Swiss National Park. *Eco.Mont, 1*, 23–34. DOI: 10.1553/eco.mont1s23

Rocha, M. I., Gonçalves, M. J., Cavaleiro, C., Cruz, M. T., Pereira, C., Moreira, P., Salgueiro, L., & Figueirinha, A. (2021). Chemical characterization and bioactive potential of Artemisia campestris L. subsp. maritima (DC) Arcang. essential oil and hydrodistillation residual water. *Journal of Ethnopharmacology*, *276*, 114146. DOI: 10.1016/j.jep.2021.114146 PMID: 33932510

Rogora, M., Frate, L., Carranza, M. L., Freppaz, M., Stanisci, A., Bertani, I., Bottarin, R., Brambilla, A., Canullo, R., Carbognani, M., Cerrato, C., Chelli, S., Cremonese, E., Cutini, M., Di Musciano, M., Erschbamer, B., Godone, D., Iocchi, M., Isabellon, M., & Matteucci, G. (2018). Assessment of climate change effects on mountain ecosystems through a cross-site analysis in the Alps and Apennines. *The Science of the Total Environment*, *624*, 1429–1442. DOI: 10.1016/j.scitotenv.2017.12.155 PMID: 29929254

Roland, J., & Matter, S. F. (2007). Encroaching forests decouple alpine butterfly population dynamics. *Proceedings of the National Academy of Sciences of the United States of America*, *104*(34), 13702–13704. DOI: 10.1073/pnas.0705511104 PMID: 17699630

Romeo, R., Manuelli, S., Geringer, M., & Barchiesi, V. (2021). *Mountain farming systems–Seeds for the future: Sustainable agricultural practices for resilient mountain livelihoods*. Food & Agriculture Org.

Rónavári, A., Igaz, N., Adamecz, D. I., Szerencsés, B., Molnar, C., Kónya, Z., Pfeiffer, I., & Kiricsi, M. (2021). Green silver and gold nanoparticles: Biological synthesis approaches and potentials for biomedical applications. *Molecules (Basel, Switzerland)*, *26*(4), 844. DOI: 10.3390/molecules26040844 PMID: 33562781

Rosbakh, S., Margreiter, V., & Jelcic, B. (2020). Seedlings of alpine species do not have better frost-tolerance than their lowland counterparts. *Alpine Botany*, *130*(2), 179–185. DOI: 10.1007/s00035-020-00237-4

Roseo, F., Celada, C., & Brambilla, M. (2025). Ski resorts threaten climate refugia for high-elevation biodiversity under current and future conditions in the Alps. *Biological Conservation*, *301*, 110890. DOI: 10.1016/j.biocon.2024.110890

Rose-Person, A., Spasojevic, M. J., Forrester, C., Bowman, W. D., Suding, K. N., Oldfather, M. F., & Rafferty, N. E. (2024). Early snowmelt advances flowering phenology and disrupts the drivers of pollinator visitation in an alpine ecosystem. *Alpine Botany*, *134*(2), 141–150. DOI: 10.1007/s00035-024-00315-x

Ross, I. A. (2005). Medicinal plants of the world, volume 3: Chemical constituents, traditional and modern medicinal uses: Springer.

Ross, A., Sherman, K. P., Snodgrass, J. G., Delcore, H. D., & Sherman, R. (2016). *Indigenous peoples and the collaborative stewardship of nature: knowledge binds and institutional conflicts*. Routledge. DOI: 10.4324/9781315426617

Rousk, J., & Bengtson, P. (2014). Microbial regulation of global biogeochemical cycles. *Frontiers in Microbiology, 5*. Advance online publication. DOI: 10.3389/fmicb.2014.00103 PMID: 24672519

Rowland, J. A., Walsh, J. C., Beitzel, M., Brawata, R., Brown, D., Chalmers, L., Evans, L., Eyles, K., Gibbs, R., Grover, S., Grundy, S., Harris, R. M. B., Haywood, S., Hilton, M., Hope, G., Keaney, B., Keatley, M., Keith, D. A., Lawrence, R., & Moore, J. L. (2023). Setting research priorities for effective management of a threatened ecosystem: Australian alpine and subalpine peatland. *Conservation Science and Practice, 5*(3), e12891. DOI: 10.1111/csp2.12891

Roychoudhury, S., Das, D., Das, S., Jha, N. K., Pal, M., Kolesarova, A., Kesari, K. K., Kalita, J. C., & Slama, P. (2022). Clinical Potential of Himalayan Herb Bergenia ligulata: An Evidence-Based Study. *Molecules (Basel, Switzerland), 27*(20), 7039. DOI: 10.3390/molecules27207039 PMID: 36296631

Roychoudhury, S., Sinha, B., Choudhury, B. P., Jha, N. K., Palit, P., Kundu, S., Mandal, S. C., Kolesarova, A., Yousef, M. I., Ruokolainen, J., Slama, P., & Kesari, K. K. (2021). Scavenging properties of plant-derived natural biomolecule para-coumaric acid in the prevention of oxidative stress-induced diseases. *Antioxidants, 10*(8), 1205. DOI: 10.3390/antiox10081205 PMID: 34439453

Rueda-Uribe, C., Chautá, A., Woodman, T. L., Lasso, E., Pelayo, R. C., Manrique-Garzón, L. M.,...Bocedi, G. J. B. R. (2025). Pollination ecology in the tropical Andes: moving towards a cross-scale approach.

Rueda-Uribe, C., Chautá, A., Woodman, T. L., Lasso, E., Pelayo, R. C., Manrique-Garzón, L. M., & Travis, J. M. (2025). Pollination ecology in the tropical Andes: Moving towards a cross-scale approach. *Biological Reviews of the Cambridge Philosophical Society, 100*(6), 2312–2345.

Ruhí, A., Datry, T., & Sabo, J. L. (2017). Interpreting beta-diversity components over time to conserve metacommunities in highly dynamic ecosystems. *Conservation Biology, 31*(6), 1459–1468. DOI: 10.1111/cobi.12906 PMID: 28188969

Ruiz-Gonzalez, A. (2024). Ion-Selective Electrodes in the Food Industry: Development Trends in the Potentiometric Determination of Ionic Pollutants. *Electrochem, 5*(2), 178–212. DOI: 10.3390/electrochem5020012

Ruiz-Molina, N., Ortega-Bedoya, I., & Arias-Zabala, M. (2019). Protonema Suspension Cultures of *Polytrichum juniperinum* as a Potential Production Platform for Bioactive Compounds. *Journal of Herbs, Spices & Medicinal Plants, 25*(2), 114–127. DOI: 10.1080/10496475.2019.1577321

Rumpf, S. B., Gravey, M., Brönnimann, O., Luoto, M., Cianfrani, C., Mariethoz, G., & Guisan, A. (2022). From white to green: Snow cover loss and increased vegetation productivity in the European Alps. *Science, 376*(6597), 1119–1122. DOI: 10.1126/science.abn6697 PMID: 35653482

Rundel, P., & Millar, C. (2016). Alpine ecosystems. *In: Zavaleta, E.; Mooney, H., eds. Ecosystems of California. Berkeley, California: University of California Press: 613-634. Chapter 29.*, 613-634.

Rundel, P. W., & Millar, C. I. (2016). Alpine ecosystems. In Zavaleta, E., & Mooney, H. (Eds.), *Ecosystems of California* (pp. 613–634). University of California Press.

Saddiqa, A., Yaseen, R., Hassan, M., Aftab, K., & Iqbal, R. J. R. B. d. F. (2025). Bridging the Gap Between Herbs and Pharmaceutics: The Science Behind Herbs' Medicinal Properties and Potential in Modern Drugs. 1-17.

Saffari, A. M., Kevan, P. G., & Atkinson, J. L. (2004). A promising pollen substitute for honey bees. *American Bee Journal, 144*(3), 230–231.

Sah, P., Sharma, S., Latwal, A., & Shaik, R. (2023). Timberline and climate in the Indian western Himalayan region: Changes and impact on timberline elevations. In Shukla, B., Chakraborty, A., & Katiyar, N. K. (Eds.), *Climate change and urban environment sustainability* (pp. 205–225). Springer. DOI: 10.1007/978-981-19-7618-6_12

Said, A., Naeem, N., Siraj, S., Khan, T., Javed, A., Rasheed, H. M., Sajjad, W., Shah, K., & Wahid, F. (2020). Mechanisms underlying the wound healing and tissue regeneration properties of *Chenopodium album. 3 Biotech, 10*(10), 452. DOI: 10.1007/s13205-020-02436-6

Sáinz-Bariáin, M., Zamora-Munoz, C., Soler, J. J., Bonada, N., Sáinz-Cantero, C. E., & Alba-Tercedor, J. (2016). Changes in Mediterranean high mountain Trichoptera communities after a 20-year period. *Aquatic Sciences, 78*(4), 669–682. DOI: 10.1007/s00027-015-0457-9

Saklani, S., Kawra, M., Chandra, S., & Coutinho, H. D. M. (2025). GC-MS Analysis, Phytochemical Profiling and Pharmacological Assessment of *Abies pindrow* from Uttarakhand. *Food and Humanity, 100863*, 100863. Advance online publication. DOI: 10.1016/j.foohum.2025.100863

Saleem, M., Shazmeen, N., Nazir, M., Riaz, N., Zengin, G., Ataullah, H. M., Nisar, F., Mukhtar, M., & Tousif, M. I. (2021). Investigation on the phytochemical composition, antioxidant and enzyme inhibition potential of *Polygonum plebeium* R. Br: A comprehensive approach to disclose new nutraceutical and functional food ingredients. *Chemistry & Biodiversity, 18*(12), e2100706. DOI: 10.1002/cbdv.202100706 PMID: 34636484

Saleem, S., Rather, J. A., Ahmad, S., Ahmed, R., & Hajam, L. A. (2024). Assessment of livestock carrying capacity in the alpine grasslands of the Kashmir Himalayas. *Pastoralism: Research. Policy & Practice, 14*, 13541.

Saleh, I. A., BiBi, A., Bibi, S., Abu-Dieyeh, M., & Al-Ghouti, M. A. (2023). Waste-to-value: Guidelines for the potential applications of *Prosopis juliflora. Bioresource Technology Reports, 24*, 101678. DOI: 10.1016/j.biteb.2023.101678

Samant, S. P. S. (2007). Assessment of plant diversity and prioritization of communities for conservation in Mornaula Reserve Forest. *Applied Ecology and Environmental Research, 5*(2), 123–138. DOI: 10.15666/aeer/0502_123138

Sana, S., Aftab, T., Naeem, M., Jha, P. K., & Prasad, P. (2025). Production of secondary metabolites under challenging environments: Understanding functions and mechanisms of signalling molecules. *Frontiers in Plant Science, 16*, 1569014. DOI: 10.3389/fpls.2025.1569014 PMID: 40860743

Sandvik, S. M., & Odland, A. (2014). Changes in alpine snowbed-wetland vegetation over three decades in northern Norway. *Nordic Journal of Botany, 32*(3), 377–384. DOI: 10.1111/j.1756-1051.2013.00249.x

Saqib, Z., & Sultan, A. (2005). Ethnobotany of palas valley, Pakistan. *Ethnobotanical Leaflets, 2004*(1), 11.

Sarasan, V., Cripps, R., Ramsay, M. M., & Atherton, C. McMICHEN, M., Prendergast, G., & Rowntree, J. K. (2006). Conservation in vitro of threatened plants—progress in the past decade. *In Vitro Cellular & Developmental Biology-Plant, 42*(3), 206-214.

Sax, D. F., Early, R., & Bellemare, J. (2013). Niche syndromes, species extinction risks, and management under climate change. *Trends in Ecology & Evolution, 28*(9), 517–523. DOI: 10.1016/j.tree.2013.05.010 PMID: 23769712

Saxena, H., & Brahman, M. (1996). The Flora of Orissa" 1-4 Orissa Forest Development Corporation Ltd. *Bhubaneswar, India.*

Scherrer, D., & Koerner, C. (2010). Infra-red thermometry of alpine landscapes challenges climatic warming projections. *Global Change Biology, 16*(9), 2602–2613. DOI: 10.1111/j.1365-2486.2009.02122.x

Schickhoff, U., Bobrowski, M., Böhner, J., Bürzle, B., Chaudhary, R. P., Gerlitz, L., & Schwab, N. (2016). Climate change and treeline dynamics in the Himalaya. In *Climate change, glacier response, and vegetation dynamics in the Himalaya: Contributions toward future earth initiatives* (pp. 271–306). Springer. DOI: 10.1007/978-3-319-28977-9_15

Schippmann, U., Leaman, D. J., & Cunningham, A. B. (2002). Impact of cultivation and gathering of medicinal plants on biodiversity: global trends and issues. *Biodiversity and the ecosystem approach in agriculture, forestry and fisheries.*

Schippmann, U. (2001). *Medicinal plants significant trade study*. Bundesamt für Naturschutz.

Schirpke, U., Leitinger, G., Tasser, E., Schermer, M., Steinbacher, M., & Tappeiner, U. (2013). Multiple ecosystem services of a changing Alpine landscape: Past, present and future. *The International Journal of Biodiversity Science, Ecosystem Services & Management*, *9*(2), 123–135. DOI: 10.1080/21513732.2012.751936 PMID: 31491038

Schirpke, U., Timmermann, F., Tappeiner, U., & Tasser, E. (2016). Cultural ecosystem services of mountain regions: Modelling the aesthetic value. *Ecological Indicators*, *69*, 78–90. DOI: 10.1016/j.ecolind.2016.04.001 PMID: 27482152

Schirpke, U., Wang, G., & Padoa-Schioppa, E. (2021). Editorial: Mountain landscapes: Protected areas, ecosystem services, and future challenges. *Ecosystem Services*, *49*, 101302. DOI: 10.1016/j.ecoser.2021.101302

Schmeller, D. S., Urbach, D., Bates, K., Catalan, J., Cogălniceanu, D., Fisher, M. C., Friesen, J., Füreder, L., Gaube, V., Haver, M., Jacobsen, D., Le Roux, G., Lin, Y.-P., Loyau, A., Machate, O., Mayer, A., Palomo, I., Plutzar, C., Sentenac, H., & Ripple, W. J. (2022). Scientists' warning of threats to mountains. *The Science of the Total Environment*, *853*, 158611. DOI: 10.1016/j.scitotenv.2022.158611 PMID: 36087665

Schmithüsen, F. J. (2013). Three hundred years of applied sustainability in forestry. *Working papers/Forest Policy and Forest Economics Department of Forest Sciences. International series, 2013*(1).

Schmitt, M., T., Haensel, M., Kaim, A., Lee, H., Reinermann, S., & Koellner, T. (. (2024). Recreation and its synergies and trade-offs with other ecosystem services of Alpine and pre-Alpine grasslands. *Regional Environmental Change*, *24*(2), 57. DOI: 10.1007/s10113-024-02213-8

Schroeder, L., Robles, V., Jara-Arancio, P., Lapadat, C., Hobbie, S. E., Arroyo, M. T. K., & Cavender-Bares, J. (2024). Drivers of plant diversity, community composition, functional traits, and soil processes along an alpine gradient in the central Chilean Andes. *Ecology and Evolution, 14*(2), e10888. DOI: 10.1002/ece3.10888 PMID: 38343572

Schuldt, A., Wang, Z., Zhou, H., & Assmann, T. (2009). Integrating highly diverse invertebrates into broad-scale analyses of cross-taxon congruence across the Palaearctic. *Ecography, 32*(6), 1019–1030. DOI: 10.1111/j.1600-0587.2009.05973.x

Schunko, C., Lechthaler, S., & Vogl, C. R. (2019). Conceptualising the factors that influence the commercialisation of non-timber forest products: The case of wild plant gathering by organic herb farmers in South Tyrol (Italy). *Sustainability (Basel), 11*(7), 2028. DOI: 10.3390/su11072028

Scott, J. M., Davis, F., Csuti, B., Noss, R., Butterfield, B., Groves, C., & Edwards, T. C.Jr. (1993). Gap analysis: A geographic approach to protection of biological diversity. *Wildlife Monographs*, •••, 3–41.

Sears, C. E. (2024). *Botanical Bar Craft: A Guide to the Art of Apothecary Cocktails and Herbal Tonic Elixirs*. Chelsea Green Publishing.

Seidl, A. (1999). *Population demographics and the conservation status of the uncompahgre fritillary Boloria acrocnema*. Nymphalidae.

Sekar, K. C., Pandey, A., & Giri, L. (2015). Invasive alien plants of Himachal Pradesh, India. *Indian Forester, 141*, 520–527.

Sekar, S., Arumugasamy, S. K., Lee, S., Sekar, S., Sadhasivam, S., Vaithilingam, S., Srinivasan, N., Krishnan, E., Chang, S.-C., & Manikandan, R. (2025). Magnetosome-Based Immunosensors for Foodborne Pathogen Detection. *Chemosensors (Basel, Switzerland), 13*(9), 355. DOI: 10.3390/chemosensors13090355

Selvaraj, P., De Britto, A. J., & Sahayaraj, K. (2005). Phytoecdysone of *Pteridium aquilinum* (L) Kuhn (Dennstaedtiaceae) and its pesticidal property on two major pests. *Archiv für Phytopathologie und Pflanzenschutz, 38*(2), 99–105. DOI: 10.1080/0323540040007517

Semenchuk, P. R., Gillespie, M. A., Rumpf, S. B., Baggesen, N., Elberling, B., & Cooper, E. J. (2016). High Arctic plant phenology is determined by snowmelt patterns but duration of phenological periods is fixed: An example of periodicity. *Environmental Research Letters, 11*(12), 125006. DOI: 10.1088/1748-9326/11/12/125006

Seth, P., & Sebastian, J. (2024). Plants and global warming: Challenges and strategies for a warming world. *Plant Cell Reports, 43*(1), 27. DOI: 10.1007/s00299-023-03083-w PMID: 38163826

Shadid, K. A., Shakya, A. K., Naik, R. R., Jaradat, N., Farah, H. S., Shalan, N., Khalaf, N. A., & Oriquat, G. A. (2021). Phenolic content and antioxidant and antimicrobial activities of *Malva sylvestris* L., *Malva oxyloba* Boiss., *Malva parviflora* L., and *Malva aegyptia* L. leaves extract. *Journal of Chemistry, 2021*(1), 8867400. DOI: 10.1155/2021/8867400

Shahar, B., Dolma, N., Indira, A., & Chongtham, N. (2025). Nutritional and phytochemical characterization of *Allium carolinianum* Redoute, a wild garlic species from cold Desert, Ladakh. *Proceedings of the Indian National Science Academy*, 1-14. DOI: 10.1007/s43538-025-00606-9

Shaheen, S., Harun, N., Ijaz, R., Mukhtar, N., Ashfaq, M., Bibi, F., . . . Khalid, Z. J. S. (2023). Sustainability issues in conservation of traditional medicinal herbs and their associated knowledge: A case study of District Lahore, Punjab, Pakistan. *15*(9), 7343.

Shaheen, H., Aziz, S., Nasar, S., Waheed, M., Manzoor, M., Siddiqui, M. H., Alamri, S., Haq, S. M., & Bussmann, R. W. (2023). Distribution patterns of alpine flora for long-term monitoring of global change along a wide elevational gradient in the Western Himalayas. *Global Ecology and Conservation, 48*, e02702. https://doi.org/https://doi.org/10.1016/j.gecco.2023.e02702. DOI: 10.1016/j.gecco.2023.e02702

Shah, S. B., Sartaj, L., Hussain, S., Ullah, N., Idrees, M., Shaheen, A., Javed, M. S., & Aslam, M. K. (2020a). *In-vitro* evaluation of antimicrobial, antioxidant, alpha-amylase inhibition and cytotoxicity properties of *Cannabis sativa. Advances in Traditional Medicine, 20*(2), 181–187. DOI: 10.1007/s13596-019-00414-9

Shah, Z., Hassan, S., Shaheen, K., Khan, S. A., Gul, T., Anwar, Y., Al-Shaeri, M. A., Khan, M., Khan, R., Haleem, M. A., & Suo, H. (2020b). Synthesis of AgNPs coated with secondary metabolites of *Acacia nilotica*: An efficient antimicrobial and detoxification agent for environmental toxic organic pollutants. *Materials Science and Engineering C, 111*, 110829. DOI: 10.1016/j.msec.2020.110829 PMID: 32279826

Shahzadi, S., Fatima, S., Shafiq, Z., & Janjua, M. R. S. A. (2025). A review on green synthesis of silver nanoparticles (SNPs) using plant extracts: A multifaceted approach in photocatalysis, environmental remediation, and biomedicine. *RSC Advances, 15*(5), 3858–3903. DOI: 10.1039/D4RA07519F PMID: 39917042

Shameem, N., Kamili, A. N., Parray, J. A., Hamid, R., & Bandh, S. A. (2015). Antimicrobial and antioxidant activity of methanol extracts of *Arnebia benthamii* (Wall ex. G. Don) Johnston—A critically endangered medicinal plant of North western Himalaya. *Journal of Analytical Science and Technology*, *6*(1), 36. DOI: 10.1186/s40543-015-0076-z

Sharma, A., Chandel, V., & Roy, S. (2025). Exploring the bioactive properties and diverse uses of *Bauhinia variegata*: A comprehensive review. *Discover Plants*, *2*(1), 86. DOI: 10.1007/s44372-025-00167-7

Sharma, H., & Shrivastav, A. (2022). HPLC Purification of bio-active *Cleome viscosa* and It's Anticancer Activity against Breast Cancer Cell Lines. *Oriental Journal of Chemistry*, *38*(6), 1505–1513. Advance online publication. DOI: 10.13005/ojc/380624

Sharma, M. K. (2024). Exploring the biochemical profiles of medicinal plants cultivated under stressful environmental conditions. *Current Agriculture Research Journal*, *12*(1), 81–103. DOI: 10.12944/CARJ.12.1.07

Sharma, S., Rasal, V. P., Patil, P. A., & Joshi, R. K. (2018). *Mentha arvensis* essential oil suppressed airway changes induced by histamine and ovalbumin in experimental animals. *Natural Product Research*, *32*(4), 468–472. DOI: 10.1080/14786419.2017.1311891 PMID: 28385073

Shazdehahmadi, F., Pournajaf, A., Kazemi, S., & Ghasempour, M. (2023). Determining the antibacterial effect of *Mentha longifolia* essential oil on cariogenic bacteria and its compounds: An *in vitro* study. *Journal of Dentistry*, *24*(1, Suppl), 146. DOI: 10.30476/dentjods.2022.92992.1688 PMID: 37051493

Shen, J., Li, S., Liu, L., Liang, Z., Wang, Y., Wang, H., & Wu, S. (2021). Uncovering the relationships between ecosystem services and social-ecological drivers at different spatial scales in the Beijing-Tianjin-Hebei region. *Journal of Cleaner Production*, *290*, 125193. DOI: 10.1016/j.jclepro.2020.125193

Shinwari, A., Ahmad, M., Alsahli, A., Israr, A., Shinwari, Z., Shinwari, S., Zhang, L., Gafforov, Y., & Enes Yilmaz, M. (2024). Himalayan medicinal plants wealth, threats and conservation for securing the future of biodiversity. *Pakistan Journal of Botany*, *56*(6). Advance online publication. DOI: 10.30848/PJB2024-6(42)

Shrivastava, R., Kushwaha, P., Bhutia, Y. C., & Flora, S. (2016). Oxidative stress following exposure to silver and gold nanoparticles in mice. *Toxicology and Industrial Health*, *32*(8), 1391–1404. DOI: 10.1177/0748233714562623 PMID: 25548373

Sickel, H., Bilger, W., & Ohlson, M. (2012). High levels of α-tocopherol in Norwegian alpine grazing plants. *Journal of Agricultural and Food Chemistry*, *60*(31), 7573–7580. DOI: 10.1021/jf301756j PMID: 22775115

Sieniawska, E., Baj, T., Dudka, J., Gieroba, R., Swiatek, L., Rajtar, B., Glowniak, K., & Polz-Dacewicz, M. (2013a). Cytotoxicity, antioxidant activity and an effect on CYP3A4 and CYP2D6 of *Mutellina purpurea* L. extracts. *Food and Chemical Toxicology*, *52*, 188–192. DOI: 10.1016/j.fct.2012.11.017 PMID: 23182739

Sieniawska, E., Baj, T., Los, R., Skalicka-Wozniak, K., Malm, A., & Glowniak, K. (2013b). Phenolic acids content, antioxidant and antimicrobial activity of *Ligusticum mutellina* L. *Natural Product Research*, *27*(12), 1108–1110. DOI: 10.1080/14786419.2012.698413 PMID: 22724452

Sierra, K., Naranjo, L., Carrillo-Hormaza, L., Franco, G., & Osorio, E. (2022). Spearmint (*Mentha spicata* L.) phytochemical profile: Impact of pre/post-harvest processing and extractive recovery. *Molecules (Basel, Switzerland)*, *27*(7), 2243. DOI: 10.3390/molecules27072243 PMID: 35408640

Singh, C. P., Mohapatra, J., & Dharaiya, N. (2015). Remote sensing of alpine treeline dynamics. *International Journal of Remote Sensing*, *21*(8), 106–113.

Singh, H., Husain, T., Agnihotri, P., Pande, P., & Khatoon, S. (2014). An ethnobotanical study of medicinal plants used in sacred groves of Kumaon Himalaya, Uttarakhand, India. *Journal of Ethnopharmacology*, *154*(1), 98–108. DOI: 10.1016/j.jep.2014.03.026 PMID: 24685588

Singh, M., Pandey, N., Agnihotri, V., Singh, K., & Pandey, A. (2017). Antioxidant, antimicrobial activity and bioactive compounds of *Bergenia ciliata* Sternb.: A valuable medicinal herb of Sikkim Himalaya. *Journal of Traditional and Complementary Medicine*, *7*(2), 152–157. DOI: 10.1016/j.jtcme.2016.04.002 PMID: 28417084

Smita, R., Sangeeta, R., Kumar, S. S., Soumya, S., & Deepak, P. (2012). An ethnobotanical survey of medicinal plants in Semiliguda of Koraput District, Odisha, India. *Research Journal of Recent Sciences*, *5*, 97–107.

Snoussi, M., Noumi, E., Trabelsi, N., Flamini, G., Papetti, A., & De Feo, V. (2015). *Mentha spicata* essential oil: Chemical composition, antioxidant and antibacterial activities against planktonic and biofilm cultures of Vibrio spp. strains. *Molecules (Basel, Switzerland)*, *20*(8), 14402–14424. DOI: 10.3390/molecules200814402 PMID: 26262604

Soemarwoto, R. (2007). Kasepuhan Rice Landrace Diversity, Risk Management and Agricultural. *Modern crises and traditional strategies: local ecological knowledge in island Southeast Asia, 6*, 84.

Soloveva, M. I., & Kuzmina, S. S. (2023). Antioxidant and antimicrobial activity of extracts of *Cetraria islandica* (L.) Ach. and *Cladonia arbuscula* (Wallr.) Flot. *Acta Biologica Sibirica, 9*, 139–146-139–146. DOI: 10.5281/zenodo.7751287

Soni, R. (2024). Green synthesis and characterization of metallic nanoparticles of Aconitum heterophyllum as antifungal agents. *Frontiers in Health Informatics, 13*(7).

Son, S.-H., Kwon, S.-J., Im, J.-H., Kim, S.-K., Kong, D., & Choi, J.-Y. (2021). Aquatic Macrophytes Determine the Spatial Distribution of Invertebrates in a Shallow Reservoir. *Water (Basel), 13*(11), 1455. DOI: 10.3390/w13111455

Sõukand, R., Stryamets, N., Fontefrancesco, M. F., & Pieroni, A. (2020). The importance of tolerating interstices: Babushka markets in Ukraine and Eastern Europe and their role in maintaining local food knowledge and diversity. *Heliyon, 6*(1), e03222. DOI: 10.1016/j.heliyon.2020.e03222 PMID: 32095643

Sousa, D. S. M., Ruiz, L. S., Pereira, A. F. M., Sani, A. A., Zapata, T. B., Monari, G. P. M., Rosa, C. F., Fernandes, A., Rall, V. L. M., & Prado, D. G. (2025). *Plantago major* leaf extract against bacteria and fungi of medical importance. *Brazilian Journal of Pharmaceutical Sciences, 61*, e24143. DOI: 10.1590/s2175-97902025e24143

Spehn, E. M., Rudmann-Maurer, K., & Körner, C. (2011). Mountain biodiversity. *Plant Ecology & Diversity, 4*(4), 301–302. DOI: 10.1080/17550874.2012.698660

Stanisavljević, D. M., Stojičević, S. S., Đorđević, S. M., Zlatković, B. P., Veličković, D. T., Karabegović, I. T., & Lazić, M. L. (2012). Antioxidant activity, the content of total phenols and flavonoids in the ethanol extracts of *Mentha longifolia* (L.) Hudson dried by the use of different techniques. *Chemical Industry & Chemical Engineering Quarterly, 18*(3), 411–420. DOI: 10.2298/CICEQ110919017S

Stanton, M. L., & Galen, C. (1989). Consequences of flower heliotropism for reproduction in an alpine buttercup (Ranunculus adoneus). *Oecologia, 78*(4), 477–485. DOI: 10.1007/BF00378737 PMID: 28312176

Steger, C., Kotlarski, S., Jonas, T., & Schär, C. (2013). Alpine snow cover in a changing climate: A regional climate model perspective. *Climate Dynamics, 41*(3), 735–754. DOI: 10.1007/s00382-012-1545-3

Steinbauer, M. J., Grytnes, J.-A., Jurasinski, G., Kulonen, A., Lenoir, J., Pauli, H., & Barni, E. (2018). Accelerated increase in plant species richness on mountain summits is linked to warming. *Nature*, *556*(7700), 231–234. DOI: 10.1038/s41586-018-0005-6 PMID: 29618821

Steinger, T., Körner, C., & Schmid, B. (1996). Long-term persistence in a changing climate: DNA analysis suggests very old ages of clones of alpine Carex curvula. *Oecologia*, *105*(1), 94–99. DOI: 10.1007/BF00328796 PMID: 28307127

Sterling, E. J., Betley, E., Sigouin, A., Gomez, A., Toomey, A., Cullman, G., & Blair, M. (2017). Assessing the evidence for stakeholder engagement in biodiversity conservation. *Biological Conservation*, *209*, 159–171. DOI: 10.1016/j.biocon.2017.02.008

Stern, N. (2006). Stern Review: The economics of climate change.

Stöcklin, J., Bosshard, A., Klaus, G., Rudmann-Maurer, K., & Fischer, M. (2007). Landnutzung und biologische Vielfalt in den Alpen. *Fakten, Perspektiven, Empfehlungen. Synthese NFP, 48.*

Stoecklin, J., & Armbruster, G. F. (2016). Environmental filtering, not local adaptation of established plants, determines the occurrence of seed-and bulbil-producing Poa alpina in a local flora. *Basic and Applied Ecology*, *17*(7), 586–595. DOI: 10.1016/j.baae.2016.06.004

Stotten, R., Ambrosi, L., Tasser, E., & Leitinger, G. (2021). Social-ecological resilience in remote mountain communities: Toward a novel framework for an interdisciplinary investigation. *Ecology and Society*, *26*(3), art29. DOI: 10.5751/ES-12580-260329

Streb, P., Aubert, S., & Bligny, R. (2003). High temperature effects on light sensitivity in the two high mountain plant species *Soldanella alpina* (L.) and *Rannunculus glacialis* (L.). *Plant Biology*, *5*(04), 432–440. DOI: 10.1055/s-2003-42713

Studer, S., Appenzeller, C., & Defila, C. (2005). Inter-annual variability and decadal trends in alpine spring phenology: A multivariate analysis approach. *Climatic Change*, *73*(3), 395–414. DOI: 10.1007/s10584-005-6886-z

Suárez, E., Encalada, A. C., Chimbolema, S., Jaramillo, R., Hofstede, R., & Riveros-Iregui, D. (2023). On the Use of "Alpine" for High-Elevation Tropical Environments. *Mountain Research and Development*, *43*(1). Advance online publication. DOI: 10.1659/mrd.2022.00024

Sugier, P., Sugier, D., Miazga-Karska, M., Nurzyńska, A., Król, B., Sęczyk, Ł., & Kowalski, R. (2025). Effect of Different *Arnica montana* L. Plant Parts on the Essential Oil Composition, Antimicrobial Activity, and Synergistic Interactions with Antibiotics. *Molecules (Basel, Switzerland)*, *30*(18), 3812. DOI: 10.3390/molecules30183812 PMID: 41011702

Sultana, K., Ahmad, B., Ahmad, Z., Rauf, A., Al-Awthan, Y. S., Bahattab, O., Bawazeer, S., Khattak, A., & Ramadan, M. F. (2022). Effect of *Bistorta affinis* (D. Don) and *Malcolmia cabulica* (Boiss) essential oils on human and foodborne pathogens, insects and termites. *Journal of Pure & Applied Microbiology*, *5*(4), 14–21. DOI: 10.22207/JPAM.16.4.45

Sumner, E. E., & Venn, S. E. (2022). Thermal tolerance and growth responses to in situ soil water reductions among alpine plants. *Plant Ecology & Diversity*, *15*(5-6), 297–308. DOI: 10.1080/17550874.2022.2160674

Sumner, E. E., Williamson, V. G., Gleadow, R. M., Wevill, T., & Venn, S. E. (2022). Acclimation to water stress improves tolerance to heat and freezing in a common alpine grass. *Oecologia*, *199*(4), 831–843. DOI: 10.1007/s00442-022-05245-1 PMID: 35974110

Sumner, E., & Venn, S. (2021). Plant responses to changing water supply and availability in high elevation ecosystems: A quantitative systematic review and meta-analysis. *Land (Basel)*, *10*(11), 1150. DOI: 10.3390/land10111150

Sundararajan, B., & Kumari, B. R. (2017). Novel synthesis of gold nanoparticles using Artemisia vulgaris L. leaf extract and their efficacy of larvicidal activity against dengue fever vector Aedes aegypti L. *Journal of Trace Elements in Medicine and Biology*, *43*, 187–196. DOI: 10.1016/j.jtemb.2017.03.008 PMID: 28341392

Sun, J., Ni, C., & Wang, M. (2023). Analysis of Water Conservation Trends and Drivers in an Alpine Region: A Case Study of the Qilian Mountains. *Remote Sensing (Basel)*, *15*(18), 4611. DOI: 10.3390/rs15184611

Suozzi, R. M. (1995). Dizionario delle erbe medicinali: oltre duecentocinquanta specie vegetali esaminate nei loro effetti curativi e nella loro utilizzazione pratica: Grandi tascabili economici Newton.pp.1-416.

Swathi, K., Suganya, K., Aswathy, R., Angel, C., Sudha, B., Sumathi, S., & Mandal, S. K. (2025). Drugs Derived from High Altitude Plants. In *High Altitude Medicinal Plants: Botany, Conservation and Cultivation* (pp. 235-268). Springer.

Tabassam, S., Anwar, M. A., Gulfraz, M., Arshad, M., Sabitaliyevich, U. Y., Nurmurzayevich, S. B., & Ahmad, M. S. (2019). Bioactivity evaluation and HPLC UV-VIS based quantification of antioxidant secondary metabolites from extract and fractions of *Bistorta amplexicaulis* rhizome. *Cellular and Molecular Biology, 65*(1), 19–26. DOI: 10.14715/cmb/2019.65.1.4 PMID: 30782289

Tabesh, M., Sh, M. E., Etemadi, M., Naddaf, F., Heidari, F., & Alizargar, J. (2022). The antibacterial activity of *Nasturtium officinale* extract on common oral pathogenic bacteria. *Nigerian Journal of Clinical Practice, 25*(9), 1466–1475. DOI: 10.4103/njcp.njcp_1887_21 PMID: 36149206

Tahir, M. F., Khan, M. Z., Attacha, S., Asim, N., Tayyab, M., Ali, A., Militky, J., & Tomková, B. (2022). The comparative performance of phytochemicals, green synthesised silver nanoparticles, and green synthesised copper nanoparticles-loaded textiles to avoid nosocomial infections. *Nanomaterials (Basel, Switzerland), 12*(20), 3629. DOI: 10.3390/nano12203629 PMID: 36296819

Tailang, M. (2014). Ethnomedicinal plants used by some Indian tribe. *World Journal of Pharmaceutical Research, 4*, 549–558.

Tang, R. (2012). *An analysis of traditional ecological knowledge's status and its conservation options.* PhD Thesis. Victoria University of Wellington.pp.1-397.

Tang, R., & Gavin, M. C. (2010). Traditional ecological knowledge informing resource management: Saxoul conservation in inner Mongolia, China. *Society & Natural Resources, 23*(3), 193–206. DOI: 10.1080/08941920903375503

Tang, R., & Gavin, M. C. (2016). A classification of threats to traditional ecological knowledge and conservation responses. *Conservation & Society, 14*(1), 57–70. DOI: 10.4103/0972-4923.182799

Tang, S., Liang, J., Li, O., Shao, N., Jin, Y., Ni, J., Fei, X., & Li, Z. (2023). Morphology-Tailored Hydroxyapatite Nanocarrier for Rhizosphere-Targeted Phosphorus Delivery. *Small, 19*(14), 2206954. DOI: 10.1002/smll.202206954 PMID: 36599675

Tang, Y., Zhao, W., Zhu, G., Tan, Z., Huang, L., Zhang, P., Gao, L., & Rui, Y. (2024). Nano-Pesticides and Fertilizers: Solutions for Global Food Security. *Nanomaterials (Basel, Switzerland), 14*(1), 90. DOI: 10.3390/nano14010090 PMID: 38202545

Tapalsky, D., Petrenev, D., Khramchenkova, O., & Doroshkevich, A. (2017). Antimicrobial and antifungal activity of lichens prevalent in Belarus. *Journal of microbiology, epidemiology and immunobiology, 94*(2), 60-65. DOI: 10.36233/0372-9311-2017-2-60-65

Tappeiner, U., & Tasser, E. I. (2019). *Historical Land Uses and Their Changes in the European Alps*. Oxford Bibliographies. DOI: 10.1093/obo/9780199363445-0114

Tariq, A., Shu, H., Siddiqui, S., Munir, I., Sharifi, A., Li, Q., & Lu, L. (2022). Spatio-temporal analysis of forest fire events in the Margalla Hills, Islamabad, Pakistan using socio-economic and environmental variable data with machine learning methods. *Journal of Forestry Research*, *33*(1), 183–194. DOI: 10.1007/s11676-021-01354-4

Tasser, E., Leitinger, G., Tappeiner, U., & Schirpke, U. (2024). Shaping the European Alps: Trends in landscape patterns, biodiversity and ecosystem services. *Catena*, *235*, 107607. DOI: 10.1016/j.catena.2023.107607

Tello-García, E., Huber, L., Leitinger, G., Peters, A., Newesely, C., Ringler, M.-E., & Tasser, E. (2020). Drought-and heat-induced shifts in vegetation composition impact biomass production and water use of alpine grasslands. *Environmental and Experimental Botany*, *169*, 103921. DOI: 10.1016/j.envexpbot.2019.103921

Testolin, R., Attorre, F., Borchardt, P., Brand, R. F., Bruelheide, H., Chytrý, M., De Sanctis, M., Dolezal, J., Finckh, M., Haider, S., Hemp, A., Jandt, U., Kessler, M., Korolyuk, A. Y., Lenoir, J., Makunina, N., Malanson, G. P., Montesinos-Tubée, D. B., Noroozi, J., & Jiménez-Alfaro, B. (2021). Global patterns and drivers of alpine plant species richness. *Global Ecology and Biogeography*, *30*(6), 1218–1231. DOI: 10.1111/geb.13297

Testolin, R., Attorre, F., & Jiménez-Alfaro, B. (2020). Global distribution and bioclimatic characterization of alpine biomes. *Ecography*, *43*(6), 779–788. DOI: 10.1111/ecog.05012

Testolin, R., Carmona, C. P., Attorre, F., Borchardt, P., Bruelheide, H., Dolezal, J., Finckh, M., Haider, S., Hemp, A., Jandt, U., Korolyuk, A. Y., Lenoir, J., Makunina, N., Malanson, G. P., Mucina, L., Noroozi, J., Nowak, A., Peet, R. K., Peyre, G., & Jiménez-Alfaro, B. (2021). Global functional variation in alpine vegetation. *Journal of Vegetation Science*, *32*(2), e13000. DOI: 10.1111/jvs.13000

Tewari, D., Gupta, P., Bawari, S., Sah, A. N., Barreca, D., Khayatkashani, M., & Khayat Kashani, H. R. (2021a). Himalayan *Ficus palmata* L. Fruit extract showed in vivo central and peripheral analgesic activity involving COX-2 and Mu opioid receptors. *Plants*, *10*(8), 1685. DOI: 10.3390/plants10081685 PMID: 34451731

Tewari, D., Zengin, G., Ak, G., Sinan, K. I., Cziáky, Z., Mishra, S. T., & Jekő, J. (2021b). Phenolic profiling, antioxidants, multivariate, and enzyme inhibitory properties of wild Himalayan Fig (*Ficus palmata* Forssk.): A potential candidate for designing innovative nutraceuticals and related products. *Analytical Letters*, *54*(9), 1439–1456. DOI: 10.1080/00032719.2020.1804395

Thakur, P., Kumar, R., Choudhary, N., Sharma, R., & Chaudhary, A. (2023). Network pharmacology on mechanistic role of *Thymus linearis* Benth. against gastrointestinal and neurological diseases. *Phytomedicine, 121*, 155098. DOI: 10.1016/j.phymed.2023.155098 PMID: 37757710

Thakur, R., Singh, R., Saxena, P., & Mani, A. (2014). Evaluation of antibacterial activity of *Prosopis juliflora* (SW.) DC. leaves. *African Journal of Traditional, Complementary, and Alternative Medicines : AJTCAM, 11*(3), 182–188. DOI: 10.4314/ajtcam.v11i3.26 PMID: 25371581

Thakur, S., Negi, V. S., Dhyani, R., Bhatt, I., & Yadava, A. (2022). Influence of environmental factors on tree species diversity and composition in the Indian western Himalaya. *Forest Ecology and Management, 503*, 119746. DOI: 10.1016/j.foreco.2021.119746

Thakur, S., Negi, V. S., Pathak, R., Dhyani, R., Durgapal, K., & Rawal, R. S. (2020). Indicator based integrated vulnerability assessment of community forests in Indian west Himalaya. *Forest Ecology and Management, 457*, 117674. DOI: 10.1016/j.foreco.2019.117674

Thapliyal, N., Bhojak, P., Mehta, P., & Sekar, K. C. (2024). Threats and Conservation of Plant Diversity in the Himalayan Biodiversity Hotspot. In (pp. 459-511). DOI: 10.1201/9781003455202-18

Theurillat, J. P., & Guisan, A. (2001). Potential impact of climate change on vegetation in the European Alps: A review. *Climatic Change, 50*(1), 77–109. DOI: 10.1023/A:1010632015572

Thorn, C. E., Darmody, R. G., & Dixon, J. C. (2011). Rethinking weathering and pedogenesis in alpine periglacial regions: Some Scandinavian evidence. *Special Publication - Geological Society of London, 354*(1), 183–193. DOI: 10.1144/SP354.11

Thuiller, W., Saillard, A., Abdulhak, S., Augé, V., Birck, C., Bonet, R., Choler, P., Delestrade, A., Kunstler, G., Leccia, M.-F., Lienard, B., Poulenard, J., Valay, J.-G., Bayle, A., Bonfanti, N., Brousset, L., Bizard, L., Calderón-Sanou, I., Dentant, C., & Renaud, J.The Orchamp Consortium. (2024). ORCHAMP: An observation network for monitoring biodiversity and ecosystem functioning across space and time in mountainous regions. *Comptes Rendus Biologies, 347*(G1), 223–247. DOI: 10.5802/crbiol.165 PMID: 39774921

Tian, Y., Liu, Z., Fan, Y., Li, Y., Tao, H., Han, C., Ao, X., & Chen, R. (2025). Patterns and Drivers of Surface Energy Flux in the Alpine Meadow Ecosystem in the Qilian Mountains, Northwest China. *Plants, 14*(2), 155. DOI: 10.3390/plants14020155 PMID: 39861508

Titley, M. A., Snaddon, J. L., & Turner, E. C. (2017). Scientific research on animal biodiversity is systematically biased towards vertebrates and temperate regions. *PLoS One*, *12*(12), e0189577. DOI: 10.1371/journal.pone.0189577 PMID: 29240835

Tittensor, D. P., Walpole, M., Hill, S. L. L., Boyce, D. G., Britten, G. L., Burgess, N. D., Butchart, S. H. M., Leadley, P. W., Regan, E. C., Alkemade, R., Baumung, R., Bellard, C., Bouwman, L., Bowles-Newark, N. J., Chenery, A. M., Cheung, W. W. L., Christensen, V., Cooper, H. D., Crowther, A. R., & Ye, Y. (2014). A mid-term analysis of progress toward international biodiversity targets. *Science*, *346*(6206), 241–244. DOI: 10.1126/science.1257484 PMID: 25278504

Tiwari, P. C., & Joshi, B. (2014). Environmental changes and their impact on rural water, food, livelihood, and health security in Kumaon Himalayas. *Journal of Urban and Regional studies on contemporary India, 1*(1), 1-12.

Tomczyk, M., & Latté, K. P. (2009). Potentilla—A review of its phytochemical and pharmacological profile. *Journal of Ethnopharmacology*, *122*(2), 184–204. DOI: 10.1016/j.jep.2008.12.022 PMID: 19162156

Totland, Ø. (1993). Pollination in alpine Norway: Flowering phenology, insect visitors, and visitation rates in two plant communities. *Canadian Journal of Botany*, *71*(8), 1072–1079. DOI: 10.1139/b93-124

Treben, M. (2018). *La salute dalla farmacia del Signore*. Ennsthaler.

Tribsch, A., & Schönswetter, P. (2003). Patterns of endemism and comparative phylogeography confirm palaeo-environmental evidence for Pleistocene refugia in the Eastern Alps. *Taxon*, *52*(3), 477–497. DOI: 10.2307/3647447

Tsing, A. (2005). *Friction: An anthropology of global connection*. Princeton University Press.

Turan, K., Akcan, M., & Kalfa, O. M. (2021). Preparation, Characterization and Antibacterial Activity of *Malva sylvestris* L. Seed Extract Containing Novel Electrospun PVA Nanofibers. *Gazi University Journal of Science*, *34*(4), 987–998. DOI: 10.35378/gujs.794638

Tvrda, E., Straka, P., Galbavy, D., & Ivanic, P. (2019). Epicatechin provides antioxidant protection to bovine spermatozoa subjected to induced oxidative stress. *Molecules (Basel, Switzerland)*, *24*(18), 3226. DOI: 10.3390/molecules24183226 PMID: 31491847

Tyagi, P. K., Arya, A., Ramniwas, S., & Tyagi, S. (2023). Editorial: Recent trends in nanotechnology in precision and sustainable agriculture. *Frontiers in Plant Science*, *14*, 1256319. DOI: 10.3389/fpls.2023.1256319 PMID: 37636083

Uba, A. I., Zengin, G., Montesano, D., Cakilcioglu, U., Selvi, S., Ulusan, M. D., Caprioli, G., Sagratini, G., Angeloni, S., Jugreet, S., Hasan, M. M., & Mahoomodally, M. F. (2022). Antioxidant and enzyme inhibitory properties, and HPLC–MS/MS profiles of different extracts of *Arabis carduchorum* Boiss.: An endemic plant to Turkey. *Applied Sciences (Basel, Switzerland)*, *12*(13), 6561. DOI: 10.3390/app12136561

Ullah, A., Munir, S., Badshah, S. L., Khan, N., Ghani, L., Poulson, B. G., Emwas, A.-H., & Jaremko, M. (2020). Important flavonoids and their role as a therapeutic agent. *Molecules (Basel, Switzerland)*, *25*(22), 5243. DOI: 10.3390/molecules25225243 PMID: 33187049

Ullah, H., & Badshah, L. (2023). Nutritional and mineral analysis of the ultimate wild food plants of Lotkuh, Chitral, the Eastern Hindukush Pakistan. *Heliyon*, *9*(3), e14449. DOI: 10.1016/j.heliyon.2023.e14449 PMID: 36950657

Ullah, M. A., Gul, F. Z., Khan, T., Bajwa, M. N., Drouet, S., Tungmunnithum, D., Giglioli-Guivarc'h, N., Liu, C., Hano, C., & Abbasi, B. H. (2021). Differential induction of antioxidant and anti-inflammatory phytochemicals in agitated micro-shoot cultures of *Ajuga integrifolia* Buch. Ham. ex D. Don with biotic elicitors. *AMB Express*, *11*(1), 137. DOI: 10.1186/s13568-021-01297-3 PMID: 34661766

Ullah, M., Mehmood, S., Ali, M., Bussmann, R. W., Aldosari, A., Khan, R. A., & Shah, M. A. R. (2019). An ethnopharmacological study of plants used for treatment of diabetes in the Southern and Tribal regions of Khyber Pakhtunkhwa province, Pakistan. *Ethnobotany Research and Applications*, *18*, 1–20. DOI: 10.32859/era.18.8.1-20

Umair, M., Altaf, M., Bussmann, R. W., & Abbasi, A. M. (2019). Ethnomedicinal uses of the local flora in Chenab riverine area, Punjab province Pakistan. *Journal of Ethnobiology and Ethnomedicine*, *15*(1), 7. DOI: 10.1186/s13002-019-0285-4 PMID: 30709360

Usman, M., Hasnain, M., Banaras, S., Akram, M., Abbas, Q., Shah, J. A.,…Khan, M. N. (2022). Potential emerging constraints and management strategies of different honeybee species in Pakistan: A review. *CABI Reviews*(2022).

Valdez, J. (2020). *Arthropods as Vertebrate Predators: A Review of Global Patterns*. BIOLOGY. https://doi.org/DOI: 10.20944/preprints202002.0119.v1

Vandebroek, I., & Balick, M. J. (2012). Globalization and loss of plant knowledge: Challenging the paradigm. *PLoS One*, *7*(5), e37643. DOI: 10.1371/journal.pone.0037643 PMID: 22662184

Vanneste, T., Michelsen, O., Graae, B. J., Kyrkjeeide, M. O., Holien, H., Hassel, K., Lindmo, S., Kapás, R. E., & De Frenne, P. (2017). Impact of climate change on alpine vegetation of mountain summits in Norway. *Ecological Research*, *32*(4), 579–593. DOI: 10.1007/s11284-017-1472-1

Vargas-Chaves, I., Lopez-Oliva, J., & Alarcon-Pena, A. (2025). The Intersections of Intellectual Property and Traditional Ecological Knowledge: A Critical Analysis within the Nagoya Protocol Framework. *Contemporary Readings in Law and Social Justice*, *17*, 267.

Varshney, N., Kumari, D., Janmeda, P., Meena, M., & Singh, D. (2025). *Salvadora oleoides* (Decne.): A systematic review and comprehensive botanical assessment. *Discover Plants*, *2*(1), 140. DOI: 10.1007/s44372-025-00150-2

Vashistha, R. K., Rawat, N., Chaturvedi, A. K., Nautiyal, B., Prasad, P., & Nautiyal, M. (2009). An exploration on the phenology of different growth forms of an alpine expanse of North-West Himalaya, India. *New York Science Journal*, *2*(6), 29–41.

Venditti, A., Serafini, M., Nicoletti, M., & Bianco, A. (2015). Terpenoids of *Linaria alpina* (L.) Mill. from Dolomites, Italy. *Natural Product Research*, *29*(21), 2041–2044. DOI: 10.1080/14786419.2015.1018252 PMID: 25738749

Venn, S. E., & Morgan, J. W. (2007). Phytomass and phenology of three alpine snowpatch species across a natural snowmelt gradient. *Australian Journal of Botany*, *55*(4), 450–456. DOI: 10.1071/BT06003

Verhagen, W., Van Teeffelen, A. J., Baggio Compagnucci, A., Poggio, L., Gimona, A., & Verburg, P. H. (2016). Effects of landscape configuration on mapping ecosystem service capacity: A review of evidence and a case study in Scotland. *Landscape Ecology*, *31*(7), 1457–1479. DOI: 10.1007/s10980-016-0345-2

Verma, P., Dangar, R., Shah, K., Gandhi, D. M., & Suhagia, B. (2011). Pharmacognostical Potential of *Capparis decidua* Edgew. *Journal of Applied Pharmaceutical Science*(Issue), 06-11.

Verrall, B. (2023). *Assessing alpine vegetation dynamics using long-term ecological monitoring amidst rapid climate change*.

Verrall, B., & Pickering, C. M. (2020). Alpine vegetation in the context of climate change: A global review of past research and future directions. *The Science of the Total Environment*, *748*, 141344. DOI: 10.1016/j.scitotenv.2020.141344 PMID: 32814293

Viktória, B., Stenger-Kovacs, C., Török, P., & Török-Krasznai, E. (2021). Are recent protection strategies sufficient for maintaining diverse freshwater benthic diatom assemblages? *Ecological Indicators, 127*, 107782. DOI: 10.1016/j.ecolind.2021.107782

Vincent, C., Fernandes, R. F., Cardoso, A. R., Broennimann, O., Di Cola, V., D'Amen, M., Ursenbacher, S., Schmidt, B. R., Pradervand, J.-N., Pellissier, L., & Guisan, A. (2019). Climate and land-use changes reshuffle politically-weighted priority areas of mountain biodiversity. *Global Ecology and Conservation, 17*, e00589. DOI: 10.1016/j.gecco.2019.e00589

Vishwakarma, R. K., Negi, A., & Negi, D. S. (2022). Anti-oxidant, COX inhibition activity of different extract of bark of *Betula utilis* and molecular docking analysis of its phytochemicals against COX-1 and COX-2 isoenzyme. *Materials Today: Proceedings, 57*, 251–258. DOI: 10.1016/j.matpr.2022.02.497

Vitalini, S., Iriti, M., Puricelli, C., Ciuchi, D., Segale, A., & Fico, G. (2013). Traditional knowledge on medicinal and food plants used in Val San Giacomo (Sondrio, Italy)—An alpine ethnobotanical study. *Journal of Ethnopharmacology, 145*(2), 517–529. DOI: 10.1016/j.jep.2012.11.024 PMID: 23220197

Vitasse, , YUrsenbacher, , SKlein, , GBohnenstengel, , TChittaro, , YDelestrade, , AMonnerat, , CRebetez, , MRixen, , CStrebel, , NSchmidt, , B.RWipf, , SWohlgemuth, , TYoccoz, , N.GLenoir, , J. (2021), Phenological and elevational shifts of plants, animals and fungi under climate change in the European Alps. *Biological Reviews, 96*(5), 1816-1835. https://doi.org/10.1111/brv.12727

Viviroli, D., Archer, D. R., Buytaert, W., Fowler, H. J., Greenwood, G. B., Hamlet, A. F., Huang, Y., Koboltschnig, G., Litaor, M., López-Moreno, J. I., Lorentz, S., Schädler, B., Schreier, H., Schwaiger, K., Vuille, M., & Woods, R. (2011). Climate change and mountain water resources: Overview and recommendations for research, management and policy. *Hydrology and Earth System Sciences, 15*(2), 471–504. DOI: 10.5194/hess-15-471-2011

Vohra, A., & Kaur, H. (2011). Chemical investigation of medicinal plant *Ajuga bracteosa. J Nat Prod Plant Resour, 1*(1), 37–45.

Voisin, C., Choler, P., Vacher, C., Perrier, C., Renaud, J., Rioux, D., & Boucher, F. C. (2025). Deciphering the drivers of speciation in Alpine Noccaea (Brassicaceae) reveal an evolution predominantly shaped by geography. *Molecular Phylogenetics and Evolution, 211*, 108381. DOI: 10.1016/j.ympev.2025.108381 PMID: 40456477

von Büren, R. S., & Hiltbrunner, E. (2022). Low winter temperatures and divergent freezing resistance set the cold range limit of widespread alpine graminoids. *Journal of Biogeography, 49*(8), 1562–1575. DOI: 10.1111/jbi.14455

Von Glasenapp, M., & Thornton, T. F. (2011). Traditional ecological knowledge of Swiss alpine farmers and their resilience to socioecological change. *Human Ecology: an Interdisciplinary Journal, 39*(6), 769–781. DOI: 10.1007/s10745-011-9427-6

Vorkauf, M., Kahmen, A., Körner, C., & Hiltbrunner, E. (2021). Flowering phenology in alpine grassland strongly responds to shifts in snowmelt but weakly to summer drought. *Alpine Botany, 131*(1), 73–88. DOI: 10.1007/s00035-021-00252-z

Vučić, D., Čomić, L., Petković, M., & Stefanović, O. (2009). Antibacterial effects of leaves of *Vaccinium vitis-idaea* L. *Planta Medica, 75*(09), PJ58. DOI: 10.1055/s-0029-1234863

Vysochina, G., & Khramova, E. (2010). Component composition of flavonols and their content in *Aconogonon alpinum* (All.) Schur Growing in the Altay. *Chem Sustain Dev, 18*, 477–483.

Waghchoure-Camphor, E., & Martin, S. (2008). Beekeeping in Pakistan: A bright future in a troubled land. *American Bee Journal, 148*(8), 726–728.

Waheed, M., Haq, S. M., Arshad, F., Bussmann, R. W., Pieroni, A., Mahmoud, E. A., Casini, R., Yessoufou, K., & Elansary, H. O. (2023). Traditional wild food plants gathered by ethnic groups living in semi-arid region of Punjab, Pakistan. *Biology (Basel), 12*(2), 269. DOI: 10.3390/biology12020269 PMID: 36829546

Wakgari, M., & Yigezu, G. (2021). Honeybee keeping constraints and future prospects. *Cogent Food & Agriculture, 7*(1), 1872192. DOI: 10.1080/23311932.2021.1872192

Walker, D., Halfpenny, J. C., Walker, M. D., & Wessman, C. A. (1993). Long-term studies of snow-vegetation interactions. *Bioscience, 43*(5), 287–301. DOI: 10.2307/1312061

Wang, W., Du, J., He, Z., Miao, C., Wu, J., Ma, D., & Zhao, P. (2024). Pollinator peaking earlier than flowering is more detrimental to plant fecundity. *Science of the Total Environment, 917*, 170458.[1]

Wang, J., Zhao, H., Guo, Q., & Ding, H. (2022). Identification and antibacterial activity of *Thamnolia vermicularis* and *Thamnolia subuliformis*. *Journal of Microbiological Methods, 203*, 106628. DOI: 10.1016/j.mimet.2022.106628 PMID: 36442696

Wang, L., Pan, X., Jiang, L., Chu, Y., Gao, S., Jiang, X., Zhang, Y., Chen, Y., Luo, S., & Peng, C. (2022). The biological activity mechanism of chlorogenic acid and its applications in food industry: A review. *Frontiers in Nutrition, 9*, 943911. DOI: 10.3389/fnut.2022.943911 PMID: 35845802

Wang, W.-T., Guo, W.-Y., Jarvie, S., & Svenning, J.-C. (2021). The fate of Meconopsis species in the Tibeto-Himalayan region under future climate change. *Ecology and Evolution, 11*(2), 887–899. DOI: 10.1002/ece3.7096 PMID: 33520173

Wang, Y., Xiao, J., Ma, Y., Ding, J., Chen, X., Ding, Z., & Luo, Y. (2023). Persistent and enhanced carbon sequestration capacity of alpine grasslands on Earth's Third Pole. *Science Advances, 9*(20), eade6875. DOI: 10.1126/sciadv.ade6875 PMID: 37196073

Wang, Z., Barrow, C. J., Dunshea, F. R., & Suleria, H. A. (2021). A comparative investigation on phenolic composition, characterization and antioxidant potentials of five different australian grown pear varieties. *Antioxidants, 10*(2), 151. DOI: 10.3390/antiox10020151 PMID: 33498549

Wani, S. A., Mugal, M. A., Dar, F. A., Reddy, C. S., Rashid, I., & Khuroo, A. A. (2023). Biodiversity data synthesis on trees of Indian Himalayan Region: Policy and management implications. *Global Ecology and Conservation, 48*, e02698. DOI: 10.1016/j.gecco.2023.e02698

Wani, Z. A., Pant, S., & Sharma, V. (2021). Diversity, Distribution, and Status of Weed Species of Northwest Himalaya. In *Bioremediation using weeds* (pp. 3–34). Springer. DOI: 10.1007/978-981-33-6552-0_1

Warren, R., Price, J., VanDerWal, J., Cornelius, S., & Sohl, H. (2018). The implications of the United Nations Paris Agreement on climate change for globally significant biodiversity areas. *Climatic Change, 147*(3), 395–409. DOI: 10.1007/s10584-018-2158-6

Watts, S. H., Mardon, D. K., Mercer, C., Watson, D., Cole, H., Shaw, R. F., & Jump, A. S. (2022). Riding the elevator to extinction: Disjunct arctic-alpine plants of open habitats decline as their more competitive neighbours expand. *Biological Conservation, 272*, 109620. DOI: 10.1016/j.biocon.2022.109620

Waykar, B., & Baviskar, R. K. (2015). Diversity of bee foraging flora and floral calendar of Paithan taluka of Aurangabad district (Maharashtra), India. *Journal of Applied Horticulture, 17*(2), 155–159.

Wei, G. (2013). Green technology for nanoparticles in biomedical applications. In *Green biosynthesis of nanoparticles: mechanisms and applications* (pp. 1–12). CABI Wallingford UK.

Wesche, K., Miehe, G., & Kaeppeli, M. (2000). The significance of fire for afroalpine ericaceous vegetation. *Mountain Research and Development, 20*(4), 340–347. DOI: 10.1659/0276-4741(2000)020[0340:TSOFFA]2.0.CO;2

Wester, P., Mishra, A., Mukherji, A., & Shrestha, A. B. (2019). *The Hindu Kush Himalaya assessment: mountains, climate change, sustainability and people.* Springer Nature.

Weston, P. A., Gurusinghe, S., Birckhead, E., Skoneczny, D., Quinn, J. C., & Weston, L. A. (2019). Chemometric analysis of *Amaranthus retroflexus* in relation to livestock toxicity in southern Australia. *Phytochemistry, 161*, 1–10. DOI: 10.1016/j.phytochem.2019.01.016 PMID: 30776591

White, F. J., Hay, F. R., Abeli, T., & Mondoni, A. (2023). Two decades of climate change alters seed longevity in an alpine herb: Implications for ex situ seed conservation. *Alpine Botany, 133*(1), 11–20. DOI: 10.1007/s00035-022-00289-8

WHO. (2009). International Classification of Diseases-ICD. German Version. Bern: Huber.

WHO. (2015). WHO guidelines on safety monitoring of herbal medicines in pharmacovigilance systems. 2004. In: World Health Organization vii Geneva.

Wilkes, M. A., Carrivick, J. L., Castella, E., Ilg, C., Cauvy-Fraunié, S., Fell, S. C., Füreder, L., Huss, M., James, W., Lencioni, V., Robinson, C., & Brown, L. E. (2023). Glacier retreat reorganizes river habitats leaving refugia for Alpine invertebrate biodiversity poorly protected. *Nature Ecology & Evolution, 7*(6), 841–851. DOI: 10.1038/s41559-023-02061-5 PMID: 37142743

Woodford, D. J., Ivey, P., Novoa, A., Shackleton, R., Richardson, D., Weyl, O., & Zengeya, T. (2017). Managing conflict-generating invasive species in South Africa: Challenges and trade-offs. *Bothalia - African Biodiversity &amp. Conservation, 47*(2), 1–11. DOI: 10.4102/abc.v47i2.2160

World Meteorological Organization. (2025, January 10). *WMO confirms 2024 as warmest year on record at about 1.55 °C above pre-industrial level.* https://wmo.int/news/media-centre/wmo-confirms-2024-warmest-year-record-about-155degc-above-pre-industrial-level

Wróblewska-Łuczka, P., Cabaj, J., Bargieł, J., & Łuszczki, J. J. (2023). Anticancer effect of terpenes: Focus on malignant melanoma. *Pharmacological Reports, 75*(5), 1115–1125. DOI: 10.1007/s43440-023-00512-1 PMID: 37515699

Wu, H., Chen, J., Xu, J., Zeng, G., Sang, L., Liu, Q., Yin, Z., Dai, J., Yin, D., Liang, J., & Ye, S. (2019). Effects of dam construction on biodiversity: A review. *Journal of Cleaner Production, 221*, 480–489. DOI: 10.1016/j.jclepro.2019.03.001

Xue, N., Li, K., Chen, K., Li, P., Ji, X., Ma, Z., & Ji, W. (2024). Predicting climate change impacts on distribution and conservation of critically endangered *Picea neoveitchii* using MaxEnt. *Frontiers in Forests and Global Change*, *7*, 1472857. DOI: 10.3389/ffgc.2024.1472857

Yadav, E., Singh, D., Yadav, P., & Verma, A. (2018). Antioxidant and anti-inflammatory properties of *Prosopis cineraria* based phenolic rich ointment in wound healing. *Biomedicine and Pharmacotherapy*, *108*, 1572–1583. DOI: 10.1016/j.biopha.2018.09.180 PMID: 30372859

Yadav, S., Kumari, A., & Kumar, V. (2012). Nanotechnology: A tool to enhance therapeutic values of natural plant products. *Trends in Medical Research*, *7*(2), 34–42. DOI: 10.3923/tmr.2012.34.42

Yaglioglu, A. S., Yaglioglu, M. S., Tosyalıoglu, N., Adem, S., & Demirtas, I. (2022). Chemical profiling, in vitro biological activities and Pearson correlation between chemical profiling and anticancer activities of four Abies species from Turkey. *South African Journal of Botany*, *151*, 600–613. DOI: 10.1016/j.sajb.2022.08.005

Yahdjian, L., Campana, S., Tognetti, P. M., Alberti, J., Graff, P., Molina, C., Borer, E. T., Seabloom, E. W., Prober, S. M., MacDougall, A. S., Risch, A. C., Price, J. N., Power, S. A., Barrio, I. C., Hersch-Green, E., Fay, P. A., Bagchi, S., Bakker, J. D., Blumenthal, D., & Virtanen, R. (2026). Insights on global rangeland ecosystem services shaped by grazing and fertilization. *Frontiers in Ecology and the Environment*, *24*(3), 70022. DOI: 10.1002/fee.70022

Yang, H., Gou, X., Xue, B., Ma, W., Kuang, W., Tu, Z., Gao, L., Yin, D., & Zhang, J. (2023). Research on the change of alpine ecosystem service value and its sustainable development path. *Ecological Indicators*, *146*, 109893. DOI: 10.1016/j.ecolind.2023.109893

Yang, Y., Zhang, Y., Zhang, B., Bao, Y., Xu, S., Tang, X., Zhao, Q., Li, J., & Li, R. (2023). *In vitro* antioxidative activity of *Fritillaria cirrhosa* D. Don straw ethanolic extract and its effect on lipid, protein oxidation, and quality of Chinese-style sausage. *Journal of Food Science*, *88*(11), 4745–4772. DOI: 10.1111/1750-3841.16757 PMID: 37751083

Yang, Z., You, J., Li, J., Zhao, W., Xing, M., Zhang, Y., Cui, M., Gong, Y., Zhao, Y., & Chen, X. (2025). Divergent functional traits and gene expression profiles in native and encroaching plant species across an alpine elevational gradient. *Frontiers in Plant Science*, *16*, 1656812. DOI: 10.3389/fpls.2025.1656812 PMID: 41127076

Yasir, M., Sultana, B., & Anwar, F. (2018). LC–ESI–MS/MS based characterization of phenolic components in fruits of two species of *Solanaceae*. *Journal of Food Science and Technology*, *55*(7), 2370–2376. DOI: 10.1007/s13197-017-2702-9 PMID: 30042551

Yatoo, M. I., Dimri, U., Gopalakrishnan, A., Saxena, A., Wani, S. A., & Dhama, K. (2018). *In vitro* and *in vivo* immunomodulatory potential of *Pedicularis longiflora* and *Allium carolinianum* in alloxan-induced diabetes in rats. *Biomedicine and Pharmacotherapy*, *97*, 375–384. DOI: 10.1016/j.biopha.2017.10.133 PMID: 29091887

Yazar, M., Sevindik, M., Uysal, I., & Polat, A. O. (2025). Effects of Caffeic Acid on Human Health: Pharmacological and Therapeutic Effects, Biological Activity and Toxicity. *Pharmaceutical Chemistry Journal*, *59*(1), 1–7. DOI: 10.1007/s11094-025-03363-7

Yazhiniprabha, M., Banu, S., Ishwarya, R., Vinotha, V., Govindarajan, M., Wadaan, M. A., Mahboob, S., Nicoletti, M., & Vaseeharan, B. (2022). Biomimetically synthesized *Physalis minima* fruit extract-based zinc oxide nanoparticles as eco-friendly biomaterials for biological applications. *Journal of Drug Delivery Science and Technology*, *73*, 103475. DOI: 10.1016/j.jddst.2022.103475

Yohan, D., & Chithrani, B. D. (2014). Applications of nanoparticles in nanomedicine. *Journal of Biomedical Nanotechnology*, *10*(9), 2371–2392. DOI: 10.1166/jbn.2014.2015 PMID: 25992462

Yuan, H., Ma, Q., Ye, L., & Piao, G. (2016). The traditional medicine and modern medicine from natural products. *Molecules (Basel, Switzerland)*, *21*(5), 559. DOI: 10.3390/molecules21050559 PMID: 27136524

Yuce, K., & Ozkan, A. I. (2020). Cys2His2 zinc finger proteins boost survival ability of plants against stress conditions. In *Plant stress physiology*. IntechOpen., DOI: 10.5772/intechopen.92590

Yunusa, U. M., & Urek, R. O. (2023). Determination of phenolic compounds in *Diospyros lotus* by RP-HPLC-DAD and evaluation of antioxidant and cytotoxic properties. *Journal of Food Measurement and Characterization*, *17*(2), 1413–1427. DOI: 10.1007/s11694-022-01696-5

Zachos, F. E., & Habel, J. C. (2011). *Biodiversity hotspots: distribution and protection of conservation priority areas*. Springer Science & Business Media. DOI: 10.1007/978-3-642-20992-5

Zafar, M., Abbasi, A. M., Abbas, Z., Narzullayeva, U., Gafforov, Y., Filimban, F. Z., & Azimova, N. (2025). Health Risk Assessment and Bioactive Potential in Wild Food Plants from Deosai: A High-Altitude Ecosystem in Northern Pakistan: M. Zafar et al. *Biological trace element research*, 1-26.

Zafar, S., & Rana, N. (2025). The convergence of nanotechnology and artificial intelligence: Unlocking future innovations. *Recent Innovations in Chemical Engineering*, *18*(2), 85–99. DOI: 10.2174/0124055204359215250127071256

Zahara, K., Tabassum, S., Sabir, S., Arshad, M., Qureshi, R., Amjad, M. S., & Chaudhari, S. K. (2014). A review of therapeutic potential of Saussurea lappa-An endangered plant from Himalaya. *Asian Pacific Journal of Tropical Medicine*, *7*, S60–S69. DOI: 10.1016/S1995-7645(14)60204-2 PMID: 25312191

Zahra, A., & Talal, M. (2008). Impact of pollen supplements and vitamins on the development of hypopharyngeal glands and on brood number in honey bees. *Journal of Apicultural Science*, *52*(2).

Zain El Shandidy, M. (2023). The power of intangible heritage in sustainable development. *International Journal of Advanced Studies in World Archaeology*, *6*(2), 92–97. DOI: 10.21608/ijaswa.2024.275163.1043

Zaman, S., Al-Joufi, F. A., Zafar, M., & Zahoor, M. (2022). Phytochemical, Antimicrobial and Cytotoxic Activities of *Gaultheria trichophylla* Royle. *Applied Sciences (Basel, Switzerland)*, *12*(14), 6921. DOI: 10.3390/app12146921

Zaman, W., Ayaz, A., & Park, S. (2025). Climate change and medicinal plant biodiversity: Conservation strategies for sustainable use and genetic resource preservation. *Genetic Resources and Crop Evolution*, *72*(6), 6275–6308. DOI: 10.1007/s10722-025-02410-2

Zambare, V. P., & Christopher, L. P. (2012). Biopharmaceutical potential of lichens. *Pharmaceutical Biology*, *50*(6), 778–798. DOI: 10.3109/13880209.2011.633089 PMID: 22471936

Zaric, N. M., Deljanin, I., Ilijević, K., Stanisavljević, L., Ristić, M., & Gržetić, I. (2018). Assessment of spatial and temporal variations in trace element concentrations using honeybees (*Apis mellifera*) as bioindicators. *PeerJ*, *6*, e5197. DOI: 10.7717/peerj.5197 PMID: 30038861

Zehnder, M., Pfund, B., Svoboda, J., Marty, C., Vitasse, Y., Alexander, J., Hille Ris Lambers, J., & Rixen, C. (2025). Snow Height Sensors Reveal Phenological Advance in Alpine Grasslands. *Global Change Biology*, *31*(5), e70195. DOI: 10.1111/gcb.70195 PMID: 40346003

Zemp, M. (2008). *Global glacier changes: facts and figures*. UNEP/Earthprint.

Zhakipbekov, K., Turgumbayeva, A., Issayeva, R., Kipchakbayeva, A., Kadyrbayeva, G., Tleubayeva, M., Akhayeva, T., Tastambek, K., Sainova, G., Serikbayeva, E., Tolenova, K., Makhatova, B., Anarbayeva, R., Shimirova, Z., & Tileuberdi, Y. (2023). Antimicrobial and other biomedical properties of extracts from *Plantago major*, Plantaginaceae. *Pharmaceuticals (Basel, Switzerland), 16*(8), 1092. DOI: 10.3390/ph16081092 PMID: 37631007

Zhang, B., Sun, X., Tang, W., Jiang, L., Liu, J., Wang, Y., Li, Y., Li, Y., & Ma, X. (2025). An effective high-performance liquid chromatography method for fingerprint and multi-component quantification of *Oxalis corniculata. Current Pharmaceutical Analysis, 21*(9), 434–438. Advance online publication. DOI: 10.1016/j.cpan.2025.09.005

Zhang, J., Ren, X., Xu, E., Evans, A. M., Jing, W., Wang, R., Jia, X., Bi, M., Amoah, I. D., Pohlmann, M., Mecha, C., & Smith, C. K. (2025). Soil Chemical Properties Along an Elevational Gradient in the Alpine Shrublands of the Northeastern Tibetan Plateau. *Soil Systems, 9*(3), 95. DOI: 10.3390/soilsystems9030095

Zhang, Q., Chen, X., Zhou, X., Nie, X., Liu, G., Zhuang, G., Zheng, G., Fortin, D., & Ma, A. (2024). Tibetan Plateau grasslands might increase sequestration of microbial necromass carbon under future warming. *Communications Biology, 7*(1), 686. DOI: 10.1038/s42003-024-06396-y PMID: 38834864

Zhang, Y., Sun, J., Lu, Y., & Song, X. J. L. (2022).. . *Influence of Anthropogenic Activities and Major Natural Factors on Vegetation Changes in Global Alpine Regions., 11*(7), 1084.

Zhang, Y., Wang, Y., Yang, H., Hull, V., Zhang, J., Wang, F., Zhao, Z., & Liu, J. (2024). Heterogeneous and interactive effects of payments for ecosystem services on household income across giant panda nature reserves. *Heliyon, 10*(15), e34866. DOI: 10.1016/j.heliyon.2024.e34866 PMID: 39145020

Zhao, R. (2023). The influence of cushion plants on soil fertility islands under their canopies with ontogenesis in an extreme alpine ecosystem. *Plant and Soil, 484*(1), 379–392. DOI: 10.1007/s11104-022-05798-3

Zhao, Z.-H., Zhang, H.-J., Cao, X.-J., Wu, B., Ge, Y.-C., & Wang, K.-W. (2025). Preparation, characterization and biological activity of *Pteridium aquilinum* polysaccharide-Cu (II)/Zn (II) complexes in vitro. *Food Bioscience, 107355*, 107355. Advance online publication. DOI: 10.1016/j.fbio.2025.107355

Zhen, L., Du, B., Wei, Y., Xiao, Y., & Sheng, W. (2018). Assessing the effects of ecological restoration approaches in the alpine rangelands of the Qinghai-Tibetan Plateau. *Environmental Research Letters, 13*(9), 095005. DOI: 10.1088/1748-9326/aada51

Zhou, C., Dai, S., Yin, L., Zhuo, Y., Zhao, N., Li, D., & Sun, C. (2025). Antibacterial Activity and Mechanism of *Pinus pumila* (Pall.) Pinecone Essential Oil Against *Escherichia coli. Waste and Biomass Valorization, 16*(9), 1–16. DOI: 10.1007/s12649-025-02961-6

Zhou, J., Cai, W., Qin, Y., Lai, L., Guan, T., Zhang, X., & Cong, Z. (2016). Alpine vegetation phenology dynamic over 16 years and its covariation with climate in a semi-arid region of China. *The Science of the Total Environment, 572*, 119–128. DOI: 10.1016/j.scitotenv.2016.07.206 PMID: 27494658

Zhuo, Z., Xu, D., Li, Y., Pu, B., & Ye, M. (2021). Fingerprint analysis of *Zanthoxylum armatum* DC. by HPLC. *Journal of Food Composition and Analysis, 96*, 103736. DOI: 10.1016/j.jfca.2020.103736

Zia-Ul-Haq, M., Ćavar, S., Qayum, M., Imran, I., & de Feo, V. (2011). Compositional studies: Antioxidant and antidiabetic activities of *Capparis decidua* (Forsk.) Edgew. *International Journal of Molecular Sciences, 12*(12), 8846–8861. DOI: 10.3390/ijms12128846 PMID: 22272107

Zierl, B., & Bugmann, H. (2005). Global change impacts on hydrological processes in Alpine catchments. *Water Resources Research, 41*(2), 2004WR003447. DOI: 10.1029/2004WR003447

Zimmer, A., Beach, T., Klein, J. A., & Recharte Bullard, J. (2022). The need for stewardship of lands exposed by deglaciation from climate change. *Wiley Interdisciplinary Reviews: Climate Change, 13*(2), e753. DOI: 10.1002/wcc.753

Žitek, T., Postružnik, V., Knez, Ž., Golle, A., Dariš, B., & Knez Marevci, M.Supercritical Extraction Optimization for Antibiotic and Anticancer Activity. (2022). *Arnica montana* L. supercritical extraction optimization for antibiotic and anticancer activity. *Frontiers in Bioengineering and Biotechnology, 10*, 897185. DOI: 10.3389/fbioe.2022.897185 PMID: 35620474

Zoderer, B. M., Lupo Stanghellini, P. S., Tasser, E., Walde, J., Wieser, H., & Tappeiner, U. (2016). Exploring socio-cultural values of ecosystem service categories in the Central Alps: The influence of socio-demographic factors and landscape type. *Regional Environmental Change, 16*(7), 2033–2044. DOI: 10.1007/s10113-015-0922-y

Zoderer, B. M., Tasser, E., Erb, K.-H., Lupo Stanghellini, P. S., & Tappeiner, U. (2016). Identifying and mapping the tourists perception of cultural ecosystem services: A case study from an Alpine region. *Land Use Policy*, *56*, 251–261. DOI: 10.1016/j.landusepol.2016.05.004

About the Contributors

Muhammad Nauman Khan holds a Ph.D. in Botany from Islamia College Peshawar, Pakistan, and is currently serving as an Assistant Professor at the University of Chakwal, Punjab, Pakistan. He is an accomplished researcher with more than 150 peer-reviewed international research articles and 40 book chapters published in reputed journals, achieving a cumulative impact factor of nearly 250. His academic activities include participation in international conferences through abstracts, oral presentations, and poster sessions. Dr. Khan is actively involved in scholarly publishing as a Subject Editor of Ethnobotany Research and Applications and an editorial board member of multiple journals. He has also served on organizing committees of major international conferences. Dr. Khan is a lifetime member of the Pakistan Botanical Society, the Weed Science Society of Pakistan, and the Khwarizmi Science Society.

***

Nouman Ahmad is an undergraduate Biotechnology student at the University of Mianwali, Punjab, Pakistan, and completed his intermediate education at PAEC Boys College Chashma. His main research interests are drug design and vaccine design using computational and molecular biotechnology approaches. He plans to expand his research into epigenetics and other areas of medical biotechnology in the future. Nouman is engaged in academic research writing and aims to pursue advanced studies and a research career in biomedical and molecular sciences.

Nevin Aslan graduated with Ph.D. in Biology at Harran University in 2024. She developed expertise in biological sciences, natural products, and their biological effects. Through her research, she has focused on antimicrobial, antioxidant, and anticancer properties of natural compounds, while also exploring interdisciplinary fields such as nanotechnology and applying bioinformatics to complex biological data. She has published articles, book chapters, and conference papers in international peer-reviewed journals. She has deep academic interests in biological sciences, with a particular focus on natural products and their diverse biological effects. Her academically interests on omics and emerging interdisciplinary fields such as nanotechnology. She actively investigates biological activities including antimicrobial, antioxidant, and anticancer properties, aiming to contribute to health and therapeutic development.

Aleena Gul is a student of BS Biotechnology at the University of Mianwali, she has a strong research interest in medical biotechnology and biomedical sciences. Her academic focus includes the application of modern biotechnological tools, such as molecular regulatory strategies, gene-based approaches, and bioinformatics techniques, to understand disease mechanisms and develop innovative therapeutic and diagnostic solutions. She is particularly interested in how targeted biological interventions can contribute to improved healthcare outcomes and precision medicine. Her contribution to this work reflects a strong commitment to integrating biotechnology with advanced medical research and sustainable biomedical innovations.

Majid Iqbal is a researcher and educator specializing in plant sciences, ecology, and soil–plant interactions. He completed his undergraduate studies at the University of Peshawar and his postgraduate degree from Quaid-i-Azam University, Islamabad. He was awarded the ANSO Young Talent Scholarship to pursue his PhD at the University of Chinese Academy of Sciences, Beijing, China. He has published over 50 peer-reviewed international papers, 15 book chapters, and 25 conference abstracts. His research focuses on plant–environment interactions, biodiversity, and ecological monitoring. He has attended more than 40 national and international workshops and conferences. Currently, he serves as a Research Officer at Qarshi Industries (PVT) Ltd, Pakistan, working on natural cosmetics, herbal products, and nutraceuticals, with a focus on sustainability and wellness.

Haris Ahmed Khan is a Lecturer at the University of Mianwali, where he is committed to advancing higher education through dedicated teaching, academic supervision, and research engagement. Known for his disciplined approach and clear instructional style, he actively contributes to developing students' conceptual understanding and scientific thinking. At the University of Mianwali, Haris plays a key role in delivering quality education within his department, guiding students in coursework, practical learning, and research-based activities. His academic interests span modern biological sciences, biotechnology, and emerging interdisciplinary fields, and he continuously seeks to integrate contemporary knowledge, computational tools, and innovative teaching methods into the classroom. Driven by a passion for academic excellence and student mentorship, Haris Ahmed Khan remains focused on contributing to Pakistan's growing scientific and educational landscape.

Javed Khan is a plant nutrition and soil science researcher specializing in Nitrogen Use Efficiency (NUE) and crop–rhizosphere interactions in wheat. He earned his MSc in Plant Nutrition from Sichuan Agricultural University, Chengdu, China, where his research focused on root and leaf trait evaluation under low-nitrogen conditions and the use of genome-wide association studies (GWAS) to identify candidate genes linked to nitrogen acquisition. He is currently enrolled in the first year of a PhD at the College of Resources, Sichuan Agricultural University, Chengdu, China. He has published research in international journal, and several of his manuscripts are currently under review. His research interests include sustainable nutrient management, plant-microbe interactions, rhizosphere dynamics, and the integration of physiological and molecular approaches to improve crop performance under low-input conditions.

Zahida Nasreen is an Assistant Professor of Zoology at the University of Mianwali, specializing in Environmental Microbiology and Industrial Biotechnology. She earned her PhD from the University of the Punjab, focusing on composting biotechnology and soil-fertility–promoting microbes. Dr. Nasreen has published extensively in national and international journals and has presented her research at numerous scientific conferences. She regularly organizes academic workshops and seminars and is an active member of the USA Composting Council. Her research interests include microbial biotechnology, waste bioprocessing, environmental health, and sustainable resource management.

Ahmet İsmail Özkan graduated with a Ph.D. in Biology from Dicle University in 2022, where he developed expertise in siRNA delivery via magnetic liposomes, and their biological effects. His academic contributions include articles, book chapters, and conference papers published in international peer reviewed journals, reflecting his works in scientific community. He has focused on antimicrobial, antioxidant, and anticancer properties of both natural and synthetic compounds, while also exploring interdisciplinary fields such as nanotechnology and siRNA delivery via liposomes. His research integrates classical biological sciences with modern interdisciplinary approaches, aiming to translate fundamental discoveries into innovative strategies that support human health and therapeutic advancement.

Shivansu Sachan is Pursuing Ph.D. in Management, a Master's holder from Integral University, Lucknow, Bachelor's from IHM, Lucknow. Academic and professional accomplishments of 6 years with renowned organizations and institutes. His Teaching Career has been very satisfying and have continuously upgraded the skills and knowledge by attending many courses and training programs, seminars, conferences organized by different organizations and also delivered many lectures as a Hospitality sector. He is working as an Assistant Professor at Chhatrapati Shahu ji Maharaj University Kanpur .

Abdul Bari Shah is currently serving as an Associate Professor and Researcher of Natural Product Chemistry at the Faculty of Chemistry and Chemical Technology and the Research Center for Medicinal Plants, Al-Farabi Kazakh National University, Almaty, Kazakhstan. Prior to this appointment, he worked as a Research Professor at the College of Pharmacy, Korea University, South Korea. He completed his Postdoctoral Fellowship at the Natural Products Research Institute, College of Pharmacy, Seoul National University, and earned his PhD in Organic and Natural Product Chemistry from Gyeongsang National University, South Korea. Dr. Shah's research focuses on the discovery of bioactive small molecules from natural sources, including plants, microorganisms, and human microbiome–derived probiotics. His work emphasizes the development of advanced dereplication strategies through spectral database construction and integrates NMR and LC–MS techniques with biological assays, molecular docking, GNPS-based molecular networking, and network pharmacology to elucidate mechanisms of action and facilitate the development of novel therapeutics. He serves as an Editorial Board Member of Scientific Reports, PLOS ONE, Frontiers in Microbiology, and Current Bioactive Compounds.

Syed Nasar Shah completed his undergraduate studies in Botany with honours at the University of Peshawar, Pakistan, in 2015. He earned his M.Phil. in Plant Systematics and Biodiversity from Quaid-i-Azam University, Islamabad, Pakistan, in 2017. He is currently serving as a Senior Science Teacher (SST SCI 1) at Government Higher Secondary School, Olandar, District Shangla, Khyber Pakhtunkhwa, Pakistan. His research focuses on multiple aspects of plant science, including the ecology, taxonomy, and systematics of lower vascular plants (Pteridophytes), angiosperm taxonomy, medicinal plant biodiversity, and desiccation-tolerant ferns with their ecological and micromorphological adaptations. His current work encompasses plant systematics and taxonomy, plant ecophysiology, plant morphology (including micromorphology, anatomy, and palynology), plant ecology, adaptation to environmental change, medicinal plants, and ethnobotany. Shah has authored over 40 journal articles with a cumulative impact factor exceeding 100, a co-authored book Ferns of Swat Revisited with renowned fern taxonomist Chris Fraser-Jenkins, several book chapters, and numerous conference and symposium papers. He is particularly recognized for his research on the systematics and ecology of ferns in Malakand Division and Northern Pakistan. He has extensive field experience in the alpine and subalpine regions of the Pakistani Himalaya. His thesis herbarium is preserved at Quaid-i-Azam University, Islamabad.

Kemal Yüce graduated with Ph.D. research on mesenchymal stem cells at Selçuk University in 2019. He was assigned as an assistant professor in 2025. He has strong academic interests in health and biological sciences, with a particular emphasis on natural products and their biological effects, as well as emerging interdisciplinary fields such as nanotechnology. His research focuses on investigating the biological activities of natural compounds, including antimicrobial, antioxidant, and anticancer properties, aiming to reveal their mechanisms of action and potential applications in health sciences. By combining experimental approaches with interdisciplinary perspectives, he contributes to advancing knowledge at the interface of biology, chemistry, and technology, positioning his work within both fundamental and applied biomedical research.

Index

A

Alpine biodiversity 53, 63, 82, 83, 84, 91, 92, 94, 104, 138, 139, 140, 144, 149, 150, 153, 159, 169, 207, 216, 219, 221, 229, 230, 341, 342, 344, 345, 348, 356, 361, 363
alpine ecosystems 2, 3, 10, 11, 12, 14, 15, 16, 18, 19, 20, 29, 31, 33, 34, 36, 37, 49, 52, 53, 55, 56, 58, 59, 60, 61, 63, 64, 67, 68, 70, 71, 76, 80, 81, 82, 84, 85, 86, 87, 92, 93, 95, 98, 99, 104, 138, 139, 142, 143, 149, 153, 168, 172, 192, 208, 215, 216, 228, 234, 251, 268, 269, 280, 281, 286, 345, 347, 351, 353, 355, 359, 360, 362, 370, 383, 384, 387, 388, 389, 391
Alpine landscapes 20, 81, 82, 87, 88, 90, 97, 98, 161, 242
Alpine medicinal plants 12, 43, 191, 196, 207, 215, 217, 219, 221, 222, 223, 224, 225, 227, 228, 231, 233, 241, 273, 341, 351, 355, 357, 360
Alpine vegetation 27, 29, 30, 33, 36, 38, 40, 49, 51, 54, 62, 63, 65, 70, 83, 84, 108, 141, 142, 165, 169, 173, 191, 196, 216, 219, 250, 287, 311, 316, 334, 336, 345, 346, 374
Alpine Zone 43, 104, 113, 115, 133, 139, 166, 213, 280, 286, 287, 326
anthropogenic activities 51, 53, 56, 59, 61, 62, 73, 77, 116
Anthropogenic threats 112, 114, 125, 345, 367
Apiculture 376, 377, 378, 379, 380, 382, 389

B

Biodiversity Conservation 1, 3, 4, 6, 9, 18, 67, 93, 105, 129, 133, 138, 143, 144, 148, 159, 162, 168, 170, 177, 195, 210, 215, 217, 219, 228, 230, 350, 369
Biological Effects 243, 279, 280, 285, 286, 287, 290, 291, 292, 299, 304, 305

C

Climate Change Effects 45, 70, 82, 105, 207
Climate change impacts 31, 36, 44, 50, 102, 133, 206, 368
Conservation strategies 19, 40, 69, 70, 74, 92, 95, 104, 112, 126, 129, 133, 138, 139, 146, 147, 148, 202, 205, 213, 214, 220, 351, 352

D

driving factors 1, 5, 7, 9, 16, 18
Drug discovery 181, 186, 197, 205, 211, 234, 270, 271, 342, 363, 365

E

Ecological resilience 90, 102, 107, 138, 353
Ecosystem resilience 97, 103, 130, 153
ecosystem services 1, 2, 3, 4, 5, 6, 7, 9, 10, 11, 13, 14, 15, 16, 17, 18, 19, 21, 22, 23, 24, 33, 58, 69, 79, 81, 87, 91, 92, 95, 97, 98, 101, 102, 105, 106, 109, 132, 138, 143, 155, 162, 169, 170, 268
Endemic species 31, 37, 52, 55, 61, 65, 84, 113, 114, 116, 117, 121, 122, 123, 124, 125, 127, 128, 142, 155, 169, 200, 280, 286, 349, 350, 372
Environmental degradation 28, 62, 64, 179, 181
Environmental Threats 14, 354, 369
Ethnobotany 165, 166, 168, 171, 176, 180, 182, 183, 184, 185, 186, 187, 188, 190, 204, 205, 211, 212, 213, 223, 225, 227, 275, 326
Ex situ conservation 75, 76, 137, 145, 147, 152, 211

G

Global change 3, 19, 20, 25, 42, 43, 45, 50, 76, 77, 99, 103, 108, 131, 155, 156, 161, 187, 369

Green nanotechnology 215, 217, 220, 221, 223, 224, 226, 227, 228, 229, 231, 233, 234

H

Habitat fragmentation 5, 8, 59, 91, 98, 113, 132, 145, 152, 153, 170, 191, 216, 349, 351, 384
Habitat protection 69, 144, 192
Honey bee flora 375, 376, 378, 380, 381

I

Indicator Species 114, 126, 127, 128, 129
In situ conservation 73, 137, 144, 146, 148, 203

L

land use change 5, 8, 19, 20

M

Medicinal Plants 2, 4, 7, 12, 13, 17, 43, 65, 66, 77, 91, 143, 160, 161, 169, 170, 173, 174, 175, 176, 179, 180, 181, 184, 185, 186, 187, 188, 191, 192, 194, 195, 196, 198, 199, 200, 201, 203, 204, 207, 209, 211, 212, 213, 214, 215, 217, 219, 221, 222, 223, 224, 225, 227, 228, 231, 233, 239, 241, 244, 273, 275, 276, 277, 313, 322, 324, 330, 341, 342, 344, 351, 355, 357, 360, 361, 362, 363, 365, 370
Modern Application 177

N

Nanobiotechnology 225, 364, 372
Nanoparticles 215, 217, 220, 221, 222, 223, 224, 225, 226, 227, 228, 229, 230, 232, 233, 234, 235, 236, 237, 238, 239, 240, 314, 325, 338, 343, 355, 356, 357, 358, 359, 360, 367, 371, 373, 374
National parks 65, 69, 137, 144, 145, 152, 153, 203, 206, 352

P

Phenological Changes 27, 32, 41, 348
Phytochemical Composition 232, 250, 251, 279, 280, 286, 290, 328, 331
Phytochemicals 197, 216, 217, 222, 223, 227, 235, 239, 240, 243, 249, 251, 252, 253, 265, 267, 292, 299, 305, 306, 336, 354, 355, 357, 358, 359, 362, 363, 364
pollen substitutes 375, 386, 387, 395

S

Secondary metabolites 197, 199, 216, 222, 241, 243, 244, 247, 248, 249, 250, 251, 252, 253, 255, 262, 266, 268, 275, 276, 279, 285, 286, 290, 291, 292, 293, 299, 306, 307, 308, 312, 316, 322, 327, 330, 331, 332, 334, 342, 345, 367
spatial analysis 8
Species Range Shifts 95
Sustainable use 19, 148, 153, 206, 214, 216, 219, 245, 268, 269, 270, 341, 364

T

trade-offs and synergies 5, 6, 9, 10
Traditional Medicine 67, 166, 171, 186, 190, 192, 194, 205, 216, 243, 264, 266, 292, 331, 344, 349, 374

W

Wild Food Plants 185, 241, 242, 244, 245, 246, 247, 248, 251, 252, 257, 264, 268, 273, 275, 277, 279, 280, 286, 304, 317, 336, 337

www.ingramcontent.com/pod-product-compliance
Lightning Source LLC
LaVergne TN
LVHW062337300926
849161LV00007B/1727